DYNAMICS OF ROTATING MACHINES

This book equips the reader to understand every important aspect of the dynamics of rotating machines. Will the vibration be large? What influences machine stability? How can the vibration be reduced? Which sorts of rotor vibration are the worst? The book develops this understanding initially using extremely simple models for each phenomenon, in which (at most) four equations capture the behavior. More detailed models are then developed based on finite element (FE) analysis, to enable the accurate simulation of the relevant phenomena for real machines. Analysis software compatible with MATLAB® is available for download from the book's Web site, www.cambridge.org/friswell, and novices to rotordynamics can expect to make good predictions of critical speeds and rotating mode shapes within days. The book is structured more for self-study than as a reference handbook and, as such, provides readers with more than 100 worked examples and more than 100 problems and solutions.

Professor Michael I. Friswell joined Aston University as a Lecturer in 1987, after five years with the Admiralty Research Establishment in Portland. He moved to Swansea in 1993 and was promoted to a personal chair in 2000. Between 2002 and 2008, he was the Sir George White Professor of Aerospace Engineering at Bristol University before returning to Swansea in 2009 as Professor of Aerospace Structures. He received an EPSRC Advanced Research Fellowship (1996–2001), a Royal Society–Wolfson Research Merit Award (2002–2007), and an EC Marie Curie Excellence Grant (2005–2008). Professor Friswell has a wide range of research interests, primarily involving rotordynamics and structural dynamics, including inverse methods, condition monitoring, damping, nonlinear dynamics, and model-reduction methods. Professor Friswell's recent associate editorships include the *Journal of Intelligent Material Systems and Structures, Structural Health Monitoring*, and the *Journal of Vibration and Acoustics*. He is a Fellow of the Institute of Mathematics and Its Applications and the Institute of Physics and a Member of the American Society of Mechanical Engineers.

Professor John E. T. Penny served an apprenticeship with the English Electric Co. and worked for that company as a development engineer for three years. He then joined the staff at Aston University, initially as a Research Fellow, then as a Lecturer and Senior Lecturer, and became Head of the Mechanical and Electrical Engineering Department. Following this, Professor Penny became Director of Research at the School of Engineering and Applied Science. He has taught bachelor- and master's-level students in vibration and rotordynamics and related topics, such as numerical analysis and instrumentation. His research interests include topics in structural dynamics and rotordynamics. He has published in journals including the *Journal of Sound and Vibration, Mechanical Systems and Signal Processing*, and *AIAA Journal*. He is now an Emeritus Professor at Aston University but is still teaching and doing research. Professor Penny is a Fellow of the Institute of Mathematics and Its Applications.

Professor Seamus D. Garvey began his career with six years at GEC Large Electrical Machines Ltd., Rugby, and his first rotordynamics experience was acquired there. When he left the company in 1990, he was Principal Engineer for Mechanical Analysis and had written the computer program that has been used ever since for rotordynamics analysis. He then spent 10 years at Aston University, after which he joined Nottingham University as a Professor of Dynamics. He remains active in rotordynamics research – especially in the areas of active control and developing control forces through the airgaps of electrical machines – and serves on the organizing committees of both the IFToMM Rotordynamics conference and the IMechE Conference on Vibrations in Rotating Machines. He is currently Director of the Rolls-Royce University Technology Centre in Gas Turbine Transmissions at Nottingham University. Professor Garvey is a Fellow of the Institution of Mechanical Engineers and a Member of the Institute of Engineering and Technology.

Professor Arthur W. Lees has spent most of his career in the power-generation industry. After completing his PhD in physics, he joined the Central Electricity Generating Board, initially developing FE codes and later resolving plant problems. After a sequence of positions, he was appointed head of the Turbine Group for Nuclear Electric Plc. He moved to Swansea University in 1995, where his position was jointly funded by British Energy Plc and BNFL until August 2000. He was then appointed to a permanent chair within Swansea University. He is a regular reviewer of many technical journals and is currently on the editorial boards of the *Journal of Sound and Vibration* and *Communications on Numerical Methods in Engineering*. His research interests include structural dynamics, rotordynamics, and heat transfer. Professor Lees is a Fellow of the Institution of Mechanical Engineers and a Fellow of the Institute of Physics.

Cambridge Aerospace Series

Editors: Wei Shyy and Michael J. Rycroft

Dynamics of Rotating Machines

Michael I. Friswell
Swansea University

John E. T. Penny
Aston University

Seamus D. Garvey
Nottingham University

Arthur W. Lees
Swansea University

CAMBRIDGE
UNIVERSITY PRESS

CAMBRIDGE
UNIVERSITY PRESS

32 Avenue of the Americas, New York NY 10013-2473, USA

Cambridge University Press is part of the University of Cambridge.

It furthers the University's mission by disseminating knowledge in the pursuit of
education, learning and research at the highest international levels of excellence.

www.cambridge.org
Information on this title: www.cambridge.org/9780521850162

First published 2010

A catalogue record for this publication is available from the British Library

Library of Congress Cataloguing in Publication data

Dynamics of rotating machines / Michael Friswell ... [et al.].
    p.   cm. – (Cambridge aerospace series ; 26)
Includes bibliographical references and index
ISBN 978-0-521-85016-2 Hardback
1. Rotors – Dynamics.   2. Rotors – Vibration.   I. Friswell, M.   I. Title
II. Series.
TJ1058.F86 2010
621.82–dc22 2009042020

ISBN 978-0-521-85016-2 Hardback

Additional resources for this publication at www.cambridge.org/friswell

Contents

Preface

This book addresses the dynamics of rotating machines, and its purpose may be considered threefold: (1) to inform readers of the various dynamic phenomena that may occur during the operation of machines; (2) to provide an intuitive understanding of these phenomena at the most basic level using the simplest possible mathematical models; and (3) to elucidate how detailed modeling may be achieved. This is an engineering textbook written for engineers and students studying engineering at undergraduate and postgraduate levels. Its aim is to allow readers to learn and gain a comprehensive understanding of the dynamics of rotating machines by reading, problem solving, and experimenting with rotor models in software.

The book deliberately eschews any detailed historical accounts of the development of thinking within the dynamic analysis of rotating machines, focusing exclusively on modern matrix-based methods of numerical modeling and analysis. The structure of the book (described in Chapter 1) is driven largely by the desire to introduce the subject in terms of matrix formulations, beginning with the exposition of the necessary matrix algebra. All of the authors are avid devotees of matrix-based approaches to dynamics problems and all are constantly inspired by the intricacy and detail that emerge from even relatively simple numerical models. The emergence of software packages such as MATLAB that enable what would once have been considered large matrix computations to be conducted easily on a personal computer is one of the most exciting and important innovations in dynamics in the past two decades. With such a package, sophisticated models of machines can be assembled "from scratch" using only a few prewritten functions, which are available from the Web site associated with this book.

This book was written in a period of several years and, during that time, the single remark that emerged most often among the authors is this: "There is always more to discover about the dynamics of rotating machines"; this remark is usually exclaimed in wonder. It has been a pleasure to write this book and we hope that this pleasure is visible to and shared by readers. We thank our respective wives, Wendy, Wendy, Antonia, and Rita, for their patience, and the publishers for their considerable forbearance. During the preparation of the manuscript, we drew on the knowledge and insight of many other seasoned practitioners in the field – too many to thank individually – but a collective acknowledgment is entirely appropriate because it is heartfelt.

Acronyms

BSF   ball spin frequency

DFT   discrete Fourier transform

FE   finite element

FEA   finite element analysis

FEM   finite element method

FFT   fast Fourier transform

FRF   frequency response function

FTF   fundamental train frequency

IRS   Improved Reduced System

ISO   International Organization for Standardization

MMF   magneto-motive force

ODE   ordinary differential equation

ODS   operating deflection shape

SEREP   System Equivalent Reduction Expansion Process

UMP   unbalanced magnetic pull

1 Introduction

1.1 Overview

The aim of this book is to introduce readers to modern methods of modeling and analyzing rotating machines to determine their dynamic behavior. This is usually referred to as *rotordynamics*. The text is suitable for final-year undergraduates, postgraduates, and practicing engineers who require both an understanding of modern techniques used to model and analyze rotating systems and an ability to interpret the results of such analyses.

Before presenting a text on the dynamics of rotating machines, it is appropriate to consider why one would wish to study this subject. Apart from academic interest, it is an important practical subject in industry, despite the forbidding appearance of some of the mathematics used. There are two important application areas for the techniques found in the following pages. First, when designing the rotating parts of a machine, it is clearly necessary to consider their dynamic characteristics. It is crucial that the design of a machine is such that while running up to and functioning at its operating speed(s), vibration does not exceed safe and acceptable levels. An unacceptably high level of rotor vibration can cause excessive wear on bearings and may cause seals to fail. Blades on a rotor may come into contact with the stationary housing with disastrous results. An unacceptable level of vibration might be transmitted to the supporting structure and high levels of vibration could generate an excessive noise level. The second aspect of importance is the understanding of a machine's behavior when circumstances change, implying that a fault has occurred in the rotating parts of the machine. This understanding is needed for the diagnosis of the fault and for the formulation of repair strategies involving important decisions, such as "Is it safe to run?" and "How long can it be run?" These questions concern personnel and machine safety, legal issues, and, in many cases, very large sums of money.

After accepting that the dynamics of rotors is an important subject worthy of study, it is necessary to explain why a book on this topic is required. Are not the dynamics of rotors simply particular cases of the more general dynamics of structures? In fact, whereas in many respects, a rotor system behaves dynamically like a fixed structure, there are some important differences because of the rotation. The most fundamental of these differences is that fixed structures do not have inherent forcing, whereas rotating machines do. The following is a list of some, but by no means all, of the phenomena unique to rotating systems:

- When a rotor spins, lateral forces and moments may be generated. These so-called unbalance forces and moments are always present due to limitations in machining and assembly accuracy. These forces and moments give rise to vibration at the same frequency as the rotational speed.
- Gyroscopic moments also act on the spinning rotor and cause its natural frequencies to change with rotational speed.
- The stiffness and damping properties of some types of bearings vary with rotor speed; these changes also influence the system's natural frequencies.
- Centrifugal forces acting on a blade attached to a spinning rotor cause the blade stiffness to increase with rotational speed.
- Errors in gear profiles generate forces on the rotor, which are generated by the imposed motion introduced by geometric errors.
- Not all rotors are perfectly symmetric, and even minor asymmetries can have significant effects. When an asymmetric rotor spins, its stiffness changes periodically at the rotational speed or a multiple thereof, when viewed in a fixed set of coordinates. This can cause instability.
- The damping in rotating machines, in some circumstances, may be relatively high compared with that found in most fixed structures. In other circumstances, the effective damping may be negative, causing instability.

Thus, there is a variety of respects in which the analysis of rotating machines differs from that of a normal fixed structure. Study of the behavior of such systems is the science of rotordynamics and, whereas great reliance is placed on techniques of analysis from structural dynamics, rotordynamic analysis represents a considerable extension in scope and complexity. Given the wide application of rotating machinery, the field is one of great practical importance.

1.2 Rotating Machine Components

Any attempt to describe a typical rotating machine is inevitably fraught with difficulty in view of the wide range of machine sizes, duties, and speeds. Nevertheless, the techniques of modeling and analysis described in the following chapters can be applied to a wide variety of machines. The methods are equally applicable to small motors with rotors of, for instance, 40 mm in length with a mass of 0.04 kg (or smaller), or to turbo-generators, in which a typical large 660 MW machine has a total length of 50 m and a mass of 250,000 kg.

Every rotating machine consists of three principal components – namely, the rotor, the bearings, and their supporting structure. Some type of rationalization may be reached by considering the features that are pertinent to each component, which has its own properties that range enormously in complexity. Despite this, however, there are some generic methods to aid analysis that can be applied to gain insight into the behavior of a machine. It is important to emphasize this concept of insight: It is frequently more important to gain an understanding of a machine's behavior than it is to calculate precise numerical values of its response. Very often, the level of forces in service are unknown or known only approximately; therefore, detailed response calculations cannot be performed. Nevertheless, an understanding of the behavior in such circumstances can lead an analyst to propose appropriate courses of action to ensure the safe operation of a machine. We now consider each component in turn.

1.2.1 Features of Rotors

The rotor is at the heart of any rotating machine; therefore, the first part of any analysis must address the dynamic properties of the rotating element. In most cases, the rotor is relatively simple insofar as it often can be represented as a single beam or as a series of beam elements and rigid disks. The rotor, of course, may be complicated by numerous changes of section and the need for some treatment of effects, such as the shrink-fitting of disks. However, because the rotor is conceptually simple, it is a straightforward matter to check the accuracy of the rotor model by comparing the predicted behavior of the rotor alone with measured data, for example, by performing an impulse test on the rotor freely suspended.

The rotor of a small machine is usually rigid but, with increasing machine size, it generally becomes more flexible, and this must be accounted for in the analysis. However, there are exceptions. For example, the large electrical motors that drive rolling mills have rotors that are essentially rigid.

In many machines, the lateral stiffness of the rotor is the same in all planes containing the axis of rotation; in other cases, this is not so. For example, the rotors of many two-pole electrical machines have a lateral stiffness in one plane that is lower than in a perpendicular plane because of groups of slots that are cut into the rotor to carry the electrical conductors. Another feature of rotors is that they sometimes carry components or subassemblies with dynamic characteristics of their own. For example, a helicopter rotor consists of a relatively short, rigid vertical shaft carrying three very flexible blades. Finally, a rotor may have internal damping; contrary to intuition, internal damping can cause instability.

1.2.2 Features of Bearings and Rotor–Stator Interactions

The rotor is connected to the supporting structure by means of bearings, which may be of several types. For small machines with light loads, the bearing may take the form of a simple bush in which the rotor runs. However, as bearing loads increase, such a simple arrangement becomes inadequate; rotors are then mounted on ball or rolling-element bearings. These provide greater load capacity and stiffness but because they have internal moving components, they can make significant contributions to the dynamics of the overall machine. For large, heavy machines, such as turbo-generators, the bearings are almost invariably of the journal type, of which there are several forms. In a journal, or hydrodynamic, bearing, there is a film of oil in a small clearance between the static and rotating elements. The rotor creates a hydrodynamic pressure distribution within the oil film, which supports the weight and unbalance forces of the rotor.

In the past two decades, magnetic bearings have been introduced for some machine types. In these bearings, the rotor and stator are held apart by a magnetic field. The advantage of this system is the complete elimination of contact and the consequent very low values of effective friction; the disadvantage is the need to generate the magnetic field necessary to support the rotor.

Although the discussion so far has focused implicitly on the bearings that locate the rotor laterally, similar considerations are made in the selection of thrust bearings, which locate the rotor axially. Many machines – including jet engines, pumps, and turbines – generate considerable axial thrust; indeed, in the case of jet engines, this is their principal purpose. The function of the thrust bearing, therefore, is to react this force to the casing structure and to maintain the relative position of the casing and the rotor. This is clearly crucial in determining the axial dynamics of the system. It can also have substantial influence on the lateral vibrations by contributing rotational stiffness terms. The design and use of each bearing type is a substantial topic in its own right and is described extensively in the literature. In this book, bearings are treated simply as a local rotor–stator interaction with their own damping and stiffness properties; however, in some cases, these properties are speed-dependent. Using these properties, methods are provided to predict the behavior of a rotor under a variety of conditions.

The bearing is the most obvious device that couples the moving part of a machine (i.e., the rotor) to the stationary parts (i.e., the foundations, or the stator). However, there are other forms of interaction. Seals and glands also cause forces between the rotor and the stator. In steam and water turbines, the steam and water act on both the moving and fixed parts of the machine, thereby coupling them. Finally, in electrical machines, magnetic forces between the rotor and stator play a similar role in coupling the stator and rotor.

1.2.3 Stators and Foundations

The last component to consider is the structure supporting the bearings. In general, this structure consists of two major components: the *stator* and the *foundation*. For the purposes of this book, the stator is an integral part of the machine and does not rotate. In contrast, the foundation is the supporting structure and serves only to hold the machine in place. When considering the dynamics of the rotor, there is no distinction between the stator and the foundation. However, we make the distinction because we may want to know the forces transmitted by the machine to the foundations. In situations in which these forces are not of interest, we may refer to the combined stator and foundation as the foundation.

Although it is commonly thought that the modeling of the bearings represents the greatest challenge in the description of machinery, in many cases, there is an even more difficult problem. Although small machines may be supported on simple structures, which are relatively stiff compared to the properties of the machine, many large turbo-machines are mounted on complex structures that are relatively flexible. In such cases, the properties of the foundation have a significant influence on the dynamics of the complete machine and, therefore, should be represented within the model of the complete unit. This may be difficult due to problems in determining the parameters of the supporting structure model. In the case of a large turbine, the supporting foundation may be a complex structure consisting of many components and welded joints, to which a number of items of ancillary equipment are connected. It has been found that due to build tolerances, even machines and structures built to the same design show significant differences in vibration behavior. In practice, the approach frequently used is simply to select parameters that give adequate agreement with experimental data. In addition, the foundations can couple the horizontal and vertical motion, and they might introduce coupling between the bearing supports.

1.3 Aspects of Rotating Machine Behavior

In discussing the dynamic behavior of rotating machinery throughout this book, considerable reliance is placed on mathematical descriptions of the physical processes involved. Before embarking on this discussion, however, it is worth reflecting on the motion of a machine rotor. A rotor can vibrate in three ways: axially, torsionally, and laterally. *Axial vibrations* occur along the axis of the rotor, and *torsional vibrations* cause the rotor to twist about its axis of rotation. *Lateral vibrations* cause displacements of the rotor in both the horizontal and vertical directions (for a horizontal rotor). These motions combine to produce an orbital motion of the rotor in a plane perpendicular to its axis of rotation. In some systems, the three types of vibration are independent of one another, and they can be analyzed separately. However, in other systems, there is coupling among all three forms of vibration (see Chapter 10).

The analysis and, indeed, the manner in which a machine performs in service are determined largely by whether the rotors are rigid or flexible, the type of bearings used, and the nature of the supporting structure. This is true whatever the size or complexity of the machine. In all machines, there are features that must be modeled; however, in individual instances, it may be possible to demonstrate that they have a small or negligible influence on the dynamics of the overall system. In these instances, detailed modeling is not required; however, this cannot be assumed without investigation.

1.3.1 Lateral Vibrations

Rotor lateral vibration (sometimes called *transverse* or *flexural vibration*) is perpendicular to the axis of the rotor and is the largest vibration component in most high-speed machinery. Understanding and controlling this lateral vibration is important because excessive lateral vibration leads to bearing wear and, ultimately, failure. In extreme cases, lateral vibration also can cause the rotating parts of a machine to come into contact with stationary parts, with potentially disastrous consequences.

Lateral vibration is generally caused by lateral forces, the most common of which are unbalance forces that are present in all rotating machines, despite efforts to minimize or eliminate them. In subsequent chapters, we discuss the effects of rotor unbalance and the methods for balancing real machines, but the balance will never be perfect.

As in all elastic systems, a machine has natural frequencies of lateral vibration determined by the lateral stiffness and mass distribution of the rotor–bearing–foundation system. When the rotational speed – and, hence, the frequency of the out-of-balance forces – is equal to any of these natural frequencies, the vibration response becomes large and the rotor is considered to be rotating at a critical speed. When a machine is accelerated from rest to its operating speed, it might have to pass through one or more of these critical speeds. For most classes of machine, it is important that it is not permitted to operate at or close to a critical speed for any length of time.

Because the rotor can vibrate laterally in two mutually perpendicular directions, the vibrations combine to create an orbit for the rotor motion. If the supporting structure of the bearings of a horizontal rotor has identical stiffness and damping properties in both the horizontal and vertical directions, then this orbit is circular and the bending stresses in the rotor are constant. In many instances, however, the structure supporting the bearings is stiffer in the vertical than in the horizontal direction. In such a situation, the rotor orbit is elliptical and the bending stress in the rotor varies at twice the rotational speed.

In the discussion thus far, the role of dissipative or damping forces on the motion has not been mentioned. As in structural dynamics, damping has a major influence close to the resonant frequencies. Although it might be anticipated that damping always tends to reduce vibration, this is not always the case. If the damping forces arise in the supporting structure, then the effects are invariably beneficial and may be treated in much the same way as damping in any structural system. Problems arise, however, when there is damping in the rotor itself. Far from being beneficial, this type of damping can be destabilizing.

1.3.2 Axial Vibrations

The ultimate function of a jet engine is to produce thrust in the axial direction. A thrust bearing must be fitted to transmit this thrust to the housing and, hence, the aircraft to which it is attached. Without this thrust bearing, the rotor would simply be propelled away from the engine housing and, therefore, would be ineffective! Of course, there is some time-varying fluctuation about the mean level of thrust, which gives rise to axial vibrations of the rotor, with this motion having its own set of resonance frequencies. In contrast to the lateral motion of the rotor, stresses arising from axial vibration are uniform across a complete cross-section of the rotor. There may be cross-coupling between axial and lateral vibrations – for example, in helical and bevel gear meshes.

1.3.3 Torsional Vibrations

The third type of vibration is torsional vibration, or a twisting motion of the rotor about its own axis. In some respects, this is relatively straight-forward to model because bearings and supporting structures have little influence on the natural frequencies. There is also a practical problem: lateral and, to a lesser extent, axial vibrations become obvious by their effects on the machine and its surroundings, enabling the deployment of appropriate effort to resolve developing problems. In complete contrast, torsional problems can go unnoticed without special instrumentation. Furthermore, because little motion is transmitted to components other than the rotor, torsional modes often have low damping. During this undetected phase, however, considerable damage may be caused to a machine.

Figure 1.1. The quietrevolution QR5 vertical-axis wind turbine. This photograph is reproduced with the permission of quietrevolution ltd, copyright ©quietrevolution ltd 2009.

1.4 Examples of Rotating Machines

Rotating machines can vary enormously in size, complexity, and general configuration. The basic rotor–bearing–foundation system can be found in numerous products and systems ranging from small electrical motors in refrigerators and washing machines to turbo-generator sets. The list of examples is almost endless but includes centrifuges and vacuum pumps running at up to 90,000 rev/min, gyroscopes, machine spindles, helicopters, reciprocating gasoline and diesel engines, rotary and reciprocating compressors, and gas and steam turbines of all sizes. Figure 1.1 shows a typical vertical-axis wind turbine and Figure 1.2 shows an example of a turbocharger. In the following sections, we provide a short description of five types of rotating machines, including two of the largest and most powerful: namely, turbo-generators and aircraft gas turbines.

1.4.1 Electrical Machines

Rotating electrical machines convert mechanical power to electrical power or vice versa. These are the most numerous of all rotating machines, with several tens of them being included in every modern car (hundreds in luxury cars) and similar numbers in most modern households. Depending on which way the power conversion is made, the machines are described as motors or generators. The flexibility and control offered by electrical power transmission is such that most prime-movers drive electrical machines and most loads are driven by electrical machines. In many cases, electrical motors are integrated with a load to form more complex machines, such as the integrated motor and compressor shown in Figure 1.3; it is meaningless in

Figure 1.2. A cutaway view of a turbocharger. Photograph courtesy of NASA.

these cases to discuss the rotordynamics of the electrical motor in its own right. This integration is most common at small scales, and one example (i.e., vacuum pumps) is discussed in Section 1.4.4. At larger scales, electrical machines tend to be separate items coupled to the load or prime-mover only through a shaft-coupling and, indirectly, through the stator mountings. One example concerns turbo-generators, which are discussed in Section 1.4.2. When the electrical machine rotor is coupled rigidly to a load or prime-mover, the rotordynamics of the combination must be considered as a whole. Often, however, flexible couplings are provided between the

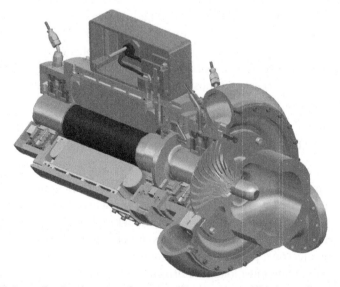

Figure 1.3. Schematic of an integrated motor and compressor. This image is reproduced with the permission of Corac Group plc, copyright ©Corac Group plc 2009.

electrical machine and its load or prime-mover; the lateral and axial dynamic behavior of the rotating machine then can be considered reasonably accurately in isolation. Rotating electrical machines present at least two specific interesting features with regard to their dynamics: (1) the rotors are often laminated and/or contain insulated conductors, which gives rise to uncertainty about the rotor stiffness and to rotor damping; and (2) electrical machines experience unbalanced magnetic pull (UMP), which can significantly influence the rotordynamics (see Chapters 5 and 7). Certain electrical machines can themselves produce strong torsional forcing and, in the presence of significant static eccentricity, most alternating current (AC) motors produce strong transverse forcing at twice the frequency of the electrical supply.

Electrical machines come in a vast range of sizes and types spanning sub-millimeter-scale machines delivering fractions of a Watt to machines delivering hundreds of MW. Torques vary from a few mNm in true micro scale machines (usually electrostatic rather than electromagnetic) to several MNm (e.g., for ship-propulsion machines). The types of electrical machine commonly used at present include the following:

- *Direct current (DC) machines*. These machines, which still dominate automotive applications, have brushes to conduct current onto the rotating part. The most efficient and most expensive DC machines include permanent magnets to provide the main magnetic field.
- *Universal machines*. These are essentially DC motors that also can work from AC electricity at any frequency. Universal machines dominate the white-goods market.
- *Induction motors*. These are inexpensive and robust and the most common industrial electrical machine (approximately 35 percent of all electricity generated in Europe in 2000 was consumed in induction machines).
- *Permanent-magnet synchronous machines*. Permanent-magnet synchronous machines often are used as servo-motors when very accurate and high-bandwidth control of rotor angular position is critical. Applications include packaging and printing machinery; specialist manufacturing machines for folding, cutting, and stitching; and so forth. Permanent-magnet machines are also commonly used in cordless power tools because of their high efficiency. They are usually used with a power-electronic drive that requires a sensor to establish rotor angular position.
- *Wound-rotor synchronous machines*. Large generators are almost invariably of this construction because it enables separate control over the phase relationships between voltage and current.
- *Stepper-motors*. These are used in many lower-grade servo-control applications where open-loop control is acceptable. Stepper-motors are particularly common in printers and photocopiers because they can achieve position control with acceptably small error and low cost.
- *Switched-reluctance machines*. These machines have no single major established niche at present but they compete with induction and permanent-magnet machines in numerous applications. Switched-reluctance machines are often more efficient than their induction-machine counterparts because there are no currents on the rotor; however, they are reputed to be much noisier. Like

Figure 1.4. A typical steam-turbine rotor. This photograph is reproduced with the permission of Alstom, copyright ©Alstom 2009.

permanent-magnet synchronous machines, they require some sensing provision to detect rotor angular position. However, they offer an advantage over permanent-magnet machines in that they do not generate high voltages if the machine rotates quickly and if, for any reason, the stator circuits are open-circuit.

1.4.2 Turbo-Generator Sets

Turbo-generator sets, used in the generation of electrical power, are examples of large and complex rotating machines. Figure 1.4 shows a typical rotor of one of these machines. The total system typically consists of up to eight individual rotors coupled together, resulting in a typical total rotor length of 50 m, and supported on 16 oil-film journal bearings. The rotational speed is 3,000 rev/min (3,600 rev/min in the United States) and the power output is typically 660 MW, but there are machines of 1,300 MW. It is worth noting that for a 660 MW machine, the mass of the composite rotor is about 250 tonnes. This implies that if the mass axis center of gravity is displaced from the geometric center by 25 μm, then the forces exerted at each bearing are of the order of 44 kN. The bearings are supported on a foundation consisting of a steel or concrete structure, which – although appearing massive and rigid – is, in fact, often relatively flexible. In addition to unbalance, many different forces act on these machines. Steam forces act on the turbine rotors and an unbalanced magnetic force acts on the generator. If the rotor suffers any type of deterioration in service (e.g., a bend or rubbing against its housing), then new forces arise that influence the dynamic behavior of the machine. Nonuniform heating can result from

Figure 1.5. The Rolls-Royce Trent 900 gas-turbine engine. This photograph is reproduced with the permission of Rolls-Royce plc, copyright ©Rolls-Royce plc 2009.

faults in the electrical windings, which result in bends that develop as load is added to the machine. There are other thermal effects that sometimes result from changes of alignment due to expansion in the supporting structures. Care is required in modeling these machines, but the understanding gained from such models has proved to be an excellent investment for many machine operators.

1.4.3 Gas Turbines

Aero-engines (Figures 1.5 and 1.6) are physically smaller than turbo-generators and have a very high power-to-weight ratio. Typically, they produce power outputs of 50 MW with a rotational speed of up to 21,000 rev/min. There are usually two or three co-axial rotors, the longest of which is typically 3 m long, and they are supported by rolling-element bearings. Thin diaphragms support the bearings so that the bearing supports are laterally stiff but axially flexible. Forces are exerted on the rotor by a series of squeeze-film dampers comprising an oil-filled clearance between the rotor and a housing that can *float* (being connected to a soft spring). As the name implies, these devices are included to provide damping rather than any further stiffness effect. Because these engines operate at such high rotational speeds, it is clear that unbalance forces can be very high. In addition, because the engines are so light with very high energy density, fluid force is significant and, consequently, rotor stability is a concern: hence, the provision of additional damping.

1.4.4 Vacuum Pumps

The need to create vacuums is becoming more important as the number of scientific instruments (e.g., mass spectrometers) increases and the demand for semiconductors expands. Some vacuum pumps consist of a short rotor with a number of

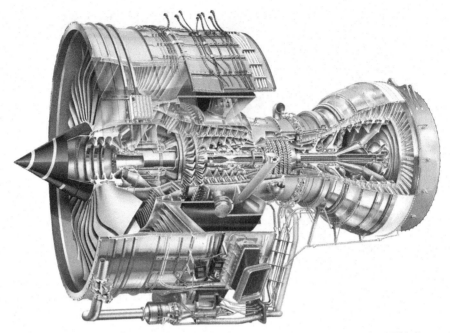

Figure 1.6. A cutaway view of the Rolls-Royce Trent 900 gas-turbine engine. This image is reproduced with the permission of Rolls-Royce plc, copyright ©Rolls-Royce plc 2009.

bladed disks toward one end and a motor at the other end, as shown in Figure 1.7. Generally, rolling-element bearings are used on either side of the motor, and these bearings are mounted on elastomer supports to reduce vibration transmission to the foundation. Rolling-element bearings cannot be used at the low-pressure end of the rotor because the grease would cause contamination. Thus, the rotor is often

Figure 1.7. A cutaway view of the Edwards nEXT300 Turbomolecular pump. This image is reproduced with the permission of Edwards Limited, copyright ©Edwards Limited 2009.

overhung, although the free end may be supported by passive or active magnetic bearings for high-value applications. The rotors run at high speed – typically, up to 90,000 rev/min – and because these pumps are often fitted to sensitive equipment, detailed consideration of the pump dynamics is vital. Often, the rotors are rigid and the pumps operate just below their second critical speed. The pumps must safely pass through the first critical speed to reach the operating speed, which should be designed to be sufficiently removed from the running speed. Furthermore, the vibration transmission to the pump foundation should be low, which is achieved by accurately balancing the machine after manufacture. There is an increasing need to develop the use of flexible rotors in these machines as current designs approach limiting-performance factors. This trend intensifies the need for accurate analysis of the machine dynamics.

1.4.5 Vertical-Axis Pumps

The design of a vertical-axis pump (sometimes referred to as a downhole pump) is relatively simple. It normally consists of an electrical motor driving one or more pumps placed vertically below the motor. The number of pumps used depends on the required head. When designing a system to pump fluid, it is necessary to ensure that the net head is high enough to prevent cavitation. One way to achieve this is to increase the suction head by placing the pump unit(s) below ground level. Thus, the drive shaft connecting the electrical motor at ground level to the pump(s) can be long and subject to vibration. A typical system might run at 3,600 rev/min with the pumps placed 5 m below ground level. Vertical-axis pumps are used in a diverse range of applications, such as pumping water in utility companies or pumping hydrocarbons in the oil and gas industry. Figure 1.8 shows a typical vertical-axis pump.

1.5 Scope and Structure of the Book

This book focuses on the linear analysis of rotor–bearing–foundation systems to determine their free vibration characteristics and their response to lateral, torsional, and axial forces. The differential equations describing the motion are linear and their coefficients generally are constant, although in some circumstances, the coefficients may vary with time.

Following this introduction, Chapter 2 is an overview of basic vibration theory. It covers all the essential theory that is required for the dynamic analysis of simple structures. Chapter 3 introduces readers to the free lateral response of some simple rotor–bearing systems and describes how gyroscopic moments affect both natural frequencies and mode shapes of the system. However, to model complex structures, finite element analysis (FEA) or some other numerical method must be used. We advocate the use of finite elements for the analysis of rotor–bearing systems, and the relevant theory is reviewed in Chapter 4. Chapter 5 builds on Chapters 3 and 4 and shows how the free lateral response of more complex rotor–bearing systems can be computed and interpreted.

Chapter 6 is devoted to the lateral response of rotor–bearing systems to out-of-balance and other forces and introduces critical speeds. In Chapter 7, asymmetric rotors and other causes of lateral instability are analyzed. Each chapter

Figure 1.8. The Floway® Pumps vertical turbine pump. This photograph is reproduced with the permission of Weir Floway Inc., Weir Minerals Division, and The Weir Group plc, ©Weir Minerals Division – The Weir Group plc 2009.

begins by considering simple models of uncomplicated systems and then applies the finite element method (FEM) to create models of more complex systems. Chapter 8 presents the topic of rotor balancing. Although not strictly part of rotor–bearing–foundation analysis, balancing is an important requirement in the commissioning of any rotor; therefore, we make no apology, for including the topic in a text primarily concerned with modeling and analysis.

In Chapter 9, axial and torsional vibrations of rotors are considered. Although these vibrations are often of less significance than lateral components, there are instances in which their understanding is crucial. This situation arises predominantly in geared systems. Finally, Chapter 10 briefly discusses more complex dynamic models of rotating machines.

At the beginning of this chapter, we state that rotordynamics is an important subject of study and that the dynamics of rotating machinery has many features not found in the dynamic analysis of fixed structures. It is hoped that this book helps to clarify the issues involved.

1.6 Required Background Knowledge

We assume that readers have some prior knowledge of topics related to this text. In particular, we expect that they have a basic knowledge of dynamics, including the concept of the free-body diagram and Newton's laws of motion, energy, and momentum. In mathematics, we expect readers to be familiar with complex numbers, differentiation and integration, linear differential equations, root structure of polynomial equations, and – most important – elementary matrix algebra, including eigenvalues, eigenvectors, and the concepts of symmetry and skew-symmetry.

1.7 Developing a Course of Instruction Using this Book

This book aims to provide a comprehensive view of the dynamics of rotating machinery. Consequently, it contains a great amount of material, and it may be necessary to make selections to prepare for an undergraduate or postgraduate course.

An introductory course for undergraduates might include Chapter 2 (if required) to provide a background in structural dynamics. Rotordynamics begins in Chapter 3 with a discussion of the free vibration of simple rotor systems and to a large extent, sets out the basic ideas required for the analysis of simple rotors. The first half of Chapter 6 extends the calculations to give forced response, so – coupled with Chapter 3 – this can form the core of the course with emphasis on understanding rotordynamic phenomena. The more elementary parts of Chapter 8 (balancing) and Chapter 9 (axial and torsional vibrations) also can be an appropriate part of an introductory course, if time allows.

An advanced course might include Chapter 4 to provide a background in FEA (if required). Chapter 5 and the second half of Chapter 6 provide the modeling and analysis of complex rotors using FEA. The more advanced parts of Chapters 8 and 9 are also appropriate for this course.

A more advanced course can be based on the material in Chapter 7, in which a range of sources of rotor instabilities is discussed. Such a course would refer to the more advanced topics briefly covered in Chapter 9.

It is hoped, however, that students will read each chapter separately, enabling them to gain insight into the general concepts of the dynamics of rotating machines.

1.8 Software

Many of the problems provided in this book are too complex to be solved manually and rotordynamic analysis software must be used. For those who have access to MATLAB, a software package designed to accompany this book is available.

This package consists of several MATLAB functions and a user guide, and it can be downloaded from www.rotordynamics.info. The software requires the user to have at least MATLAB Version 6; no extra MATLAB toolboxes are required. The SCILAB software package has functionality similar to MATLAB and is free. The purpose of providing this software – and the practice problems that use it – is to allow readers to solve the set problems and variants of them, to obtain solutions, and to then make engineering judgments about them. It is important to use the software to explore variants of the problems. What, for example, is the effect modifying the bearings? Doing this allows the user to rapidly gain experience of the manner in which rotor systems behave. Using the software should be more than an exercise in following instructions and making keystrokes: It should be an exercise in experimentation, interpretation, and insight.

2 Introduction to Vibration Analysis

2.1 Introduction

In this chapter, we briefly examine the dynamic characteristics and properties of elastic systems composed of discrete components. In subsequent chapters, we extend and apply these ideas to the more complex problems arising from the dynamic analysis of continuous components, rotors, stators, and rotor–bearing–stator systems. Readers introduced to this material for the first time may need to consult other textbooks that develop these ideas in more detail and at a more measured pace (e.g., Inman, 2008; Meirovitch, 1986; Newland, 1989; Rao, 1990; Thompson, 1993). Here, the basic theory is reviewed in a manner suitable for those who already have some familiarity with it. For such readers, this chapter provides both a revision and a concise summary.

The purpose of analyzing any elastic system or structure is to determine the static or dynamic displacements (or strains) and to find the internal forces (or stress) in the system or structure. To determine displacements, we require a frame of reference from which to measure them. Before we can begin the analysis, however, we must create a mathematical model of our system that may be very simple – perhaps devised intuitively – and easy to analyze but provides information of limited accuracy. Conversely, it may be a very complex model that requires significant computation but provides relatively accurate information.

In a real structure such as a rotor or an aircraft wing, material is distributed continuously and the system likewise has distributed inertial, flexibility, and damping properties. However, within a system, it may be that some parts or components have a large mass and negligible distortion under load, whereas other parts may distort under load but have a negligible mass. Thus, in some instances, we may readily approximate the actual system by an assemblage of discrete and idealized components. Other components are possible, such as viscous dampers that have negligible mass or stiffness but provide a force proportional to the relative velocity of their ends. These assumptions and simplifications lead to the concept of a discrete system – that is, a system built up from discrete components. Further simplification results from the assumption that there is a linear relationship between applied forces and resultant displacements. This assumption is generally valid provided that the

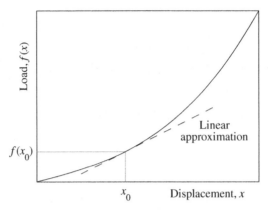

Figure 2.1. A typical load-deflection curve for a nonlinear spring.

systems vibrate with only small oscillations. Some types of bearings, for example, have nonlinear characteristics; however, for small oscillations, linear models with speed-dependent properties may be obtained.

For many systems, the inertia, damping, and stiffness properties do not vary with time; these are called *time-invariant vibrating systems*. The analysis of time-invariant systems is easier than the analysis of time-variant systems because closed-form solutions based on the system's eigenvalues and eigenvectors often exist. Section 7.4 discusses an example of a time-variant system consisting of an asymmetric rotor supported by anisotropic bearing.

The simple models studied in this chapter are *static* in the sense that they do not contain rotating parts. We begin by examining the behavior of the simplest possible dynamic system – namely, a mass–spring–damper oscillator that requires only a single coordinate to specify the displacement of the mass from its equilibrium position. We then analyze discrete systems with many masses, springs, and dampers.

Throughout this chapter, complex numbers are used extensively for eigenvalues and also to derive frequency response functions and time responses. Although the system response may be obtained using only real arithmetic, there are occasions when the use of complex functions considerably simplifies the calculation.

2.2 Linear Systems

The development discussed in most of this book is concerned with linear models of systems. A linear model is best illustrated by considering the load-deflection relationship for a spring. If a force is applied to a spring, it deforms, and the resulting displacement may be written as a function of the force. Alternatively, the spring may be extended by a fixed amount and the restoring force measured. This second approach is more convenient for incorporation into equations of motion of systems and machines, and it assumes that the force in the spring is a function of spring extension. Of course, the force also may depend on the rate of spring extension and possibly past values of the extension and its derivatives, but these possibilities are ignored for now. Figure 2.1 shows a typical load-deflection curve, where the force, f, is given as a function of the spring extension, x. In a vibrating system, the displacement often is relatively small and occurs about some equilibrium position, x_0.

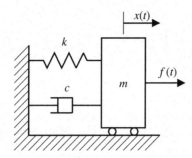

Figure 2.2. Single degree of freedom example.

Thus, for small deflections, the Taylor series of $f(x)$ is

$$f(x) = f(x_0 + \delta x)$$
$$= f(x_0) + f'(x_0)\delta x + \frac{1}{2}f''(x_0)\delta x^2 + \text{higher-order terms} \tag{2.1}$$

where the prime denotes differentiation with respect to x. Neglecting the second- and higher-order terms, we have

$$f(x) = f(x_0 + \delta x) \approx f(x_0) + f'(x_0)\delta x = f(x_0) + k_0\delta x \tag{2.2}$$

where $k_0 = f'(x_0)$ is constant and is called the *stiffness coefficient*. The approximation given by Equation (2.2) is also shown in Figure 2.1. For vibration about the equilibrium point, the change in force as a function of the change in displacement is required, giving

$$\delta f = f(x) - f(x_0) = k_0\delta x \tag{2.3}$$

This linearized force-displacement relationship is called *Hooke's law* and is used to model the spring.

2.3 Single Degree of Freedom Systems

We begin by considering the mass–spring–damper system shown in Figure 2.2. This discrete system consists of a mass, a spring, and a viscous damper. The mass is assumed to be infinitely rigid and only has the property of inertia. The linear spring is massless and only has the properties of stiffness. Stiffness is defined (by Hooke's law) as the force required to produce a unit axial extension of the spring (see Section 2.2). The linear viscous damper is also massless and only has the property of damping. The damping coefficient is defined as the force required to produce a unit axial velocity across the damper. In fact, a viscous damper does not model commonly occurring mechanisms for energy dissipation particularly accurately, but it has the advantage of simplicity. In the system, the mass component moves on a frictionless surface and, because we are going to assume that its motion is restricted to the x direction only, a single coordinate is sufficient to specify the displacement of the system mass at any instant. The system is said to have only one *degree of freedom*. A dynamic force $f(t)$ is applied to the mass.

Sometimes it is necessary to combine stiffnesses of more than one spring into a single equivalent spring. Consider the case of two springs connected in parallel,

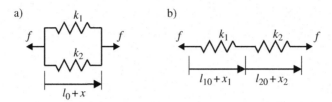

Figure 2.3. The springs connected in parallel and series: l_0, l_{10}, and l_{20} are the undeformed lengths of the springs; and x, x_1, and x_2 are the extensions due to the force f.

as shown in Figure 2.3(a). The total force is shared between the springs, but the displacement of the springs is identical. Thus,

$$f = k_1 x + k_2 x = (k_1 + k_2)\, x \qquad (2.4)$$

and the effective stiffness of the combined springs is

$$k_{eff} = k_1 + k_2 \qquad (2.5)$$

For the case of two springs connected in series, as shown in Figure 2.3(b), the force in both springs is equal, and the total displacement is the sum of the displacement in the two springs. Thus,

$$f = k_1 x_1 = k_2 x_1 \qquad (2.6)$$

Hence,

$$x_1 = f/k_1 \quad \text{and} \quad x_2 = f/k_2, \qquad (2.7)$$

$$x = x_1 + x_2 = f(1/k_1 + 1/k_2), \qquad (2.8)$$

and the effective stiffness of the combined springs is given by

$$1/k_{eff} = 1/k_1 + 1/k_2 \qquad (2.9)$$

These results are the reverse of those obtained for resistors in electrical-circuit theory. Equivalent results may be obtained for damping elements, and they have the same form.

2.3.1 The Equation of Motion

To develop the equation of motion for the system in Figure 2.2, we use Newton's second law, which states that the rate of change of momentum of a free body is equal to the sum of the forces acting on it. For a body with a constant mass, the rate of change of momentum is equal to the mass multiplied by the acceleration of the body. Referring to the mass in the free-body diagram for the system, Figure 2.4, we have

$$m\ddot{x} = f(t) - f_s - f_d \qquad (2.10)$$

where f_s and f_d are the forces acting on the mass due to the spring and the damper, respectively. Following the usual convention, a single dot indicates the derivative with respect to time; two dots indicate the second derivative with respect to time.

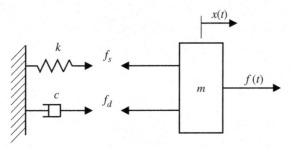

Figure 2.4. Free-body diagram for the single degree of freedom example.

The stiffness, k, of a linear spring is defined as the force required to obtain a unit extension in the spring. Thus, using the notation of this application, $f_s = kx$. The damping coefficient, c, for a viscous damper is defined as the force required to obtain a unit velocity across the damper. Hence, we have $f_d = c\dot{x}$. Substituting for f_s and f_d in Equation (2.10) and rearranging gives

$$m\ddot{x} + c\dot{x} + kx = f(t) \qquad (2.11)$$

Dividing this ordinary differential equation (ODE) by m and letting $c/m = 2\zeta\omega_n$ and $k/m = \omega_n^2$, we have

$$\ddot{x} + 2\zeta\omega_n\dot{x} + \omega_n^2 x = f(t)/m \qquad (2.12)$$

Before solving Equation (2.12), we consider the system in Figure 2.5, which consists of a disk connected to a shaft and a torsional damper. An external torque, $\tau(t)$, is applied to the disk. The motion $\theta(t)$ is such that the shaft is twisted and behaves as a torsional spring. The shaft and torsional damper are assumed to have a negligible polar moment of inertia compared to the disk. The free-body diagram for the disk is shown in Figure 2.6. In the case of rotation about a fixed axis, Newton's second law may be stated as follows: The sum of the torques acting on a body rotating about a fixed axis is equal to the rate of change of angular momentum. The rate of change of angular momentum is the product of the polar moment of inertia of the body about its axis of rotation and the angular acceleration. The sum of the torques acting on

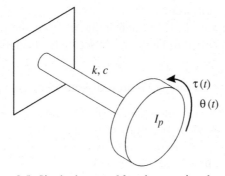

Figure 2.5. Single degree of freedom torsional example.

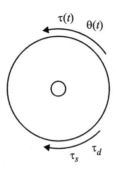

Figure 2.6. Free-body diagram for the disk in the single degree of freedom torsional example.

the disk is $\tau(t) - \tau_s - \tau_d$ (see Figure 2.6); thus,

$$I_p \ddot{\theta} = \tau(t) - \tau_s - \tau_d \qquad (2.13)$$

where I_p is the polar moment of inertia of the disk and τ_s and τ_d are the torques due to the stiffness and damping. The shaft behaves as a torsional spring with a torsional stiffness $k_r = \tau_s / \theta$. For a circular shaft, we can calculate this stiffness from the shaft dimensions because $\tau_s / J = G\theta / L$, where J is the polar second moment of area of the shaft, L is the length of the shaft, and G is a material property (i.e., the modulus of rigidity). The damping coefficient is defined for a torsional viscous damping device by $c_r = \tau_d / \dot{\theta}$. Thus, $\tau_s = k_r \theta$ and $\tau_d = c_r \dot{\theta}$. Substituting into Equation (2.13) and rearranging gives

$$I_p \ddot{\theta} + c_r \dot{\theta} + k_r \theta = \tau(t) \qquad (2.14)$$

Dividing this equation by I_p gives

$$\ddot{\theta} + 2\zeta \omega_n \dot{\theta} + \omega_n^2 \theta = \tau(t)/I_p \qquad (2.15)$$

where $c_r / I_p = 2\zeta \omega_n$ and $k_r / I_p = \omega_n^2$.

The two systems in Figures 2.2 and 2.5 are physically different, but they are described by differential equations with an identical form: Equations (2.12) and (2.15). In the axial system, Equation (2.12), the displacement x is related to the applied force/unit system mass, f/m. In the torsional system, Equation (2.15), the rotation θ is related to the applied torque/unit system polar moment of inertia, τ/I_p. In all other respects, the equations are identical and the two systems are dynamically identical although physically different.

We now solve Equation (2.12) to determine the free and forced vibrations of a single degree of freedom system. The procedure to solve Equation (2.15) is identical.

2.3.2 Free Vibrations of a Single Degree of Freedom System

So far, the parameters ω_n and ζ of Equations (2.12) and (2.15) have not been given a physical meaning. However, from a study of the free vibration of a system, physical meanings can be given to these parameters. By free vibration, we mean the vibration that occurs due to an initial disturbance of the system alone. Thus, to study free vibrations, we remove the dynamic force $f(t)$ and the dynamic torque $\tau(t)$ from the axial and torsional models, respectively. For convenience, we only consider the free vibration of the axial model; the results for the torsional model are analogous.

To remove $f(t)$ from our model, we set it to zero; Equation (2.12) becomes the homogeneous equation

$$\ddot{x} + 2\zeta\omega_n\dot{x} + \omega_n^2 x = 0 \tag{2.16}$$

This equation can be solved by assuming a solution of the form $x(t) = x_0 e^{st}$ for some complex constants x_0 and s. Although assuming a complex solution for the real response $x(t)$ may seem inconsistent, it will be demonstrated that these solutions must occur in complex conjugate pairs. On introducing real initial conditions, the actual solution is then the sum of these complex functions, and this sum produces a real response. Equation (2.16) becomes

$$\left(s^2 + 2\zeta\omega_n s + \omega_n^2\right) x_0 e^{st} = 0 \tag{2.17}$$

Because $x_0 e^{st} \neq 0$, then $s^2 + 2\zeta\omega_n s + \omega_n^2 = 0$. Thus,

$$s = \omega_n\left(-\zeta \pm \sqrt{\zeta^2 - 1}\right) \tag{2.18}$$

Generally, there are two roots, which we call s_1 and s_2; if these roots are complex, they occur as a complex conjugate pair.

We begin by considering the case when the damping is zero, so that $c = 0$; hence, $\zeta = 0$. Thus, $s = \pm j\omega_n$, where ω_n is real and positive. The time response then becomes

$$x(t) = x_0 e^{j\omega_n t} + \bar{x}_0 e^{-j\omega_n t} \tag{2.19}$$

where x_0 is a complex constant that depends on the initial conditions, $\bar{x}_0$ denotes the complex conjugate of x_0, and $j = \sqrt{-1}$. Recalling that the sine and cosine functions can be expressed as sums and differences of complex exponentials, the previous equation can be rearranged to give

$$x(t) = a_0 \cos \omega_n t + b_0 \sin \omega_n t \tag{2.20}$$

where a_0 and b_0 are real constants determined from the initial conditions. Alternatively, we can write Equation (2.20) as

$$x(t) = c_0 \cos(\omega_n t - \phi) \tag{2.21}$$

where the magnitude, $c_0 = \sqrt{a_0^2 + b_0^2}$, and the phase angle, ϕ, is such that $\tan\phi = b_0/a_0$. This equation shows that the system vibrates freely with a frequency ω_n, called the *natural frequency* of the system. The time to complete one oscillation of the system is called the *period*, T. The period is related to the natural frequency by $T = 2\pi/\omega_n$. Figure 2.7 is a plot of the solution of Equation (2.16) with $\zeta = 0$ when the mass is given an initial displacement, x_0, and zero initial velocity. The amplitude of the oscillations does not change with time.

Consider now the situation that pertains when $0 < \zeta < 1$. Because $c/m = 2\zeta\omega_n$, then c must lie in the range $0 < c < 2m\omega_n$. In this case, because $\zeta < 1$, Equation (2.18) may be rearranged as

$$s = \omega_n\left(-\zeta \pm j\sqrt{1 - \zeta^2}\right) = -\zeta\omega_n \pm j\omega_d \tag{2.22}$$

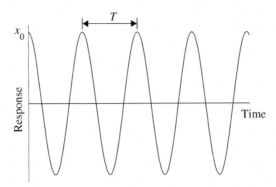

Figure 2.7. Response of the mass for the undamped single degree of freedom example ($\zeta = 0$). The initial displacement is x_0 and the time period is T.

where $\omega_d = \omega_n\sqrt{1 - \zeta^2}$. The solution of Equation (2.16) is obtained using Equation (2.22) and is

$$x(t) = x_0 e^{(-\zeta\omega_n + J\omega_d)t} + \bar{x}_0 e^{(-\zeta\omega_n - J\omega_d)t} \tag{2.23}$$

Rearranging gives

$$x(t) = e^{-\zeta\omega_n t}\left(a_0 \cos\omega_d t + b_0 \sin\omega_d t\right) \tag{2.24}$$

where a_0 and b_0 are real constants, determined from the initial conditions. Alternatively, Equation (2.24) can be written as

$$x(t) = c_0 e^{-\zeta\omega_n t} \cos\left(\omega_d t - \phi\right) \tag{2.25}$$

Figure 2.8 is a plot of the solution of Equation (2.16) with the same initial conditions as in Figure 2.7 but with $\zeta = 0.1$. Comparing Equations (2.21) and (2.25) with the aid of Figures 2.7 and 2.8, we see that the effect of the introduction of damping is twofold: (1) an exponential decay term has been introduced causing the amplitude of the oscillation to decay exponentially to zero; and (2) the frequency of oscillation has changed from ω_n to ω_d, the *damped natural frequency*. In many practical situations, ζ is small so that ω_d is very close to ω_n. The period of oscillation, T, is now given by $T = 2\pi/\omega_d$.

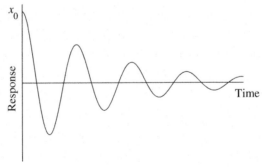

Figure 2.8. Response of the mass for the single degree of freedom example, $\zeta = 0.1$. The initial displacement is x_0.

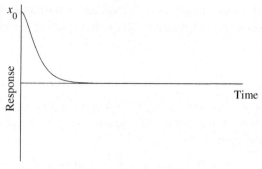

Figure 2.9. Response of the mass for the single degree of freedom example, $\zeta = 1$. The initial displacement is x_0.

We now consider the situation that exists if $\zeta = 1$; that is, $c = 2m\omega_n$. Equation (2.22) is

$$s = -\omega_n \quad \text{(twice)} \tag{2.26}$$

Here, there is only one root and the solution to Equation (2.16) (Inman, 2008) is given by

$$x(t) = (a_0 + b_0 t)\, e^{-\omega_n t} \tag{2.27}$$

Figure 2.9 is a plot of the solution of Equation (2.16) with $\zeta = 1$ and the same initial conditions as in Figure 2.7. The system no longer oscillates and $\zeta = 1$ is a critical value that just prevents oscillations from occurring. The damping is called the *critical damping*, c_c, where $c_c = 2m\omega_n$. Because, in general, $\zeta = c/2m\omega_n$, then $\zeta = c/c_c$. Thus, ζ is the ratio between the actual damping in a system and the damping required to just prevent vibrations from occurring; ζ is called the *damping ratio* or *damping factor*.

Sometimes ζ can be greater than 1 (e.g., in machines with fluid bearings). The system is now *overdamped*. In this case, Equation (2.22) gives two real and negative roots, s_1 and s_2, and the solution to Equation (2.16) is given by

$$x(t) = a_0 e^{s_1 t} + b_0 e^{s_2 t} \tag{2.28}$$

The transient response is similar to the case when $\zeta = 1$, shown in Figure 2.9.

2.3.3 Forced Vibrations

We now consider the situation that arises when a dynamic force is applied to the system. Although the forcing function can take various forms, here we only consider harmonic (i.e., sinusoidal) forces for several reasons. Harmonic forces are widely used in vibration testing and they cause a response that is easily calculated and interpreted. Harmonic and periodic forces arise naturally in rotordynamics, and periodic forces can be decomposed into the sum of harmonic forces using Fourier analysis.

Suppose that the force applied to the system is harmonic with frequency ω, so that $f(t) = f_0 \cos \omega t$. From Equation (2.12), the equation of motion of the system is now

$$\ddot{x} + 2\zeta \omega_n \dot{x} + \omega_n^2 x = \frac{f_0}{m} \cos \omega t = \Re \left(\frac{f_0}{m} e^{J\omega t} \right) \tag{2.29}$$

where $\Re$ denotes the real part of a complex number or quantity. For a linear system, the *steady-state response* is harmonic with the same frequency as the excitation force. The solution is of the form

$$x_s(t) = \Re \left(\alpha(\omega) f_0 e^{J\omega t} \right) = f_0 (A_s \cos \omega t + B_s \sin \omega t) \tag{2.30}$$

where $\alpha(\omega)$ is the frequency response function (FRF), also called the *receptance*. Substituting the complex trial solution into Equation (2.29) using

$$\dot{x}_s(t) = \Re \left(J\omega \alpha(\omega) f_0 e^{J\omega t} \right), \qquad \ddot{x}_s(t) = \Re \left(-\omega^2 \alpha(\omega) f_0 e^{J\omega t} \right) \tag{2.31}$$

gives

$$\alpha(\omega) = \frac{1/m}{-\omega^2 + 2\zeta \omega_n J\omega + \omega_n^2} = \frac{1/(m\omega_n^2)}{1 - r^2 + 2\zeta Jr} \tag{2.32}$$

where r is the frequency ratio, given by $r = \omega/\omega_n$. Equation (2.32) may be written in terms of A_s and B_s, using Equation (2.30), as

$$A_s = \frac{(1 - r^2)/(m\omega_n^2)}{(1 - r^2)^2 + (2\zeta r)^2}, \qquad B_s = \frac{2\zeta r/(m\omega_n^2)}{(1 - r^2)^2 + (2\zeta r)^2} \tag{2.33}$$

The solution also may be expressed as

$$x_s(t) = f_0 C_s \cos(\omega t - \phi_s) \tag{2.34}$$

where the magnitude, C_s, and phase shift, ϕ_s, are

$$C_s = \frac{1/(m\omega_n^2)}{\sqrt{(1 - r^2)^2 + (2\zeta r)^2}}, \qquad \phi_s = \tan^{-1} \frac{2\zeta r}{1 - r^2} \tag{2.35}$$

and

$$\alpha(\omega) = C_s e^{-J\phi_s} \tag{2.36}$$

The general solution for the motion of the system is the sum of the steady-state response given previously and the transient response, which has the same form as Equation (2.25) for an underdamped system. Thus,

$$x(t) = f_0 C_s \cos(\omega t - \phi_s) + C_t e^{-\zeta \omega_n t} \cos(\omega_d t - \phi_t) \tag{2.37}$$

The unknown constants in the transient response, C_t and ϕ_t, are now obtained from the initial conditions of the system. If the system is damped, then the transient response becomes small as t increases, and we may consider the steady-state response alone. The lighter the damping, the longer it takes for the steady-state response to dominate the transient response. Figure 2.10 shows the magnitude and phase for a force of unit magnitude per unit mass for a number of damping ratios. For light damping, the magnitude of the response approaches a maximum as the excitation frequency approaches the natural frequency. The maximum magnitude

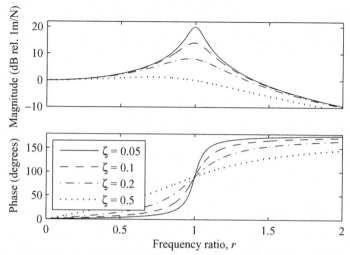

Figure 2.10. Receptance for the single degree of freedom example, for damping ratios, ζ, of 0.05, 0.1, 0.2, and 0.5.

occurs at $\omega = \omega_n\sqrt{1 - 2\zeta^2}$ and is $f_0/(2m\zeta\omega_n^2)$. Because the magnitude at zero frequency is $f_0/(m\omega_n^2)$, we may define the following ratio

$$Q = \frac{\text{maximum } C_s}{C_s \text{ at zero frequency}} = \frac{1}{2\zeta} \qquad (2.38)$$

The ratio Q is called the *Q factor*, or *dynamic magnifier*. If, for example, a system has a damping ratio of 0.005 (or 0.5 percent, which is typical of metallic structures), then $Q = 100$ and the displacement caused by a dynamic force can be up to 100 times that of a static force of the same magnitude. Furthermore, the force in the spring can be 100 times the applied force. For heavily damped systems (e.g., $\zeta = 0.5$, shown in Figure 2.10), the resonant peak has disappeared altogether. Also notice that the phase shift, given in Equation (2.35), crosses through 90° as the excitation frequency passes the undamped natural frequency. This crossing point defines the resonance for the damped case. An alternative definition of the frequency of resonance is the frequency of maximum response. For small damping ratios, these two definitions of resonance frequency produce similar results, as shown in Figure 2.10.

Although the parameter Q is defined by Equation (2.38), it can be shown that for a lightly damped system, an equivalent definition for Q is

$$Q = \omega_{pk}/\omega_{bw} \qquad (2.39)$$

where ω_{pk} and ω_{bw} are defined in Figure 2.11. The *half-power bandwidth* is defined as the difference in the frequencies at the half-power points – that is, at a power level of one half the peak-power level of the signal. Thus, the bandwidth is defined at a response level of $1/\sqrt{2}$ of the peak level of response. This definition is frequently used in the measurement of Q because it is relatively easy to estimate ω_{pk} and ω_{bw}, providing that the frequency resolution is sufficiently small for lightly damped systems.

The force acting on the system is given by the right side of Equation (2.29) and the response is given by Equation (2.34). If the excitation frequency is 2 Hz (so that

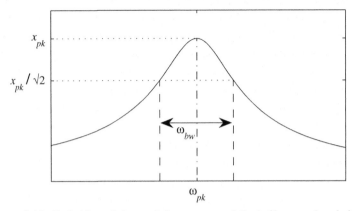

Figure 2.11. Definition of the peak frequency and the half-power bandwidth.

$T = 0.5\,\text{s}$) and the phase angle, ϕ_s, is $60°$, then the force and response harmonic waves are shown in Figure 2.12. The response is said to be *lagging* the force by $60°$; this is shown in Figure 2.12. Because we read from left to right and, hence, we equate this with progress from left to right, it is easy to delude ourselves into thinking that the response is leading the force because it is to the right of the force. This is a wholly incorrect interpretation. More careful consideration shows that the response reaches a peak value at a later time than the force; hence, it is lagging the force. Alternatively, we can say that the force is *leading* the response by $60°$. If a harmonic wave lags the reference wave by more than $180°$, then we normally express it as leading; that is, a lag of $200°$ is a lead of $160°$.

2.3.4 Nonviscous Damping

Thus far, only viscous damping has been considered. Viscous damping is simple to apply in mathematical models, although it does not always model the actual energy dissipation particularly well. A frequency-dependent viscous damping model is used to model bearing damping in Chapter 5. Damping, in general, consists of a range of complex mechanisms that are difficult to model and, including some dissipation,

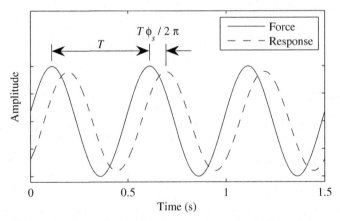

Figure 2.12. The interpretation of the phase angle, ϕ_s, in the forced time response.

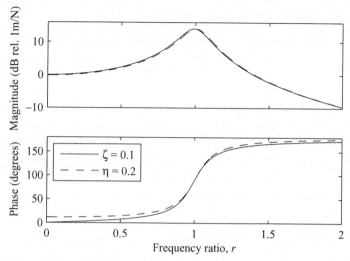

Figure 2.13. Comparison of the receptance for systems with viscous and hysteretic damping.

does allow a first approximation of the response to be calculated. In structural dynamics, *hysteretic* (also called *structural*) damping is often used for modeling purposes. Hysteretic damping is valid only for harmonic excitation and may be considered as a frequency-dependent viscous damping coefficient, in which the damping force is $f_d = c\dot{x}/\omega$. An equivalent modeling approach assumes a complex stiffness, given by $k(1 + j\eta)$, with $c = \eta k$. The time response to a harmonic excitation is given by Equation (2.30), where now

$$A_s = \frac{(1 - r^2)/(m\omega_n^2)}{(1 - r^2)^2 + \eta^2}, \qquad B_s = \frac{\eta/(m\omega_n^2)}{(1 - r^2)^2 + \eta^2} \qquad (2.40)$$

Figure 2.13 compares the magnitude and phase of the response with viscous and hysteretic damping, with $\zeta = 0.1$ and $\eta = 0.2$. There is little difference near resonance, which may be reduced by adjusting slightly the equivalent natural frequencies. There is more difference away from resonance; in particular, the static response ($r = 0$) for a hysteretically damped system does not have a zero phase.

Another form of damping that can arise is *dry friction*, or *Coulomb damping*. Here, the damping force is assumed to be constant and opposes the direction of motion. Thus, $f_{dry} = f_0\dot{x}/|\dot{x}|$, where f_{dry} is a constant force due to friction. The free-vibration problem is readily solved by determining the points where the velocity is zero and the constant force changes sign. The decay of the free response is linear rather than exponential in the case of viscous damping. Coulomb damping is nonlinear, and the equations of motion for forced vibrations must be solved numerically or by some approximation method. However, if the friction force is small compared to the harmonic force acting on the system, we can linearize the effect of the damping by determining an equivalent viscous-damping coefficient by equating the work done on the damper when it is displaced harmonically. For a harmonic displacement $x(t) = x_0 \sin(\omega t)$, the work done in one period on a Coulomb damper is $W = 4f_{dry}x_0$. By comparison, the work done on a viscous damper during one period is $W = \pi c\omega x_0^2$. Equating the work done, the equivalent viscous damping is $c_e = 4f_{dry}/(\pi\omega x_0)$.

2.3.5 Forced Vibration: Periodic Excitation

The analysis so far has concentrated on harmonic excitation – that is, sinusoidal forcing. Other forms of forcing also may occur. Often, the excitation is periodic – for example, excitation due to gear meshing. General, nonperiodic excitation may be modeled approximately as a periodic excitation when the period is very long. Indeed, this approach is adopted for measuring force or response when measurements are performed during a limited period. Analysis of the response to such forcing is made easier by two features: (1) the linear superposition principle, and (2) the Fourier decomposition of periodic force signals. Any periodic signal may be written as a series of sinusoidal terms, called a *Fourier series* (Newland, 1984). Thus, a force $f(t)$, which is periodic, with period T may be written as

$$f(t) = a_0 + \sum_{n=1}^{\infty} (a_n \cos \omega_0 nt + b_n \sin \omega_0 nt) = \Re \left(\sum_{n=0}^{\infty} c_n e^{J\omega_0 nt} \right) \qquad (2.41)$$

where $\omega_0 = 2\pi/T$ is the fundamental frequency of the force, and

$$a_n = \frac{2}{T} \int_0^T f(t) \cos \omega_0 nt \, dt, \qquad b_n = \frac{2}{T} \int_0^T f(t) \sin \omega_0 nt \, dt \qquad (2.42)$$

for $n \geq 1$, and

$$a_0 = \frac{1}{T} \int_0^T f(t) \, dt, \qquad b_0 = 0 \qquad (2.43)$$

Combining the real constants gives the complex constants

$$c_n = a_n - j b_n \qquad (2.44)$$

Once the force is written as the Fourier series, Equation (2.41), the response to each term in the series may be computed using the techniques described in Sections 2.3.3 and 2.3.4. The superposition principle for linear systems means that the response to a sum of forces is equal to the sum of the response to each force individually. Thus, from Equation (2.30), the steady-state response to the force in Equation (2.41) is

$$x_s(t) = \Re \left(\sum_{n=0}^{\infty} \alpha(\omega_0 n) c_n e^{J\omega_0 nt} \right) \qquad (2.45)$$

EXAMPLE 2.3.1. Find the Fourier series of a square wave force with a minimum of 0 and a maximum of 1 N, with a fundamental frequency of 1 Hz. Find the response to this force of a single degree of freedom system with mass $m = 1$ kg, damping ratio $\zeta = 0.06$, and natural frequency $= 7$ Hz.

Solution. For this example, $T = 1$ s and $\omega_0 = 2\pi$ rads. With an appropriate choice of time origin, the force is an even function of time; therefore, the b_n

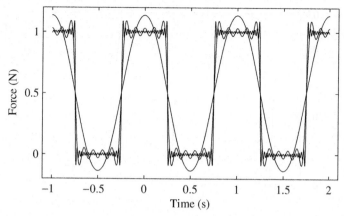

Figure 2.14. The Fourier series of a square wave force for 2, 10, and 40 terms (Example 2.3.1). The larger the number of terms, the better is the approximation to the square wave.

terms are all zero. The a_n coefficients are computed from Equations (2.42) and (2.43) as $a_0 = 0.5$, and

$$a_n = \frac{2}{T} \int_0^T f(t) \cos \omega_0 nt \, dt = 2 \int_0^{0.25} \cos 2\pi nt \, dt + 2 \int_{0.75}^1 \cos 2\pi nt \, dt$$

$$= \begin{cases} \dfrac{2(-1)^{(n-1)/2}}{\pi n} & n \text{ odd} \\ 0 & n \text{ even} \end{cases}$$

Figure 2.14 shows the Fourier series approximation to the force for increasing numbers of terms in Equation (2.41). The "overshoot" before and after each discontinuity in the square wave is known as the *Gibbs phenomenon*.

Figure 2.15 shows the response to the square wave force obtained by summing the response in Equation (2.45). The natural frequency of the system is close to a harmonic of the forcing (at 7 Hz). A *harmonic* is a component frequency of a signal that is an integer multiple of the fundamental frequency. The response is a combination of the transient response at this natural frequency and the steady-state response at the forcing frequency. Although the force is poorly approximated using 40 terms, the response is accurately approximated using this number because the system response is small at the higher frequencies.

2.3.6 Forced Vibration: Arbitrary Excitation

Suppose that the system is excited by a force $f(t)$, so that the equation of motion is

$$\ddot{x} + 2\zeta \omega_n \dot{x} + \omega_n^2 x = \frac{1}{m} f(t) \tag{2.46}$$

The resulting response may be written as the *convolution* or *Duhamel integral* (Inman, 2008)

$$x(t) = \int_0^t f(\tau) h(t - \tau) \, d\tau \tag{2.47}$$

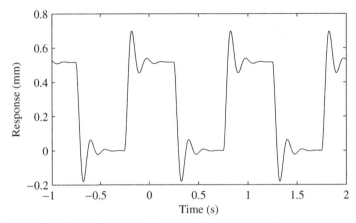

Figure 2.15. The response of a single degree of freedom system to a square wave force (with 40 terms) (Example 2.3.1).

where the *impulse response function* is

$$h(t) = \frac{1}{m\omega_d} e^{-\zeta \omega_n t} \sin \omega_d t \tag{2.48}$$

For a given system and excitation force, this integral may be computed either analytically or numerically. An alternative approach is to integrate the equation of motion, Equation (2.46), directly using a numerical procedure. Another approach to calculate the response to an arbitrary force uses Fourier transforms. Section 2.6 discusses these latter two issues in more detail.

2.4 Multiple Degrees of Freedom Systems

Although single degree of freedom models can provide insight into the dynamics of some real systems, many systems require more complex models with more degrees of freedom. This introduces complexity in the analysis and calculations because these systems feature multiple natural frequencies, each with a corresponding mode of vibration. Here, we consider only systems with discrete components.

2.4.1 System Equations

For systems involving discrete components, the equations of motion may be written in terms of the displacements of the system masses: $q_1, q_2, \ldots, q_n$. This choice of coordinates is not unique and is chosen for simplicity. Other choices may be more appropriate for different systems, although the coordinates must be independent of one another. These are called *generalized coordinates*; the number required is fixed and is called the *number of degrees of freedom*. Generalized coordinates are discussed in more detail in Section 4.2. The approach adopted here to generate the governing equations draws the necessary free-body diagrams for each mass, specifies the forces acting on each mass, and applies Newton's second law. For large, complex problems and for approximating continuous structures, energy methods are more appropriate (see Chapter 4).

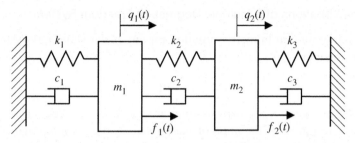

Figure 2.16. The two degrees of freedom discrete example.

Figures 2.16 and 2.17 show a two degrees of freedom discrete system and the associated free-body diagrams. Applying Newton's second law to each mass in turn gives

$$
\begin{aligned}
m_1 \ddot{q}_1 &= f_1(t) - f_{s1} + f_{s2} - f_{d1} + f_{d2} \\
&= f_1(t) - k_1 q_1 + k_2 (q_2 - q_1) - c_1 \dot{q}_1 + c_2 (\dot{q}_2 - \dot{q}_1) \\
m_2 \ddot{q}_2 &= f_2(t) - f_{s2} + f_{s3} - f_{d2} + f_{d3} \\
&= f_2(t) - k_2 (q_2 - q_1) + k_3 q_2 - c_2 (\dot{q}_2 - \dot{q}_1) + c_3 \dot{q}_2
\end{aligned}
\tag{2.49}
$$

or, in matrix form, after rearranging

$$
\begin{aligned}
\begin{bmatrix} m_1 & 0 \\ 0 & m_2 \end{bmatrix} \begin{Bmatrix} \ddot{q}_1 \\ \ddot{q}_2 \end{Bmatrix} + \begin{bmatrix} c_1 + c_2 & -c_2 \\ -c_2 & c_2 + c_3 \end{bmatrix} \begin{Bmatrix} \dot{q}_1 \\ \dot{q}_2 \end{Bmatrix} & \\
+ \begin{bmatrix} k_1 + k_2 & -k_2 \\ -k_2 & k_2 + k_3 \end{bmatrix} \begin{Bmatrix} q_1 \\ q_2 \end{Bmatrix} &= \begin{Bmatrix} f_1(t) \\ f_2(t) \end{Bmatrix}
\end{aligned}
\tag{2.50}
$$

which is in the standard form for a multiple degrees of freedom structural dynamics problem. In Equation (2.49), Hooke's law is used to relate the stiffness force to the displacement and the viscous damping law is used to relate the damping force to the velocity.

Whenever it is necessary to develop the equations of motion of a system from discrete components using Newton's laws of motion (in contrast to an energy method), the procedure that must always be adopted draws free-body diagrams and specifies the unknown forces of interaction and then eliminates them by using Hooke's law or similar constitutive relations. However, in this book, we frequently do not show the full process and merely quote the resulting equations.

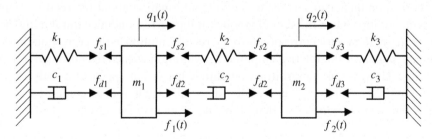

Figure 2.17. Free-body diagrams for the two degrees of freedom discrete example.

2.4.2 Free Vibrations of a Multiple Degrees of Freedom System

The equation of motion in matrix notation for any linear stationary system with n degrees of freedom is

$$\mathbf{M}\ddot{\mathbf{q}} + \mathbf{C}\dot{\mathbf{q}} + \mathbf{K}\mathbf{q} = \mathbf{Q}(t) \qquad (2.51)$$

where $\mathbf{M}$, $\mathbf{C}$, and $\mathbf{K}$ are $n \times n$ matrices that are usually symmetric.[†] These matrices are usually referred to as the *mass* or *inertia matrix*, the *damping matrix*, and the *stiffness matrix*, respectively. The generalized displacement, $\mathbf{q}(t)$, and the generalized force, $\mathbf{Q}(t)$, are vectors of length n. Equation (2.50) is a particular case of this equation. For free vibration, $\mathbf{Q}(t) = \mathbf{0}$; the undamped case is given by $\mathbf{C} = \mathbf{0}$.

We now assume a solution for free undamped vibration of the form $\mathbf{q}(t) = \mathbf{u}e^{st}$, where in this case $\mathbf{u}$ is a constant real vector. Thus

$$\ddot{\mathbf{q}} = s^2 \mathbf{u}e^{st} \qquad (2.52)$$

and substituting into the equation of motion gives

$$\left[s^2\mathbf{M} + \mathbf{K}\right]\mathbf{u} = \mathbf{0} \quad \text{or} \quad \omega_n^2\mathbf{M}\mathbf{u} = \mathbf{K}\mathbf{u} \qquad (2.53)$$

where $s = \pm \jmath\omega_n$ and ω_n is a natural frequency of the multiple degrees of freedom system and is real and positive. The exponential term has been canceled from these equations because it is generally non-zero.

Equation (2.53) is an eigenvalue problem. It can be rearranged into several forms, such as

$$\mathbf{D}\mathbf{u} = \lambda\mathbf{u} \qquad (2.54)$$

where $\lambda = 1/\omega_n^2 = -1/s^2$ and $\mathbf{D} = \mathbf{K}^{-1}\mathbf{M}$ if the stiffness matrix is nonsingular. $\mathbf{D}$ is sometimes referred to as the *dynamical matrix* (Meirovitch, 1986). There are cases in which the stiffness matrix is singular and, for certain displacement vectors $\mathbf{u}$, $\mathbf{K}\mathbf{u} = \mathbf{0}$. These vectors represent *rigid-body modes*, which are natural modes of the system with a resonance frequency of zero. Although the dynamical matrix cannot be calculated in this case, alternative methods of analysis are available.

For an undamped system with n degrees of freedom, $\mathbf{D}$ is an $n \times n$ matrix, and the complete solution for the eigenvalue problem consists of n values of λ (the eigenvalues) and n corresponding values of $\mathbf{u}$ (the eigenvectors). The ith eigenvalue corresponds to two values of s, such that $\lambda_i = -1/s_i^2$. The values of s_i occur in complex conjugate pairs, given by $s_i = +\jmath\omega_i$ and $s_{n+i} = -\jmath\omega_i$, such that each pair corresponds to one of the n real and positive natural frequencies, ω_i, for $i = 1, 2, \ldots, n$. The corresponding eigenvectors, $\mathbf{u}_i$, provide information about the relative amplitudes of vibration when the system vibrates at the corresponding natural frequency. Each entry in the vector corresponds to the relative motion at the corresponding coordinate, and $u_{\ell i}$ is the relative amplitude of vibration at the ℓth coordinate for the ith mode. However, $\mathbf{u}_i$ has arbitrary amplitude because it appears on both sides of Equation (2.54); therefore, if $\mathbf{u}_i$ is a solution, so also is $\beta\mathbf{u}_i$ for any complex scalar β. The scaling may be chosen so that all of the elements of $\mathbf{u}_i$ are real. Thus, if one

---

[†] Systems involving fluid–structure interactions or systems with rotating parts may lead to nonsymmetric damping and stiffness matrices. The latter case is considered from Chapter 3 onward.

element is chosen arbitrarily, the remaining elements are fixed and $\mathbf{u}_i$ gives the spatial shape of the vibrations. This shape is called the *mode shape* or the *normal mode of vibration*. These mode shapes are as important as the natural frequencies in the characterization of a system.

The eigenvalues may be obtained by recognizing that Equations (2.53) and (2.54) are equivalent to

$$\left[s^2 \mathbf{M} + \mathbf{K}\right] \mathbf{u} = \mathbf{0} \quad \text{or} \quad [\mathbf{D} - \lambda \mathbf{I}] \mathbf{u} = \mathbf{0} \tag{2.55}$$

For nontrivial solutions to Equation (2.55), the coefficient matrix must be singular, and the eigenvalues are obtained as the solution of the characteristic equation

$$\det\left[s^2 \mathbf{M} + \mathbf{K}\right] = 0 \quad \text{or} \quad \det[\mathbf{D} - \lambda \mathbf{I}] = 0 \tag{2.56}$$

These determinants give a polynomial of degree n in either s^2 or λ. The terms *root* and *eigenvalue* are interchangeable because they are essentially the same: the eigenvalue is obtained from the eigenvalue problem and the root is obtained from the characteristic equation, which are equivalent. The mode shape, $\mathbf{u}_i$, is then obtained as any nonvanishing column of the adjoint matrix (Horn and Johnson, 1985) given by

$$\text{adj}\left[s_i^2 \mathbf{M} + \mathbf{K}\right] \quad \text{or} \quad \text{adj}[\mathbf{D} - \lambda_i \mathbf{I}] \tag{2.57}$$

Alternatively, one element of the vector $\mathbf{u}$ may be fixed in Equation (2.55), with $s = s_i$ or $\lambda = \lambda_i$, and the remaining elements of $\mathbf{u}$ are obtained from the solution of any $n - 1$ linear equations from the n available (Meirovitch, 1986).

If the number of degrees of freedom in the analysis, n, is three or less, then the eigenvalues may be readily obtained from the characteristic equation. When n is larger, a computer is used to solve the eigenvalue problem directly. There are many procedures to solve the equations, and choosing the best procedure depends on several factors: the size of the problem (because some procedures are more efficient when solving large problems); whether all of the eigenvalues are required or only those in a range of frequencies; whether the eigenvectors are required as well as the eigenvalues; and so on. Such methods are beyond the scope of this book, and Newland (1989), Petyt (1990), and Wilkinson (1965) should be consulted for further details.

The mode shapes satisfy the so-called *orthogonality relationships*. It can be proved[†] that

$$\mathbf{u}_i^\top \mathbf{M} \mathbf{u}_\ell = \mathbf{u}_i^\top \mathbf{K} \mathbf{u}_\ell = 0, \quad \text{for } i, \ell = 1, 2, \dots, n, \text{ and } i \neq \ell \tag{2.58}$$

When $i = \ell$, the products $\mathbf{u}_i^\top \mathbf{M} \mathbf{u}_i$ and $\mathbf{u}_i^\top \mathbf{K} \mathbf{u}_i$ are generally not zero. Thus,

$$\mathbf{u}_i^\top \mathbf{M} \mathbf{u}_i = m_i \quad \text{and} \quad \mathbf{u}_i^\top \mathbf{K} \mathbf{u}_i = k_i \quad \text{for } i = 1, 2, \dots, n \tag{2.59}$$

In Equation (2.59), $\mathbf{M}$ is positive-definite; hence, m_i are positive constants. $\mathbf{K}$ is positive-semidefinite; hence, k_i are non-negative constants. However, these constants are not unique because $\mathbf{u}_i$ is not unique. The ratio of k_i and m_i is unique and

---

[†] Equation (2.58) may not hold in cases of repeated eigenvalues; however, even when the eigenvalues are repeated, often the eigenvectors may be chosen to ensure orthogonality and certainly when the matrices are symmetric.

$k_i/m_i = \omega_i^2$. Because $\mathbf{u}_i$ is not unique, we can scale it to make m_i any value; in particular, we can make m_i equal to 1. Under these conditions, we say that $\mathbf{u}_i$ is *mass normalized* and write it as $\mathbf{u}_{Ni}$. Thus,

$$\mathbf{u}_{Ni}^{\top}\mathbf{M}\mathbf{u}_{Ni} = 1 \quad \text{and} \quad \mathbf{u}_{Ni}^{\top}\mathbf{K}\mathbf{u}_{Ni} = \omega_i^2 \quad \text{for} \quad i = 1, 2, \ldots, n \tag{2.60}$$

The orthogonality equations, of course, are still valid in terms of $\mathbf{u}_{Ni}$. To determine $\mathbf{u}_{Ni}$, it is necessary to first calculate m_i from $\mathbf{u}_i^{\top}\mathbf{M}\mathbf{u}_i$ and then divide each element of $\mathbf{u}_i$ by $\sqrt{m_i}$. Thus, $\mathbf{u}_{Ni} = \mathbf{u}_i/\sqrt{m_i}$. In the analysis that follows, it is not essential to use mass normalized vectors but their use makes the analysis clearer; for that reason, they are used here.

The mass normalized mode shapes may be combined to give the normalized modal matrix $\mathbf{U}_N$ as

$$\mathbf{U}_N = [\mathbf{u}_{N1} \quad \mathbf{u}_{N2} \quad \ldots \quad \mathbf{u}_{Nn}] \tag{2.61}$$

Thus, to create this matrix, we simply place the n eigenvectors in an array. Because each vector is n elements long, the array is $n \times n$. The spectral matrix, $\mathbf{\Lambda}$, is defined as

$$\mathbf{\Lambda} = \begin{bmatrix} \ddots & & \\ & \omega_i^2 & \\ & & \ddots \end{bmatrix} = \begin{bmatrix} \omega_1^2 & 0 & \vdots & 0 \\ 0 & \omega_2^2 & \vdots & 0 \\ \cdots & \cdots & \ddots & 0 \\ 0 & 0 & \vdots & \omega_n^2 \end{bmatrix} \tag{2.62}$$

As a consequence of the orthogonality conditions and the method of mass normalization of the eigenvectors,

$$\mathbf{U}_N^{\top}\mathbf{M}\mathbf{U}_N = \mathbf{I} \quad \text{and} \quad \mathbf{U}_N^{\top}\mathbf{K}\mathbf{U}_N = \mathbf{\Lambda} \tag{2.63}$$

The free response of the structure may be written in terms of the natural frequencies and mode shapes as the sum of the trial functions

$$\mathbf{q}(t) = \sum_{i=1}^{2n} c_i e^{-J\phi_i} \mathbf{u}_{Ni} e^{s_i t} = \sum_{i=1}^{n} c_i \mathbf{u}_{Ni} \left(e^{-J\phi_i} e^{J\omega_i t} + e^{J\phi_i} e^{-J\omega_i t} \right) \tag{2.64}$$

where the real constants c_i and ϕ_i are calculated from the initial displacement and velocity vectors. The second n roots, $s_{n+i} = -J\omega_i$, are complex conjugates of the first n roots, $s_i = J\omega_i$; hence, $c_{n+i} = c_i$ and $\phi_{n+i} = -\phi_i$.

An alternative approach to calculate the free response is to write Equation (2.64) as

$$\mathbf{q}(t) = \sum_{i=1}^{n} \mathbf{u}_{Ni} p_{ci} \cos(\omega_i t) + \sum_{i=1}^{n} \mathbf{u}_{Ni} p_{si} \sin(\omega_i t) \tag{2.65}$$

where p_{ci} and p_{si} are $2n$ unknown coefficients. Differentiating Equation (2.65) gives

$$\dot{\mathbf{q}}(t) = -\sum_{i=1}^{n} \mathbf{u}_{Ni} p_{ci} \omega_i \sin(\omega_i t) + \sum_{i=1}^{n} \mathbf{u}_{Ni} p_{si} \omega_i \cos(\omega_i t) \tag{2.66}$$

If the initial conditions are $\mathbf{q}(0) = \mathbf{q}_0$ and $\dot{\mathbf{q}}(0) = \mathbf{v}_0$, then

$$\mathbf{q}_0 = \sum_{i=1}^{n} \mathbf{u}_{Ni} p_{ci} = \mathbf{U}_N \mathbf{p}_c \tag{2.67}$$

and

$$\dot{\mathbf{q}}_0 = \mathbf{v}_0 = \sum_{i=1}^{n} \mathbf{u}_{Ni} p_{si} \omega_i = \mathbf{U}_N \boldsymbol{\omega} \mathbf{p}_s \tag{2.68}$$

where $\boldsymbol{\omega}$ is a diagonal matrix of the natural frequencies and $\mathbf{p}_c$ and $\mathbf{p}_s$ are vectors of coefficients p_{ci} and p_{si}. Thus,

$$\mathbf{p}_c = \mathbf{U}_N^{-1} \mathbf{q}_0 = \mathbf{U}_N^{\top} \mathbf{M} \mathbf{q}_0 \tag{2.69}$$

and

$$\mathbf{p}_s = \boldsymbol{\omega}^{-1} \mathbf{U}_N^{-1} \mathbf{v}_0 = \boldsymbol{\omega}^{-1} \mathbf{U}_N^{\top} \mathbf{M} \mathbf{v}_0 \tag{2.70}$$

Knowing $\mathbf{p}_c$ and $\mathbf{p}_s$, we can substitute these values into Equation (2.65) to determine the free response of the system to the initial conditions.

EXAMPLE 2.4.1. Calculate the free response of the two degrees of freedom system shown in Figure 2.16, where each mass is 1 kg and each spring has a stiffness of 10 kN/m. The masses are initially at rest with the four initial displacements, $\mathbf{q}(0) = \left\{ \begin{matrix} 1 \\ 1 \end{matrix} \right\}, \left\{ \begin{matrix} 1 \\ -1 \end{matrix} \right\}, \left\{ \begin{matrix} 1 \\ 0 \end{matrix} \right\}$, and $\left\{ \begin{matrix} 0 \\ 1 \end{matrix} \right\}$ mm. Ignore the effect of the dampers.

Solution. The equations of motion are given by Equation (2.50), where the mass and stiffness matrices are

$$\mathbf{M} = \begin{bmatrix} 1 & 0 \\ 0 & 1 \end{bmatrix} \text{kg}, \quad \mathbf{K} = \begin{bmatrix} 2 & -1 \\ -1 & 2 \end{bmatrix} \times 10^4 \text{ N/m}$$

The characteristic equation, Equation (2.56), is then

$$\det \left[s^2 \mathbf{M} + \mathbf{K} \right] = \left(s^2 + 2 \times 10^4 \right)^2 - 10^8$$
$$= s^4 + 4 \times 10^4 s^2 + 3 \times 10^8$$
$$= \left(s^2 + 10^4 \right) \left(s^2 + 3 \times 10^4 \right) = 0$$

giving roots, s_i, of $\pm 100_J$ and $\pm \sqrt{3} \times 100_J$. The adjoint matrices, Equation (2.57), for these two pairs of roots (or eigenvalues) are

$$\begin{bmatrix} -1 & -1 \\ -1 & -1 \end{bmatrix} \times 10^4 \quad \text{and} \quad \begin{bmatrix} 1 & -1 \\ -1 & 1 \end{bmatrix} \times 10^4$$

Each column of the adjoint matrices represents the same mode. In the case of the second mode, the second column is obtained from the first by multiplying by -1. The factor of 10^4 has no significance. The mode shapes derived from these adjoint matrices may be normalized to give the natural frequencies and mass normalized mode shapes as

$$\omega_1 = 100\,\text{rad/s}, \qquad\qquad \mathbf{u}_{N1} = \frac{1}{\sqrt{2}}\begin{Bmatrix}1\\1\end{Bmatrix}\,\text{kg}^{-1/2}$$

$$\omega_2 = 100 \times \sqrt{3} = 173.2\,\text{rad/s}, \qquad \mathbf{u}_{N2} = \frac{1}{\sqrt{2}}\begin{Bmatrix}1\\-1\end{Bmatrix}\,\text{kg}^{-1/2}$$

In this case, $\mathbf{u}_N$ has the dimension $1/[M]^{0.5}$.

The time response of the two masses from Equation (2.64) is

$$\mathbf{q}(t) = c_1 \begin{Bmatrix}1\\1\end{Bmatrix}\cos\left(100t - \phi_1\right) + c_2\begin{Bmatrix}1\\-1\end{Bmatrix}\cos\left(173.2t - \phi_2\right)$$

for some real constants, c_1, ϕ_1, c_2, and ϕ_2. Because the masses are initially at rest, $\dot{\mathbf{q}}(0) = \mathbf{0}$; then, ϕ_1 and ϕ_2 are zero. Furthermore, if both masses are initially displaced equally – that is, in proportion to the first mode – so that $\mathbf{q}(0) = \begin{Bmatrix}1\\1\end{Bmatrix}$ mm, then

$$\mathbf{q}(t) = \begin{Bmatrix}1\\1\end{Bmatrix}\cos\left(100t\right)\,\text{mm}$$

which implies that the masses continue to oscillate in-phase at the first natural frequency of 100 rad/s. If the masses are initially displaced equally but out-of-phase – that is, in proportion to the second mode – so that $\mathbf{q}(0) = \begin{Bmatrix}1\\-1\end{Bmatrix}$ mm, then

$$\mathbf{q}(t) = \begin{Bmatrix}1\\-1\end{Bmatrix}\cos\left(173.2t\right)\,\text{mm}$$

and the masses oscillate at the second natural frequency of 173.2 rad/s. For any other initial displacements, the masses oscillate simultaneously at both natural frequencies. For example, if $\mathbf{q}(0) = \begin{Bmatrix}1\\0\end{Bmatrix}$ mm, then

$$\mathbf{q}(t) = \frac{1}{2}\begin{Bmatrix}1\\1\end{Bmatrix}\cos\left(100t\right) + \frac{1}{2}\begin{Bmatrix}1\\-1\end{Bmatrix}\cos\left(173.2t\right)\,\text{mm}$$

Figure 2.18 shows the response of the discrete mass-spring system for a selection of different initial conditions.

Meirovitch (1967) provides a formal procedure to determine the response to an arbitrary set of initial conditions.

2.4.3 The Influence of Damping on the Free Response

The previous analysis does not include the effect of damping, but its effect can easily be included. Viscous damping produces a force proportional to the velocity and requires the retention of all the terms in Equation (2.51), which is repeated here:

$$\mathbf{M}\ddot{\mathbf{q}} + \mathbf{C}\dot{\mathbf{q}} + \mathbf{K}\mathbf{q} = \mathbf{Q}(t)$$

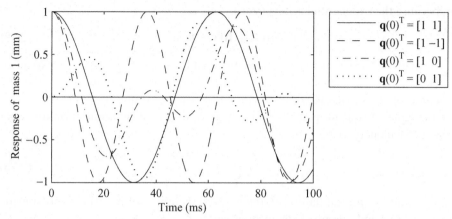

Figure 2.18. Time response for mass 1 of the two degrees of freedom discrete system (Example 2.4.1).

We consider the response in two parts: namely, the transient or unforced solution and the steady-state or forced solution. For free vibration, $\mathbf{Q}(t) = \mathbf{0}$ and the approach adopted for the undamped system in Section 2.4.2 leads to a quadratic eigenvalue problem that is difficult to solve directly. For convenience, the transient solution is obtained by converting Equation (2.51) into $2n$ first-order differential equations, in terms of $\mathbf{q}$ and $\dot{\mathbf{q}}$ (Meirovitch, 1967). Thus,

$$\begin{bmatrix} \mathbf{C} & \mathbf{M} \\ \mathbf{M} & \mathbf{0} \end{bmatrix} \frac{d}{dt} \begin{Bmatrix} \mathbf{q} \\ \dot{\mathbf{q}} \end{Bmatrix} + \begin{bmatrix} \mathbf{K} & \mathbf{0} \\ \mathbf{0} & -\mathbf{M} \end{bmatrix} \begin{Bmatrix} \mathbf{q} \\ \dot{\mathbf{q}} \end{Bmatrix} = \begin{Bmatrix} \mathbf{0} \\ \mathbf{0} \end{Bmatrix} \tag{2.71}$$

The lower set of equations is a dummy set that enforces the constraint that the lower partition of the coordinates is velocity and the upper partition is displacement. Other representations equivalent to Equation (2.71) are possible, although Equation (2.71) has the advantage that the matrices are symmetric – given that $\mathbf{M}$, $\mathbf{C}$, and $\mathbf{K}$ are symmetric – and the coefficient matrix of the derivative is nonsingular when the mass matrix is nonsingular. The first-order differential equation in the *state vector* $\mathbf{x} = \begin{Bmatrix} \mathbf{q} \\ \dot{\mathbf{q}} \end{Bmatrix}$ is

$$\mathbf{A}\dot{\mathbf{x}} + \mathbf{B}\mathbf{x} = \mathbf{0} \tag{2.72}$$

where the definition of the matrices follows from Equation (2.71). Equation (2.72) is called a *state–space* form of the equations of motion. In a manner similar to the undamped case, solutions are sought of the form

$$\mathbf{x}(t) = \mathbf{v}e^{st} \tag{2.73}$$

and $\mathbf{v}$ is a complex vector and s is a complex scalar. If the trial function is a solution to the differential equation, then so also is the complex conjugate of this function because $\mathbf{A}$ and $\mathbf{B}$ are real matrices. Therefore, these two solutions are added together to produce a real solution for $\mathbf{x}(t)$. Substituting Equation (2.73) into Equation (2.72) and rearranging gives

$$[s\mathbf{A} + \mathbf{B}]\mathbf{v}e^{st} = \mathbf{0} \tag{2.74}$$

or, because the exponential is never zero,

$$[s\mathbf{A} + \mathbf{B}]\mathbf{v} = \mathbf{0} \qquad (2.75)$$

Equation (2.75) is an eigenvalue problem and has $2n$ solutions that, as explained previously, occur as complex conjugate pairs. Comparing Equation (2.73) to the trial function for the single degree of freedom case used in Equation (2.17) shows that each complex conjugate pair of eigenvalues, s_i, s_{n+i}, is of the same form as Equation (2.18); namely

$$s_i, s_{n+i} = \omega_i \left(-\zeta_i \pm \sqrt{\zeta_i^2 - 1} \right) \qquad (2.76)$$

where ω_i and ζ_i are the natural frequency and damping ratio, respectively, for the ith mode. For underdamped eigenvalues where $\zeta_i < 1$

$$s_i, s_{n+i} = \omega_i \left(-\zeta_i \pm J\sqrt{1 - \zeta_i^2} \right) = -\zeta_i \omega_i \pm J\omega_{di} \qquad (2.77)$$

For complex eigenvalues, the corresponding eigenvectors also occur in complex conjugate pairs, $\mathbf{v}_{n+i} = \bar{\mathbf{v}}_i$, and the transient response can be computed from the sum

$$\mathbf{x}(t) = \sum_{i=1}^{2n} v_i \mathbf{v}_i e^{s_i t} \qquad (2.78)$$

where the v_i are complex constants depending on the initial displacement and velocity. They occur in complex conjugate pairs corresponding to the eigenvalues and eigenvectors.

EXAMPLE 2.4.2. Calculate the eigenvalues and eigenvectors for the two degrees of freedom system shown in Figure 2.16 with $m_i = 1\,\text{kg}$, $k_i = 10\,\text{kN/m}$, $c_1 = 10\,\text{Ns/m}$, and $c_2 = c_3 = 0$.

Solution. The damping matrix is

$$\mathbf{C} = \begin{bmatrix} 10 & 0 \\ 0 & 0 \end{bmatrix} \text{Ns/m}$$

The stiffness and damping matrices are identical to those in Example 2.4.1. Using a computer to solve the eigenvalue problem, Equation (2.75), the eigenvalues and eigenvectors are

$$s_1 = -2.50 + 100.03J, \qquad \mathbf{v}_1 = \left\{ \begin{array}{c} 1 \\ 0.9975 + 0.0500J \\ s_1 \\ (0.9975 + 0.0500J)\,s_1 \end{array} \right\}$$

$$s_2 = -2.50 + 173.08J, \qquad \mathbf{v}_2 = \left\{ \begin{array}{c} 1 \\ -0.9975 + 0.0866J \\ s_2 \\ (-0.9975 + 0.0866J)\,s_2 \end{array} \right\}$$

$$s_3 = \bar{s}_1, \quad \mathbf{v}_3 = \bar{\mathbf{v}}_1, \qquad\qquad s_4 = \bar{s}_2, \quad \mathbf{v}_4 = \bar{\mathbf{v}}_2$$

where the overbar denotes the complex conjugate. In terms of the natural frequencies and damping ratios, the eigenvalues may be written as

$$\omega_1 = 100.06\,\text{rad/s}, \qquad \zeta_1 = 0.0250, \qquad \omega_{d1} = 100.03\,\text{rad/s},$$

$$\omega_2 = 173.10\,\text{rad/s}, \qquad \zeta_2 = 0.0144, \qquad \omega_{d2} = 173.08\,\text{rad/s}$$

Because the damping is low in this example, the damping ratios are small and the natural frequencies are close to those of the undamped system, Example 2.4.1.

EXAMPLE 2.4.3. Calculate the eigenvalues and eigenvectors for the two degrees of freedom system shown in Figure 2.16 with $m_i = 1\,\text{kg}$, $k_i = 10\,\text{kN/m}$, $c_1 = 1000\,\text{Ns/m}$, $c_2 = 0$, and $c_3 = 100\,\text{Ns/m}$.

Solution. The damping matrix is

$$\mathbf{C} = \begin{bmatrix} 1000 & 0 \\ 0 & 100 \end{bmatrix} \text{Ns/m}$$

The stiffness and mass matrices are identical to those in Example 2.4.1. The eigenvalues and associated eigenvectors are now

$$s_1 = -14.883, \qquad \mathbf{v}_1 = \begin{Bmatrix} 1 \\ 0.5338 \\ s_1 \\ 0.5338 s_1 \end{Bmatrix}$$

$$s_2 = -52.708 + 133.40 J, \qquad \mathbf{v}_2 = \begin{Bmatrix} 1 \\ -4.7726 + 11.934 J \\ s_2 \\ (-4.7726 + 11.934 J) s_2 \end{Bmatrix}$$

$$s_3 = -979.70, \qquad \mathbf{v}_3 = \begin{Bmatrix} 1 \\ 0.0113 \\ s_3 \\ 0.0113 s_3 \end{Bmatrix}$$

$$s_4 = \bar{s}_2, \qquad \mathbf{v}_4 = \bar{\mathbf{v}}_2$$

From s_2 and s_4, we have $\omega_2 = 143.43\,\text{rad/s}$, $\zeta_2 = 0.3675$, and $\omega_{d2} = 133.40\,\text{rad/s}$. The damping is now so large that one pair of complex conjugate eigenvalues has become a pair of real eigenvalues and will not produce an oscillatory response.

2.4.4 Forced Vibrations of a Multiple Degrees of Freedom System

We now consider the response of a multidegrees of freedom undamped system when acted on by one or more dynamic forces. The equation of motion for a multiple degrees of freedom system is given in Equation (2.51) and is repeated here for the convenience of readers:

$$\mathbf{M\ddot{q}} + \mathbf{C\dot{q}} + \mathbf{Kq} = \mathbf{Q}(t)$$

The undamped case is given when $\mathbf{C} = \mathbf{0}$. We now introduce an important idea in vibration measurement and analysis that is rooted in electrical-circuit theory. Receptance is a measure of the responsiveness of a system to harmonic excitation applied at some point in the system. Suppose that $\mathbf{Q}(t) = \mathbf{Q}_0 \cos \omega t = \Re (\mathbf{Q}_0 e^{j\omega t})$; then, $\mathbf{q}(t) = \Re (\mathbf{q}_0 e^{j\omega t})$ and Equation (2.51) becomes

$$\left[-\omega^2 \mathbf{M} + j\omega \mathbf{C} + \mathbf{K} \right] \mathbf{q}_0 = \mathbf{Q}_0 \quad \text{or} \quad \mathbf{q}_0 = \left[-\omega^2 \mathbf{M} + j\omega \mathbf{C} + \mathbf{K} \right]^{-1} \mathbf{Q}_0 \qquad (2.79)$$

The matrix $\mathbf{D}(\omega) = \left[-\omega^2 \mathbf{M} + j\omega \mathbf{C} + \mathbf{K} \right]$ is called the *dynamic stiffness matrix* and its inverse, $\boldsymbol{\alpha}(\omega) = [\mathbf{D}(\omega)]^{-1}$, is the *receptance matrix*. It is clear that both the receptance and the dynamic stiffness are functions of frequency, and these matrices are symmetric because $\mathbf{M}$, $\mathbf{C}$, and $\mathbf{K}$ are symmetric. Thus, Equation (2.79) may be written as

$$\mathbf{q}_0 = \boldsymbol{\alpha}(\omega) \mathbf{Q}_0 \qquad (2.80)$$

If we expand Equation (2.80) for a system with n degrees of freedom, we have

$$\begin{Bmatrix} q_{01} \\ q_{02} \\ \vdots \\ q_{0n} \end{Bmatrix} = \begin{bmatrix} \alpha_{11}(\omega) & \alpha_{12}(\omega) & \vdots & \alpha_{1n}(\omega) \\ \alpha_{21}(\omega) & \alpha_{22}(\omega) & \vdots & \alpha_{2n}(\omega) \\ \dots & \dots & \ddots & \dots \\ \alpha_{n1}(\omega) & \alpha_{n2}(\omega) & \vdots & \alpha_{nn}(\omega) \end{bmatrix} \begin{Bmatrix} Q_{01} \\ Q_{02} \\ \vdots \\ Q_{0n} \end{Bmatrix} \qquad (2.81)$$

Consider how we might measure or define the value of $\alpha_{i\ell}(\omega)$, where i and ℓ can take any values in the range $1, 2, \ldots, n$. The ith equation in Equation (2.81) is

$$q_{0i} = \alpha_{i1}(\omega) Q_{01} + \alpha_{i2}(\omega) Q_{02} + \ldots + \alpha_{i\ell}(\omega) Q_{0\ell} + \ldots + \alpha_{in}(\omega) Q_{0n} \qquad (2.82)$$

If a unit amplitude harmonic force of frequency ω is applied at coordinate ℓ only, then

$$q_{0i} = \alpha_{i\ell}(\omega) \qquad (2.83)$$

From this, we can see that the receptance $\alpha_{i\ell}(\omega)$ – that is, the receptance value relating locations i and ℓ at excitation frequency ω – is defined as follows:

$\alpha_{i\ell}(\omega)$ represents the amplitude and phase of the harmonic displacement at coordinate i in response to the application of a unit harmonic force, of frequency ω, applied at coordinate ℓ only. No other forces are applied to the system.

We can obtain $\alpha_{i\ell}(\omega)$ by experiment from the previous definition. The experiment must be carried out over the range of frequencies of interest. In an experiment, we do not necessarily use a unit force, but if our system is linear, then $\alpha_{i\ell}(\omega) = q_{0i}/Q_{0\ell}$. Note from Equation (2.79) that as the harmonic excitation frequency ω tends to zero, then $\boldsymbol{\alpha}$ tends to $\mathbf{K}^{-1}$, the system flexibility matrix. Thus, we may obtain the system receptance (or flexibility) matrix element by element; the same cannot be done for the dynamic (or static) stiffness matrix because force would have to be applied to all degrees of freedom. If the dynamic-stiffness matrix is required, the complete receptance matrix must be measured and then inverted. Similarly, the static-stiffness matrix may be derived from the complete flexibility matrix.

The receptance matrix relates the harmonic excitation force vector to the system-displacement responses. We also can define matrices relating the harmonic excitation force vector to either the velocity or acceleration. Because the motion is harmonic, the velocity and acceleration response of the system are of the form $\mathbf{v}(t) = \Re\,(\mathbf{v}_0 e^{\jmath\omega t})$ and $\mathbf{a}(t) = \Re\,(\mathbf{a}_0 e^{\jmath\omega t})$, respectively. The *mobility matrix*, $\mathbf{Y}(\omega)$, and the *accelerance* or *inertance matrix*, $\mathbf{A}(\omega)$ (not related to matrix $\mathbf{A}$ of the state–space equations), are defined as

$$\mathbf{v}_0 = \mathbf{Y}(\omega)\mathbf{Q}_0 \tag{2.84}$$

and

$$\mathbf{a}_0 = \mathbf{A}(\omega)\mathbf{Q}_0 \tag{2.85}$$

Differentiating the expressions for velocity and acceleration gives

$$\mathbf{v}_0 = \jmath\omega\mathbf{q}_0 \quad \text{and} \quad \mathbf{a}_0 = -\omega^2\mathbf{q}_0 \tag{2.86}$$

The velocity has a phase shift of $90°$ relative to the displacement indicated by the operator $\jmath$. Substituting the velocity from Equation (2.86) into Equation (2.84) and the acceleration from Equation (2.86) into Equation (2.85), the mobility and inertance are related to receptance by

$$\mathbf{Y}(\omega) = \jmath\omega\boldsymbol{\alpha}(\omega) \quad \text{and} \quad \mathbf{A}(\omega) = -\omega^2\boldsymbol{\alpha}(\omega) \tag{2.87}$$

Alternatively, we state that the modulus of the mobility matrix, $\left|\mathbf{Y}(\omega)\right|$, is equal to $\left|\omega\boldsymbol{\alpha}(\omega)\right|$ and $\phi_Y = \phi_\alpha + 90°$, where ϕ denotes the phase angle. Accelerometers are an inexpensive and convenient method of measuring acceleration; therefore, in practice, inertance is often obtained. Laser systems usually measure velocity and, in this case, mobility is obtained. Once inertance or mobility has been computed, it can easily be converted to receptance if required.

2.4.5 Computing the Receptance of an Undamped System by Modal Decomposition

Computing the receptance via the inverse matrix in Equation (2.79) is inefficient if this calculation must be performed for many distinct frequencies. An alternative is to use *modal decomposition* (sometimes referred to as *modal analysis*) to effect a solution. To use modal decomposition for an undamped system, it is necessary to solve the corresponding eigenvalue problem so that $\mathbf{U}_N$ and $\boldsymbol{\Lambda}$, given by Equations (2.61) and (2.62), are determined. We begin by defining the coordinate transformation $\mathbf{q} = \mathbf{U}_N\mathbf{p}$, where $\mathbf{p}$ is the vector of modal or principal coordinates that determines the contribution of each mode to the response. Using this transformation in Equation (2.51) and premultiplying by $\mathbf{U}_N^{\mathsf{T}}$ gives

$$\mathbf{U}_N^{\mathsf{T}}\mathbf{M}\mathbf{U}_N\ddot{\mathbf{p}} + \mathbf{U}_N^{\mathsf{T}}\mathbf{K}\mathbf{U}_N\mathbf{p} = \mathbf{U}_N^{\mathsf{T}}\mathbf{Q}(t) \tag{2.88}$$

Using the mode orthogonality, Equation (2.63), we have

$$\ddot{\mathbf{p}} + \boldsymbol{\Lambda}\mathbf{p} = \mathbf{P}(t), \quad \text{where} \quad \mathbf{P}(t) = \mathbf{U}_N^{\mathsf{T}}\mathbf{Q}(t) \tag{2.89}$$

and where Λ is the diagonal matrix of the squares of the natural frequencies. Matrix Equation (2.89) represents a set of *uncoupled* ODEs and, because Λ is diagonal, each equation can be solved independently of the others. Once we have determined $\mathbf{p}$, we can obtain $\mathbf{q}$ because $\mathbf{q} = \mathbf{U}_N \mathbf{p}$.

Equation (2.89) may be used to calculate efficiently the receptance of the system. Thus,

$$\mathbf{p} = \left[-\omega^2 \mathbf{I} + \Lambda\right]^{-1} \mathbf{P} = \left[-\omega^2 \mathbf{I} + \Lambda\right]^{-1} \mathbf{U}_N^\top \mathbf{Q} \tag{2.90}$$

and, hence

$$\mathbf{q} = \mathbf{U}_N \left[-\omega^2 \mathbf{I} + \Lambda\right]^{-1} \mathbf{U}_N^\top \mathbf{Q} \tag{2.91}$$

The receptance is then

$$\boldsymbol{\alpha}(\omega) = \mathbf{U}_N \left[-\omega^2 \mathbf{I} + \Lambda\right]^{-1} \mathbf{U}_N^\top \tag{2.92}$$

The inverse in Equation (2.92) is easy to calculate because the matrix is diagonal. Furthermore, in terms of the modes of the structure, the receptance may be written as

$$\boldsymbol{\alpha}(\omega) = \sum_{i=1}^{n} \frac{\mathbf{u}_{Ni} \mathbf{u}_{Ni}^\top}{\omega_i^2 - \omega^2} \tag{2.93}$$

Note that Equation (2.93) is equivalent to Equation (2.92). The contribution of a particular mode to the receptance matrix is highest when the frequencies of interest are close to the corresponding natural frequency. Thus, some terms dominate the summation in Equation (2.93) and other terms often can be neglected.

When the force and response are measured or calculated at the same point, the receptance is called a *point receptance*; if the force and response are at different locations, it is called a *transfer receptance*. It is usual to plot the magnitude of receptance (and mobility and inertance) against frequency on a log or log-log scale. The receptance has peaks, called *resonances*, at the natural frequencies, indicating a large amplitude of vibration when a small force is applied to the system. With zero damping, these peaks are infinitely high – although, in practice, a small amount of damping is always present. There are also frequencies, called *anti-resonances*, at which the response is zero. The resonances of a particular system are global features that appear in most, if not all, receptances of the system, whereas the frequencies of the anti-resonances change depending on the degrees of freedom considered.

The total time response is obtained by adding the steady-state and transient parts given in Equation (2.64). Notice that with zero damping, the transient, Equation (2.64), does not decay.

EXAMPLE 2.4.4. Calculate the point and transfer receptances for the two degrees of freedom discrete system shown in Figure 2.16. The discrete masses have a mass of 1 kg and the springs have a stiffness of 10 kN/m. Damping is assumed to be zero.

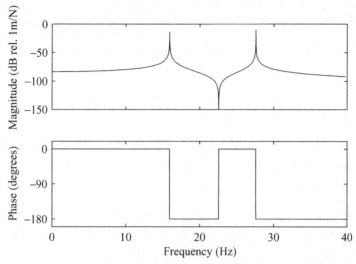

Figure 2.19. Point receptance for the two degrees of freedom discrete system (Example 2.4.4).

Solution. The natural frequencies and mode shapes were calculated in Example 2.4.1 and may be substituted into Equation (2.93) to give the receptance as

$$\boldsymbol{\alpha}(\omega) = \frac{1}{2\left(10^4 - \omega^2\right)} \begin{Bmatrix} 1 \\ 1 \end{Bmatrix} \begin{bmatrix} 1 & 1 \end{bmatrix} + \frac{1}{2\left(3 \times 10^4 - \omega^2\right)} \begin{Bmatrix} 1 \\ -1 \end{Bmatrix} \begin{bmatrix} 1 & -1 \end{bmatrix}$$

$$= \frac{1}{2\left(10^4 - \omega^2\right)} \begin{bmatrix} 1 & 1 \\ 1 & 1 \end{bmatrix} + \frac{1}{2\left(3 \times 10^4 - \omega^2\right)} \begin{bmatrix} 1 & -1 \\ -1 & 1 \end{bmatrix} \text{ m/N} \qquad (2.94)$$

Figure 2.19 shows the point receptances for both masses, which are identical because the system is symmetric. Note that the resonances and anti-resonances alternate as the frequency increases. This is not the case for the transfer receptance shown in Figure 2.20, which does not contain an anti-resonance. The resonance frequencies of 100 and 173.2 rad/s are identical in both the point and transfer receptances. Because there is no damping, the receptance is real. Thus, the phase is either $0°$ or $180°$ (which is equivalent to $-180°$), and the phase changes occur at both resonances and anti-resonances of the system.

2.4.6 Computing the Receptance of a Damped System by Modal Decomposition

In the steady-state case, zero damping implies that when $\omega = \omega_i$ (i.e., the excitation frequency equals one of the system natural frequencies), the denominator of the receptance expression becomes zero [see Equation (2.93)] and the receptance tends to infinity. Including the effect of damping allows the denominator to become small but not zero and the receptance becomes large but remains finite. Note the similarity with the behavior of a single degree of freedom mass–spring–damper system. The forced response is based on the equation of motion, Equation (2.51) – namely

$$\mathbf{M\ddot{q}} + \mathbf{C\dot{q}} + \mathbf{Kq} = \mathbf{Q}(t)$$

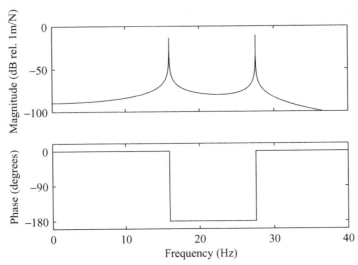

Figure 2.20. Transfer receptance for the two degrees of freedom discrete system (Example 2.4.4).

We now suppose that $\mathbf{Q}(t) = \Re\left(\mathbf{Q}_0 e^{J\omega t}\right)$ and that the response is then of the form $\mathbf{q}(t) = \Re\left(\mathbf{q}_0 e^{J\omega t}\right)$. Equation (2.51) becomes

$$\left[-\omega^2\mathbf{M} + J\omega\mathbf{C} + \mathbf{K}\right]\mathbf{q}_0 = \mathbf{Q}_0 \quad \text{or} \quad \mathbf{q}_0 = \left[-\omega^2\mathbf{M} + J\omega\mathbf{C} + \mathbf{K}\right]^{-1}\mathbf{Q}_0 \qquad (2.95)$$

where the receptance matrix is $\boldsymbol{\alpha}(\omega) = \left[-\omega^2\mathbf{M} + J\omega\mathbf{C} + \mathbf{K}\right]^{-1}$. Although this may be calculated directly for different frequencies, as in Section 2.4.4, it is more efficient to diagonalize (Horn and Johnson, 1985; Inman, 2006) the equations in state–space form, Equation (2.72), using the eigenvectors in a manner similar to the undamped case. The eigenvectors, $\mathbf{v}_N$, are given by the solutions to the eigenvalue problem, Equation (2.75), and they may be normalized for symmetric $\mathbf{A}$ and $\mathbf{B}$ so that

$$\mathbf{v}_{Ni}^\top \mathbf{A}\mathbf{v}_{Ni} = 1 \quad \text{and} \quad \mathbf{v}_{Ni}^\top \mathbf{B}\mathbf{v}_{Ni} = -s_i \qquad (2.96)$$

Even when the damping matrix is zero, this normalization is not equivalent to Equation (2.60) for the undamped system. Then, in a procedure similar to the development of Equation (2.93), we obtain

$$\boldsymbol{\alpha}(\omega) = \begin{bmatrix} \mathbf{I} & \mathbf{0} \end{bmatrix} \sum_{i=1}^{2n} \frac{\mathbf{v}_{Ni}\mathbf{v}_{Ni}^\top}{J\omega - s_i} \begin{bmatrix} \mathbf{I} \\ \mathbf{0} \end{bmatrix} \qquad (2.97)$$

A number of differences between Equations (2.93) and (2.97) should be highlighted. The eigenvectors are now generally complex and they occur in complex conjugate pairs. Furthermore, they have length $2n$ and there are $2n$ of them. For undamped systems, the roots, s_i, are purely imaginary and have a magnitude equal to the natural frequencies. Thus, in this case, the complex conjugate pairs combine to give the terms in Equation (2.93). The $\begin{bmatrix} \mathbf{I} & \mathbf{0} \end{bmatrix}$ premultiplying the sum selects the coordinates corresponding to displacements rather than velocity. The $\begin{bmatrix} \mathbf{I} & \mathbf{0} \end{bmatrix}^\top$ term postmultiplying the sum indicates that the force is applied to the first set of Equation (2.71). Recognizing that the coordinates are displacements and velocities means that the eigenvectors are of the form

$$\mathbf{v}_{Ni} = \begin{Bmatrix} \mathbf{w}_{Ni} \\ s_i \mathbf{w}_{Ni} \end{Bmatrix} \tag{2.98}$$

where $\mathbf{w}_{Ni}$ are the eigenvectors of the quadratic eigenvalue problem, defined in physical coordinates. Then, Equation (2.97) may be written as

$$\boldsymbol{\alpha}(\omega) = \sum_{i=1}^{2n} \frac{\mathbf{w}_{Ni}\mathbf{w}_{Ni}^{\top}}{J\omega - s_i} \tag{2.99}$$

Because the mass, damping, and stiffness matrices are real, any complex eigenvalues and eigenvectors occur in complex conjugate pairs and for these

$$\mathbf{w}_{N(n+i)} = \bar{\mathbf{w}}_{Ni} \quad \text{and} \quad s_{(n+i)} = \bar{s}_i \tag{2.100}$$

where the bar denotes the complex conjugate. If all of the eigenvalues are complex, then combining conjugate pairs of eigenvalues (i.e., terms i and $n+i$) in Equation (2.99) gives

$$\boldsymbol{\alpha}(\omega) = \sum_{i=1}^{n} \frac{2J\omega\Re\left(\mathbf{w}_{Ni}\mathbf{w}_{Ni}^{\top}\right) - 2\Re\left(\bar{s}_i\mathbf{w}_{Ni}\mathbf{w}_{Ni}^{\top}\right)}{-\omega^2 - 2J\omega\Re\left(s_i\right) + |s_i|^2} \tag{2.101}$$

The eigenvalues may be written in terms of the natural frequencies, ω_i, and damping ratios, ζ_i, as $s_i = -\zeta_i\omega_i + J\omega_i\sqrt{1-\zeta_i^2}$. Thus, the denominator in Equation (2.101) becomes $-\omega^2 + 2\zeta_i\omega_i J\omega + \omega_i^2$, which is equivalent to the single degree of freedom case, Equation (2.32). Even when there are real eigenvalues, the response still can be decomposed into a summation of responses of single degree of freedom systems, some of which are overdamped.

For high frequencies, the ω^2 term dominates the denominator and, hence,

$$\boldsymbol{\alpha}(\omega) \to -\frac{2J}{\omega}\Re\left[\sum_{i=1}^{n}\mathbf{w}_{Ni}\mathbf{w}_{Ni}^{\top}\right] + \frac{2}{\omega^2}\Re\left[\sum_{i=1}^{n}\bar{s}_i\mathbf{w}_{Ni}\mathbf{w}_{Ni}^{\top}\right] \tag{2.102}$$

Using an analysis parallel to that given by Garvey et al. (1998), it can be shown that the first term on the right side is zero and the matrix in the second term is constant. Thus, as ω tends to ∞, α is proportional to $1/\omega^2$.

EXAMPLE 2.4.5. Calculate the receptance for the discrete system shown in Figure 2.16 with $m_i = 1\,\text{kg}$, $k_i = 10\,\text{kN/m}$, and

(a) $c_1 = 1\,\text{Ns/m}$, $c_2 = c_3 = 0$
(b) $c_2 = 10\,\text{Ns/m}$, $c_1 = c_3 = 0$
(c) $c_2 = 50\,\text{Ns/m}$, $c_1 = c_3 = 0$

Solution

(a) Figure 2.21 shows the point receptance for mass 1 with $c_1 = 1\,\text{Ns/m}$ and $c_2 = c_3 = 0$. The receptance is similar to that of the undamped system, shown in Figure 2.19, except that now the peaks in the response are finite. The peaks are still large because the level of damping is low. Furthermore, the phase is now no longer exactly $0°$ or $\pm 180°$. The anti-resonance is very sharp in this case, which

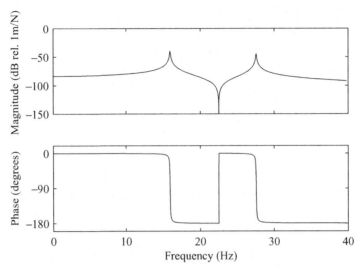

Figure 2.21. Point receptance for mass 1 in the two degrees of freedom system with $c_1 = 1 \, \text{Ns/m}$ (Example 2.4.5).

causes a jump in the phase from $-180°$ to $0°$. Often, damping causes the minimum response to be non-zero; this case is unusual because the damping arises from a grounded damper at mass 1.

(b) If the damper is placed between the two masses so that $c_2 = 10 \, \text{Ns/m}$ and $c_1 = c_3 = 0$, then consideration of the undamped modes shows that there is no damping in mode 1. The receptance for mass 1 is shown in Figure 2.22 and confirms this observation. The receptance also highlights that the second mode is highly damped.

(c) Figure 2.23 shows the effect of increasing c_2 to $50 \, \text{Ns/m}$. The peak corresponding to the second mode has now almost disappeared.

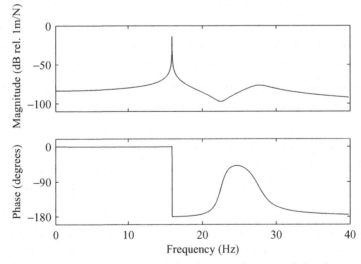

Figure 2.22. Point receptance for mass 1 in the two degrees of freedom system with $c_2 = 10 \, \text{Ns/m}$ (Example 2.4.5).

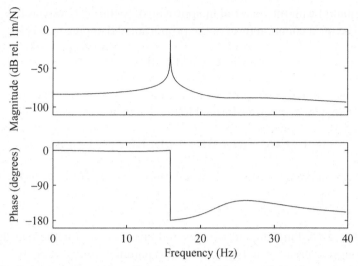

Figure 2.23. Point receptance for mass 1 in the two degrees of freedom system with $c_2 = 50\,\text{Ns/m}$ (Example 2.4.5).

2.4.7 Modal and Proportional Damping

In general, damping is difficult to model, although it is often important to provide some allowance for damping to produce reasonable estimates of the response of a structure, especially close to resonances. Two approximate methods to make some allowance for the damping are modal damping and proportional damping. General viscous damping is often termed *nonproportional damping*.

Modal damping assumes that the mode shapes of the undamped structure are also mode shapes of the damped structure. Thus, the full set of equations may be decoupled using the mass normalized mode shapes of the undamped model. The damping matrix is then transformed to the diagonal matrix

$$\mathbf{U}_N^\top \mathbf{C} \mathbf{U}_N = \begin{bmatrix} 2\zeta_1\omega_1 & 0 & \vdots & 0 \\ 0 & 2\zeta_2\omega_2 & \vdots & 0 \\ \cdots & \cdots & \ddots & 0 \\ 0 & 0 & \vdots & 2\zeta_n\omega_n \end{bmatrix} \tag{2.103}$$

The transformed equation of motion represents n single degree of freedom systems with natural frequency ω_i and damping ratio ζ_i. Modal damping assumes that the damping ratio in each undamped mode may be specified independently; often, the measured damping ratio in each mode is used. Caughey and O'Kelly (1965) give the conditions for modal damping in terms of the mass, damping, and stiffness matrices as

$$\mathbf{K}\mathbf{M}^{-1}\mathbf{C} = \mathbf{C}\mathbf{M}^{-1}\mathbf{K} \tag{2.104}$$

Proportional viscous damping assumes that the viscous damping matrix may be written as a linear combination of the mass and stiffness matrices. The main justification of this approach is that the energy-dissipation mechanism may be assumed to be distributed with the material in approximately the same way as the mass and stiffness. The most common form of the proportional damping matrix is

$$\mathbf{C} = \alpha\mathbf{M} + \beta\mathbf{K} \tag{2.105}$$

for some scalars α and β. Proportional damping is a special case of modal damping, and the damping matrix is transformed to the diagonal matrix

$$\mathbf{U}_N^T\mathbf{C}\mathbf{U}_N = \alpha\mathbf{U}_N^T\mathbf{M}\mathbf{U}_N + \beta\mathbf{U}_N^T\mathbf{K}\mathbf{U}_N = \alpha\mathbf{I} + \beta\mathbf{\Lambda} \tag{2.106}$$

Thus, the ith damping ratio is obtained in terms of the ith natural frequency as

$$\zeta_i = \frac{1}{2\omega_i}\alpha + \frac{\omega_i}{2}\beta.$$

EXAMPLE 2.4.6. Find the eigenvalues and eigenvectors for the two degrees of freedom system shown in Figure 2.16, with parameters $m_i = 1\,\text{kg}$, $k_i = 10\,\text{kN/m}$, $c_2 = 100\,\text{Ns/m}$, and $c_1 = c_3 = 0$.

Solution. The damping matrix is

$$\mathbf{C} = 100 \times \begin{bmatrix} 1 & -1 \\ -1 & 1 \end{bmatrix} \text{Ns/m} = -100 \times \mathbf{M} + 10^{-2} \times \mathbf{K}$$

Although α is negative, the presence of the stiffness term ensures that $\mathbf{C}$ is always positive-semidefinite and the damping ratios are always positive. The eigenvalues and eigenvectors are

$$s_1 = 100\jmath, \qquad\qquad \mathbf{v}_1 = \begin{Bmatrix} 1 \\ 1 \\ s_1 \\ s_1 \end{Bmatrix}$$

$$s_2 = -100 + 100 \times \sqrt{2}\jmath, \qquad\qquad \mathbf{v}_2 = \begin{Bmatrix} 1 \\ -1 \\ s_2 \\ -s_2 \end{Bmatrix}$$

$$s_3 = \bar{s}_1, \quad \mathbf{v}_3 = \bar{\mathbf{v}}_1, \qquad\qquad s_4 = \bar{s}_2, \quad \mathbf{v}_4 = \bar{\mathbf{v}}_2$$

In terms of the natural frequencies and damping ratios, the eigenvalues may be written as

$$\omega_1 = 100\,\text{rad/s} \qquad \zeta_1 = 0 \qquad \omega_{d1} = 100\,\text{rad/s}$$
$$\omega_2 = 100 \times \sqrt{3}\,\text{rad/s} \qquad \zeta_2 = 1/\sqrt{3} \qquad \omega_{d2} = 100 \times \sqrt{2}\,\text{rad/s}$$
$$= 173.2\,\text{rad/s} \qquad\quad = 0.5774 \qquad\quad = 141.4\,\text{rad/s}$$

Because the damping is proportional, the magnitude of the eigenvalues are equal for the damped and undamped cases. Also, the upper halves of the eigenvectors are real.

2.4.8 Operating Deflection Shapes

The *operating deflection shape* (*ODS*) is the deflected shape of the structure under a particular set of operating conditions. The ODS is different from the mode shapes, which are solely a property of the structure. In contrast, the ODS is not only a function of the structure, it also depends on the force. Suppose that the force is harmonic, at a fixed frequency, of the form $\mathbf{Q}(t) = \Re\left(\mathbf{Q}_0 e^{j\omega t}\right)$; then, Equation (2.95) gives the steady-state response as $\mathbf{q}(t) = \Re\left(\mathbf{q}_0 e^{j\omega t}\right)$, where

$$\mathbf{q}_0 = \boldsymbol{\alpha}(\omega)\mathbf{Q}_0 = \left[-\omega^2\mathbf{M} + j\omega\mathbf{C} + \mathbf{K}\right]^{-1}\mathbf{Q}_0 \qquad (2.107)$$

Then, $\mathbf{q}_0$ defines the spatial deformation of the structure at the forcing frequency and therefore defines the ODS. Alternatively, in terms of the complex eigenvectors, $\mathbf{w}_{Ni}$, and eigenvalues, s_i, from Equation (2.99)

$$\mathbf{q}_0 = \sum_{i=1}^{2n} \frac{\mathbf{w}_{Ni}\mathbf{w}_{Ni}^\top\mathbf{Q}_0}{j\omega - s_i} = \sum_{i=1}^{2n} \frac{\mathbf{w}_{Ni}\, f_i}{j\omega - s_i} \qquad (2.108)$$

where $f_i = \mathbf{w}_{Ni}^\top\mathbf{Q}_0$ denotes the force in the ith mode. It is clear that the ODS is a combination of the eigenvectors. For a lightly damped structure excited near a resonance, the denominator term $j\omega - s_i$ becomes small for one mode; therefore, the corresponding eigenvector dominates the ODS. However, away from resonances, the ODS is a combination of eigenvectors.

EXAMPLE 2.4.7. Find the ODSs for the undamped system given in Example 2.4.4, where the excitation is harmonic with magnitude 5 N at frequencies 95, 130, and 170 rad/s and the force is applied to mass 1 only.

Solution. The receptance is given by Equation (2.94). The force is $\mathbf{Q}_0 = \begin{Bmatrix} 5 \\ 0 \end{Bmatrix}$ N. Then,

$$\mathbf{q}_0(\omega) = \frac{5}{2\left(10^4 - \omega^2\right)}\begin{Bmatrix} 1 \\ 1 \end{Bmatrix} + \frac{5}{2\left(3 \times 10^4 - \omega^2\right)}\begin{Bmatrix} 1 \\ -1 \end{Bmatrix}$$

Thus,

$$\mathbf{q}_0(95) = \begin{Bmatrix} 2.683 \\ 2.445 \end{Bmatrix} \text{mm}, \quad \mathbf{q}_0(130) = \begin{Bmatrix} -0.1715 \\ -0.5532 \end{Bmatrix} \text{mm},$$

$$\mathbf{q}_0(170) = \begin{Bmatrix} 2.140 \\ -2.405 \end{Bmatrix} \text{mm}$$

For excitation frequencies 95 and 170 rad/s, the ODS approximates the corresponding modes; however, for 130 rad/s, the response is a combination of both modes.

2.5 Imposing Constraints and Model Reduction

A *constraint* is a restriction or loss of freedom. Either one degree of freedom may be fixed (i.e., constrained to be zero) or a constraint equation may be introduced that links two or more degrees of freedom. Thus, one of the generalized coordinates is set to zero or may be written in terms of the remaining coordinates and is not required to specify the configuration of the system. The number of degrees of freedom in the system – that is, the number of independent generalized coordinates – is therefore reduced by one.

Suppose that the constrained degrees of freedom are given by $\mathbf{q}_c$. Denote the free degrees of freedom by $\mathbf{q}_f$ and partition the undamped equations of motion as

$$\begin{bmatrix} \mathbf{M}_{ff} & \mathbf{M}_{fc} \\ \mathbf{M}_{cf} & \mathbf{M}_{cc} \end{bmatrix} \begin{Bmatrix} \ddot{\mathbf{q}}_f \\ \ddot{\mathbf{q}}_c \end{Bmatrix} + \begin{bmatrix} \mathbf{K}_{ff} & \mathbf{K}_{fc} \\ \mathbf{K}_{cf} & \mathbf{K}_{cc} \end{bmatrix} \begin{Bmatrix} \mathbf{q}_f \\ \mathbf{q}_c \end{Bmatrix} = \begin{Bmatrix} \mathbf{Q}_f \\ \mathbf{Q}_c \end{Bmatrix} \tag{2.109}$$

where $\mathbf{Q}_c$ is the unknown force required to enforce the constraints. Using the constraint $\mathbf{q}_c = \mathbf{0}$, the first block row in Equation (2.109) gives the constrained equations of motion as

$$\mathbf{M}_{ff}\ddot{\mathbf{q}}_f + \mathbf{K}_{ff}\mathbf{q}_f = \mathbf{Q}_f \tag{2.110}$$

The second block row in Equation (2.109) gives the forces required to enforce the constraints as

$$\mathbf{Q}_c = \mathbf{M}_{cf}\ddot{\mathbf{q}}_f + \mathbf{K}_{cf}\mathbf{q}_f \tag{2.111}$$

An alternative approach, which is useful when two or more degrees of freedom are linked, introduces a transformation between the unconstrained and constrained degrees of freedom. Thus, for some transformation $\mathbf{T}$

$$\mathbf{q} = \mathbf{T}\mathbf{q}_r \tag{2.112}$$

where $\mathbf{q}_r$ is a generalized coordinate vector of the constrained system. This transformation is substituted into the equations of motion of the unconstrained system, which is then premultiplied by $\mathbf{T}^\top$. The reduced, or constrained, mass and stiffness matrices are then

$$\mathbf{M}_r = \mathbf{T}^\top \mathbf{M}\mathbf{T} \quad \text{and} \quad \mathbf{K}_r = \mathbf{T}^\top \mathbf{K}\mathbf{T} \tag{2.113}$$

and the force in the constrained equation of motion is $\mathbf{Q}_r = \mathbf{T}^\top \mathbf{Q}$. If the transformation matrix is chosen as columns of the identity matrix, then $\mathbf{q}_r = \mathbf{q}_f$, $\mathbf{M}_r = \mathbf{M}_{ff}$, and $\mathbf{K}_r = \mathbf{K}_{ff}$. For example, if the first coordinate of a five degrees of freedom system is grounded (i.e., constrained to be zero), then

$$\mathbf{T} = \begin{bmatrix} 0 & 0 & 0 & 0 \\ 1 & 0 & 0 & 0 \\ 0 & 1 & 0 & 0 \\ 0 & 0 & 1 & 0 \\ 0 & 0 & 0 & 1 \end{bmatrix}$$

The more general case of applying constraints is considered in detail in Section 9.6.

2.5.1 Model Reduction

Model reduction is used to reduce the computational effort required in analyzing stationary and rotating systems. Obviously, it is impossible to emulate the behavior of a full system with a reduced model, and every reduction transformation sacrifices accuracy for speed in some way. One of the oldest and most popular reduction methods is static, or Guyan, reduction (Guyan, 1965), in which the inertia and damping terms associated with the discarded degrees of freedom are neglected.

In Guyan reduction, the deflection and force vectors, $\mathbf{q}$ and $\mathbf{Q}$, and the mass and stiffness matrices, $\mathbf{M}$ and $\mathbf{K}$, are reordered and partitioned into separate quantities relating to master (i.e., retained) and slave (i.e., discarded) degrees of freedom. Although not strictly necessary, we choose the slaves from the set of unforced degrees of freedom. Assuming that the damping is negligible, the equation of motion of the structure becomes

$$\begin{bmatrix} \mathbf{M}_{mm} & \mathbf{M}_{ms} \\ \mathbf{M}_{sm} & \mathbf{M}_{ss} \end{bmatrix} \begin{Bmatrix} \ddot{\mathbf{q}}_m \\ \ddot{\mathbf{q}}_s \end{Bmatrix} + \begin{bmatrix} \mathbf{K}_{mm} & \mathbf{K}_{ms} \\ \mathbf{K}_{sm} & \mathbf{K}_{ss} \end{bmatrix} \begin{Bmatrix} \mathbf{q}_m \\ \mathbf{q}_s \end{Bmatrix} = \begin{Bmatrix} \mathbf{Q}_m \\ \mathbf{0} \end{Bmatrix} \qquad (2.114)$$

The subscripts "m" and "s" relate to the master and slave coordinates, respectively. By neglecting the inertia terms in the second set of equations, the slave degrees of freedom may be eliminated so that

$$\begin{Bmatrix} \mathbf{q}_m \\ \mathbf{q}_s \end{Bmatrix} = \begin{bmatrix} \mathbf{I} \\ -\mathbf{K}_{ss}^{-1}\mathbf{K}_{sm} \end{bmatrix} \{\mathbf{q}_m\} = \mathbf{T}_s \mathbf{q}_m \qquad (2.115)$$

where $\mathbf{T}_s$ denotes the static transformation between the full state vector and the master coordinates (i.e., constrained degrees of freedom, denoted as $\mathbf{q}_r$ herein). The reduced mass and stiffness matrices are then given by Equation (2.113). Any response generated by the reduced matrices in Equation (2.115) is exact only at zero frequency – hence, the name *static reduction*. As the excitation frequency increases, the inertia terms neglected in Equation (2.114) become more significant. The process of choosing the master degrees of freedom may be automated in static reduction by considering the magnitude of the ratio of elastic to inertia terms (Henshell and Ong, 1975). The procedure is iterative and a single degree of freedom is eliminated at each iteration. This has the advantage that the inversion of $\mathbf{K}_{sm}$ is trivial because it is now a scalar quantity. Another advantage is that after each iteration, the inertia and stiffness properties associated with the eliminated degree of freedom are redistributed to the retained degrees of freedom before the next degree of freedom to be removed is chosen.

Dynamic reduction (Paz, 1984) is an alternative to static reduction, in which the frequency at which the reduction is exact may be chosen arbitrarily. Suppose we want the reduction to be exact at ω_0; then, approximating the inertia forces for the second set of equations in Equation (2.114) by

$$-\omega_0^2 \begin{bmatrix} \mathbf{M}_{sm} & \mathbf{M}_{ss} \end{bmatrix} \begin{Bmatrix} \mathbf{q}_m \\ \mathbf{q}_s \end{Bmatrix} \qquad (2.116)$$

produces a reduction transformation

$$\begin{Bmatrix} \mathbf{q}_m \\ \mathbf{q}_s \end{Bmatrix} = \begin{bmatrix} \mathbf{I} \\ -\begin{bmatrix} \mathbf{K}_{ss} - \omega_0^2\mathbf{M}_{ss} \end{bmatrix}^{-1} \begin{bmatrix} \mathbf{K}_{sm} - \omega_0^2\mathbf{M}_{sm} \end{bmatrix} \end{bmatrix} \{\mathbf{q}_m\} = \mathbf{T}_d \mathbf{q}_m \qquad (2.117)$$

Other more complex reduction methods based on the mass and stiffness matrices also may be used, such as the Improved Reduced System (IRS) (Gordis, 1992; O'Callahan, 1989) and the iterated IRS (Friswell et al., 1995, 1998a). However, the increased accuracy of the reduced model is gained at the expense of increased computation. Although the natural frequencies are improved, the response is not guaranteed to be exact at any frequency.

Suppose a subset of the eigenvectors of the undamped full system is to be retained in the reduced model. Let this subset be denoted by $\mathbf{U}_r$, where each column represents an eigenvector that is retained. Thus, $\mathbf{U}_r$ has many more rows (i.e., degrees of freedom) than columns (i.e., modes). Using these eigenvectors as the transformation – that is, $\mathbf{T} = \mathbf{U}_r$ – produces a reduced model that reproduces the chosen natural frequencies and mode shapes. The reduced degrees of freedom are now modal coordinates rather than physical or generalized displacements; however, this approach is often useful for time simulations, in which the transformation also must be applied to the external generalized forces. An alternative method, called the System Equivalent Reduction Expansion Process (SEREP) (O'Callahan et al., 1989), gives a reduced model that reproduces the chosen eigenvalues and eigenvectors but is based on physical degrees of freedom. The eigenvector subset of the full system, $\mathbf{U}_r$, is partitioned into master (i.e., retained) and slave (i.e., discarded) degrees of freedom

$$\mathbf{U}_r = \begin{bmatrix} \mathbf{U}_m \\ \mathbf{U}_s \end{bmatrix} \tag{2.118}$$

For model reduction, $\mathbf{U}_m$ is usually square and the reduction transformation is then

$$\mathbf{T}_{\text{SEREP}} = \begin{bmatrix} \mathbf{U}_m \\ \mathbf{U}_s \end{bmatrix} \mathbf{U}_m^{-1} \tag{2.119}$$

The same techniques for model reduction discussed for discrete systems also may be applied to systems modeled with FEA (see Chapter 4). The methods were applied to undamped and stationary structures, although the transformations determined also may be applied to systems with damping or gyroscopic effects (Rades, 1998). Qu (2004) discusses methods to reduce systems with high damping. Inman (2006) considers methods from control engineering, such as the balanced realization approach, based on the state–space equations.

The transformations defined for model reduction also may be used for the expansion of mode shapes or displacements. If analysis has been performed using the reduced model, then the full response of the system is obtained by using Equation (2.112). Alternatively, if a response or deformation is measured at a limited number of degrees of freedom, then the response or deformation may be estimated at all degrees of freedom using the transformation. The dynamic reduction transformation in Equation (2.117) is frequency-dependent and produces an exact expansion of the mode shapes or ODSs by appropriately setting the reference frequency, ω_0, for each shape.

EXAMPLE 2.5.1. Consider a mass–spring chain system consisting of 20 masses shown in Figure 2.24 with $n = 20$. Each mass is 1 kg and each spring is 1 MN/m. Estimate the first five natural frequencies of the full model and then of a reduced model, where only the degrees of freedom relating to masses 4, 8, 12, 16, and 20 are retained.

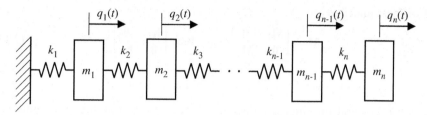

Figure 2.24. The mass-spring chain system for n masses (Example 2.5.1).

Solution. Table 2.1 shows the first five natural frequencies of the full model. The model is then reduced to five degrees of freedom using static and dynamic reduction and SEREP. Static reduction gives the following reduced mass and stiffness matrices:

$$\mathbf{M}_r = \frac{1}{8}\begin{bmatrix} 22 & 5 & 0 & 0 \\ 5 & 22 & 5 & 0 \\ 0 & 5 & 22 & 5 \\ 0 & 0 & 5 & 15 \end{bmatrix} \text{ kg}, \quad \mathbf{K}_r = \frac{1}{4}\begin{bmatrix} 2 & -1 & 0 & 0 \\ -1 & 2 & -1 & 0 \\ 0 & -1 & 2 & -1 \\ 0 & 0 & -1 & 2 \end{bmatrix} \text{ MN/m} \quad (2.120)$$

The first five natural frequencies for static reduction and dynamic reduction based on a frequency of 36 Hz are also shown in Table 2.1. For static reduction, the lower modes are clearly the most accurate; whereas, for dynamic reduction, the modes nearest to 36 Hz are the most accurate. As expected, SEREP exactly reproduces the lowest five natural frequencies.

2.5.2 Component Mode Synthesis

Component mode synthesis reduces the degrees of freedom in individual substructures or subsystems and then uses those reduced subsystem models to build a model of the complete system. The approach is useful because computing the eigensystems of a number of smaller models is faster than computing a single eigensystem of a large model. Only a summary is provided here; Craig (1981), Craig and Bampton (1968), and Petyt (1990) provide further details.

The basis of the approach is to recognize that when two structures are joined, the interface degrees of freedom are common. Suppose that there are two subsystems, labeled 1 and 2, with mass and stiffness matrices $\mathbf{M}_1$, $\mathbf{M}_2$, $\mathbf{K}_1$, and $\mathbf{K}_2$ and degrees of freedom $\mathbf{q}_1$ and $\mathbf{q}_2$. When the substructures are unconnected, the mass

Table 2.1. *Natural frequencies for reduced models of a mass-spring chain system (Example 2.5.1)*

Mode	Natural frequencies (Hz)			
	Full model	Static reduction	Dynamic reduction	SEREP
1	12.1921	12.2347	14.6229	12.1921
2	36.5049	37.6369	36.5058	36.5049
3	60.6034	65.5622	62.8965	60.6034
4	84.3462	95.6392	92.8585	84.3462
5	107.5941	120.8658	119.3026	107.5941

and stiffness matrices are

$$\mathbf{M} = \begin{bmatrix} \mathbf{M}_1 & \mathbf{0} \\ \mathbf{0} & \mathbf{M}_2 \end{bmatrix}, \qquad \mathbf{K} = \begin{bmatrix} \mathbf{K}_1 & \mathbf{0} \\ \mathbf{0} & \mathbf{K}_2 \end{bmatrix} \qquad (2.121)$$

where the vector of generalized coordinates of the *unconnected* system is $\mathbf{q}_u = \begin{Bmatrix} \mathbf{q}_1 \\ \mathbf{q}_2 \end{Bmatrix}$. The constraint that the interface degrees of freedom, denoted $\mathbf{q}_i$, are common to both subsystems may be imposed using the constraint equation

$$\mathbf{q}_i = \mathbf{E}_1^\top \mathbf{q}_1 = \mathbf{E}_2^\top \mathbf{q}_2 \qquad (2.122)$$

where $\mathbf{E}_1$ and $\mathbf{E}_2$ select the interface degrees of freedom of subsystems 1 and 2, respectively. Thus, the constraint on the degrees of freedom of the unconnected system is

$$\mathbf{E}^\top \mathbf{q}_u = \mathbf{0}, \quad \text{where} \quad \mathbf{E}^\top = \begin{bmatrix} \mathbf{E}_1^\top & -\mathbf{E}_2^\top \end{bmatrix} \qquad (2.123)$$

A transformation, $\mathbf{T}$, is now derived that enforces the constraint and gives the generalized coordinates of the coupled system as $\mathbf{q}_u = \mathbf{T}\mathbf{q}_r$. The requirements on $\mathbf{T}$ are that

$$\mathbf{E}^\top \mathbf{T} = \mathbf{0} \qquad (2.124)$$

and $\mathbf{Y} = [\mathbf{E} \quad \mathbf{T}]$ satisfies the condition that $\mathbf{Y}\mathbf{Y}^\top$ is nonsingular. The imposition of constraints in geared systems is considered in Section 9.6.2, and a general method to compute a suitable $\mathbf{T}$ is given in Section 9.6.3. The mass and stiffness matrices of the *coupled* structure are then obtained by applying the transformation matrix in a way similar to Equation (2.113) as

$$\mathbf{M}_r = \mathbf{T}^\top \begin{bmatrix} \mathbf{M}_1 & \mathbf{0} \\ \mathbf{0} & \mathbf{M}_2 \end{bmatrix} \mathbf{T}, \qquad \mathbf{K}_r = \mathbf{T}^\top \begin{bmatrix} \mathbf{K}_1 & \mathbf{0} \\ \mathbf{0} & \mathbf{K}_2 \end{bmatrix} \mathbf{T} \qquad (2.125)$$

Often, it is important to calculate the forces required to impose the constraints at the interface. The equation of motion in terms of the unconstrained degrees of freedom is

$$\begin{bmatrix} \mathbf{M}_1 & \mathbf{0} \\ \mathbf{0} & \mathbf{M}_2 \end{bmatrix} \begin{Bmatrix} \ddot{\mathbf{q}}_1 \\ \ddot{\mathbf{q}}_2 \end{Bmatrix} + \begin{bmatrix} \mathbf{K}_1 & \mathbf{0} \\ \mathbf{0} & \mathbf{K}_2 \end{bmatrix} \begin{Bmatrix} \mathbf{q}_1 \\ \mathbf{q}_2 \end{Bmatrix} = \begin{Bmatrix} \mathbf{Q}_1 \\ \mathbf{Q}_2 \end{Bmatrix} + \begin{Bmatrix} \mathbf{Q}_{i1} \\ \mathbf{Q}_{i2} \end{Bmatrix} \qquad (2.126)$$

where $\mathbf{Q}_1$ and $\mathbf{Q}_2$ are the external forces on the substructures and $\mathbf{Q}_{i1}$ and $\mathbf{Q}_{i2}$ are the forces required to ensure the constraint in Equation (2.123). By applying the transformation

$$\mathbf{q}_u = \begin{Bmatrix} \mathbf{q}_1 \\ \mathbf{q}_2 \end{Bmatrix} = \mathbf{T}\mathbf{q}_r \qquad (2.127)$$

and premultiplying Equation (2.126) by $\mathbf{T}^\top$, the equation of motion becomes

$$\mathbf{M}_r \ddot{\mathbf{q}}_r + \mathbf{K}_r \mathbf{q}_r = \mathbf{T}^\top \begin{Bmatrix} \mathbf{Q}_1 \\ \mathbf{Q}_2 \end{Bmatrix} \qquad (2.128)$$

where the forces at the interface must satisfy

$$\mathbf{T}^\top \begin{Bmatrix} \mathbf{Q}_{i1} \\ \mathbf{Q}_{i2} \end{Bmatrix} = \mathbf{0} \qquad (2.129)$$

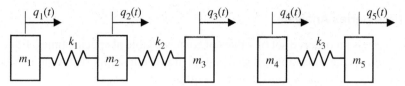

Figure 2.25. The two mass-spring chain systems (Example 2.5.2).

For a given external force, Equation (2.128) is used to obtain the response of the coupled system. The response in terms of the uncoupled degrees of freedom is obtained using Equation (2.127) through the transformation $\mathbf{T}$. The forces at the interface, $\mathbf{Q}_{i1}$ and $\mathbf{Q}_{i2}$, are then obtained from Equation (2.126) because all of the other quantities are known.

The interface degrees of freedom, $\mathbf{q}_i$, are fixed by the coupling between the subsystems. However, there is an opportunity to reduce the number of internal degrees of freedom in the subsystems, thereby reducing the dimension of the eigenvalue problem for the coupled system. Section 2.5.1 discusses various model-reduction methods for complete systems. The modal transformation is sometimes used (called the *free-interface method*), although it may perform poorly if the dynamics of the subsystem with free boundaries is not a good approximation to the motion of the subsystem within the full system. Craig and Bampton (1968) proposed the *fixed-interface method*. Here, the transformation is based on two sets of modal properties: (1) the modes of the subsystem with the interface degrees of freedom fixed; and (2) the constraint modes, which are equivalent to the static reduction transformation with the masters taken as the interface degrees of freedom. The reduced model is based on the interface degrees of freedom and a smaller set of internal degrees of freedom.

EXAMPLE 2.5.2. Figure 2.25 shows two mass-spring chain systems that are to be joined at masses m_3 and m_4. Derive the constraint matrix $\mathbf{E}$ and a suitable transformation matrix $\mathbf{T}$.

Solution. The interface constraint is given by $q_3 = q_4$; therefore,

$$\mathbf{E}^\top = \begin{bmatrix} 0 & 0 & 1 & -1 & 0 \end{bmatrix}$$

A suitable transformation matrix is easy to obtain in this case, based on a vector of generalized coordinates of the constrained system given by $\mathbf{q}_r^\top = \begin{bmatrix} q_1 & q_2 & q_3 & q_5 \end{bmatrix}$ as

$$\mathbf{T} = \begin{bmatrix} 1 & 0 & 0 & 0 \\ 0 & 1 & 0 & 0 \\ 0 & 0 & 1 & 0 \\ 0 & 0 & 1 & 0 \\ 0 & 0 & 0 & 1 \end{bmatrix}$$

It is easy to verify that $\mathbf{E}$ and $\mathbf{T}$ satisfy the required conditions; that is, $\mathbf{E}^\top \mathbf{T} = \mathbf{0}$ and $[\,\mathbf{E}\ \mathbf{T}\,]$ is nonsingular.

2.6 Time Series Analysis

A *time series* is an ordered set of values of some variable at specific times. For example, it could be the force applied to a system at given instants. The time increments do not have to be equally spaced, but it is often convenient to do so.

2.6.1 Simulation of a System Response

The steady-state response of a linear system to a set of harmonic forces can be determined by assuming a harmonic solution and obtaining a solution in the frequency domain (see Section 2.4.4). If the forcing functions are arbitrary, we can either take the Fourier transform of the forcing function (see Section 2.6.3) or the equations of motion can be uncoupled (using modal analysis). Then, it may be possible to express the response in the form of convolution, or Duhamel integrals (see Section 2.3.6), if the forcing function can be expressed in a closed form. This requires considerable algebraic manipulation and some computation. An alternative approach simulates the system by solving the equation of motion in the time domain using a numerical procedure. This has the advantage that any arbitrary forcing function can be handled; furthermore, the system equations can be nonlinear. The difficulty with this approach is that solving a large set of differential equations can be demanding in terms of computation.

   Numerical methods predict the system response only at discrete time intervals Δt apart, and the variation of displacement, velocity, and acceleration must be assumed with each time step Δt. The simplest algorithms use a fixed time step Δt and values are computed at times t_0, $t_0 + \Delta t$, $t_0 + 2\Delta t$, and so on. More advanced algorithms use variable time steps in which the step size is automatically adjusted to maintain a relative and/or absolute accuracy previously specified by the user. Large time steps are used when the solution varies slowly with respect to time and small time steps are used when the solution varies rapidly. Variable time-step algorithms tend to be more reliable than the fixed-step versions and ensure that numerical instability does not occur. Although there are many numerical integration schemes (Lindfield and Penny, 2000), the Runge-Kutta methods are reliable and widely used. These methods derive the current displacement from the previously determined values of displacement, velocity, and acceleration, starting from the initial conditions. Most numerical-integration schemes use the state–space form of the equations of motion. The Newmark-β method integrates the second-order form of the equations directly (Cook et al., 2001; Newmark, 1959; Petyt, 1990; Zienkiewicz et al., 2005).

   An important consideration that affects the time taken to solve a problem is the *stiffness* of the system of ODEs. For a linear system, this is the ratio of the highest to the lowest eigenvalue. A stiff system of equations requires a small step size to be taken in the solution process – relative to the time span of the simulation – to maintain stability and accuracy. This is illustrated in Example 2.6.1. To reduce the time taken to solve a system of equations, it is advantageous to use a model-reduction technique (e.g., Guyan reduction) to reduce the number of degrees of freedom in the model and, hence, the number of equations in the system. This reduction has two benefits: by removing the stiffest degrees of freedom from the model, we reduce the

Table 2.2. *The relative solution time for Example 2.6.1 (time relative to* $k_2 = 5 \times 10^4 \, N/m$)

k_2 (N/m)	Eigenvalue ratio	Relative time taken
5×10^4	3.56	1
5×10^5	10.7	2.2
5×10^6	33.6	8.2
5×10^7	106	27.8
5×10^8	335	82.7
Single Degree of Freedom	–	0.72

ratio of the highest to the lowest eigenvalue and we also have fewer equations to solve.

EXAMPLE 2.6.1. The two degrees of freedom system of Figure 2.16 has the following parameters: $m_1 = 2\,\mathrm{kg}$, $m_2 = 1\,\mathrm{kg}$, $c_1 = 3\,\mathrm{Ns/m}$, $c_2 = 0$, $c_3 = 6\,\mathrm{Ns/m}$, $k_1 = k_3 = 10\,\mathrm{kN/m}$, and k_2 varies from 5×10^4 to $5 \times 10^8\,\mathrm{N/m}$. A harmonic force, $\cos \omega t$, of magnitude $150\,\mathrm{N}$ with a frequency of $12\,\mathrm{Hz}$ acts at coordinate 2. The system is also given initial displacements of 0.02 and 0.01 m at coordinates 1 and 2, respectively, and initial velocities of 0.01 m/s at both coordinates. Determine the relative times taken to find the solution for $t = 0$ to 5 s using the fourth-order Runge-Kutta method with variable step size to give a consistent accuracy.

Solution. The excitation frequency is close to the first natural frequency, which varies from 12.92 to 12.99 Hz as k_2 varies from 5×10^4 to $5 \times 10^8\,\mathrm{N/m}$. The two second-order differential equations must be rearranged into four first-order differential equations to use the Runge-Kutta method. Rearranging Equation (2.51) with $\mathbf{Q}(t) = \mathbf{Q}_0 \cos \omega t$ gives

$$\frac{\mathrm{d}}{\mathrm{d}t} \begin{Bmatrix} \mathbf{q} \\ \dot{\mathbf{q}} \end{Bmatrix} = \begin{bmatrix} \mathbf{0} & \mathbf{I} \\ -\mathbf{M}^{-1}\mathbf{K} & -\mathbf{M}^{-1}\mathbf{C} \end{bmatrix} \begin{Bmatrix} \mathbf{q} \\ \dot{\mathbf{q}} \end{Bmatrix} + \begin{Bmatrix} \mathbf{0} \\ \mathbf{M}^{-1}\mathbf{Q}_0 \end{Bmatrix} \cos \omega t \qquad (2.130)$$

where the matrix is called the *state matrix*.

The stiffness of k_2 has little effect on the response of the system because the frequency of the force is so close to the first natural frequency, which is insensitive to this stiffness. The relative times taken to solve the system for $t = 0$ to 5 s are shown in Table 2.2. The variable time step for the integration is chosen to give the same estimated accuracy for the different stiffness values. The table shows that as the eigenvalue ratio increases, the time required to solve the system for the time period 0 to 5 s increases significantly. The excitation frequency is such that the higher mode is barely excited; yet, its presence has a significant effect on the solution time. If this two degrees of freedom model is reduced to a single degree of freedom model, the relative computation time falls to 0.72. Although the relative times taken for the integration change slightly for different computer systems, integration algorithms, and error estimators, the same trend occurs.

A ratio of 335 between the highest and lowest eigenvalues is by no means an unusually high ratio. For example, in a typical model of a rotor consisting of 18 elements (i.e., 76 degrees of freedom), the eigenvalue ratio is approximately 2,200. Reducing this model to 20 degrees of freedom reduced the ratio of the largest to the smallest eigenvalue to approximately 100.

2.6.2 The Fourier Transform

The Fourier series is defined in Section 2.3.5. Using the Fourier series, a periodic signal or function can be expressed as (or decomposed into) an infinite series of sine and cosine terms or, alternatively, an infinite series of complex exponential terms. The frequencies of these terms are integer multiples of the fundamental frequency of the original signal or function and are called *harmonics*. It is important to understand that this is more than just a mathematical operation. The Fourier series reveals the presence and size of frequency components present in a periodic function or signal. This information is important to an engineer, particularly in rotor-dynamics, in which the frequency of a dominant component in a signal can help to locate its source.

If the function to be decomposed is nonperiodic, then the Fourier series cannot be used; instead, the Fourier transform is used. A nonperiodic signal can be considered a periodic signal with an infinitely long period, so that it never repeats. Equations (2.41), (2.42), and (2.43) then become

$$x(t) = \frac{1}{2\pi} \int_{-\infty}^{\infty} X(\omega) e^{j\omega t} d\omega \qquad (2.131)$$

$$X(\omega) = \int_{-\infty}^{\infty} x(t) e^{-j\omega t} dt \qquad (2.132)$$

These two equations are called the *Fourier transform pair*. Some authors apply a different scaling to the transformed signal. Equation (2.132) defines the Fourier transform of $x(t)$, denoted by $X(\omega)$. Conversely, knowing $X(\omega)$, we can determine the original signal or function $x(t)$ from Equation (2.131). Both $x(t)$ and $X(\omega)$ are continuous functions. $X(\omega)$ is also called the *frequency spectrum* of $x(t)$ and it contains all the frequency components of $x(t)$.

EXAMPLE 2.6.2. Determine the Fourier transform of a rectangular pulse defined by

$$x(t) = \begin{cases} A & -T_0 < t < T_0 \\ A/2 & t = \pm T_0 \\ 0 & t < -T_0 \quad \text{or} \quad t > T_0 \end{cases}$$

and plot the Fourier transform of a rectangular pulse with $T_0 = 1$ and $A = 2$.

Solution. The Fourier transform of a rectangular pulse of amplitude A existing from time $-T_0$ to $+T_0$ is given by

$$X(\omega) = \int_{-T_0}^{T_0} A e^{-j\omega t} dt = A \int_{-T_0}^{T_0} \cos(\omega t) dt - j A \int_{-T_0}^{T_0} \sin(\omega t) dt$$

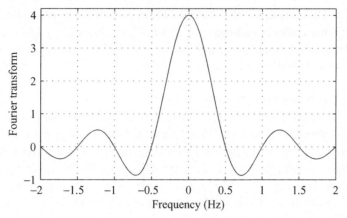

Figure 2.26. The Fourier transform of a square pulse with $T_0 = 1$ and $A = 2$.

Integrating this equation and noting that the second term is zero gives

$$X(\omega) = \frac{A}{\omega}\left[\sin(\omega t)\right]_{-T_0}^{T_0} = 2AT_0\frac{\sin(T_0\omega)}{T_0\omega} \tag{2.133}$$

A plot of the Fourier transform with $T_0 = 1$ and $A = 2$ is shown in Figure 2.26; note that the frequencies were converted from rad/s to Hz. Although the Fourier transform might appear undefined at $\omega = 0$ in Equation (2.133), because $\sin x/x \to 1$ as $x \to 0$, $X(0) = 4$. All frequencies are present in the Fourier transform of a rectangular pulse, except when the transform magnitude is zero at $\omega = \pm\pi/T_0, \pm2\pi/T_0, \pm3\pi/T_0, \ldots$ ($\pm0.5\,\text{Hz}, \pm1.0\,\text{Hz}, \pm1.5\,\text{Hz}, \ldots$ in Figure 2.26).

2.6.3 The Discrete Fourier Transform

Suppose $x(t)$ is not known as a continuous function but rather as a series of discrete real values. This situation can arise when a signal is sampled and recorded digitally or when $x(t)$ is the output from a time simulation. In such cases, we cannot obtain the frequency content of $x(t)$ using the Fourier transform; the discrete Fourier transform (DFT) is used instead, which is defined as

$$X_k = \sum_{r=0}^{n-1} x_r e^{-j2\pi kr/n}, \qquad k = 0, 1, 2, \ldots, n-1 \tag{2.134}$$

and the inverse DFT is

$$x_r = \frac{1}{n}\sum_{k=0}^{n-1} X_k e^{j2\pi kr/n}, \qquad r = 0, 1, 2, \ldots, n-1 \tag{2.135}$$

X is the DFT of x and Equations (2.134) and (2.135) form the transform pair. Some authors place the factor $1/n$ of Equation (2.135) in Equation (2.134).

If n equispaced samples are taken of a function at a time interval Δt, then the frequency components in the DFT are at intervals $\Delta f = 1/(n\Delta t)$. Note that this and subsequent frequencies have a frequency unit of Hz. The frequency spectrum covers only the range 0 to $(n/2)\,\Delta f$ (or 0 to $1/(2\Delta t)$) and consists of $n/2 + 1$ frequency

components. This is because the components of the DFT are not independent because X_{n-k} is the complex conjugate of X_k for $k = 1, \ldots, n/2 - 1$ and thus provides
no extra information. Some important features of the DFT should be noted. To distinguish closely spaced frequency components (i.e., small Δf), we require samples
during a long time period ($n\Delta t$). To obtain a high frequency spectrum, we also require a fast sample rate (i.e., small Δt).

Using Equation (2.134) to determine the DFT for n data points requires n^2 complex multiplications. For large datasets, this would be a slow process. However, the
fast Fourier transform (FFT) algorithm was developed by Cooley and Tukey (1965).
This ingenious algorithm to determine the DFT reduces the number of complex
multiplications required to $(n/2) \log_2 n$, provided n is an integer power of 2. Subsequent developments introduced efficient algorithms that do not require n to be
an integer power of 2. Also, datasets for which n is not an integer power of 2 can
be padded by adding zeros to increase the size of the dataset so that n becomes an
integer power of 2.

The DFT transforms a finite set of discrete data samples; consequently, it im-
poses a periodicity on the data – the period being the time during which the data
are sampled, $n\Delta t$. The DFT is periodic and the highest frequency that can be represented is $1/(2\Delta t)$, which is called the *Nyquist frequency*. A frequency component
in the original data higher than the Nyquist frequency appears in the spectrum at a
different frequency, which is lower than the Nyquist frequency, and is called *aliasing*. Another problem that arises when using the DFT is that if the data contain
a frequency component that does not precisely match a frequency component in
the DFT frequency spectrum, the true frequency component will be smeared over
several frequency components in the DFT. This effect is called *leakage*. It can be reduced by multiplying the data by window functions such as the Hamming, Hanning,
or exponential windows (see Example 2.6.4).

EXAMPLE 2.6.3. The function $y = 2\sin(2\pi f_1 t) + \cos(2\pi f_2 t)$, where $f_1 = 3.125\,\text{Hz}$
and $f_2 = 6.25\,\text{Hz}$, is sampled at equal time intervals during the following time
periods: (a) 3.2 s, (b) 3.6 s, (c) 1.6 s, and (d) 6.4 s. In each case, 64 data samples
are taken. Determine the DFT from the sampled data and plot the frequency
spectrum for each case.

Solution

(a) Given 64 data samples taken over 3.2 s, then $\Delta t = 3.2/64 = 0.05\,\text{s}$
and the Nyquist frequency is $1/(2\Delta t) = 1/(2 \times 0.05) = 10\,\text{Hz}$. $\Delta f = 1/3.2 =$
$0.3125\,\text{Hz}$. Hence, the frequency components in the data, 3.125 and 6.25 Hz,
coincide exactly with the 11th and 21st frequency components in the DFT spectrum. An equivalent statement is that the sampling period is exactly an integer
number of cycles of the frequency components in the data – in this case, exactly
10 and 20 cycles of the lower and higher frequency components, respectively.
The frequency spectrum gives the correct result, as shown in Figure 2.27(a).

(b) Given 64 data samples taken over 3.6 s, then $\Delta t = 3.6/64 = 0.05625\,\text{s}$ and
the Nyquist frequency is $1/(2\Delta t) = 1/(2 \times 0.05625) = 8.889\,\text{Hz}$. $\Delta f = 1/3.6 =$
$0.2778\,\text{Hz}$. The frequency components in the data no longer coincide exactly
with the frequency components in the DFT and leakage occurs; that is, the

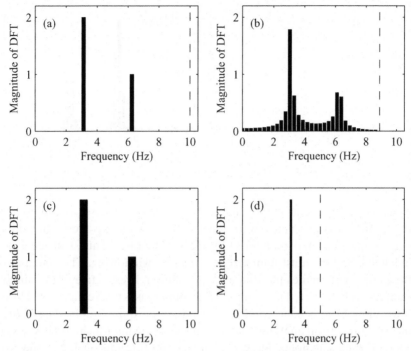

Figure 2.27. The DFT of a signal consisting of the sum of two sinusoids (Example 2.6.3) for sample periods (a) 3.2 s, (b) 3.6 s, (c) 1.6 s, and (d) 6.4 s. The dashed vertical lines indicate the Nyquist frequency.

frequency components in the data spread over several frequency components in the DFT. In this case, the sampling period has 11.25 and 22.5 cycles of the lower and higher frequency components in the data, respectively. Figure 2.27(b) shows the effect of leakage. We still can clearly see the two frequency components in the spectrum, but their amplitudes are no longer exact and energy has leaked into adjacent frequencies.

(c) Given 64 data samples taken over 1.6 s, then $\Delta t = 1.6/64 = 0.025$ s and the Nyquist frequency is $1/(2\Delta t) = 1/(2 \times 0.025) = 20$ Hz. $\Delta f = 1/1.6 = 0.625$ Hz. Because the samples are taken during a shorter time period than in (a) or (b), the frequency resolution is lower (i.e., Δf is larger) and the Nyquist frequency – the maximum frequency in the spectrum – is 20 Hz. In Figure 2.27(c), the spectrum is plotted only to 10.5 Hz to be consistent with the other plots.

(d) Given 64 data samples taken over 6.4 s, then $\Delta t = 6.4/64 = 0.1$ s and the Nyquist frequency is $1/(2\Delta t) = 1/(2 \times 0.1) = 5$ Hz. $\Delta f = 1/6.4 = 0.15625$ Hz. Thus, the maximum frequency in the spectrum is 5 Hz. Because the 6.25 Hz frequency component in the data is 1.25 Hz *above* the Nyquist frequency, aliasing causes it to appear in the spectrum at 1.25 Hz *below* the Nyquist frequency (i.e., at 3.75 Hz). This is shown in Figure 2.27(d). Although there are no data above 5 Hz, the spectrum is plotted to 10.5 Hz to be consistent with the other plots.

This example illustrates several pitfalls in determining the frequency spectrum from sampled data. Frequently, we do not know the values of all of the

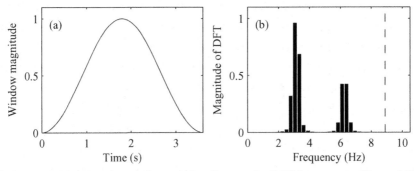

Figure 2.28. The Hanning window and its effect on the DFT (compare to Figure 2.27).

frequency components in the data, so it is unlikely that case (a) applies. Even if the frequency components in the data were known in advance, many data loggers sample at a small number of fixed sampling rates. Thus, leakage is usually inevitable and we must minimize its effect by windowing. The time period of the sample determines the frequency resolution, irrespective of the number of samples, as highlighted in a comparison among cases (a), (c), and (d). However, if we allow Δt to become too large, then we are exposed to the danger of aliasing. Most spectrum analyzers have an anti-aliasing filter included in the system. These low-pass analogue filters are set to a cutoff frequency below the Nyquist frequency so that the data to be analyzed cannot contain frequencies above the Nyquist frequency and aliasing cannot occur. Thus, the frequency spectrum obtained contains no spurious components.

EXAMPLE 2.6.4. Apply a Hanning window to Example 2.6.3(b).

Solution. The Hanning window for a time signal between $t = 0$ and $t = T$ is

$$w(t) = 0.5 - 0.5 \cos\left(\frac{2\pi t}{T}\right)$$

and this function is shown in Figure 2.28(a). Multiplying the time signal in Example 2.6.3(b) by this time window and computing the DFT gives the result shown in Figure 2.28(b). The DFT at frequencies away from the signal frequencies is now much smaller; of course, the exact frequency content is not recovered. Because none of the DFT frequencies corresponds to either of the signal frequencies, the energy must be split among neighboring frequencies. Also, the magnitude of the DFT is reduced because the window reduces the signal amplitude at the start and end of the time period. For continuous signals, such as this example, the magnitude reduction is approximately 50 percent.

2.7 Nonlinear Systems

Section 2.2 describes how a nonlinear spring can be linearized by considering only small displacements – that is, by considering small levels of vibration. Similarly, nonlinear Coulomb damping can be linearized by introducing an equivalent viscous-damping coefficient. Sometimes systems cannot be linearized because we want to study certain phenomena that appear only in the nonlinear system models. Nonlinear systems often give responses that are surprising and cannot be explained using a linear model. This section introduces some of the phenomena associated with nonlinear dynamics; Moon (2004) and Thomsen (1997) provide more comprehensive introductions.

Consider a system with a nonlinear cubic stiffening spring (i.e., Duffing's equation) whose equation of motion is

$$m\ddot{x} + c\dot{x} + kx + hx^3 = f_0 \cos(\omega t) \tag{2.136}$$

Once the transient dynamics have decayed, a linear system excited with a sinusoidal force responds only at the excitation frequency. This is not true for a nonlinear dynamic system, and the response may contain sub- or super-harmonics of the excitation frequency, or it may be chaotic.

The FRF for a nonlinear system excited by a sinusoidal force is defined as the ratio found by dividing the response at the excitation frequency by the excitation itself. The FRF may be calculated from the system model, it may be obtained from a time-response simulation, or it may be measured using a slow sweep of the excitation frequency. An alternative is to use the *harmonic-balance* method to determine the steady-state response of the system. The assumption in the harmonic-balance method is that the response is periodic and therefore may be written as a Fourier series. The fundamental period is often taken as the period of the excitation, which restricts the response to contain only super-harmonics of the excitation frequency. This assumption often predicts steady-state solutions; however, a characteristic of nonlinear systems is that multiple steady-state solutions can exist – in particular, the response may contain sub-harmonics. The approach is demonstrated by the Duffing oscillator given by Equation (2.136), in which we assume the response is of the form

$$x(t) = a_0 + \sum_{n=1}^{\infty} a_n \cos(\omega n t + \phi_n) \tag{2.137}$$

This series is equivalent to those in Section 2.3.5; however, using a definition based on amplitude and phase is more convenient for the example used here. The approach substitutes this expression into Equation (2.136), expands the x^3 term, uses trigonometric formulae to simplify the expressions, and equates the lower harmonic terms to generate sufficient nonlinear algebraic equations in a_n and b_n (Friswell and Penny, 1994; Thomsen, 1997). In practice, the number of terms in the Fourier series must be truncated and the number of algebraic equations limited. The approach is demonstrated by assuming a single-term solution to the Duffing oscillator of the form

$$x(t) = a \cos(\omega t + \phi) \tag{2.138}$$

Using standard trigonometric formulae

$$x^3 = (a \cos(\omega t + \phi))^3 = \frac{a^3}{4} \cos 3(\omega t + \phi) + \frac{3a^3}{4} \cos(\omega t + \phi) \tag{2.139}$$

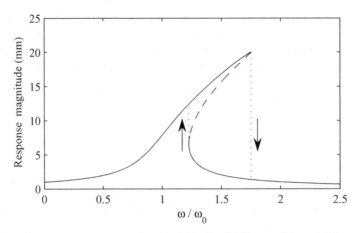

Figure 2.29. The frequency response function for the Duffing oscillator with $m = 1\,\text{kg}$, $c = 10\,\text{Ns/m}$, $k = 10\,\text{kN/m}$, $h = 50\,\text{MN/m}^3$, and $f_0 = 20\,\text{N}$. The dashed line denotes the unstable solution branch. The dotted lines and arrows denote the jumps that occur for an increasing or decreasing excitation frequency.

and, because we require the forcing function to be in terms of $\cos(\omega t + \phi)$ and $\sin(\omega t + \phi)$, we also expand $\cos \omega t$ as

$$f_0 \cos(\omega t) = f_0 \left(\cos \phi \cos(\omega t + \phi) + \sin \phi \sin(\omega t + \phi) \right) \qquad (2.140)$$

Substituting this response into Equation (2.136), neglecting the term $\cos 3(\omega t + \phi)$ arising from x^3, and equating sine and cosine terms gives the two equations

$$(-m\omega^2 + k)a + \frac{3}{4}ha^3 = f_0 \cos \phi \quad \text{and} \quad -c\omega a = f_0 \sin \phi \qquad (2.141)$$

Squaring and adding these equations to eliminate the phase angle ϕ gives a cubic equation in a^2 – that is, the response amplitude squared – as

$$\left[-m\omega^2 + k + \frac{3}{4}ha^2 \right]^2 a^2 + c^2\omega^2 a^2 = f_0^2 \qquad (2.142)$$

For a linear system, the FRF has only one response amplitude and phase for each excitation frequency; whereas, for nonlinear systems, multiple solutions can exist. This is clear from Equation (2.142) for the Duffing oscillator, in which the cubic can have one or three real solutions for a^2. When two or more real solutions are positive, the solutions obtained for response magnitude are not unique. A typical FRF is shown in Figure 2.29, in which the excitation frequency is nondimensionalized by the natural frequency of the linear system, $\omega_n = \sqrt{k/m}$. Over a range of frequencies near to the resonance, the system can take one of two possible response levels with an unstable region between them. Thus, the system can exhibit a *jump phenomenon* because the response can "jump" between the different levels. For example, if the excitation frequency is slowly increased from a low value, the response follows the higher-amplitude response solution until the nondimensional excitation frequency reaches approximately 1.75. For further increases in excitation frequency, there is only one solution; thus, the response suddenly jumps to the low-amplitude solution. When the excitation is slowly reduced in frequency from a high value, the solution remains on the lower-amplitude branch until the nondimensional

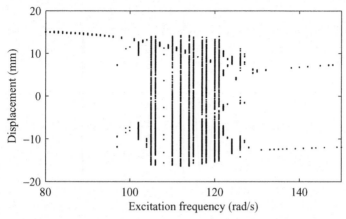

Figure 2.30. The bifurcation diagram of the Duffing oscillator, with $m = 1\,\text{kg}$, $c = 10\,\text{Ns/m}$, $k = -5\,\text{kN/m}$, $h = 50\,\text{MN/m}^3$, and $f_0 = 30\,\text{N}$, as the excitation frequency, ω, varies.

excitation frequency is approximately 1.22, at which point the solution jumps to the higher-amplitude branch. The excitation frequencies in which the jumps occur denote parameter values, where a qualitative change in the behavior of the system occurs. Such qualitative changes are often called *bifurcations*.

Duffing's equation can give a wide variety of responses for different system parameters, some of which are not periodic. The bifurcation diagram shows the qualitative nature of the solution for a range of parameter values of the system. Bifurcation plots are used to display nonlinear and chaotic effects in systems encountered in many fields, such as population growth, plant–herbivore evolution, and mechanical systems, to name just three (Thompson and Stewart, 1986). Figure 2.30 is an example bifurcation diagram for the Duffing oscillator as the frequency of excitation is changed in steps of 1 rad/s. The transients are allowed to decay and then the displacement is sampled at the same frequency as the excitation and plotted as points. The linear stiffness for this example is negative; physically, this corresponds to the (snap-through) bucking of a thin beam (Moon, 2004). If the response is a sinusoid at the same frequency as the excitation, as occurs in the steady-state response of a linear system, then only a single point occurs on the bifurcation diagram. This occurs in Figure 2.30 for frequencies below $\omega = 96$ rad/s and above $\omega = 129$ rad/s. Although superficially there appears to be two points for frequencies above $\omega = 129$ rad/s, careful examination of Figure 2.30 shows that there is only one displacement point *for each frequency*. For other excitation frequencies, the response is periodic but contains sub-harmonics of the excitation frequency. This occurs at an excitation frequency of $\omega = 100$ rad/s, for example, and the response is shown in Figure 2.31. Also shown is the DFT of the time response, which clearly highlights the significant response at one third of the excitation frequency. At other excitation frequencies – for example, at $\omega = 120$ rad/s – the response is not periodic and may be quasiperiodic or chaotic. Figure 2.32 shows the response for an excitation frequency of $\omega = 120$ rad/s, and the broadband response shows that the response is chaotic. One method used to analyze such a response is the *Poincaré map*, shown in Figure 2.33 for the same parameter values as used in Figure 2.32. To generate a Poincaré map, the system is simulated in the time domain until the transient response has decayed. The

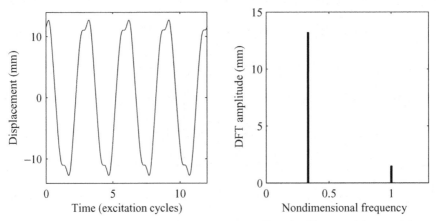

Figure 2.31. The sub-harmonic response of the Duffing oscillator, with $m = 1\,\mathrm{kg}$, $c = 10\,\mathrm{Ns/m}$, $k = -5\,\mathrm{kN/m}$, $h = 50\,\mathrm{MN/m^3}$, $f_0 = 30\,\mathrm{N}$, and $\omega = 100\,\mathrm{rad/s}$.

simulation is then continued and the resulting displacement, x, and velocity, $\dot{x}$, are sampled at the same frequency as the excitation. These data are then plotted on the phase plane; that is, the velocity is plotted against the displacement. If the response is periodic, then the Poincaré map consists of a finite number of discrete points. Figure 2.33 shows a chaotic response, which has no periodicity.

Another feature of stable damped linear systems is that the steady-state response of an unforced system is zero, which is not necessarily true for a nonlinear system. Consider a system described by the Van der Pol equation

$$m\ddot{x} + c(1 - \gamma x^2)\dot{x} + kx = 0 \tag{2.143}$$

where the damping coefficient, c, is negative. If the system is displaced by a small amount from the equilibrium position with zero displacement and velocity, the vibrations grow because the damping is negative. This is called a *self-excited system*.

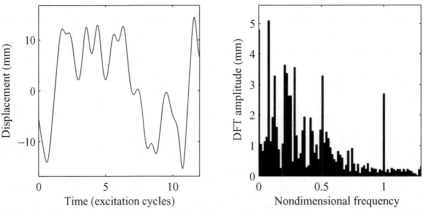

Figure 2.32. The chaotic response of the Duffing oscillator, with $m = 1\,\mathrm{kg}$, $c = 10\,\mathrm{Ns/m}$, $k = -5\,\mathrm{kN/m}$, $h = 50\,\mathrm{MN/m^3}$, $f_0 = 30\,\mathrm{N}$, and $\omega = 120\,\mathrm{rad/s}$.

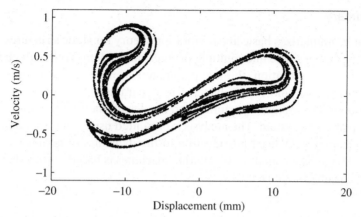

Figure 2.33. The Poincaré map of the Duffing oscillator, with $m = 1\,\mathrm{kg}$, $c = 10\,\mathrm{Ns/m}$, $k = -5\,\mathrm{kN/m}$, $h = 50\,\mathrm{MN/m^3}$, $f_0 = 30\,\mathrm{N}$, and $\omega = 120\,\mathrm{rad/s}$.

If the system has a large displacement, the vibration levels are limited by the positive damping. Thus, the vibration neither grows nor decays and the system reaches a steady-state motion called a *limit cycle*. Figure 2.34 shows the limit cycle for the Van der Pol equation in the phase plane as well as the transient response for initial conditions inside and outside the limit cycle.

It is clear that the range of phenomena that may occur in dynamics systems with a significant nonlinearity is vast. The situation is further complicated because there are many forms of nonlinearity, and different strategies often are required to analyze these different forms. The majority of this book considers linear systems or nonlinear systems that may be linearized. Chapter 10 provides examples of rotating machinery that require nonlinear dynamic analysis.

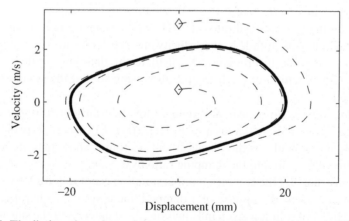

Figure 2.34. The limit cycle and transient response of the Van der Pol equation in the phase plane, with $m = 1\,\mathrm{kg}$, $c = -40\,\mathrm{Ns/m}$, $\gamma = 10^4\,\mathrm{m^{-2}}$, and $k = 10\,\mathrm{kN/m}$. The solid line denotes the limit cycle and the dashed lines represent two transient responses, with initial conditions denoted by the diamonds.

2.8 Summary

This chapter summarizes basic linear vibration theory for static structures. We analyze systems with a single and multiple degrees of freedom, both with and without damping. For free vibration, the key parameters are the system natural frequencies, damping ratios, and mode shapes. These parameters are important because they provide both a system description and an elegant method for determining the forced response of a system. The analysis is presented using matrix notation to clarify the mathematics. A brief introduction to the dynamics of nonlinear systems is provided. The dynamic analysis of rotating machines is based on this theory and is developed in subsequent chapters.

2.9 Problems

2.1 Write the constants a_0 and b_0 in Equation (2.24) in terms of the initial displacement and velocity, x_0 and $\dot{x}_0$. If $m = 1\,\text{kg}$, $c = 0$, $k = 9\,\text{N/m}$, $x_0 = 1\,\text{mm}$, and $\dot{x}_0 = 0$, calculate the resulting transient response. Calculate the response again if the damping is $c = 1\,\text{Ns/m}$. Sketch the response in both cases.

2.2 A steel disk of 400 mm diameter and 10 mm thickness is mounted on one end of a steel shaft of 10 mm diameter and 150 mm length, whose other end is fixed. Assuming that the system may be modeled as a single degree of freedom torsional system, as shown in Figure 2.5, calculate the natural frequency in Hz.

   The torsional stiffness of the shaft is $k = GJ/\ell$, where G is the material shear modulus, $J = \pi d^4/32$ is the second moment of area of the shaft, and ℓ and d are the shaft length and diameter. The polar moment of inertia of the disk is $I = \rho\pi h D^4/32$, where D and h are the disk diameter and thickness and ρ is the material density. For steel, assume $G = 80\,\text{GPa}$ and $\rho = 7,800\,\text{kg/m}^3$.

2.3 A car may be crudely modeled as a single degree of freedom system, in which the tires and suspension are assumed to act collectively as a single spring. If the mass of the car is 700 kg and the measured natural frequency is 1 Hz, what is the approximate effective stiffness of the tires and suspension?

2.4 Suppose that a system vibrates at 50 Hz with a magnitude of 1 mm. Calculate the maximum magnitude of the velocity and acceleration.

2.5 For the car given in Problem 2.3, calculate the damping coefficient required for the combination of the tires and suspension so that the system is critically damped.

2.6 The *logarithmic decrement*, δ, is defined for the transient response of a single degree of freedom system as $\delta = \ln\left(x(t)/x(t + T)\right)$, where T is the period of the oscillation and x is the displacement response. Show that the logarithmic decrement is related to the damping ratio by $\delta = 2\pi\zeta/\sqrt{1 - \zeta^2}$.

2.7 Demonstrate that the complex number notation is not required to compute the steady-state forced response of a damped linear system to harmonic forcing. Using the trial solution given in Equation (2.30) in terms of the sine and cosine, substitute into Equation (2.29) and show that Equation (2.33) holds.

2.8 A single degree of freedom system is forced harmonically with the same force magnitude at each frequency in turn. Show that the magnitude of the response is a maximum when the excitation frequency is $\omega = \omega_n\sqrt{1 - 2\zeta^2}$.

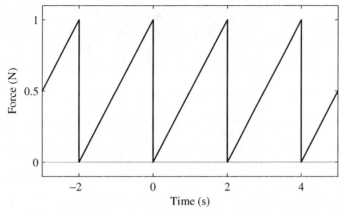

Figure 2.35. The triangular force input (Problem 2.10).

2.9 A single degree of freedom undamped system has a natural frequency of 5 Hz and is excited by a force $f(t) = f_0 \cos 11\pi t$. What is the phase of the displacement relative to the force? What happens to the magnitude and phase of the displacement response if

(a) The magnitude of the force is doubled?

(b) The phase of the force is changed by $\pi/2$?

(c) The forcing frequency is halved?

(d) Damping is introduced so that the damping ratio is 0.1?

2.10 Calculate the Fourier series of the periodic 0.5 Hz triangular force waveform shown in Figure 2.35.

2.11 For the two degrees of freedom system in Figure 2.16, the mass and stiffness coefficients are given as $m_1 = 1$ kg, $m_2 = 2$ kg, $k_1 = 1$ N/m, $k_2 = 4$ N/m, and $k_3 = 1$ N/m. The damping is assumed to be negligible. By substituting these values into Equation (2.50), calculate the natural frequencies and mode shapes for this system.

2.12 Two contra-rotating machines are mounted on a raft, as shown in Figure 2.36. Because of the synchronization of the machines, the resultant force is vertical. Assuming that the total mass of the raft and machines (including the unbalance mass) is m, the mounting stiffness is k, the unbalance mass is m_u at a radius r, and machines rotate at speed Ω rad/s, show that the equation of motion is

$$m\ddot{x} + kx = f(t) = 2m_u r \Omega^2 \cos \Omega t$$

where it is assumed that the unbalance masses are both vertical and in the same direction as x at $t = 0$.

It is now proposed to add a vibration absorber to reduce the vibration of the raft at the excitation frequency. Figure 2.37 shows a schematic of the system. Show that the natural frequencies are the roots of

$$\omega^4 - \left(\frac{k + k_a}{m} + \frac{k_a}{m_a} \right) \omega^2 + \frac{k k_a}{m m_a} = 0$$

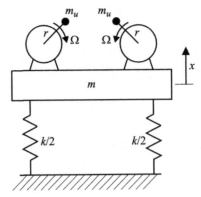

Figure 2.36. The machine (Problem 2.12).

Calculate the response of the raft and show that this is zero when $\Omega = \sqrt{k_a/m_a}$. At this frequency, show that the corresponding response of the absorber is $\dfrac{-2m_u r \Omega^2}{k_a} \cos \Omega t$.

2.13 The system shown in Figure 2.38 consists of a frame of mass 40 kg, restrained by a spring of stiffness 30 kN/m. The frame carries two masses, each of 10 kg. The masses are connected to the frame and one another by three identical springs of stiffness 10 kN/m. Determine the equation of motion for small displacements of the system in terms of the coordinates $q_1, \ldots q_3$.

Solving the eigenvalue problem for this system gives the following natural frequencies and mode shapes:

$$\omega_1 = 20.1725 \,\text{rad/s}, \quad \omega_2 = 42.9310 \,\text{rad/s}, \quad \omega_3 = 54.7723 \,\text{rad/s},$$

$$\mathbf{u}_1 = \begin{Bmatrix} 1.6861 \\ 1.6861 \\ 1.0000 \end{Bmatrix}, \quad \mathbf{u}_2 = \begin{Bmatrix} -1.1861 \\ -1.1861 \\ 1.0000 \end{Bmatrix}, \quad \mathbf{u}_3 = \begin{Bmatrix} 1 \\ -1 \\ 0 \end{Bmatrix}$$

Note that the modes are *not* normalized according to Equation (2.60). Verify that $\mathbf{u}_2$ and $\mathbf{u}_3$ are orthogonal with respect to both the mass and stiffness

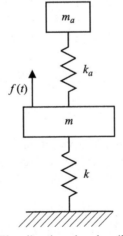

Figure 2.37. The vibration absorber (Problem 2.12).

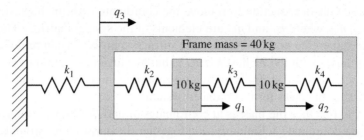

Figure 2.38. The three degrees of freedom system (Problem 2.13). The springs have stiffnesses $k_1 = 30\,\text{kN/m}$, $k_2 = k_3 = k_4 = 10\,\text{kN/m}$.

matrices; see Equation (2.58). Use the mass and stiffness matrices previously derived.

The mass-normalized first and second modes are

$$\mathbf{u}_{N1} = \begin{Bmatrix} 0.17132 \\ 0.17132 \\ 0.10161 \end{Bmatrix} \quad \text{and} \quad \mathbf{u}_{N2} = \begin{Bmatrix} -0.14369 \\ -0.14369 \\ 0.12114 \end{Bmatrix}$$

Mass-normalize the third mode using Equation (2.60) together with the mass matrix derived.

2.14 Consider the system described in Problem 2.13:

(a) The system is initially at rest (i.e., the initial velocities are zero) with displacements $q_1 = q_2 = q_3 = 5 \times 10^{-3}\,\text{m}$. Calculate the undamped free response of the system.

(b) Determine the response of the system at coordinates 1, 2, and 3 to a harmonic (i.e., sinusoidal) force of 100 N and frequency 6 Hz acting at the third coordinate. Determine the forces in each spring under these conditions.

(c) Calculate the receptance, mobility, and inertance of the system relating coordinates 1 and 3 at a frequency of 6 Hz.

2.15 For the system in Figure 2.39, derive its mass and stiffness matrices in the coordinate system shown. Solving the eigenvalue problem for this system gives the following natural frequencies: 4.8538, 13.2727, 23.5933, and 26.6981 Hz.

To reduce the model of the system to one with two degrees of freedom, the following strategies are to be tested:

(a) Use Guyan reduction to remove coordinates q_3 and q_4. What are the model natural frequencies?

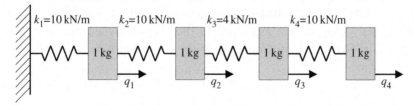

Figure 2.39. The four degrees of freedom system (Problem 2.15).

(b) Use Guyan reduction to remove the coordinate with the highest value of k_{ii}/m_{ii}, where k_{ii} and m_{ii} are elements of the system matrices. Having determined the new 3×3 mass and stiffness matrices, again remove the coordinate with the highest k_{ii}/m_{ii} value by Guyan reduction to obtain new 2×2 mass and stiffness matrices. From these matrices, determine the natural frequencies for this reduced system.

(c) Add together the masses at coordinates q_1 and q_2 to give a single mass at new coordinate 1. Similarly, add together the masses at coordinates q_3 and q_4 to give a single mass at new coordinate 2. Essentially, this implies that spring stiffnesses k_2 and k_4 are infinite. Solve the eigenvalue problem for this new system to obtain the two natural frequencies.

(d) Apply the constraints $q_1 = q_2$ and $q_3 = q_4$ using the transformation method described in Section 2.5 and show that the resulting model is identical to that obtained in (c).

Which approach gives solutions closest to the first and second natural frequencies?

2.16 By substituting the equivalent viscous-damping coefficient for Coulomb damping, $c_{eq} = 4f_{dry}/(\pi \omega x_0)$, into the equations of motion for a harmonically excited mass–spring–damper system, show that the response x_0 and phase ϕ are given by

$$x_0 = \frac{f_0\sqrt{1-\delta^2}}{k|1-r^2|} \quad \text{and} \quad \phi = \pm \tan^{-1}\left(\frac{\delta}{\sqrt{1-\delta^2}}\right)$$

where $r = \omega/\omega_n$, $\delta = 4f_{dry}/(\pi f_0)$, f_0 is the amplitude of the excitation at frequency ω, f_{dry} is the Coulomb or dry friction force, k is the spring stiffness, and ω_n is the natural frequency of the system. If $r < 1$, then ϕ is positive; if $r > 1$, then ϕ is negative.

Section 2.3.4 states that the equivalent Coulomb damping is only valid if $f_{dry} \ll f_0$. For x_0 to be real, the equation for x_0 places an upper bound on f_{dry} relative to f_0. What is it?

A mass-spring system with a Coulomb damper has a mass of 1 kg, a stiffness of 100 kN/m, and a Coulomb damping force of 10 N. Determine the system response magnitude and phase angle when a harmonic excitation force of 50 N acts on the mass at a frequency of 25 Hz. Repeat the calculation for an excitation frequency of 55 Hz.

2.17 (a) To obtain an approximate solution for an undamped nonlinear system described by Duffing's equation, assume a solution $x = x_1 \cos(\omega t)$. In the solution process, use the harmonic-balance method, discard any terms in $\cos(3\omega t)$ that arise from the x^3 term. Obtain a cubic equation in x_1. An undamped nonlinear system described by Duffing's equation (see Equation (2.136)) has the following parameters: $m = 10$ kg, $k = 1$ MN/m, and $h = 40$ MN/m^3. Estimate the amplitude of the response when the excitation frequency is 50 Hz and the amplitude of the excitation force is 30, 60, 120, and 240 N.

(b) Show that when the applied force tends to zero, the larger-magnitude solutions satisfy $\omega^2 = \omega_n^2 + \frac{3}{4}x_1^2\alpha$, where $\omega_n^2 = k/m$ and $\alpha = h/m$. The frequency ω may be interpreted as the frequency of the free response for a given displacement amplitude x_1. If $\alpha > 0$, calculate the limits of this free-response frequency. Estimate the frequency of the unforced response when the magnitude of vibration is 10 mm.

(c) Now assume a more accurate solution for the undamped Duffing oscillator of the form $x = x_1 \cos(\omega t) + x_3 \cos(3\omega t)$. Develop the two coupled equations in x_1 and x_3 using the harmonic-balance method. Numerically solve these coupled equations for an excitation force of 240 N and for excitation frequencies of 50 and 55 Hz. Use the value obtained from the cubic equation in x_1 alone as a starting value, with $x_3 = 0$.

3 Free Lateral Response of Simple Rotor Models

3.1 Introduction

In this chapter, we consider the process of creating adequate models of simple rotor systems and examine their lateral vibration in the absence of any applied forces. By "simple," we mean a rotor system that can be modeled in terms of a small number of degrees of freedom. These simple models consist of either a rigid rotor on flexible bearings and foundation or a flexible rotor on rigid bearings and foundation. Obviously, rotating machines are not designed specifically with these properties; the reality is that rotating machines are designed for a purpose. Shaft dimensions and inertias and the type and dimensions of the bearings are chosen appropriately for the machine function. It may be that the rotor is short with a large diameter, resulting in a shaft that is much stiffer than the bearing and foundations on which it is supported. In such circumstances, it might well be appropriate to model the system as a rigid rotor on flexible bearings and foundations. In these simple models, we assume that both the bearing and foundation can be represented by simple linear springs in the x and y directions. Therefore, the stiffness of the bearing and foundation can be combined and considered as a single entity, using the formula for the stiffness of springs in series, Equation (2.9). Conversely, a machine design might demand a long shaft supported on relatively stiff rolling-element bearings and a stiff foundation. In this case, the bearing and foundation stiffness relative to the shaft stiffness is very high and it may be acceptable to model the system as a flexible rotor on rigid supports. Both models are studied in this chapter.

In many situations, however, the stiffness of the bearing and/or foundation is of the same order of magnitude as the rotor stiffness, and the system can be modeled satisfactorily only if the flexibility of both parts of the system is taken into account. Models of these more complex systems are studied in Chapter 5. A key property we associate with all of the rotors modeled in Chapters 3, 5, and 6 is *symmetry*, in which the stiffness and mass properties are the same in every plane containing the axis of rotation. *Axisymmetric rotors* are solids of revolution and therefore have this symmetry property. *Asymmetric rotors* lack this symmetry and are considered further in Chapter 7. Many real machines may be considered to have symmetric rotors.

Having created the model, we then analyze it to determine the dynamic characteristics of the system: the natural frequencies, the corresponding mode shapes, and the free response of the system. In Chapters 6 and 8, we examine the effect of lateral forces and moments acting on the rotor.

When analyzing simple rotor models (or, indeed, any simple dynamic system) manually, it is essential always to work from first principles in the derivation of the equations of motion. A different model, of course, leads to a different set of equations; generally, however, the techniques for solving the equations numerically – using a calculator or simple computer program – are similar. The diversity of rotor systems encountered is so great that it is difficult to produce and correctly apply a set of formulae catering to all cases; therefore, we emphasize fundamental analysis techniques.

3.2 Coordinate Systems

To develop the equations of motion for a rotor-bearing system, it is necessary to define a coordinate system. We can use a coordinate system that is either stationary or fixed to the spinning rotor and therefore rotates. Each has its merits but, for simple systems with axisymmetric rotors, it is generally easier to use a stationary (i.e., nonrotating) coordinate system. Thus, for most of the systems described in this book, a stationary coordinate system is used, although some of the rotor models described in Chapters 7 and 10 employ a rotating coordinate system to advantage. An inertial frame of reference (in which a mass remains at rest unless a force is applied) allows the coordinate system to translate with a constant velocity and may be used instead of the stationary system.

We begin by defining a stationary coordinate system, consisting of three mutually perpendicular axes – Ox, Oy, and Oz – intersecting at the point O and fixed in space. For illustration, we describe the position of a rigid rotor, where the origin of the rotating coordinates is taken as the center of mass of the rotor. However, the same procedure may be applied to any point of interest on the shaft of a flexible rotor. We always assume that the axis of a rotor, in equilibrium, is coincident with the axis Oz. The axes Ox, Oy, and Oz, in that order, form a right-handed set, which is defined as follows:

A rotation of a right-handed screw from Ox to Oy advances it along Oz.
A rotation of a right-handed screw from Oy to Oz advances it along Ox.
A rotation of a right-handed screw from Oz to Ox advances it along Oy.

This is an accepted convention. For convenience in the following discussion, the axis of the rotor, Oz, is assumed to be horizontal and the vertical axis is assumed to be Oy (Figure 3.1).

Figure 3.2 shows side elevations and an end view of a rotor. The center of mass of the rotor is allowed to translate along axes Ox, Oy, and Oz by u, v, and w, respectively. The rotor also may rotate (by small amounts) about axes Ox and Oy by θ and ψ, respectively. Positive values of θ and ψ represent clockwise rotations about axes Ox and Oy, respectively, when viewed from O. Alternatively, we can state that the sense of the rotations is such that positive values of θ and ψ cause a right-hand screw to advance along Ox and Oy, respectively. A right-hand screw,

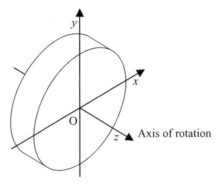

Figure 3.1. Right-handed coordinate set used in the analysis of a rotor.

which advances along Ox (or Oy), also can be represented by a screw vector acting in the direction Ox (or Oy). In Figure 3.2, a screw vector is indicated by a double-headed arrow. Because of the direction of view, the positive rotation about Ox, θ, appears clockwise in Figure 3.2. In contrast, the positive rotation about Oy, ψ, appears counterclockwise.

The rotor rotates clockwise about axis Oz with an angular displacement ϕ and an angular velocity Ω. Thus, the sense of this rotation is such that positive values of ϕ (or Ω) cause a right-hand screw to advance along axis Oz.

Visualizing a three-dimensional coordinate system can be difficult. It is helpful to make one corner of a cube from a card and mark on it the three axes and directions of rotation, as shown in Figure 3.3. This object then can be easily oriented in space to illustrate spatial relationships among the coordinates.

3.3 Gyroscopic Couples

In the dynamic analysis of rotors, it is important to include the effects of *gyroscopic couples* that act on the system. Gyroscopic couples arise because of the conservation of angular momentum in a system; these moments are perpendicular to the axis of spin. In Chapter 5, energy methods are used to develop the equations of motion of a rotating system, including gyroscopic effects. Here, we consider the angular momentum of the system to determine the gyroscopic couple.

Consider a uniform, circular disk spinning about the z-axis (i.e., the axis of rotation) with a constant angular velocity Ω. In Section 3.2, we define the sense of this

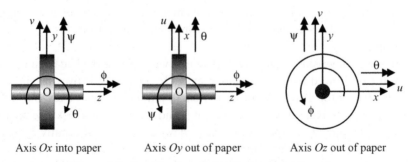

Axis Ox into paper   Axis Oy out of paper   Axis Oz out of paper

Figure 3.2. Coordinates used in the analysis of a rotor.

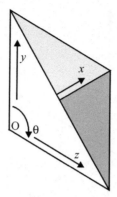

Figure 3.3. A simple aid to visualizing a right-handed coordinate set.

rotation such that a positive value of ϕ (or Ω) caused a right-hand screw to advance along axis Oz (i.e., the screw vector acts in the direction of Oz). The angular momentum of the rotor about Oz is $I_p\Omega$, where I_p is the polar moment of inertia of the rotor, defined as the moment of inertia about the longitudinal or polar axis. This angular momentum also can be represented by a screw vector acting in the direction Oz (Figure 3.4). Suppose that the disk now rotates about Oy with an angular velocity of $\dot{\psi}$. Over time δt, the disk will rotate by an angle of $\delta\psi = \dot{\psi}\delta t$; at the end of this time period, the angular momentum has magnitude $I_p\Omega$ but its direction is rotated by $\delta\psi$, as shown in Figure 3.4. For infinitesimal $\delta\psi$, the vector change in angular momentum has direction Ox and magnitude $M_x\delta t$, where M_x is the clockwise moment about the x-axis. Then

$$M_x\delta t = I_p\Omega\delta\psi \quad \text{or} \quad M_x = I_p\Omega\frac{\delta\psi}{\delta t}$$

Taking the limit as δt tends to zero, we have

$$M_x = I_p\Omega\frac{\mathrm{d}\psi}{\mathrm{d}t} = I_p\Omega\dot{\psi} \tag{3.1}$$

It is important to appreciate that Equation (3.1) defines a relationship, not a cause and effect. Thus, if a moment M_x is applied about the x-axis, the rotor

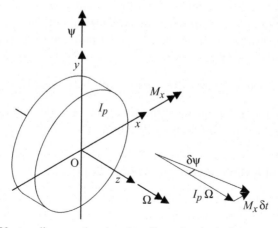

Figure 3.4. Vector diagram showing the effect of a clockwise moment about Ox.

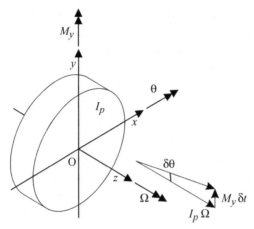

Figure 3.5. Vector diagram showing the effect of an counterclockwise moment about Oy.

(spinning about the z-axis) has angular velocity $\dot{\psi}$ about the y-axis. This velocity, $\dot{\psi}$, is called *precession*. Alternatively, if the spinning rotor is made to precess about the y-axis with an angular velocity $\dot{\psi}$, then a moment M_x must exist about the x-axis in order to close the triangle of momentum vectors and maintain equilibrium.

Consider now that the disk has a constant angular velocity of $\dot{\theta}$ about the direction Ox. Over time δt, the disk will rotate by an angle of $\delta \theta = \dot{\theta} \delta t$; the change in angular momentum is in the negative Oy direction, represented by the screw vector $-M_y \delta t$, in which M_y is the clockwise moment about the y-axis. To close the momentum-vector diagram and maintain equilibrium (Figure 3.5), we have

$$-M_y \delta t = I_p \Omega \delta \theta \quad \text{or} \quad M_y = -I_p \Omega \frac{\delta \theta}{\delta t}$$

Taking the limit as δt tends to zero, we have

$$M_y = -I_p \Omega \frac{d\theta}{dt} = -I_p \Omega \dot{\theta} \tag{3.2}$$

Note the difference in sign between Equations (3.1) and (3.2). In these equations, M_x and M_y are moments due to all the forces acting on the rotor. Because the disk can now rotate simultaneously about the x- and y-axes, the rate of change of angular momentum due to the angular acceleration about these axes also needs to be added to the contribution from the disk polar moment of inertia. Hence, Equations (3.1) and (3.2) become

$$I_d \ddot{\theta} + I_p \Omega \dot{\psi} = M_x \quad \text{and} \quad I_d \ddot{\psi} - I_p \Omega \dot{\theta} = M_y \tag{3.3}$$

where I_d is the diametral moment of inertia of the disk, defined as the moment of inertia with respect to an axis that is a diameter of the rotor.

3.4 Dynamics of a Rigid Rotor on Flexible Supports

To illustrate some of the features of the dynamics of rotating systems, we now consider two simple rotor models, each of which can be modeled using only four coordinates or degrees of freedom. The first rotor is rigid with a circular cross section,

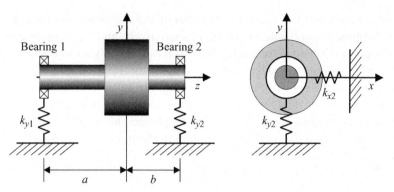

Figure 3.6. A rigid axisymmetric rotor on flexible supports.

supported by two bearings on flexible supports; the second rotor consists of a light flexible shaft carrying a single circular disk. We begin with the rigid rotor. Figure 3.6 shows a rigid rotor with a circular cross section supported by two bearings with no angular stiffness, commonly called *short bearings*. The bearings are flexibly supported in both the horizontal and vertical directions. The notation k_{y2} indicates the stiffness of the spring at bearing 2 in the y (vertical) direction. The mass of the rotor is m, its moment of inertia about axes Ox and Oy is I_d, and the polar moment of inertia of the rotor is I_p. It can be shown that $I_p \leq 2I_d$.

To develop the equations of motion for this system, we can use an energy method (e.g., Lagrange's equations) or, alternatively, directly, apply Newton's second law of motion. Here, we choose to use the latter; to do this, the free-body diagram for the system must be drawn as shown in Figure 3.7. Notice that ψ appears to be a counterclockwise rotation; however, because the Oy axis is out of the paper in the view on the right of the diagram, the rotation is clockwise about the Oy axis. Because there is no elastic coupling between the planes Oxz and Oyz, we can treat the elastic forces in these two planes separately.

The rotor has four degrees of freedom because it can translate in the directions Ox and Oy and it also can rotate about these axes. Practitioners often term the translation and rotation as *bounce* and *tilt motion*, respectively. We have chosen to

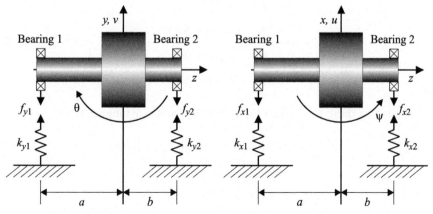

Figure 3.7. Free-body diagram for a rigid rotor on flexible supports.

describe the movement of the rotor in terms of the displacements of its center of mass in the directions Ox and Oy, u and v, respectively, and the clockwise rotations about Ox and Oy, θ and ψ, respectively. This is not the only set of coordinates that can be used to describe the displacements of the rotor. For example, we can use the displacements of the ends of the rotor along the Ox and Oy axes; again, we would have four coordinates. The resultant equations would be different but the solutions of both sets of equations would have the same physical meaning.

Applying Newton's second law of motion to the free rotor, we have:

$$\begin{aligned}
\text{Forces acting on rotor in } x \text{ direction:} \quad & -f_{x1} - f_{x2} = m\ddot{u} \\
\text{Forces acting on rotor in } y \text{ direction:} \quad & -f_{y1} - f_{y2} = m\ddot{v} \\
\text{Moments acting on rotor in } \theta \text{ direction:} \quad & -af_{y1} + bf_{y2} = I_d\ddot{\theta} + I_p\Omega\dot{\psi} \\
\text{Moments acting on rotor in } \psi \text{ direction:} \quad & af_{x1} - bf_{x2} = I_d\ddot{\psi} - I_p\Omega\dot{\theta}
\end{aligned} \tag{3.4}$$

The clockwise moments acting on the rotor about the Ox axis cause an angular acceleration about the Ox axis, $\ddot{\theta}$, and a precession about the Oy axis, $\dot{\psi}$. The clockwise moments acting on the rotor about the Oy axis cause an angular acceleration about the Oy axis, $\ddot{\psi}$, and a precession about the Ox axis, $-\dot{\theta}$. Because the Oy axis is out of the paper, the clockwise moments about Oy appear to be counterclockwise in the figure.

Let us assume that the displacements of the rotor from the equilibrium position are small, which is the case in practice (unless there is a catastrophic failure of our rotor-bearing system). This assumption means that the rotations θ and ψ are small; hence, we can replace $\sin\theta$ by θ and $\sin\psi$ by ψ. Furthermore, we assume that the spring supports are linear and that Hooke's law applies. Then, assuming no elastic coupling between the Ox and Oy directions

$$f_{x1} = k_{x1}\delta = k_{x1}(u - a\sin\psi) \approx k_{x1}(u - a\psi)$$

where δ is the deflection of the spring due to the force f_{x1}. Applying this argument to each of the forces, we have

$$\begin{aligned}
f_{x1} &= k_{x1}(u - a\psi) \\
f_{x2} &= k_{x2}(u + b\psi) \\
f_{y1} &= k_{y1}(v + a\theta) \\
f_{y2} &= k_{y2}(v - b\theta)
\end{aligned} \tag{3.5}$$

Substituting these forces into Equation (3.4) and rearranging gives

$$\begin{aligned}
m\ddot{u} + k_{x1}(u - a\psi) + k_{x2}(u + b\psi) &= 0 \\
m\ddot{v} + k_{y1}(v + a\theta) + k_{y2}(v - b\theta) &= 0 \\
I_d\ddot{\theta} + I_p\Omega\dot{\psi} + ak_{y1}(v + a\theta) - bk_{y2}(v - b\theta) &= 0 \\
I_d\ddot{\psi} - I_p\Omega\dot{\theta} - ak_{x1}(u - a\psi) + bk_{x2}(u + b\psi) &= 0
\end{aligned}$$

These equations can be further rearranged to give

$$m\ddot{u} + (k_{x1} + k_{x2})u + (-ak_{x1} + bk_{x2})\psi = 0$$

$$m\ddot{v} + (k_{y1} + k_{y2})v + (ak_{y1} - bk_{y2})\theta = 0$$

$$I_d\ddot{\theta} + I_p\Omega\dot{\psi} + (ak_{y1} - bk_{y2})v + (a^2k_{y1} + b^2k_{y2})\theta = 0$$ (3.6)

$$I_d\ddot{\psi} - I_p\Omega\dot{\theta} + (-ak_{x1} + bk_{x2})u + (a^2k_{x1} + b^2k_{x2})\psi = 0$$

Letting

$$k_{xT} = k_{x1} + k_{x2}, \qquad\qquad k_{yT} = k_{y1} + k_{y2}$$

$$k_{xC} = -ak_{x1} + bk_{x2}, \qquad\qquad k_{yC} = -ak_{y1} + bk_{y2}$$ (3.7)

$$k_{xR} = a^2k_{x1} + b^2k_{x2}, \qquad\qquad k_{yR} = a^2k_{y1} + b^2k_{y2}$$

where the subscripts T, C, and R have been chosen to indicate translational, coupling between displacement and rotation, and rotational stiffness coefficients. Then, Equation (3.6) can be written more concisely as

$$m\ddot{u} + k_{xT}u + k_{xC}\psi = 0$$

$$m\ddot{v} + k_{yT}v - k_{yC}\theta = 0$$

$$I_d\ddot{\theta} + I_p\Omega\dot{\psi} - k_{yC}v + k_{yR}\theta = 0$$ (3.8)

$$I_d\ddot{\psi} - I_p\Omega\dot{\theta} + k_{xC}u + k_{xR}\psi = 0$$

We see that there is elastic coupling between the first and fourth equations as well as the second and third equations. In addition, gyroscopic couples introduce a coupling between the third and fourth equations of Equation (3.8). Thus, all of the equations are coupled.

In summary, we derived the equations of motion for a rigid rotor with a circular cross section. Our next task is to solve the coupled differential equations given in Equation (3.8).

3.5 A Rigid Rotor on Isotropic Flexible Supports

Let us assume that the flexibility of the bearing supports is the same in both of the transverse directions; that is, the bearing supports are *isotropic*. Then, we can simplify Equation (3.8) by letting

$$k_{xT} = k_{yT} = k_T, \quad k_{xC} = k_{yC} = k_C, \quad \text{and} \quad k_{xR} = k_{yR} = k_R$$

By introducing these simplifying relationships, Equation (3.8) becomes

$$m\ddot{u} + k_Tu + k_C\psi = 0$$

$$m\ddot{v} + k_Tv - k_C\theta = 0$$

$$I_d\ddot{\theta} + I_p\Omega\dot{\psi} - k_Cv + k_R\theta = 0$$ (3.9)

$$I_d\ddot{\psi} - I_p\Omega\dot{\theta} + k_Cu + k_R\psi = 0$$

We now consider the solution to these equations under various conditions.

3.5.1 Neglecting Gyroscopic Effects and Elastic Coupling

Consider the solutions of Equation (3.9) when gyroscopic effects can be neglected; that is, the speed of rotation is low or the polar moment of inertia is small. Letting $I_p\Omega = 0$ in Equation (3.9) gives

$$m\ddot{u} + k_T u + k_C\psi = 0$$
$$m\ddot{v} + k_T v - k_C\theta = 0$$
$$I_d\ddot{\theta} - k_C v + k_R\theta = 0$$
$$I_d\ddot{\psi} + k_C u + k_R\psi = 0$$

(3.10)

We now obtain a solution for the case when $k_C = 0$. This situation arises, for example, if the shaft lengths a and b are equal and the stiffnesses of the bearings are the same at each end of the rotor. Thus, Equation (3.10) becomes

$$m\ddot{u} + k_T u = 0$$
$$m\ddot{v} + k_T v = 0$$
$$I_d\ddot{\theta} + k_R\theta = 0$$
$$I_d\ddot{\psi} + k_R\psi = 0$$

(3.11)

These equations are uncoupled and can be solved independently of one another. We look for solutions of the form

$$u(t) = u_0 e^{st}, \quad v(t) = v_0 e^{st}, \quad \theta(t) = \theta_0 e^{st}, \quad \text{and} \quad \psi(t) = \psi_0 e^{st} \quad (3.12)$$

where u_0, v_0, θ_0, and ψ_0 are complex constants. Thus, $\ddot{u} = u_0 s^2 e^{st}$, and so on. Substituting these relationships into Equation (3.11) gives

$$(ms^2 + k_T)u_0 e^{st} = 0$$
$$(ms^2 + k_T)v_0 e^{st} = 0$$
$$(I_d s^2 + k_R)\theta_0 e^{st} = 0$$
$$(I_d s^2 + k_R)\psi_0 e^{st} = 0$$

Because $e^{st} \neq 0$,

$$(ms^2 + k_T)u_0 = 0$$
$$(ms^2 + k_T)v_0 = 0$$
$$(I_d s^2 + k_R)\theta_0 = 0$$
$$(I_d s^2 + k_R)\psi_0 = 0$$

(3.13)

The nontrivial solutions of Equation (3.13) are

$$s^2 = -\frac{k_T}{m} \quad \text{(twice, from the first and second equations)}$$

$$s^2 = -\frac{k_R}{I_d} \quad \text{(twice, from the third and fourth equations)}$$

Thus,

$$s_1 = s_2 = J\sqrt{\frac{k_T}{m}}, \qquad s_3 = s_4 = J\sqrt{\frac{k_R}{I_d}},$$

$$s_5 = s_6 = -J\sqrt{\frac{k_T}{m}}, \qquad s_7 = s_8 = -J\sqrt{\frac{k_R}{I_d}}$$

(3.14)

Because the roots given by Equation (3.14) are complex conjugate pairs, the system natural frequencies are derived from $s_i = J\omega_i$, $s_{i+4} = -J\omega_i$, $i = 1, \ldots, 4$. Thus,

$$\omega_1 = \omega_2 = \sqrt{\frac{k_T}{m}} \quad \text{and} \quad \omega_3 = \omega_4 = \sqrt{\frac{k_R}{I_d}}$$

(3.15)

Here, ω_1 is the natural frequency of the bounce mode and ω_3 is the natural frequency of the tilt mode.

Depending on the value of the parameters, $\omega_3 = \omega_4$ may be larger than $\omega_1 = \omega_2$ or vice versa. By convention, the frequencies are labeled so that $\omega_1 = \omega_2 \le \omega_3 = \omega_4$; therefore, it may be necessary to interchange the definitions of $\omega_1 = \omega_2$ and $\omega_3 = \omega_4$. Substituting for s in Equation (3.12) gives the four solutions as follows:

$$u(t) = u_{01} e^{s_1 t} + u_{02} e^{s_5 t}$$

$$= a_u \sin(\omega_1 t) + b_u \cos(\omega_1 t) = c_u \cos(\omega_1 t + \alpha_u)$$

$$v(t) = v_{01} e^{s_2 t} + v_{02} e^{s_6 t}$$

$$= a_v \sin(\omega_2 t) + b_v \cos(\omega_2 t) = c_v \cos(\omega_2 t + \alpha_v)$$

$$\theta(t) = \theta_{01} e^{s_3 t} + \theta_{02} e^{s_7 t}$$

$$= a_\theta \sin(\omega_3 t) + b_\theta \cos(\omega_3 t) = c_\theta \cos(\omega_3 t + \alpha_\theta)$$

$$\psi(t) = \psi_{01} e^{s_4 t} + \psi_{02} e^{s_8 t}$$

$$= a_\psi \sin(\omega_4 t) + b_\psi \cos(\omega_4 t) = c_\psi \cos(\omega_4 t + \alpha_\psi)$$

(3.16)

The vibration of the rotor is a combination of these motions, depending on the initial conditions.

If the system is given an initial displacement $u(0)$ and $v(0)$ and an initial velocity $\dot{u}(0)$ and $\dot{v}(0)$, we can determine the subsequent motion of the rotor in the x and y directions from the first two equations in Equation (3.16). Similarly, if the system is given initial rotations and angular velocities about the Ox and Oy axes, we can determine the subsequent rotations about the two axes.

EXAMPLE 3.5.1. A uniform rigid rotor is shown in Figure 3.8. The rotor has a length of 0.5 m and a diameter of 0.2 m and is made from steel with a density of 7, 810 kg/m³. It is supported at the ends by bearings. Both bearing supports have horizontal and vertical stiffnesses of 1 MN/m. Determine the natural frequencies of this rotor. The rotor is given an initial displacement of 1 mm in the x direction and 0.5 mm in the y direction, and an initial velocity of 30 mm/s in the x direction at its center. Determine the free response of the center of the rotor.

Solution. This and most other examples in this chapter require the solution and interpretation of the equations of motion of a rotor model that is described in

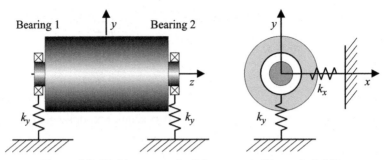

Figure 3.8. Rigid rotor on flexible supports (Example 3.5.1).

this book. In the interest of brevity, we do not develop the equations of motion from first principles in the solution because they are already developed. We simply use results that have been derived from these equations.

From Appendix 1, the mass of the rotor is

$$m = \rho \pi D^2 L/4 = 7{,}810 \times \pi \times 0.2^2 \times 0.5/4 = 122.68 \,\text{kg}$$

and the polar and diametral inertias are

$$I_p = mD^2/8 = 122.68 \times 0.2^2/8 = 0.6134 \,\text{kg m}^2,$$

$$I_d = I_p/2 + mL^2/12 = 0.6134/2 + 122.68 \times 0.5^2/12 = 2.8625 \,\text{kg m}^2$$

From Equation (3.7), the stiffness terms are

$$k_T = 1{,}000 + 1{,}000 = 2{,}000 \,\text{kN/m},$$

$$k_R = 0.25^2 \times 1{,}000 + 0.25^2 \times 1{,}000 = 125 \,\text{kNm},$$

$$k_C = -0.25 \times 1{,}000 + 0.25 \times 1{,}000 = 0$$

Because $k_C = 0$, we can use Equation (3.14); the system roots are

$$s_1 = s_2 = J\sqrt{k_T/m} = J\sqrt{2{,}000{,}000/122.68} = 127.68J \,\text{rad/s}$$

and

$$s_3 = s_4 = J\sqrt{k_R/I_d} = J\sqrt{1{,}25{,}000/2.8625} = 208.97J \,\text{rad/s}$$

From Equation (3.15), the natural frequencies are

$$\omega_1 = \omega_2 = 127.68 \,\text{rad/s} \quad \text{and} \quad \omega_3 = \omega_4 = 208.97 \,\text{rad/s}$$

These natural frequencies may be converted to Hz by dividing by 2π to give 20.32 and 33.26 Hz, respectively.

From the first equation in Equation (3.16), $u(t) = a_u \sin(\omega_1 t) + b_u \cos(\omega_1 t)$; hence, $\dot{u}(t) = a_u \omega_1 \cos(\omega_1 t) - b_u \omega_1 \sin(\omega_1 t)$. Thus, $u(0) = b_u = 1 \,\text{mm}$ and $\dot{u}(0) = a_u \omega_1 = 30 \,\text{mm/s}$. Similarly, $v(0) = b_v = 0.5 \,\text{mm}$ and $\dot{v}(0) = a_v \omega_2 = 0$. Thus, $u(t) = (30/\omega_1) \sin(\omega_1 t) + \cos(\omega_1 t) \,\text{mm}$ and $v(t) = 0.5 \cos(\omega_2 t) \,\text{mm}$. Note that $\omega_1 = \omega_2 = 127.68 \,\text{rad/s}$.

Plotting $u(t)$ versus $v(t)$ shows the orbit that the rotor executes. Because we want to control the plot of the orbit, it is more convenient to let $\omega_1 t = \omega_2 t$ be the

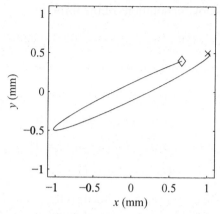

Figure 3.9. Motion of the center of mass of the rotor due to an initial disturbance (Example 3.5.1). The × indicates the start of the motion and the ◇ indicates the end.

angle β, for example, and then plot the orbit for β from 0 to 1.9π in small increments. In this way, the orbit does not close; we then can see the beginning and end of the orbit and, hence, its direction. The details of the calculations are not given but the result is shown in Figure 3.9. Due to the implied assumption that no damping is present in the system, the orbit repeats indefinitely. In practice, of course, the free orbits caused by initial displacements and velocities normally decay to zero due to damping.

Figure 3.10 shows the projection of the orbit shown in Figure 3.9 along the Ox and Oy axes, plotted against time. It shows that the rotor vibrates harmonically in the x and y directions.

3.5.2 Neglecting Gyroscopic Effects but Including Elastic Coupling

We now return to Equation (3.10) and consider the solution of this equation when the gyroscopic terms are neglected but the elastic coupling terms are included, so

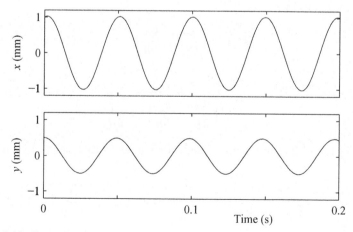

Figure 3.10. Rotor motion: Translation in the directions Ox and Oy (Example 3.5.1).

that $k_C \neq 0$. We again look for solutions of the form

$$u(t) = u_0 e^{st}, \quad v(t) = v_0 e^{st}, \quad \theta(t) = \theta_0 e^{st}, \quad \text{and} \quad \psi(t) = \psi_0 e^{st}$$

Because $e^{st} \neq 0$, substituting these expressions into Equation (3.10) gives

$$
\begin{aligned}
(ms^2 + k_T)u_0 + k_C \psi_0 &= 0 \\
k_C u_0 + (I_d s^2 + k_R)\psi_0 &= 0 \\
(ms^2 + k_T)v_0 - k_C \theta_0 &= 0 \\
-k_C v_0 + (I_d s^2 + k_R)\theta_0 &= 0
\end{aligned}
\tag{3.17}
$$

The *order* of the equations has been changed so that the original order [1234] becomes [1423]. Equation (3.17) is two pairs of coupled equations. Considering the first two equations, the roots are determined by rearranging to give

$$u_0 = -\frac{k_C \psi_0}{(ms^2 + k_T)} = -\frac{\left(I_d s^2 + k_R\right)\psi_0}{k_C} \tag{3.18}$$

and, hence

$$\left(ms^2 + k_T\right)\left(I_d s^2 + k_R\right) - k_C^2 = 0$$

Multiplying the bracketed terms and dividing each term by mI_d gives a quadratic equation in s^2; thus,

$$s^4 + \left(\frac{k_R}{I_d} + \frac{k_T}{m}\right)s^2 + \frac{k_R k_T - k_C^2}{mI_d} = 0 \tag{3.19}$$

Rearranging the second pair of equations in Equation (3.17) gives

$$v_0 = \frac{k_C \theta_0}{(ms^2 + k_T)} = \frac{\left(I_d s^2 + k_R\right)\theta_0}{k_C} \tag{3.20}$$

Multiplying the bracketed terms and dividing each term by mI_d gives a second quadratic equation in s^2, identical to Equation (3.19). The roots of these two quadratic equations in s^2 are

$$s^2 = -\left(\frac{k_R}{2I_d} + \frac{k_T}{2m}\right) \pm \gamma$$

where

$$\gamma = \sqrt{\left(\frac{k_R}{2I_d} - \frac{k_T}{2m}\right)^2 + \frac{k_C^2}{mI_d}} \tag{3.21}$$

Thus, from the two quadratic equations, the values of s that satisfy Equation (3.17) are

$$
\begin{aligned}
s_1 = s_2 &= J\sqrt{\frac{k_R}{2I_d} + \frac{k_T}{2m} - \gamma}, \qquad s_3 = s_4 = J\sqrt{\frac{k_R}{2I_d} + \frac{k_T}{2m} + \gamma}, \\
s_5 = s_6 &= -J\sqrt{\frac{k_R}{2I_d} + \frac{k_T}{2m} - \gamma}, \qquad s_7 = s_8 = -J\sqrt{\frac{k_R}{2I_d} + \frac{k_T}{2m} + \gamma}
\end{aligned}
\tag{3.22}
$$

The eight roots form two pairs of repeated complex conjugate roots with zero real parts. Thus, it is seen that when $k_C \neq 0$, the algebra is somewhat more complicated but the fundamental nature of the roots is not different from the case when $k_C = 0$. Because $s_i = j\omega_i$, $s_{i+4} = -j\omega_i$, for $i = 1, \ldots, 4$, the system natural frequencies are

$$\omega_1 = \omega_2 = \sqrt{\frac{k_R}{2I_d} + \frac{k_T}{2m} - \gamma} \quad \text{and} \quad \omega_3 = \omega_4 = \sqrt{\frac{k_R}{2I_d} + \frac{k_T}{2m} + \gamma} \qquad (3.23)$$

There is now coupling between the displacements and rotations in each plane, and the mode shape associated with each root can be computed. We found that s^2 can take only two specific values in Equation (3.18), denoted by s_1^2 and s_3^2. Thus, by replacing s by s_i ($i = 1$ or 3) in either part of Equation (3.18), we can determine the ratio between u_0 and ψ_0 for the specific value of s. Thus,

$$\left(\frac{u_0}{\psi_0}\right)^{(i)} = -\frac{I_d s_i^2 + k_R}{k_C} = -\frac{k_C}{ms_i^2 + k_T}, \quad i = 1, 3 \qquad (3.24)$$

Similarly, s^2 can take only two specific values in Equation (3.20), denoted by s_2^2 and s_4^2. Thus, by replacing s by s_i ($i = 2$ or 4) in either part of Equation (3.20), we can determine the ratio between v_0 and θ_0 for the specific value of s. Thus,

$$\left(\frac{v_0}{\theta_0}\right)^{(i)} = \frac{I_d s_i^2 + k_R}{k_C} = \frac{k_C}{ms_i^2 + k_T}, \quad i = 2, 4 \qquad (3.25)$$

Because $s_1^2 = s_2^2$, the mode shapes given by Equations (3.24) and (3.25) are identical, although in different planes, and are the mode shapes associated with the natural frequencies ω_1 or ω_2. The different sign in Equations (3.24) and (3.25) arises because of the convention in the definitions of the angular displacements, as shown in Figure 3.7. Similarly, the mode shapes associated with the roots s_3 and s_4 is identical. This mode shape is associated with the natural frequency ω_3 or ω_4. The amplitude of a mode shape is not unique and we can obtain only the ratios $(u_0/\psi_0)^{(i)}$ and $(v_0/\theta_0)^{(i)}$. Furthermore, each pair of mode shapes is not unique because each pair is associated with two repeated roots; therefore, a linear combination of these mode shapes is also a mode shape.

EXAMPLE 3.5.2. The rigid rotor of Example 3.5.1 and shown in Figure 3.8 is supported in flexibly mounted bearings. The horizontal and vertical supports of the bearings are 1 MN/m at bearing 1 and 1.3 MN/m at bearing 2. Determine the natural frequencies and the mode shape information for this system when the rotor is stationary.

Solution. From the solution of Example 3.5.1, we have

$$m = 122.68 \,\text{kg}, \quad I_p = 0.6134 \,\text{kg m}^2, \quad \text{and} \quad I_d = 2.8625 \,\text{kg m}^2$$

In this example, the bearing stiffnesses are

$$k_T = 1{,}000 + 1{,}300 = 2{,}300 \,\text{kN/m}$$

$$k_R = 0.25^2 \times 1{,}000 + 0.25^2 \times 1{,}300 = 143.75 \,\text{kNm/rad}$$

$$k_C = -0.25 \times 1{,}000 + 0.25 \times 1{,}300 = 75 \,\text{kN}$$

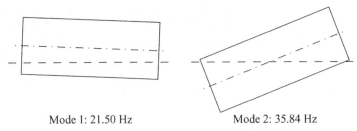

Mode 1: 21.50 Hz   Mode 2: 35.84 Hz

Figure 3.11. The bounce and tilt modes of the rigid rotor supported by flexible bearings (Example 3.5.2). The dashed line denotes zero displacement.

Thus, k_C is non-zero and we must use Equation (3.22) to determine the system roots.

From Equation (3.21), we find that $\gamma = 16{,}236\,\text{rad}^2/\text{s}^2$. Equation (3.22) gives $s_1 = s_2 = 135.08 j\,\text{rad/s}$ and $s_3 = s_4 = 225.21 j\,\text{rad/s}$. Using Equation (3.23), we have $\omega_1 = \omega_2 = 135.08\,\text{rad/s}$ and $\omega_3 = \omega_4 = 225.21\,\text{rad/s}$. Thus, the natural frequencies are $135.08/(2\pi) = 21.50\,\text{Hz}$ and $225.21/(2\pi) = 35.84\,\text{Hz}$. From Equation (3.24), the corresponding $(u_0/\psi_0)^{(i)}$ ratios are -1.2202 at $21.50\,\text{Hz}$ and 0.0191 at $35.84\,\text{Hz}$. At the lower frequency, the rotor motion is primarily in translation; if the mode is scaled to give a lateral displacement of $1\,\text{mm}$, the corresponding tilting motion of the rotor is $1.2202 \times 10^{-3}\,\text{rad}$, or approximately $0.05°$. At the higher frequency, a scaled mode with $1\,\text{mm}$ of lateral displacement corresponds to a tilting motion of about $3°$. Figure 3.11 shows these modes diagrammatically.

3.5.3 Including Gyroscopic Effects

In the analysis presented in Sections 3.5.1 and 3.5.2, the effect of gyroscopic couples is neglected; consequently, the spin speed of the rotor is immaterial. This is sometimes valid but, in many situations such as overhung rotors or rotors spinning at high speed, it is necessary to consider the effect of gyroscopic couples. In all subsequent analysis, we routinely include gyroscopic effects unless there is good evidence that they can be neglected.

To include gyroscopic effects in the analysis of a rigid rotor on isotropic supports, we begin with Equation (3.9), which is repeated here for convenience:

$$m\ddot{u} + k_T u + k_C \psi = 0$$

$$m\ddot{v} + k_T v - k_C \theta = 0$$

$$I_d\ddot{\theta} + I_p\Omega\dot{\psi} - k_C v + k_R \theta = 0$$

$$I_d\ddot{\psi} - I_p\Omega\dot{\theta} + k_C u + k_R \psi = 0$$

If $k_C = 0$, then

$$m\ddot{u} + k_T u = 0$$

$$m\ddot{v} + k_T v = 0$$

$$I_d\ddot{\theta} + I_p\Omega\dot{\psi} + k_R \theta = 0$$

$$I_d\ddot{\psi} - I_p\Omega\dot{\theta} + k_R \psi = 0$$

(3.26)

The first two equations uncouple, as previously, to give

$$s_1 = s_2 = J\sqrt{\frac{k_T}{m}}, \quad s_5 = s_6 = -J\sqrt{\frac{k_T}{m}}$$

and, hence

$$\omega_1 = \omega_2 = \sqrt{\frac{k_T}{m}} \tag{3.27}$$

The second pair of equations is coupled; letting $\theta(t) = \theta_0 e^{st}$ and $\psi(t) = \psi_0 e^{st}$ gives

$$\begin{aligned}
\left(I_d s^2 + k_R\right)\theta_0 + I_p \Omega s \psi_0 &= 0 \\
-I_p \Omega s \theta_0 + \left(I_d s^2 + k_R\right)\psi_0 &= 0
\end{aligned} \tag{3.28}$$

Eliminating θ_0 and ψ_0 leads to

$$\left(I_d s^2 + k_R\right)^2 + \left(I_p \Omega s\right)^2 = 0 \tag{3.29}$$

Moving the last term to the right side of the equation and taking the square root gives

$$I_d s^2 + k_R = \pm J I_p \Omega s \quad \text{or} \quad I_d s^2 \mp J I_p \Omega s + k_R = 0$$

The four solutions of Equation (3.28) are obtained by solving this pair of quadratic equations in s to give

$$s_{3,4} = J\left\{\pm \frac{I_p \Omega}{2 I_d} + \sqrt{\left(\frac{I_p \Omega}{2 I_d}\right)^2 + \frac{k_R}{I_d}}\right\}, \quad s_{7,8} = -s_{3,4} \tag{3.30}$$

Because $s_i = J\omega_i$, $s_{i+4} = -J\omega_i$, for $i = 3, 4$, then

$$\omega_3 = -\frac{I_p \Omega}{2 I_d} + \sqrt{\left(\frac{I_p \Omega}{2 I_d}\right)^2 + \frac{k_R}{I_d}}, \quad \omega_4 = \frac{I_p \Omega}{2 I_d} + \sqrt{\left(\frac{I_p \Omega}{2 I_d}\right)^2 + \frac{k_R}{I_d}} \tag{3.31}$$

These roots (and natural frequencies) are dependent on the speed of rotation. As this speed tends to zero, the roots become identical to the second pair of roots of Equation (3.14). The fact that ω_3 and ω_4 are dependent on the rotational speed is in contrast to the situation that prevails in all fixed structures. In a fixed structure, the natural frequencies are essentially controlled by the mass and stiffness properties of the system. Thus, because the mass and stiffness of the system are normally fixed, the natural frequencies are fixed values and vary only slightly with second-order effects such as temperature.

By rearranging Equation (3.28), we have

$$\left(\frac{\theta_0}{\psi_0}\right)^{(i)} = -\frac{I_p \Omega s_i}{I_d s_i^2 + k_R} = \frac{I_d s_i^2 + k_R}{I_p \Omega s_i} \tag{3.32}$$

The terms on the right side of Equation (3.32) are of the form $-A/B = B/A$ and, thus, A/B must equal $\pm J$. Noting that $s_i^2 = -\omega_i^2$, where the natural frequency ω_i is positive, then in Equation (3.32)

if $k_R > \omega_i^2 I_d$

$$(\theta_0/\psi_0)^{(i)} = \begin{cases} -J & \text{if } s_i = J\omega_i \\ J & \text{if } s_i = -J\omega_i \end{cases}$$

if $k_R < \omega_i^2 I_d$

$$(\theta_0/\psi_0)^{(i)} = \begin{cases} J & \text{if } s_i = J\omega_i \\ -J & \text{if } s_i = -J\omega_i \end{cases} \tag{3.33}$$

The sign of the relationship between θ_0 and ψ_0 determines the direction of rotation of the modes. The real response is obtained by adding together the contribution of the two complex conjugate modes. Because the scaling of the mode shapes is arbitrary, we may assume – without loss of generality – that $\theta_0 = 1$ for all eigenvectors. If $k_R > \omega_i^2 I_d$, then for the root $s_i = J\omega_i$, $\theta_0 = -J\psi_0$ and, hence, $\psi_0 = J$. For the complex conjugate root, $\psi_0 = -J$. Thus, the time response is

$$\begin{Bmatrix} \theta(t) \\ \psi(t) \end{Bmatrix} = \begin{Bmatrix} 1 \\ J \end{Bmatrix} e^{J\omega_i t} + \begin{Bmatrix} 1 \\ -J \end{Bmatrix} e^{-J\omega_i t} = 2 \begin{Bmatrix} \cos\omega_i t \\ -\sin\omega_i t \end{Bmatrix} \tag{3.34}$$

In the (θ, ψ) plane, the orbit is a circle. The mode rotates in a clockwise direction and, because the positive rotor spin has been defined to be counterclockwise, is called a *backward mode*. Similarly, for $k_R < \omega_i^2 I_d$

$$\begin{Bmatrix} \theta(t) \\ \psi(t) \end{Bmatrix} = \begin{Bmatrix} 1 \\ -J \end{Bmatrix} e^{J\omega_i t} + \begin{Bmatrix} 1 \\ J \end{Bmatrix} e^{-J\omega_i t} = 2 \begin{Bmatrix} \cos\omega_i t \\ \sin\omega_i t \end{Bmatrix} \tag{3.35}$$

In this case, the mode rotates in the counterclockwise direction and is called a *forward mode*. Forward and backward modes are considered in more detail in Section 3.6.1.

It is difficult to visualize the modes in the (θ, ψ) plane. However, with reference to Figure 3.7, the displacement in the x and y directions at the bearing 1 end of the rotor are

$$\begin{aligned} \text{along } Ox: \quad & u_0 - a\psi_0, \\ \text{along } Oy: \quad & v_0 + a\theta_0 \end{aligned} \tag{3.36}$$

At frequencies ω_3 and ω_4, the corresponding displacements are $u_0 = v_0 = 0$; therefore, at these frequencies, the displacement along Ox is $-a\psi_0$ and along Oy is $a\theta_0$. Because the orbit in the (θ, ψ) plane is circular, the orbit in the (x, y) plane at bearing 1 (and also at bearing 2) also must be circular.

Consider now the case when $k_C \neq 0$. The system is described by Equation (3.9), which is repeated here for convenience:

$$m\ddot{u} + k_T u + k_C \psi = 0$$

$$m\ddot{v} + k_T v - k_C \theta = 0$$

$$I_d\ddot{\theta} + I_p\Omega\dot{\psi} - k_C v + k_R \theta = 0$$

$$I_d\ddot{\psi} - I_p\Omega\dot{\theta} + k_C u + k_R \psi = 0$$

Assuming the response is of the form given by Equation (3.12) – for example, $u(t) = u_0 e^{st}$ – then

$$(ms^2 + k_T)u_0 + k_C\psi_0 = 0$$

$$(ms^2 + k_T)v_0 - k_C\theta_0 = 0$$

$$-k_C v_0 + (I_d s^2 + k_R)\theta_0 + I_p\Omega s\psi_0 = 0 \tag{3.37}$$

$$k_C u_0 - I_p\Omega s\theta_0 + (I_d s^2 + k_R)\psi_0 = 0$$

Using the first two of these equations to eliminate θ_0 and ψ_0 in the last two equations gives

$$-I_p\Omega s(ms^2 + k_T)u_0 + \left\{(ms^2 + k_T)(I_d s^2 + k_R) - k_C^2\right\}v_0 = 0$$

$$\left\{(ms^2 + k_T)(I_d s^2 + k_R) - k_C^2\right\}u_0 + I_p\Omega s(ms^2 + k_T)v_0 = 0 \tag{3.38}$$

Hence, eliminating u_0 and v_0

$$\left\{(ms^2 + k_T)(I_d s^2 + k_R) - k_C^2\right\}^2 + \left\{I_p\Omega s(ms^2 + k_T)\right\}^2 = 0 \tag{3.39}$$

and

$$(ms^2 + k_T)(I_d s^2 + k_R) - k_C^2 = \pm_J I_p\Omega s(ms^2 + k_T) \tag{3.40}$$

or, rearranging

$$\left(ms^2 + k_T\right)\left(I_d s^2 \mp_J I_p\Omega s + k_R\right) - k_C^2 = 0 \tag{3.41}$$

Multiplying the bracketed terms and dividing each term by mI_d gives

$$s^4 \mp_J \left(\frac{I_p}{I_d}\right)\Omega s^3 + \left(\frac{k_R}{I_d} + \frac{k_T}{m}\right)s^2 \mp_J \left(\frac{k_T I_p}{mI_d}\right)\Omega s + \frac{k_R k_T - k_C^2}{mI_d} = 0 \tag{3.42}$$

From the definitions of k_R, k_T, and k_C in Equation (3.7), it can be shown that $k_R k_T - k_C^2 > 0$; thus, the real terms in Equation (3.42) are always positive. In fact, the roots of this quartic equation are purely imaginary and occur in complex conjugate pairs according to the sign of the imaginary terms in Equation (3.42). It is convenient to let $s = _J\omega$ so that the coefficients of the quartic equation become real, as follows:

$$\omega^4 \mp \left(\frac{I_p}{I_d}\right)\Omega\omega^3 - \left(\frac{k_R}{I_d} + \frac{k_T}{m}\right)\omega^2 \pm \left(\frac{k_T I_p}{mI_d}\right)\Omega\omega + \frac{k_R k_T - k_C^2}{mI_d} = 0 \tag{3.43}$$

The eight roots of this pair of quartic equations occur in four pairs with opposite signs; in practice, these roots are determined numerically. The positive solutions give the four natural frequencies appropriate to a problem with four degrees of freedom.

We now determine the mode shapes for each root s_i, for $i = 1, \ldots, 8$. By the same argument used to obtain Equation (3.33) from Equation (3.32), Equations (3.38) and (3.40) give $u_0 = \pm_J v_0$. Then, from the first two equations of

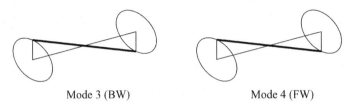

Mode 3 (BW)   Mode 4 (FW)

Figure 3.12. Mode shapes for the rigid rotor (Example 3.5.3(a)). The rotor displacements at the bearing locations are displayed. FW indicates forward whirl and BW indicates backward whirl.

Equation (3.37), $\theta_0 = \pm_J \psi_0$. Using the same argument as for the uncoupled case when $k_C = 0$, Equation (3.34), the orbits in the (u, v) and (θ, ψ) planes are circular. Furthermore

$$\left(\frac{u_0}{\psi_0}\right)^{(i)} = -\left(\frac{v_0}{\theta_0}\right)^{(i)} = -\frac{k_C}{ms_i^2 + k_T} \tag{3.44}$$

This ratio is real for purely imaginary roots, s_i, and is the same for both roots of a complex conjugate pair. This shows that for a given root, s_i, there is a fixed relationship between u_0 and ψ_0 and between v_0 and θ_0.

EXAMPLE 3.5.3. The rigid rotor of Example 3.5.1 and shown in Figure 3.8 rotates at 4,000 rev/min. Determine the natural frequencies and the mode-shape information for this system if:

(a) both bearings have horizontal and vertical support stiffnesses of 1 MN/m
(b) the horizontal and vertical support stiffnesses are 1.0 MN/m at bearing 1 and 1.3 MN/m at bearing 2

Solution. The inertia data for this problem are identical to Example 3.5.1 and are repeated here for convenience:

$$m = 122.68\,\text{kg}, \qquad I_p = 0.6134\,\text{kg m}^2, \qquad I_d = 2.8625\,\text{kg m}^2$$

The rotor spin speed of 4,000 rev/min must be converted to rad/s and is

$$\Omega = 4,000 \times 2\pi/60 = 418.88\,\text{rad/s}$$

(a) The stiffness data for this problem are identical to Example 3.5.1 and are

$$k_T = 2\,\text{MN/m}, \qquad k_R = 125\,\text{kNm}, \qquad k_C = 0$$

From Equations (3.27) and (3.31), $\omega_1 = \omega_2 = 127.68\,\text{rad/s}$, $\omega_3 = 168.85\,\text{rad/s}$, and $\omega_4 = 258.61\,\text{rad/s}$ (equivalent to 20.32, 26.87, and 41.16 Hz, respectively). The backward- and forward-whirling mode shapes for the third and fourth natural frequencies are shown in Figure 3.12. In this example, in the stationary frame, the backward-whirling modes occur at a lower frequency than the forward-whirling modes. Section 3.6.1 considers forward- and backward-whirling modes in more detail.

(b) The stiffness data for this problem are identical to Example 3.5.2 and are

$$k_T = 2.3\,\text{MN/m}, \qquad k_R = 143.75\,\text{kNm}, \qquad k_C = 75\,\text{kN}$$

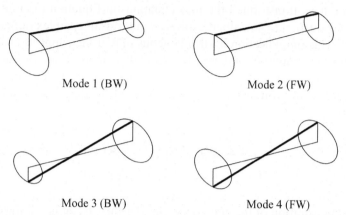

Mode 1 (BW)   Mode 2 (FW)

Mode 3 (BW)   Mode 4 (FW)

Figure 3.13. Mode shapes for the rigid rotor (Example 3.5.3(b)). The rotor displacements at the bearing locations are displayed. FW indicates forward whirl and BW indicates backward whirl.

Substituting these, parameter values in Equation (3.42) gives

$$s^4 \mp j89.76s^3 + 6{,}8966s^2 \mp j1{,}682{,}800s + 925.48 \times 10^6 = 0$$

Solving this equation gives $s = \pm134.0j$, $\pm135.6j$, $\pm185.9j$, and $\pm274.0j$ rad/s, corresponding to natural frequencies 21.33, 21.58, 29.58, and 43.62 Hz, respectively. Substituting these roots into Equation (3.38) gives the ratios u_0/v_0. From this, we can deduce that the modes at 21.33 and 29.58 Hz are clockwise- (or backward-) whirling modes and those at 21.58 and 43.62 Hz are counterclockwise (or forward-) whirling modes. Using Equation (3.44), the mode-shape ratios u_0/ψ_0 are -0.7723, -1.6795, 0.0387, and 0.0108, and the ratios v_0/θ_0 are 0.7723, 1.6795, -0.0387, and -0.0108 for frequencies $\omega_1, \ldots, \omega_4$, respectively. The mode shapes are shown in Figure 3.13.

3.5.4 Complex Coordinates

An alternative approach to determine the natural frequencies of a rigid circular rotor on isotropic supports combines the four coordinates required to describe the motion of the rotor into two complex coordinates. This has the advantage of halving the number of coordinates required, thereby allowing the natural frequencies and mode shapes to be determined from relatively simple equations. The complex coordinates, r and φ, are

$$r = u + jv \quad \text{and} \quad \varphi = \psi - j\theta \tag{3.45}$$

where $j = \sqrt{-1}$. Adding j times the second equation to the first equation of Equation (3.9) and subtracting j times the third equation from the fourth equation of Equation (3.9) and noting that $j\dot{\varphi} = j\dot{\psi} + \dot{\theta}$ gives

$$m\ddot{r} + k_T r + k_C \varphi = 0$$

$$I_d\ddot{\varphi} - jI_p\Omega\dot{\varphi} + k_C r + k_R \varphi = 0 \tag{3.46}$$

Thus, we have transformed the four equations in Equation (3.9) into the two equations of Equation (3.46), albeit expressed in terms of complex rather than real coordinates. This simplification is advantageous only if the supports are isotropic; thus, the dynamic properties of the bearings and their supports are the same in both the Ox and Oy directions. To proceed, we look for solutions of the form $r(t) = r_0 e^{st}$ and $\varphi(t) = \varphi_0 e^{st}$. Substituting these expressions into Equation (3.46) and noting that $e^{st} \neq 0$ gives

$$
\begin{aligned}
\left(ms^2 + k_T\right) r_0 + k_C \varphi_0 &= 0 \\
\left(I_d s^2 - \jmath I_p \Omega s + k_R\right) \varphi_0 + k_C r_0 &= 0
\end{aligned}
$$

$$(3.47)$$

As a demonstration of the approach, consider the case of no elastic coupling – that is, when $k_C = 0$. The first equation of Equation (3.47) immediately gives the same natural frequencies as Equation (3.27). The second equation gives a quadratic that is solved easily to give the same natural frequencies as in Equation (3.31).

It is possible to extend this approach to the more general case of anisotropic supports – that is, supports that have different properties in the x and y directions. Lee (1993) and Kessler and Kim (2001) analyzed spinning rotors in anisotropic supports in terms of a set of complex coordinates and their complex conjugates. There are advantages in this approach, but the use of complex coordinates no longer results in a reduction in the number of coordinates required. For example, in the case of a rigid rotor on flexible supports, four real coordinates are required to describe its motion. Whereas only two complex coordinates were required in the analysis of a rigid rotor on isotropic supports, four complex coordinates are required for the more general anisotropic case. For the general case, we prefer the more conventional method of keeping the coordinates in the x and y directions separate. However, complex coordinates are used to advantage in Chapter 6 to calculate the response due to unbalance.

3.6 A Rigid Rotor on Anisotropic Flexible Supports

For general flexible supports, the task is to solve Equation (3.8), repeated here for convenience:

$$
\begin{aligned}
m\ddot{u} + k_{xT} u + k_{xC} \psi &= 0 \\
m\ddot{v} + k_{yT} v - k_{yC} \theta &= 0 \\
I_d \ddot{\theta} + I_p \Omega \dot{\psi} - k_{yC} v + k_{yR} \theta &= 0 \\
I_d \ddot{\psi} - I_p \Omega \dot{\theta} + k_{xC} u + k_{xR} \psi &= 0
\end{aligned}
$$

These are the equations of motion for a flexibly supported spinning rotor with differing stiffness properties in the x and y directions. It is helpful to express these equations in matrix form as

$$\mathbf{M}\ddot{\mathbf{q}} + \Omega \mathbf{G}\dot{\mathbf{q}} + \mathbf{K}\mathbf{q} = \mathbf{0} \qquad (3.48)$$

where

$$
\mathbf{M} = \begin{bmatrix} m & 0 & 0 & 0 \\ 0 & m & 0 & 0 \\ 0 & 0 & I_d & 0 \\ 0 & 0 & 0 & I_d \end{bmatrix}, \quad \mathbf{G} = \begin{bmatrix} 0 & 0 & 0 & 0 \\ 0 & 0 & 0 & 0 \\ 0 & 0 & 0 & I_p \\ 0 & 0 & -I_p & 0 \end{bmatrix},
$$

$$
\mathbf{K} = \begin{bmatrix} k_{xT} & 0 & 0 & k_{xC} \\ 0 & k_{yT} & -k_{yC} & 0 \\ 0 & -k_{yC} & k_{yR} & 0 \\ k_{xC} & 0 & 0 & k_{xR} \end{bmatrix}
$$

(3.49)

and

$$
\mathbf{q} = \begin{bmatrix} u & v & \theta & \psi \end{bmatrix}^{\top}
$$

The mass and stiffness matrices, $\mathbf{M}$ and $\mathbf{K}$, respectively, are symmetric, positive-definite matrices. In contrast, the gyroscopic matrix, $\mathbf{G}$, is skew-symmetric.

To determine the roots of Equation (3.48), we must rearrange the equation into the form of Equation (2.71). This gives

$$
\begin{bmatrix} \Omega\mathbf{G} & \mathbf{M} \\ \mathbf{M} & \mathbf{0} \end{bmatrix} \frac{\mathrm{d}}{\mathrm{d}t} \begin{Bmatrix} \mathbf{q} \\ \dot{\mathbf{q}} \end{Bmatrix} + \begin{bmatrix} \mathbf{K} & \mathbf{0} \\ \mathbf{0} & -\mathbf{M} \end{bmatrix} \begin{Bmatrix} \mathbf{q} \\ \dot{\mathbf{q}} \end{Bmatrix} = \begin{Bmatrix} \mathbf{0} \\ \mathbf{0} \end{Bmatrix}
$$

(3.50)

Note that $\mathbf{C}$ of Equation (2.71) has been replaced by $\Omega\mathbf{G}$ in Equation (3.50). Equation (3.50) can be written as

$$
\mathbf{A}\dot{\mathbf{x}} + \mathbf{B}\mathbf{x} = \mathbf{0}
$$

(3.51)

where

$$
\mathbf{x} = \begin{Bmatrix} \mathbf{q} \\ \dot{\mathbf{q}} \end{Bmatrix}, \quad \dot{\mathbf{x}} = \frac{\mathrm{d}}{\mathrm{d}t} \begin{Bmatrix} \mathbf{q} \\ \dot{\mathbf{q}} \end{Bmatrix}, \quad \mathbf{A} = \begin{bmatrix} \Omega\mathbf{G} & \mathbf{M} \\ \mathbf{M} & \mathbf{0} \end{bmatrix}, \quad \text{and} \quad \mathbf{B} = \begin{bmatrix} \mathbf{K} & \mathbf{0} \\ \mathbf{0} & -\mathbf{M} \end{bmatrix}
$$

Looking for solutions of the form $\mathbf{x}(t) = \mathbf{x}_0 e^{st}$, then $\dot{\mathbf{x}} = s\mathbf{x}_0 e^{st}$; hence, Equation (3.51) becomes

$$
s\mathbf{A}\mathbf{x}_0 = -\mathbf{B}\mathbf{x}_0
$$

(3.52)

This is an 8×8 eigenvalue problem and it must be solved numerically. Using any appropriate software, it readily can be solved to give eight roots or eigenvalues. If the system is described by n coordinates (four in this case), then there are $2n$ roots in the form of n complex conjugate pairs. Each pair of complex conjugate roots represents one natural frequency. Thus, for the rigid rotor

$$
\begin{aligned}
s_1 &= +\jmath\omega_1, \quad s_2 = +\jmath\omega_2, \quad s_3 = +\jmath\omega_3, \quad s_4 = +\jmath\omega_4, \\
s_5 &= -\jmath\omega_1, \quad s_6 = -\jmath\omega_2, \quad s_7 = -\jmath\omega_3, \quad s_8 = -\jmath\omega_4
\end{aligned}
$$

(3.53)

where s_i and s_{4+i} form a complex conjugate pair for each i. Corresponding to these roots are n complex conjugate pairs of eigenvectors. For each pair of eigenvectors, the first n elements correspond to a mode of the system.

3.6.1 Forward and Backward Whirl

Section 3.5.3 shows that the mode shapes of a rigid rotor on isotropic supports have circular orbits that rotate either forward or backward – that is, in the direction of rotor spin or in the opposite direction, respectively. For a rigid rotor on anisotropic supports, determining the direction of rotation of the mode is not as simple; because the system properties in the x and y directions are different, the orbits are generally elliptical. The direction of whirl at a particular axial location may be determined as follows.

In general, the eigenvector $\mathbf{x}^{(i)}$ and eigenvalue s_i are complex. Here, we consider only the undamped case, $s_i = +\jmath\omega_i$, where the natural frequency ω_i is real and positive. If the eigenvalue $s_i = -\jmath\omega_i$ is chosen, then the direction of whirl based on the associated eigenvector for a given relative phase difference given in Equation (3.59) must be reversed. The damping affects only the amplitude of time response, not the shape of the orbit or the direction. The free response in this mode only is

$$\mathbf{x}(t) = \Re\left(\mathbf{x}^{(i)}\mathrm{e}^{\jmath\omega_i t}\right) \tag{3.54}$$

To determine the direction of rotation of the mode, we consider either displacements or rotations at a single node. Hence, for illustration, consider only the two rows of this equation corresponding to displacements u and v at a single node. The response may be written as

$$
\begin{aligned}
\begin{Bmatrix} u(t) \\ v(t) \end{Bmatrix} &= \Re\left(\begin{Bmatrix} r_u\mathrm{e}^{\jmath\eta_u} \\ r_v\mathrm{e}^{\jmath\eta_v} \end{Bmatrix} \mathrm{e}^{\jmath\omega_i t}\right) \\
&= \begin{Bmatrix} r_u\cos(\eta_u + \omega_i t) \\ r_v\cos(\eta_v + \omega_i t) \end{Bmatrix} = \mathbf{T}\begin{Bmatrix} \cos\omega_i t \\ \sin\omega_i t \end{Bmatrix}
\end{aligned}
\tag{3.55}
$$

where $r_u\mathrm{e}^{\jmath\eta_u}$ and $r_v\mathrm{e}^{\jmath\eta_v}$ are the elements of the ith eigenvector, $\mathbf{x}^{(i)}$, corresponding to the degrees of freedom of interest, and

$$\mathbf{T} = \begin{bmatrix} r_u\cos\eta_u & -r_u\sin\eta_u \\ r_v\cos\eta_v & -r_v\sin\eta_v \end{bmatrix} \tag{3.56}$$

From Equation (3.55),

$$\begin{Bmatrix} \cos\omega_i t \\ \sin\omega_i t \end{Bmatrix} = \mathbf{T}^{-1}\begin{Bmatrix} u(t) \\ v(t) \end{Bmatrix} \tag{3.57}$$

Hence, the orbit (u, v) forms an ellipse because

$$\begin{Bmatrix} u(t) \\ v(t) \end{Bmatrix}^{\top}\mathbf{T}^{-\top}\mathbf{T}^{-1}\begin{Bmatrix} u(t) \\ v(t) \end{Bmatrix} = \cos^2\omega_i t + \sin^2\omega_i t = 1 \tag{3.58}$$

The length of the major and minor axes of the ellipse are obtained from the eigenvalues of $\mathbf{H} = \mathbf{T}\mathbf{T}^{\top}$. The matrix $\mathbf{H}$ is symmetric and positive-definite and therefore has real eigenvalues λ_1 and λ_2 with $\lambda_1 \geq \lambda_2$, and real eigenvector matrix $\mathbf{U}$. Then, $\mathbf{H} = \mathbf{U}\boldsymbol{\Lambda}\mathbf{U}^{\top}$ and $\mathbf{U}$ represents a rotation of the ellipse. The lengths of the semiminor and semimajor axes are given by $\sqrt{\lambda_2}$ and $\sqrt{\lambda_1}$, respectively.

This analysis shows only that the response is an ellipse. To decide the direction of the mode, we must return to the original definition given by Equation (3.55). If we shift the time origin so that $t \rightarrow (t - \eta_u/\omega_i)$, then

$$\begin{Bmatrix} u(t) \\ v(t) \end{Bmatrix} = \begin{Bmatrix} r_u \cos(\omega_i t) \\ r_v \cos(\eta_v - \eta_u + \omega_i t) \end{Bmatrix} \tag{3.59}$$

It is now clear that the direction of rotation depends on the phase difference between the u and v response given by $\eta_v - \eta_u$. If $\eta_v = \eta_u$ or $\eta_v = \eta_u + \pi$, then the response is a straight line, and that point on the rotor centerline is vibrating in only one direction. If $0 < \eta_v - \eta_u < \pi$, then a backward-rotating mode exists. Alternatively, if $-\pi < \eta_v - \eta_u < 0$, then a forward-rotating mode exists. To apply these criteria, it is assumed that $-\pi < \eta_v - \eta_u < \pi$; if this is not the case, then multiples of 2π are added or subtracted to ensure that $\eta_v - \eta_u$ is within the required range. If $r_u = r_v$ and $\eta_v = \eta_u \pm \pi/2$, then the orbit is circular.

The properties of the orbit may be encoded into a single parameter κ, defined as

$$\kappa = \pm\sqrt{\lambda_2/\lambda_1} \tag{3.60}$$

where κ is positive for a forward-rotating orbit and negative for a backward-rotating orbit, as previously determined. If $\kappa = \pm 1$, the orbit is circular.

The modal vector also contains information about the angles θ and ψ. The whirling of these orbits can be addressed exactly the same way as the displacements, but the direction of whirling is more difficult to visualize. In the case of a rigid rotor, we can use the angular information to determine the orbit direction and shape at the ends of the rotor. Thus, with reference to Figure 3.7, for small displacements and rotations we have

$$\begin{aligned} u_1 &= u - a\psi, & v_1 &= v + a\theta \\ u_2 &= u + b\psi, & v_2 &= v - b\theta \end{aligned} \tag{3.61}$$

where u_1, for example, is the displacement of the rotor in the x direction at bearing 1 and u_2 is the displacement of the rotor in the x direction at bearing 2. Thus, for the ith mode, the translational and rotational response may be calculated and the response at each end calculated. This analysis then may be used to determine the direction and shape of the orbit at each end of the rotor. When the whirl orbits are circular, the backward-whirling modes occur at a lower frequency in the stationary frame than the forward-whirling modes. Often, this also occurs for elliptical-whirl orbits but is not guaranteed.

EXAMPLE 3.6.1. Determine the natural frequencies and the mode shapes of the rigid rotor of Example 3.5.1 (shown in Figure 3.8) for the following:

(a) The horizontal and vertical support stiffnesses are 1.0 MN/m at bearing 1 and 1.3 MN/m at bearing 2 and the rotor spins at 4,000 rev/min. (This is the same as (b) in Example 3.5.3.)

(b) The horizontal and vertical support stiffnesses are 1.0 and 1.1 MN/m, respectively, at bearing 1, and 1.3 and 1.4 MN/m, respectively, at bearing 2. Note that the bearings are anisotropic. Obtain the natural frequencies

Table 3.1. *The values of κ for a rotor speed of 4,000 rev/min (Example 3.6.1(a))*

Natural frequency	21.33 Hz	21.58 Hz	29.58 Hz	43.62 Hz
Bearing 1	−1	1	−1	1
Center	−1	1	−1	1
Bearing 2	−1	1	−1	1

and mode shapes when the rotor is stationary and also rotating at 4,000 and 8,000 rev/min.

Solution

(a) Mass, stiffness, and gyroscopic matrices for this system can be determined by substituting the data of Example 3.5.3 into Equation (3.49) to give

$$\mathbf{M} = \begin{bmatrix} 122.68 & 0 & 0 & 0 \\ 0 & 122.68 & 0 & 0 \\ 0 & 0 & 2.8625 & 0 \\ 0 & 0 & 0 & 2.8625 \end{bmatrix}, \quad \mathbf{G} = \begin{bmatrix} 0 & 0 & 0 & 0 \\ 0 & 0 & 0 & 0 \\ 0 & 0 & 0 & 0.6134 \\ 0 & 0 & -0.6134 & 0 \end{bmatrix},$$

$$\mathbf{K} = 10^3 \begin{bmatrix} 2,300 & 0 & 0 & 75 \\ 0 & 2,300 & -75 & 0 \\ 0 & -75 & 143.75 & 0 \\ 75 & 0 & 0 & 143.75 \end{bmatrix}$$

Forming and solving the eigenvalue problem, Equation (3.52), gives the following eigenvalues or roots (in rad/s):

$$s_1 = +134.0j, \quad s_2 = +135.6j, \quad s_3 = +185.9j, \quad s_4 = +274.0j,$$
$$s_5 = -134.0j, \quad s_6 = -135.6j, \quad s_7 = -185.9j, \quad s_8 = -274.0j$$

The roots form complex conjugate pairs and they determine the four natural frequencies as 134.0, 135.6, 185.9, and 274.0 rad/s, or 21.33, 21.58, 29.58, and 43.62 Hz. These frequencies are identical to those calculated in (b) of Example 3.5.3. The corresponding eigenvectors are approximately

$$\begin{bmatrix} 1 & 1 & 1 & 1 & 1 & 1 & 1 & 1 \\ j & -j & j & -j & -j & j & -j & j \\ 1.295j & -0.595j & -28.8j & 92.2j & -1.295j & 0.595j & 28.8j & -92.2j \\ -1.295 & -0.595 & 28.8 & 92.2 & -1.295 & -0.595 & 28.8 & 92.2 \end{bmatrix}$$

The eigenvectors are normalized so that the first entry of each column is unity. From the mode shapes, we can determine u/ψ and v/θ for each mode, together with the shape and direction of the orbits. For example, for the first mode, $u^{(1)}/\psi^{(1)} = -0.772$, $v^{(1)}/\theta^{(1)} = 0.772$. These results are in agreement with those obtained in (b) of Example 3.5.3.

Using Equation (3.60), we find for the four modes of vibration that in every case, $|\kappa| = 1$, indicating that the orbit is a circle (Table 3.1). The sign of κ indicates that for the natural frequencies 21.58 and 43.62 Hz, the mode orbit is forward in the direction of rotation. At 21.33 and 29.58 Hz, the mode orbit

Table 3.2. *The values of κ for a rotor speed of 4,000 rev/min (Example 3.6.1(b)(ii))*

Natural frequency	21.44 Hz	22.43 Hz	30.26 Hz	44.40 Hz
Bearing 1	−0.1694	0.0586	−0.8630	0.9176
Center	−0.1103	0.1304	−0.9685	0.8921
Bearing 2	−0.0133	0.2405	−0.8890	0.9154

is backward, in the direction opposite to rotation. These results confirm those previously obtained in (b) of Example 3.5.3.

(b)(i) *Stationary rotor on anisotropic bearings.* Forming and solving the eigenvalue problem gives the following eigenvalues (in rad/s):

$$s_1 = +135.08j, \quad s_2 = +141.13j, \quad s_3 = +225.21j, \quad s_4 = +234.62j,$$
$$s_5 = -135.08j, \quad s_6 = -141.13j, \quad s_7 = -225.21j, \quad s_8 = -234.62j$$

The eigenvalues consist of four complex conjugate pairs. The natural frequencies are deduced from the eigenvalues and are 135.08, 141.13, 225.21, and 234.62 rad/s, or 21.50, 22.46, 35.84, and 37.34 Hz. The corresponding eigenvectors are real and given by

$$\begin{bmatrix} 1 & 0 & 1 & 0 & 1 & 0 & 1 & 0 \\ 0 & 1 & 0 & 1 & 0 & 1 & 0 & 1 \\ 0 & 0.756 & 0 & -56.7 & 0 & 0.756 & 0 & -56.7 \\ -0.820 & 0 & 52.3 & 0 & -0.820 & 0 & 52.3 & 0 \end{bmatrix}$$

The pairs of complex conjugate roots have identical (real) eigenvectors in this case. For every mode, $\kappa = 0$ because the modes do not have a circular or elliptical orbit. Motion is either in the Oxz plane ($\omega = 135.08$ and 225.21 rad/s, equivalent to 21.50 and 35.84 Hz) or the Oyz plane ($\omega = 141.13$ and 234.62 rad/s, equivalent to 22.46 and 37.34 Hz).

(b)(ii) *Rotor rotating at 4,000 rev/min on anisotropic bearings.* Forming and solving the eigenvalue problem gives the following four pairs of complex conjugate eigenvalues (in rad/s):

$$s_1 = +134.74j, \quad s_2 = +140.94j, \quad s_3 = +190.15j, \quad s_4 = +278.94j,$$
$$s_5 = -134.74j, \quad s_6 = -140.94j, \quad s_7 = -190.15j, \quad s_8 = -278.94j$$

Hence, we can determine the natural frequencies as 134.74, 140.94, 190.15, and 278.94 rad/s, or 21.44, 22.43, 30.26, and 44.40 Hz. The corresponding eigenvectors are:

$$\begin{bmatrix} 1 & 1 & 1 & 1 & 1 & 1 & 1 & 1 \\ -0.110j & 7.67j & -0.968j & 1.12j & 0.110j & -7.67j & 0.968j & -1.12j \\ -0.401j & 6.45j & 25.0j & -105j & 0.401j & -6.45j & -25.0j & 105j \\ -0.968 & 1.83 & 28.5 & 96.6 & -0.968 & 1.83 & 28.5 & 96.6 \end{bmatrix}$$

The corresponding values of κ are shown in Table 3.2. From this table, we see that for the first and third natural frequencies, the direction of the mode orbit is backward, in the opposite direction of the rotor spin. For the second

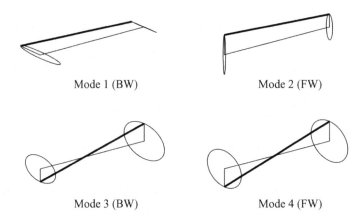

Figure 3.14. Mode shapes for the rigid rotor (Example 3.6.1(b)) at 4,000 rev/min. The rotor displacements at the bearing locations are displayed. FW indicates forward whirl and BW indicates backward whirl.

and fourth natural frequencies, the direction of the mode orbit is forward, in the same direction as the rotor spin. The mode orbits are not circular (i.e., $\kappa \neq \pm 1$) but rather elliptical. Figure 3.14 shows the mode shapes.

(b)(iii). *The rotor spins at 8,000 rev/min.* Forming and solving the eigenvalue problem gives the four pairs of complex conjugate eigenvalues with zero real parts (in rad/s) as:

$$s_1 = +132.78j, \quad s_2 = +140.16j, \quad s_3 = +161.05j, \quad s_4 = +336.07j,$$
$$s_5 = -132.78j, \quad s_6 = -140.16j, \quad s_7 = -161.05j, \quad s_8 = -336.07j$$

Hence, we can determine the four natural frequencies for the system as 132.87, 140.16, 161.05, and 336.07 rad/s, or 21.13, 22.30, 25.63, and 53.49 Hz. The corresponding eigenvectors follow and again we see that they form complex conjugate pairs:

$$\begin{bmatrix} 1 & 1 & 1 & 1 & 1 & 1 & 1 & 1 \\ 0.311j & -2.36j & 1.19j & -1.06j & -0.311j & 2.36j & -1.19j & 1.06j \\ 1.40j & -2.83j & -10.8j & 160j & -1.40j & 2.83j & 10.8j & -160j \\ -1.83 & 1.47 & 11.8 & 154 & -1.83 & 1.47 & 11.8 & 154 \end{bmatrix}$$

Table 3.3 gives the values of κ for the four modes. It shows that the direction of the orbits for the second and fourth modes is forward and for the third mode is backward. The first mode has mixed orbits. The orbits at bearing 1 and the center are backward and the orbit at bearing 2 is forward. Although this is counterintuitive, it can occur when a system is supported by anisotropic bearings.

The orbits of the rotor studied in this example show that in some modes, the direction of the rotor orbit is forward and, in some modes, the direction of the rotor orbit is backwards; in one case, it is a combination of forward and backward motion.

Table 3.3. *The values of κ for a rotor speed of 8,000 rev/min (Example 3.6.1(b)(iii))*

Natural frequency	21.13 Hz	22.30 Hz	25.63 Hz	53.49 Hz
Bearing 1	−0.4540	0.2066	−0.7783	0.9649
Center	−0.3115	0.4242	−0.8433	0.9477
Bearing 2	+0.0709	0.8279	−0.9852	0.9640

3.7 Natural Frequency Maps

In Sections 3.5.3 and 3.6, we show that due to gyroscopic effects, the roots of the characteristic equation (i.e., the eigenvalues) vary with rotational speed. This is not the only reason why the eigenvalues vary with rotational speed. For example, hydrodynamic bearings have stiffness and damping properties that vary with rotational speed; these, in turn, affect the eigenvalues (see Chapter 5).

It is convenient to illustrate graphically the way in which the roots change with rotational speed. Graphs can be plotted that show these changes in various ways. Typically, the rotational speed is plotted on the *x*-axis and the imaginary part of the roots, or the natural frequencies are plotted on the *y*-axis. This plot is usually referred to as a *natural frequency map* and Figures 3.15 through 3.18 are examples of these maps. Not only does the map provide a considerable amount of information about the system's roots in a single diagram, it also provides a simple method to determine critical speeds (see Chapter 6).

Natural frequency maps also can illustrate the relationship between resonances and parameters other than rotational speed. In this way, the effect of varying a bearing stiffness or rotor-inertia property may be examined (see Example 3.7.2).

EXAMPLE 3.7.1. Plot the natural frequency maps for rotor spin speeds up to 20,000 rev/min for the rigid rotor described in Example 3.5.1, supported by the following bearing stiffnesses:

(a) $k_{x1} = 1.0$ MN/m, $k_{y1} = 1.0$ MN/m, $k_{x2} = 1.0$ MN/m, $k_{y2} = 1.0$ MN/m
(b) $k_{x1} = 1.0$ MN/m, $k_{y1} = 1.0$ MN/m, $k_{x2} = 1.3$ MN/m, $k_{y2} = 1.3$ MN/m
(c) $k_{x1} = 1.0$ MN/m, $k_{y1} = 1.5$ MN/m, $k_{x2} = 1.0$ MN/m, $k_{y2} = 1.5$ MN/m
(d) $k_{x1} = 1.0$ MN/m, $k_{y1} = 1.5$ MN/m, $k_{x2} = 1.3$ MN/m, $k_{y2} = 2.0$ MN/m

Solution. For (a) and (b), the rotor is supported on isotropic bearings. The motion of the rotor is described by Equation (3.9), repeated here for convenience:

$$m\ddot{u} + k_T u + k_C \psi = 0$$

$$m\ddot{v} + k_T v - k_C \theta = 0$$

$$I_d \ddot{\theta} + I_p \Omega \dot{\psi} - k_C v + k_R \theta = 0$$

$$I_d \ddot{\psi} - I_p \Omega \dot{\theta} + k_C u + k_R \psi = 0$$

(a) The natural frequency map for this system is shown in Figure 3.15. In this case, $k_C = 0$; therefore, the first and second equations of Equation (3.9) are uncoupled from the rest of the system and are independent of rotational speed.

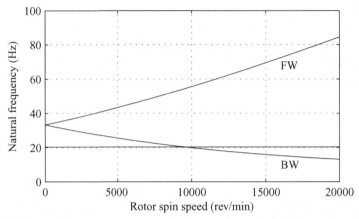

Figure 3.15. Natural frequency map for a rotor on uncoupled isotropic bearing supports (Example 3.7.1(a)). FW and BW indicate forward and backward whirl, respectively.

The stiffnesses in the x and y directions are identical; therefore, the two natural frequencies that are derived from these two equations are identical and are equal to approximately 20 Hz. At zero rotational speed, the third and fourth equations are uncoupled; hence, the two natural frequencies that are derived from these equations also are identical. Once the rotor begins to spin, the third and fourth equations become coupled due to gyroscopic effects and the two natural frequencies separate. In the map, we can see that the line of the pair of the constant natural frequencies (at approximately 20 Hz) is intersected by another natural-frequency line at about 9,000 rev/min. In this case, the intersection is a consequence of the first and second equations being uncoupled from the third and fourth and, hence, independent of one another. More complex examples show that the natural frequencies often *veer* away from each other.

(b) The natural frequency map for this system is shown in Figure 3.16. In this case, $k_C \neq 0$ and all of the equations are coupled, except when $\Omega = 0$.

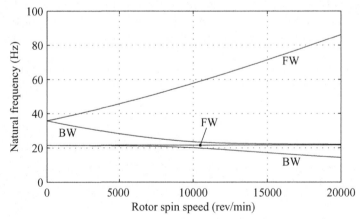

Figure 3.16. Natural frequency map for a rotor on coupled isotropic bearing supports (Example 3.7.1(b)). FW and BW indicate forward and backward whirl, respectively.

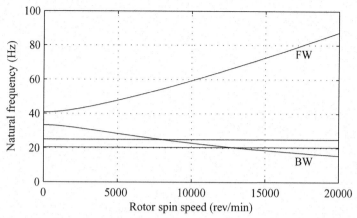

Figure 3.17. Natural frequency map for a rotor on uncoupled anisotropic bearing supports (Example 3.7.1(c)). FW and BW indicate forward and backward whirl, respectively.

All natural frequencies are influenced by gyroscopic effects. Two frequencies decrease with rotational speed; the other two frequencies increase, although the increase from approximately 20 Hz is barely perceptible. In this case, all of the equations are coupled and frequency lines do not intersect.

For (c) and (d), the rotor is supported by anisotropic bearings. The motion of the rotor is described by Equation (3.8), repeated here for convenience:

$$m\ddot{u} + k_{xT}u + k_{xC}\psi = 0$$

$$m\ddot{v} + k_{yT}v - k_{yC}\theta = 0$$

$$I_d\ddot{\theta} + I_p\Omega\dot{\psi} - k_{yC}v + k_{yR}\theta = 0$$

$$I_d\ddot{\psi} - I_p\Omega\dot{\theta} + k_{xC}u + k_{xR}\psi = 0$$

(c) The natural frequency map for this system is shown in Figure 3.17. In this example, $k_{xC} = k_{yC} = 0$. The first and second equations are uncoupled from the rest of the system and they are independent of rotational speed. The stiffnesses in the x and y directions are different; therefore, the two natural frequencies derived from these two equations are distinct. At zero rotational speed, the third and fourth equations also are uncoupled and the two natural frequencies derived from them are also distinct. When the rotor begins to spin, the map shows that one of these natural frequencies increases and the other decreases, due to gyroscopic effects. The map also shows that the path of the two constant natural-frequency lines at about 20 and 25 Hz are intersected by another natural-frequency line at about 8,000 and 13,000 rev/min. This is a consequence of the equations being uncoupled and therefore independent of one another.

(d) The natural frequency map for this system is shown in Figure 3.18. In this case, $k_{xC} \neq 0$ and $k_{yC} \neq 0$ so that all of the equations are coupled, except when $\Omega = 0$. From the natural frequency map, we see that all of the frequencies are influenced by gyroscopic couples and frequency lines do not intersect.

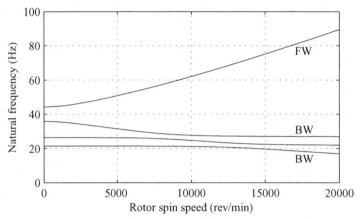

Figure 3.18. Natural frequency map for a rotor on coupled anisotropic bearing supports (Example 3.7.1(d)). FW and BW indicate forward and backward whirl, respectively. In the second mode, some positions whirl forward and some whirl backward.

EXAMPLE 3.7.2. Determine the effect of varying the stiffness of bearing 2 for the rigid rotor as described in Example 3.5.1 for a rotor spin speed of 5,000 rev/min. Assume the bearing is isotropic and vary $k_{x2} = k_{y2}$ in the range 0.4 to 2.0 MN/m. The properties of bearing 1 are

(a) $k_{x1} = 1.0$ MN/m, $k_{y1} = 1.0$ MN/m
(b) $k_{x1} = 1.0$ MN/m, $k_{y1} = 2.0$ MN/m

Solution. Figures 3.19 and 3.20 show the natural frequency maps for (a) and (b), respectively, as the stiffness of bearing 2 varies. These maps are functions of bearing stiffness, whereas other maps in this section have been functions of rotor spin speed. When both bearings are isotropic, the natural frequencies may cross (see Figure 3.19), whereas when bearing 1 is anisotropic, curve veering occurs and the natural frequencies do not cross (see Figure 3.20).

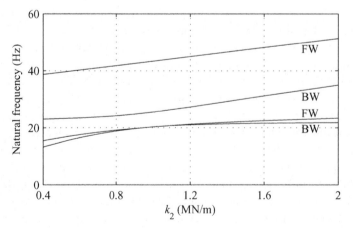

Figure 3.19. The effect of $k_{x2} = k_{y2} = k_2$ on the natural frequencies of a rotor on isotropic bearing supports at 5,000 rev/min (Example 3.7.2(a)). FW and BW indicate forward and backward whirl, respectively.

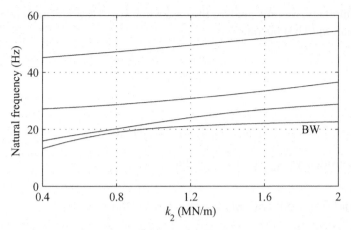

Figure 3.20. The effect of $k_{x2} = k_{y2} = k_2$ on the natural frequencies of a rotor on anisotropic bearing supports at 5,000 rev/min (Example 3.7.2(b)). BW indicates backward whirl. Modes 2, 3, and 4 do not whirl forward or backward at all locations for all stiffness values.

3.8 The Effect of Damping in the Supports

We now consider the effect of viscous damping in the bearings. Assuming that a viscous damper is placed in parallel with each spring element supporting the bearing, the forces f_{x1}, f_{x2}, f_{y1}, and f_{y2} of Equation (3.5) now become

$$f_{x1} = k_{x1}(u - a\psi) + c_{x1}(\dot{u} - a\dot{\psi})$$
$$f_{x2} = k_{x2}(u + b\psi) + c_{x2}(\dot{u} + b\dot{\psi})$$
$$f_{y1} = k_{y1}(v + a\theta) + c_{y1}(\dot{v} + a\dot{\theta})$$
$$f_{y2} = k_{y2}(v - b\theta) + c_{y2}(\dot{v} - b\dot{\theta})$$

(3.62)

where c is the viscous-damping coefficient and is defined as the force required to produce a unit velocity across the damping element. Let

$$c_{xT} = c_{x1} + c_{x2}, \qquad\qquad c_{yT} = c_{y1} + c_{y2}$$
$$c_{xC} = -ac_{x1} + bc_{x2}, \qquad\qquad c_{yC} = -ac_{y1} + bc_{y2}$$
$$c_{xR} = a^2 c_{x1} + b^2 c_{x2}, \qquad\qquad c_{yR} = a^2 c_{y1} + b^2 c_{y2}$$

Using these definitions, substituting Equation (3.62) into Equation (3.4) and rearranging these equations gives

$$m\ddot{u} + c_{xT}\dot{u} + c_{xC}\dot{\psi} + k_{xT}u + k_{xC}\psi = 0$$
$$m\ddot{v} + c_{yT}\dot{v} - c_{yC}\dot{\theta} + k_{yT}v - k_{yC}\theta = 0$$
$$I_d\ddot{\theta} + I_p\Omega\dot{\psi} - c_{yC}\dot{v} + c_{yR}\dot{\theta} - k_{yC}v + k_{yR}\theta = 0$$
$$I_d\ddot{\psi} - I_p\Omega\dot{\theta} + c_{xC}\dot{u} + c_{xR}\dot{\psi} + k_{xC}u + k_{xR}\psi = 0$$

(3.63)

For simplicity, we only analyze the case when $k_{xC} = k_{yC} = 0$ and $c_{xC} = c_{yC} = 0$. Thus, Equation (3.63) becomes

$$m\ddot{u} + c_{xT}\dot{u} + k_{xT}u = 0$$

$$m\ddot{v} + c_{yT}\dot{v} + k_{yT}v = 0$$

$$I_d\ddot{\theta} + I_p\Omega\dot{\psi} + c_{yR}\dot{\theta} + k_{yR}\theta = 0$$

$$I_d\ddot{\psi} - I_p\Omega\dot{\theta} + c_{xR}\dot{\psi} + k_{xR}\psi = 0$$

(3.64)

We now consider two specific cases of damping.

3.8.1 Rigid Rotor on Isotropic Supports with Damping

Consider a rigid rotor on isotropic supports; that is, the support stiffness and damping are the same in both the x and y directions. Thus, we can let $k_{xT} = k_{yT} = k_T$, $c_{xT} = c_{yT} = c_T$, and so on. Equation (3.64) becomes

$$m\ddot{u} + c_T\dot{u} + k_T u = 0$$

$$m\ddot{v} + c_T\dot{v} + k_T v = 0$$

$$I_d\ddot{\theta} + I_p\Omega\dot{\psi} + c_R\dot{\theta} + k_R\theta = 0$$

$$I_d\ddot{\psi} - I_p\Omega\dot{\theta} + c_R\dot{\psi} + k_R\psi = 0$$

(3.65)

The first two equations are uncoupled and may be solved by assuming solutions of the form $u(t) = u_0 e^{st}$ and $v(t) = v_0 e^{st}$ to give

$$\left(ms^2 + c_T s + k_T\right) u_0 = 0$$

$$\left(ms^2 + c_T s + k_T\right) v_0 = 0$$

(3.66)

If $c_T^2 < 4mk_T$, then Equation (3.66) has repeated complex conjugate roots given by

$$s_1 = s_2 = -\frac{c_T}{2m} + J\sqrt{\frac{k_T}{m} - \left(\frac{c_T}{2m}\right)^2}$$

$$s_5 = s_6 = -\frac{c_T}{2m} - J\sqrt{\frac{k_T}{m} - \left(\frac{c_T}{2m}\right)^2}$$

(3.67)

The last two equations of Equation (3.65) are coupled and may be solved by assuming solutions of the form $\theta(t) = \theta_0 e^{st}$ and $\psi(t) = \psi_0 e^{st}$. Thus,

$$\left(I_d s^2 + c_R s + k_R\right)\theta_0 + I_p\Omega s\psi_0 = 0$$

$$-I_p\Omega s\theta_0 + \left(I_d s^2 + c_R s + k_R\right)\psi_0 = 0$$

(3.68)

The solution is given by

$$\left(I_d s^2 + c_R s + k_R\right)^2 + \left(I_p\Omega s\right)^2 = 0$$

(3.69)

or

$$I_d s^2 + c_R s + k_R = \pm_J I_p\Omega s$$

and so

$$I_d s^2 + (c_R \mp J I_p \Omega) s + k_R = 0 \tag{3.70}$$

The roots of Equation (3.70) cannot be expressed as a simple algebraic equation with a real and an imaginary part. However, we know that the roots of Equation (3.69) are of the form

$$s_{3,7} = -\zeta_3 \omega_3 \pm J \omega_{d3} \quad \text{and} \quad s_{4,8} = -\zeta_4 \omega_4 \pm J \omega_{d4}$$

In these equations, $\omega_{di} = \omega_i \sqrt{1 - \zeta_i^2}$. These four roots are the solutions of the two quadratic equations in Equation (3.70), and the roots must pair as $\{s_3, s_8\}$ and $\{s_4, s_7\}$. Generating the quadratic equation from the pair of solutions s_4 and s_7 gives

$$(s - (-\zeta_3 \omega_3 - J \omega_{d3}))(s - (-\zeta_4 \omega_4 + J \omega_{d4})) = 0 \tag{3.71}$$

Complex conjugate solutions have not been obtained because the quadratic Equation (3.70) has a complex coefficient. Expanding Equation (3.71) gives

$$s^2 + ((\zeta_3 \omega_3 + \zeta_4 \omega_4) - J (\omega_{d4} - \omega_{d3})) s$$
$$+ ((\zeta_3 \zeta_4 \omega_3 \omega_4 + \omega_{d3} \omega_{d4}) + J (\omega_{d3} \zeta_4 \omega_4 - \omega_{d4} \zeta_3 \omega_3)) = 0 \tag{3.72}$$

Examining Equation (3.70), we see that the constant term is real. Thus in Equation (3.72), the imaginary part of the constant term must be zero. Hence,

$$\omega_{d3} \zeta_4 \omega_4 - \omega_{d4} \zeta_3 \omega_3 = 0$$

Thus, it follows that

$$\omega_3 \omega_4 \zeta_4 \sqrt{1 - \zeta_3^2} = \omega_3 \omega_4 \zeta_3 \sqrt{1 - \zeta_4^2}$$

Rearranging this equation gives

$$\frac{\zeta_3}{\sqrt{1 - \zeta_3^2}} = \frac{\zeta_4}{\sqrt{1 - \zeta_4^2}}$$

and, hence, $\zeta_3 = \zeta_4$.

This tells us that when a pair of natural frequencies separate due to gyroscopic effects, the damping factors for the two modes are identical if the bearings are isotropic.

3.8.2 Anisotropic Support Damping

We now consider the case when the support stiffnesses are identical in the x and y directions but the damping is not. Thus, in Equation (3.64), we let $k_{xT} = k_{yT} = k_T$ and $k_{xR} = k_{yR} = k_R$; Equation (3.64) becomes

$$m \ddot{u} + c_{xT} \dot{u} + k_T u = 0$$

$$m \ddot{v} + c_{yT} \dot{v} + k_T v = 0$$

$$I_d \ddot{\theta} + I_p \Omega \dot{\psi} + c_{yR} \dot{\theta} + k_R \theta = 0 \tag{3.73}$$

$$I_d \ddot{\psi} - I_p \Omega \dot{\theta} + c_{xR} \dot{\psi} + k_R \psi = 0$$

The first and second equations of Equation (3.73) are uncoupled and can be readily solved. Focusing our attention of the second pair of coupled equations, we solve these equations by letting $\theta(t) = \theta_0 e^{st}$ and $\psi(t) = \psi_0 e^{st}$. Thus, the second pair of equations of Equation (3.73) becomes

$$\left(I_d s^2 + c_{yR}s + k_R\right)\theta_0 + I_p\Omega s\psi_0 = 0$$
$$-I_p\Omega s\theta_0 + \left(I_d s^2 + c_{xR}s + k_R\right)\psi_0 = 0$$

(3.74)

The characteristic equation for Equation (3.74) is obtained by eliminating θ_0 and ψ_0 to give

$$\left(I_d s^2 + c_{yR}s + k_R\right)\left(I_d s^2 + c_{xR}s + k_R\right) + I_p^2\Omega^2 s^2 = 0$$

(3.75)

This can be rearranged to give

$$\left(I_d s^2 + c_{mR}s + k_R\right)^2 = \left(c_{dR}^2 - I_p^2\Omega^2\right)s^2$$

(3.76)

where $c_{mR} = (c_{xR} + c_{yR})/2$ and $c_{dR} = (c_{xR} - c_{yR})/2$. Now, if $\Omega^2 < c_{dR}^2/I_p^2$, the right-hand term is positive. Letting $\alpha^2 = c_{dR}^2 - I_p^2\Omega^2$ and taking the square root of Equation (3.76) gives

$$I_d s^2 + c_{mR}s + k_R = \mp \alpha s$$

or

$$I_d s^2 + (c_{mR} \pm \alpha)s + k_R = 0$$

(3.77)

It is clear that if $\Omega^2 < c_{dR}^2/I_p^2$, then the natural frequencies of the two forms of Equation (3.77) are equal but the damping coefficients – and, hence, the damping ratios – are different, depending on whether we take $(c_{mR} + \alpha)$ or $(c_{mR} - \alpha)$. When $\Omega = 0$, $(c_{mR} \mp \alpha)$ simplifies to c_{xR} and c_{yR}. As Ω increases, α tends to zero, and both damping coefficients become c_{mR}.

If $\Omega^2 > c_{dR}^2/I_p^2$, the right-hand side of Equation (3.76) is negative and we can define a transformed speed, $\hat{\Omega}$, by

$$\hat{\Omega}^2 = \Omega^2 - (c_{dR}/I_p)^2$$

Equation (3.76) may be written as

$$\left(I_d s^2 + c_{mR}s + k_R\right)^2 + I_p^2\hat{\Omega}^2 s^2 = 0$$

(3.78)

Consider again the third and fourth equations of Equation (3.65). These equations describe the case when both the support stiffness and damping are isotropic. Letting $\theta(t) = \theta_0 e^{st}$ and $\psi(t) = \psi_0 e^{st}$, we have

$$\left(I_d s^2 + c_R s + k_R\right)\theta_0 + I_p\Omega s\psi_0 = 0$$
$$-I_p\Omega s\theta_0 + \left(I_d s^2 + c_R s + k_R\right)\psi_0 = 0$$

This leads to the characteristic equation

$$\left(I_d s^2 + c_R s + k_R\right)^2 + I_p^2\Omega^2 s^2 = 0$$

(3.79)

Comparing Equations (3.78) and (3.79), we see that when $\Omega^2 > c_{dR}^2/I_p^2$, the anisotropic damping case gives the same pattern of behavior as the isotropic case,

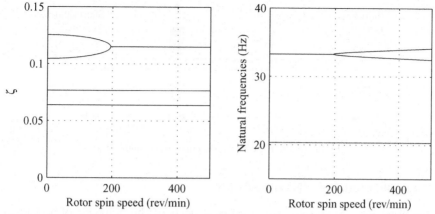

Figure 3.21. Plot of modal damping ratios and natural frequencies against rotor spin speed (Example 3.8.1).

except that the behavior is dependent on the mean rather than the actual damping coefficient, and the rotor speed is shifted from Ω to $\hat{\Omega}$.

EXAMPLE 3.8.1. The rigid rotor of Example 3.5.1 is supported by bearings with the following dynamic properties:

$$k_{x1} = k_{x2} = 1\,\text{MN/m}, \qquad\qquad k_{y1} = k_{y2} = 1\,\text{MN/m},$$

$$c_{x1} = c_{x2} = 1\,\text{kNs/m}, \qquad\qquad c_{y1} = c_{y2} = 1.2\,\text{kNs/m}.$$

Plot the variation of the damping factors, ζ, and natural frequencies with rotational speed in the range 0 to 500 rev/min.

Solution. Because the damping in the bearing supports is anisotropic, we can calculate the rotor speed at which the damping characteristic changes:

$$c_{xR} = (L/2)^2 c_{x1} + (L/2)^2 c_{x2} = 125\,\text{Ns/m},$$

$$c_{yR} = (L/2)^2 c_{y1} + (L/2)^2 c_{y2} = 150\,\text{Ns/m}$$

Thus,

$$c_{mR} = (c_{xR} + c_{yR})/2 = 137.5\,\text{Ns/m},$$

$$c_{dR} = (c_{xR} - c_{yR})/2 = -12.5\,\text{Ns/m}$$

Hence, the speed where the character of the solution changes is $\Omega = |c_{dR}|/I_p$, which in this example is 194.60 rev/min.

We can solve the eigenvalue problem for various rotor speeds to determine the roots s_i. Now, because $\{s_i, s_{4+i}\} = -\zeta_i \omega_i \pm j\omega_{di}$, we can deduce ζ_i. Figure 3.21 shows that the damping factors of the two higher modes become identical at 194.60 rev/min. The natural frequencies of the two higher modes are identical below 194.60 rev/min and then bifurcate (Friswell et al., 2001).

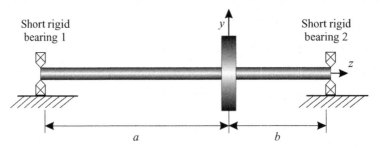

Figure 3.22. Flexible rotor, carrying a single disk.

3.9 Simple Model of a Flexible Rotor

In Sections 3.4, 3.5, and 3.7, we examine aspects of the behavior of a rigid rotor on flexible supports. Many rotors cannot be modeled as a rigid body; for instance, they are flexible because they have a small diameter relative to their length. Thus, a rotor – bearing system consisting of a flexible rotor can vibrate, even if supported by rigid bearings on rigid supports. Figure 3.22 shows such a rotor; it consists of a long, uniform, flexible shaft with circular cross section, carried in two short or self-aligning, rigidly supported bearings. The assumption of short, rigidly supported bearings means that the rotor is considered to be simply supported (or pinned) at its ends. The shaft carries a single disk and it is assumed that the mass of the shaft is small compared to that of the disk. As a consequence, the mass of the shaft can be neglected in the analysis.

If the single disk is placed at the midspan of the shaft, the rotor is often called a *Jeffcott* (Jeffcott, 1919) or a *De Laval rotor* in honor of Henry Jeffcott and Carl De Laval, who conducted some of the earliest studies of the dynamics of flexible rotors.

To analyze the dynamic behavior of the rotor shown in Figure 3.22, we must consider the displacement of the disk from the equilibrium position along and about the Ox and Oy axes. Thus, our dynamic system requires four coordinates to specify the displacements along and the rotations about the Ox and Oy axes. These are the only coordinates required; thus, the model has four degrees of freedom. The coordinate definitions are given in Figure 3.2.

For small, static displacements of the shaft, there is a linear relationship between a force applied to the shaft in the direction Ox (or Oy) and the resultant displacements and rotations. There is also a linear relationship between a moment applied to the shaft about the Oy (or Ox) axes and the resultant displacements and rotations. Thus, for a specific point on the shaft, we have

$$f_x = k_{uu}u + k_{u\psi}\psi$$
$$M_y = k_{\psi\psi}\psi + k_{\psi u}u$$

(3.80)

In these equations, f_x is a force applied to the shaft in the direction Ox and M_y is a moment applied about the Oy axis. The parameters u and ψ are defined in Figure 3.2. The coefficients k_{uu}, $k_{\psi u}$, $k_{u\psi}$, and $k_{\psi\psi}$ are stiffness coefficients at the particular location on the shaft. From Equation (3.80), we can deduce definitions for these coefficients. For example, k_{uu} is the force in the direction Ox required to produce a unit displacement u when no other displacements or rotations are allowed to occur. Similarly, $k_{\psi u}$ is the moment about the Oy axis required to produce a unit displacement u when no other displacements or rotations are allowed to occur. Note that for a conservative system, $k_{u\psi} = k_{\psi u}$.

The relationships between the forces and moments applied and the resulting displacement v and rotation θ are

$$f_y = k_{vv}v + k_{v\theta}\theta$$
$$M_x = k_{\theta\theta}\theta + k_{\theta v}v \tag{3.81}$$

Note that for a conservative system, $k_{v\theta} = k_{\theta v}$.

To determine the equations of motion for this system, we must apply Newton's second law of motion to the system. Thus, for a disk of mass m, diametral moment of inertia I_d, and polar moment of inertia I_p

Force on disk in x direction: $\qquad -f_x = m\ddot{u}$

Force on disk in y direction: $\qquad -f_y = m\ddot{v} \tag{3.82}$

Moments acting on disk in θ direction: $\qquad -M_x = I_d\ddot{\theta} + I_p\Omega\dot{\psi}$

Moments acting on disk in ψ direction: $\qquad -M_y = I_d\ddot{\psi} - I_p\Omega\dot{\theta}$

The forces and moments applied to the disk due to the elasticity of the shaft are equal and opposite to the forces acting on the shaft due to the displacements of the disk. Gyroscopic effects are also included in this analysis. Substituting Equations (3.80) and (3.81) into Equation (3.82) and rearranging gives

$$m\ddot{u} + k_{uu}u + k_{u\psi}\psi = 0$$
$$m\ddot{v} + k_{vv}v + k_{v\theta}\theta = 0$$
$$I_d\ddot{\theta} + I_p\Omega\dot{\psi} + k_{\theta v}v + k_{\theta\theta}\theta = 0 \tag{3.83}$$
$$I_d\ddot{\psi} - I_p\Omega\dot{\theta} + k_{\psi u}u + k_{\psi\psi}\psi = 0$$

The stiffness properties of a circular shaft are identical in each direction, so that $k_{uu} = k_{vv}$ and $k_{\theta\theta} = k_{\psi\psi}$. However, $k_{u\psi} = -k_{v\theta}$ because of the particular sign convention we are using (see Figure 3.2). Thus, letting $k_{u\psi} = -k_{v\theta} = k_C$, $k_{uu} = k_{vv} = k_T$, and $k_{\theta\theta} = k_{\psi\psi} = k_R$, we have

$$m\ddot{u} + k_T u + k_C\psi = 0$$
$$m\ddot{v} + k_T v - k_C\theta = 0$$
$$I_d\ddot{\theta} + I_p\Omega\dot{\psi} - k_C v + k_R\theta = 0 \tag{3.84}$$
$$I_d\ddot{\psi} - I_p\Omega\dot{\theta} + k_C u + k_R\psi = 0$$

This set of equations is identical to Equation (3.9). The shaft stiffness coefficients can be determined from the information given in Appendix 2. For a shaft of length L between short bearings (i.e., simply supported at the ends), with a disk located a distance a from one bearing and b from the other, then

$$k_T = k_{uu} = k_{vv} = \frac{3EI\left(a^3 + b^3\right)}{a^3 b^3}$$

$$k_C = k_{u\psi} = -k_{v\theta} = \frac{3EI\left(a^2 - b^2\right)}{a^2 b^2} \qquad (3.85)$$

$$k_R = k_{\psi\psi} = k_{\theta\theta} = \frac{3EI\left(a + b\right)}{ab}$$

Because Equation (3.84) is identical to Equation (3.9), it follows that the procedure for solving Equation (3.84) is identical to that used for Equation (3.9). The procedure is described in Sections 3.5.1 and 3.5.3 neglecting and including gyroscopic effects, respectively. The difference in the two systems is whether the flexibility arises in the shaft or in the supports. Another difference between these systems concerns the relative magnitudes of the polar and diametral moments of inertia. Consider the case of a uniform rigid rotor of diameter D and length h and the case of a disk of diameter D and uniform thickness h mounted on a flexible shaft. From the data given in Appendix 1, it easily can be shown that for a cylinder of diameter D and length h, $I_d < I_p$ if $h < \left(\sqrt{3}/2\right) D$. For a rigid, cylindrical rotor, it is likely that the length of the rotor is substantially greater than its diameter; therefore, for this system, $I_d > I_p$. In contrast, the thickness of a disk on a shaft is likely to be less than the disk diameter and, in this system, $I_d < I_p$ (but note that $I_d \geq I_p/2$ always).

We can analyze a disk on a light shaft supported by other combinations of bearings. In each case, the appropriate shaft-stiffness properties must be used (see Appendix 2). In every case, the model requires four coordinates; hence, it has four degrees of freedom.

EXAMPLE 3.9.1. A 38 mm-diameter solid shaft is supported in self-aligning bearings 1.1 m apart. A disk, 650 mm in diameter and 100 mm in thickness is shrunk onto the shaft 0.8 m from one bearing. The material properties of the shaft and disk are density $\rho = 7{,}810\,\text{kg/m}^3$ and modulus of elasticity $E = 211\,\text{GPa}$.

(a) Plot the natural frequency map for this system for speeds of up to 3,000 rev/min.
(b) Plot the mode shapes for this rotor at rest and at a rotational speed of 900 rev/min.

Solution

The disk mass is $m = \rho h \pi D^2/4 = 7{,}810 \times 0.1 \times \pi \times 0.65^2/4 = 259.2\,\text{kg}$. Also, $I_p = mD^2/8 = 259.2 \times 0.65^2/8 = 13.69\,\text{kg m}^2$, and $I_d = mD^2/16 + mh^2/12 = 259.2 \times 0.65^2/16 + 259.2 \times 0.1^2/12 = 7.06\,\text{kg m}^2$.

For the shaft, the second moment of area $I = \pi d^4/64 = 0.038^4 \pi/64 = 1.024 \times 10^{-7}\,\text{m}^4$. Thus, $EI = 211 \times 10^9 \times 1.024 \times 10^{-7} = 21.6\,\text{kNm}^2$.

The mass of the shaft $= \rho L \pi d^2/4 = 7{,}810 \times 1.1 \times \pi \times 0.038^2/4 = 9.74\,\text{kg}$. Note that the mass of the shaft is small compared to the mass of the disk.

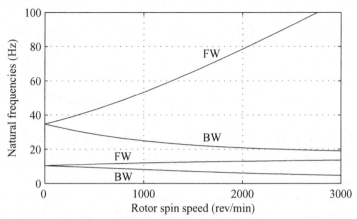

Figure 3.23. Natural frequency map for the flexible rotor (Example 3.9.1(a)). FW and BW indicate forward and backward whirl, respectively.

Given that $a = 0.8$ m and $b = 0.3$ m, using Equation (3.85) gives

$$k_T = 3 \times 21.6 \times 10^3 \times (0.8^3 + 0.3^3)/(0.3^3 \times 0.8^3) = 2.53 \times 10^6 \, \text{N/m}$$

$$k_C = 3 \times 21.6 \times 10^3 \times (0.8^2 - 0.3^2)/(0.3^2 \times 0.8^2) = 6.19 \times 10^5 \, \text{N}$$

$$k_R = 3 \times 21.6 \times 10^3 \times (0.3 + 0.8)/(0.3 \times 0.8) = 2.97 \times 10^5 \, \text{Nm}$$

(a) For a particular rotational speed Ω, we must solve Equation (3.42) or (3.43) to obtain four roots and, hence, the natural frequencies of the system. This process then must be repeated for all rotational speeds of interest. Once the roots or natural frequencies are computed at a range of rotational speeds, then the natural frequency map can be drawn as shown in Figure 3.23.

The system has four natural frequencies at a particular rotational speed. When the rotational speed is zero, the system has two pairs of repeated natural frequencies. As the rotational speed increases, these frequencies become distinct due to the gyroscopic effect. As for the rigid-rotor examples, two of the system natural frequencies correspond to forward-rotating modes and two correspond to backward-rotating modes, as described in Section 3.5.3. In Figure 3.23, the frequencies are plotted and the direction of the circular orbit of the rotor at that frequency is indicated.

(b) For a particular $s_i(= j\omega_i)$, we can obtain the ratio $u^{(i)}/\psi^{(i)} = -v^{(i)}/\theta^{(i)}$ by substituting the value of s_i into Equation (3.44). This is the mode shape for the ith mode. Thus, for the stationary rotor

$$s_1 = s_2 = +64.96j \, \text{rad/s} \quad (10.34 \, \text{Hz}), \quad u/\psi = -0.43$$
$$s_3 = s_4 = +218.16j \, \text{rad/s} \quad (34.72 \, \text{Hz}), \quad u/\psi = 0.063$$

At 900 rev/min

$$s_1 = +52.22j \, \text{rad/s} \quad (8.31 \, \text{Hz}), \quad u/\psi = -0.34$$
$$s_2 = +74.66j \, \text{rad/s} \quad (11.88 \, \text{Hz}), \quad u/\psi = -0.57$$
$$s_3 = +160.55j \, \text{rad/s} \quad (25.55 \, \text{Hz}), \quad u/\psi = +0.15$$
$$s_4 = +320.84j \, \text{rad/s} \quad (51.06 \, \text{Hz}), \quad u/\psi = +0.026$$

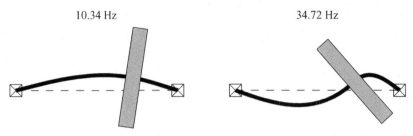

Figure 3.24. Mode shapes for the stationary rotor (Example 3.9.1(b)).

The roots $s_5, \ldots, s_8$ have not been given but, of course, $s_{i+4} = -s_i$.

Figures 3.24 and 3.25 show the rotor mode shapes for the natural frequencies for the system. The figures show that in each mode, the disk displaces and rotates about the Ox and Oy axes. We cannot determine the actual amplitude of either the displacements or the rotations because they are mode shapes. At each frequency, the value of u/ψ is used to construct the figures. The shaft line is added to the diagrams for clarity; however, determining its exact shape is not part of this calculation. If required, these deflections may be determined from the equations for static beam bending.

3.10 Summary

In this chapter, we analyze simple rotor–bearing systems to determine their natural frequencies and free lateral response. The modeling restrictions are that the rotors are circular in cross section with no internal damping and systems can be modeled with only four degrees of freedom. This restricts the analysis to rigid rotors on flexible supports or flexible rotors of negligible mass carrying a single disk.

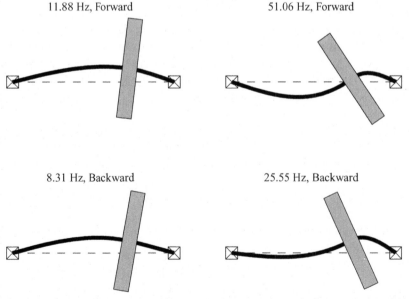

Figure 3.25. Mode shapes for the rotor spinning at 900 rev/min (Example 3.9.1(b)).

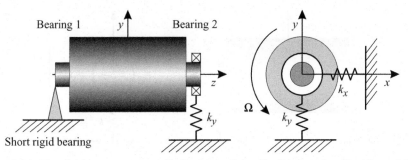

Figure 3.26. The rigid rotor on one self-aligning and one flexible bearing (Problems 3.1, 3.2, 3.3, and 3.4).

It is shown that when the flexibility in the rotor–bearing system is isotropic, the rotor orbits are circular, rotating either in the same or the opposite direction to the direction of spin of the rotor. When the rotor is supported by anisotropic bearings, the orbits are elliptical; however, rotor orbits can be in the same or the opposite direction to that of the spin of the rotor.

An important feature of the behavior of a rotor is the influence of gyroscopic couples. The simple models described in this chapter show that the rotor frequencies change with rotational speed due to the action of gyroscopic couples. These changes in frequency with rotational speed are conveniently shown using the natural frequency map.

In Chapters 5 and 6, the behavior of more complex rotordynamic systems is examined.

3.11 Problems

3.1 The rigid rotor shown in Figure 3.26 is 0.5 m long. It is carried by a rigidly supported, short, self-aligning bearing at the left end and a flexibly supported, short, self-aligning bearing at the right end. The rotor has a polar moment of inertia, I_p, of 0.6 kg m^2 and a diametral moment of inertia about the left end, I_{d1}, of 10 kg m^2. The bearing support stiffness, k, is 1 MN/m in both the x and y directions (i.e., $k_x = k_y = k$). By taking moments about the rigid bearing (i.e., bearing 1, pinned left end), show that the equations of motion for this system, including the gyroscopic couple, in terms of the rotations θ and ψ are

$$I_{d1}\ddot{\theta} + I_p\Omega\dot{\psi} + kL^2\theta = 0$$
$$I_{d1}\ddot{\psi} - I_p\Omega\dot{\theta} + kL^2\psi = 0$$

where L is the rotor length.

Find the roots and therefore the natural frequencies of the system when it is stationary and when it is rotating at 3,000 and 10,000 rev/min. Describe the motion of the rotor at bearing 2 when it vibrates in each of its natural frequencies.

Determine the direction of the orbit of the rotor in the $\theta - \psi$ and $x - y$ planes for each of its natural frequencies when the rotor is spinning at 3,000 rev/min.

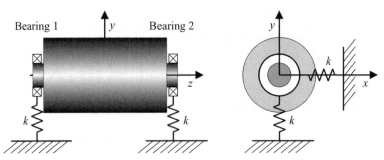

Figure 3.27. Schematic of a rigid rotor (Problem 3.5).

3.2 The rigid rotor, described in Problem 3.1, is carried in bearings as shown in Figure 3.26. The support stiffness at bearing 2 is 1 MN/m in the y direction and 1.3 MN/m in the x direction. Find the roots and therefore the natural frequencies of the system when it is stationary and when it is rotating at 3,000 and 10,000 rev/min.

3.3 The rigid rotor, described in Problem 3.1, is carried in bearings as shown in Figure 3.26. The support stiffness at bearing 2 is 1 MN/m in both the x and y directions and the damping in the bearing is 500 Ns/m in both the x and y directions.

   Find the roots and therefore the natural frequencies and the damping factors for the system when it is stationary and when it is spinning at 3,000 and 10,000 rev/min.

3.4 A rigid rotor, shown in Figure 3.26, is 0.5 m long. It is supported by a short, rigid bearing at the left end (bearing 1) and a short, flexibly supported bearing at the right end (bearing 2). The rotor has a polar moment of inertia of 0.6 kg m^2 and a diametral moment of inertia about the left end of 10 kg m^2. The bearing-support stiffness in both the x and y directions, k, is 1 MN/m. There is also a cross-coupling stiffness, k_c, between the x and y directions of 0.2 MN/m. Thus, if the force acting on the bearing in the x direction is f_x, then $f_x = ku + k_c v$ where u and v are the displacements at the bearing 2 end of the rotor in the x and y directions. Similarly, $f_y = k_c u + kv$.

   (a) Develop the differential equations of motion for this system in terms of the rotations of the rotor about the left end, θ and ψ, allowing for gyroscopic effects. Do not attempt to simplify the equations by introducing complex coordinates.

   (b) By letting $\theta = \theta_0 e^{st}$ and $\psi = \psi_0 e^{st}$, solve the equations of motion and determine the system natural frequencies when the rotor is stationary and when it rotates at 3,000 rev/min.

3.5 A rigid rotor mounted on two flexible supports, shown in Figure 3.27, has a center of gravity located midway between the supports. The rotor has a mass of 1,000 kg, a polar moment of inertia of 30 kg m^2, and a diametral moment of inertia at the center of gravity of 50 kg m^2. The rotor is supported on isotropic bearings, which are 1 m apart, with stiffnesses in both the horizontal and vertical directions of 4 MN/m. Determine the natural frequencies of the machine at 3,000 rev/min.

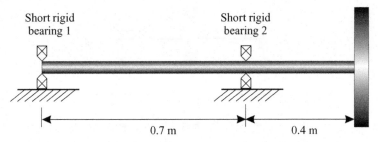

Figure 3.28. The flexible overhung rotor (Problem 3.6).

3.6 A rotating machine consists of a shaft with an overhung disk, as shown in Figure 3.28. The bearings are stiff and 0.7 m apart and the shaft extends 0.4 m beyond the bearing, as shown in the diagram. The shaft is 40 mm in diameter and is made from steel with a Young's modulus of 210 GPa. The disk has a mass of 30 kg, a diametral moment of inertia of 2 kg m², and a polar moment of inertia of 3.75 kg m². The mass of the shaft is negligible compared with the mass of a disk.

Develop the differential equations of motion for this system, allowing for gyroscopic effects, using the stiffness coefficients given in Appendix 2. Then, determine the natural frequencies of the rotor when it is stationary and when it is spinning at 1,000 rev/min. A quartic equation must be solved. If the facilities to do this are not available, substitute the solutions (provided at the end of the book) into the frequency equation and verify that the given solutions satisfy this equation.

Use the values of s for this system – both when stationary and spinning – to determine the corresponding ratios u/ψ, and sketch the mode shape of the system as it whirls (i.e., vibrates) in each mode.

3.7 A rigid steel rotor, 1 m long, is supported by plain bearings at each end, as shown in Figure 3.29. In the y direction, the effective bearing and foundation stiffnesses are k_1 and k_2 at the left and right ends, respectively. In the x direction, the left-end combined bearing and foundation stiffnesses are so high that the support can be assumed to be rigid. The right end combined bearing and foundation stiffnesses is k_0. The center of mass of the rotor is at the midspan. Develop the equations of motion for small displacements of this three degrees of freedom system.

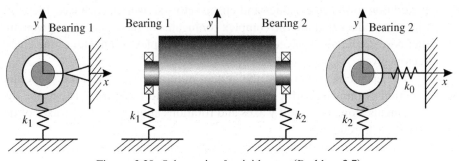

Figure 3.29. Schematic of a rigid rotor (Problem 3.7).

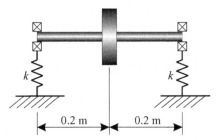

Figure 3.30. The flexible rotor on flexible supports (Problem 3.8).

Suppose $k_0 = 2.0\,\text{MN/m}$, $m = 120\,\text{kg}$, $I_p = 2\,\text{kg m}^2$, the diametral moment of inertia about the center of mass is $I_d = 8\,\text{kg m}^2$, and $a = L/2 = 0.5\,\text{m}$. Using the parallel-axis theorem, the diametral moment of inertia about the left end is $I_{d1} = 38\,\text{kg m}^2$. Determine the natural frequencies of this system under the following conditions:

(a) $\Omega = 0$, $k_1 = 1.8\,\text{MN/m}$, $k_2 = 2.2\,\text{MN/m}$

(b) $\Omega = 9{,}550\,\text{rev/min}$, $k_1 = k_2 = 2.0\,\text{MN/m}$

3.8 The simple rotor shown in Figure 3.30 is supported by two identical, short, flexible isotropic bearings. The central disk has a mass of 14 kg and a diametral moment of inertia of $0.08\,\text{kg m}^2$. The shaft is 0.4 m long with a diameter of 25 mm. It is made of steel with a modulus of elasticity of $E = 200\,\text{GPa}$ and, compared to the disk, its mass is negligible.

Ignoring gyroscopic effects, determine the first two natural frequencies of the system when the short bearings are of stiffness 50 kN/m, 1 MN/m, and 100 MN/m, in turn. The stiffness coefficients for a shaft supported by short, flexible bearings are given in Appendix 2, System 6, of Table A2.4.

Show that when the bearing stiffness is 50 kN/m, the system can be adequately modeled as a rigid rotor on flexible supports and that when the bearing stiffness is 100 MN/m, the system can be modeled as a flexible rotor on short, rigid bearings. Finally, show that when the bearing stiffness is 1 MN/m, neither of these assumptions provides an accurate estimate of the natural frequencies.

3.9 A light shaft, supported by flexible bearings at the ends, carries a disk. The disk is not at the midspan of the shaft. Table 3.4 shows the eigenvalues (s) and the corresponding eigenvectors for the system when supported on isotropic bearings (Case 1) and anisotropic bearings (Case 2). Only the elements of the eigenvectors related to the coordinates u, v, θ, and ψ (in that order) are shown. In addition to the eigenvalues and eigenvectors shown, in each case there are four eigenvalues that are the complex conjugates of the eigenvalues given and four corresponding eigenvectors that are the complex conjugates of the eigenvectors given.

For each case, determine the system natural frequencies, the shape and direction of the whirl orbit at each frequency, and the ratio between the displacements along the x and y axes and rotations about the x and y axes of the disk at each frequency.

3.10 A rigid steel rotor, 0.8 m long, is carried in flexibly supported bearings, as shown in Figure 3.31. The bearing supports at each end of the rotor are

Table 3.4. *The eigenvalues and eigenvectors for the flexible machine (Problem 3.9)*

Case 1. Isotropic bearings				
$s_i, i = 1, \ldots, 4$	297.85j	297.97j	558.55j	973.47j
	3.3574 − 0.0000j	0.0000 + 3.3561j	0.0000 + 0.0045j	0.0007 − 0.0000j
Corresponding	0.0000 + 3.3574j	3.3561 − 0.0000j	−0.0045 + 0.0000j	−0.0000 − 0.0007j
eigenvectors	−0.0000 − 0.9963j	−0.5703 + 0.0000j	−1.7903 + 0.0000j	−0.0001 − 1.0272j
	0.9963 − 0.0000j	0.0000 + 0.5703j	−0.0000 − 1.7903j	−1.0272 + 0.0001j
Case 2. Anisotropic bearings				
$s_i, i = 1, \ldots, 4$	297.91j	316.17j	575.68j	991.99j
	−0.0000 − 3.3567j	0.0240 + 0.0000j	−0.0040 + 0.0000j	0.0006 + 0.0000j
Corresponding	0.0225 + 0.0000j	−0.0000 − 3.1628j	−0.0000 − 0.0090j	0.0000 − 0.0015j
eigenvectors	−0.1995 − 0.0000j	−0.0000 + 1.6272j	−0.0000 − 1.5440j	0.0000 − 1.0081j
	0.0000 − 0.7797j	−0.4761 − 0.0000j	1.7371 − 0.0000j	−0.9423 − 0.0000j

isotropic and identical at each end of the rotor. The rotor has a mass of 30 kg and is symmetrical about all three axes. The bearing-support stiffness and the polar and the diametral moments of inertia for the rotor are unknown.

Develop the equations of motion for this system in terms of the rotor length, L; the position of the center of gravity; the bearing support stiffness, k; and the polar and the diametral moments of inertia, I_p and I_d, respectively.

The natural frequencies of the system are measured and one is found to be equal to 7.1 Hz, irrespective of the rotor speed. The second frequency is 21.2 Hz when the rotor is stationary. When the rotor begins to spin, this 21.2 Hz frequency splits into a pair of frequencies; when the rotor spins at 2,000 rev/min, the higher frequency of the pair is 46.7 Hz. Using this information, determine the stiffness of the bearing supports and the polar and diametral moments of inertia of the rotor.

3.11 Rework the analysis of Section 3.6.1 for the eigenvalue $s_i = -j\omega_i$ and determine the range of phase-difference angles for forward- and backward-whirling orbits.

3.12 A flexible shaft carrying a single mass is supported by short bearings, rigidly supported at one end and flexibly supported at the other, as shown in Figure 3.32. By letting k_1 in System 6 of Appendix 2 tend to infinity, derive the flexibility coefficients for this system.

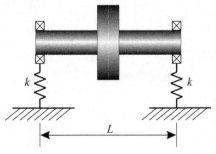

Figure 3.31. The rigid rotor on flexible supports (Problem 3.10).

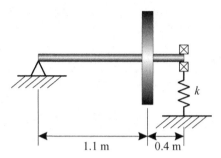

Figure 3.32. The flexible rotor on one pinned support and one flexible support (Problem 3.12).

The shaft is 1.5 m long and has an outside diameter of 60 mm and an internal diameter of 40 mm. The disk is mounted on the shaft, 1.1 m from the short, rigidly supported bearing and 0.4 m from the short, flexibly supported bearing, as shown in Figure 3.32. The bearing support stiffness is 10 MN/m. Using the stiffness coefficients that have been derived and assuming the shaft has negligible mass, determine the two natural frequencies of this system when the shaft carries (a) a disk of diameter 0.65 m and thickness 65 mm, and (b) a disk of diameter 1.2 m and thickness 120 mm. Assume $E = 200$ GPa and $\rho = 7,800$ kg/m^3 and that the rotor is stationary.

Using the reduction formula given in Appendix 2, reduce the model to a single degree of freedom and determine the first natural frequency for each of the two disk cases. Comment on the single degree of freedom natural frequency compared with the two degrees of freedom natural frequencies.

3.13 Figure 3.33 shows a simplified representation of the rotor assembly of a turbofan aircraft engine. The rotor is modeled as a uniform cylinder 600 mm long and 400 mm in diameter. The rotor mass is 100 kg with a center of gravity 800 mm from bearing 1. The fan is modeled as a uniform (short) cylinder of length 100 mm and outside diameter 1.4 m. The fan has a mass of 100 kg with a center of gravity 300 mm forward of bearing 2. The shaft is assumed to be rigid.

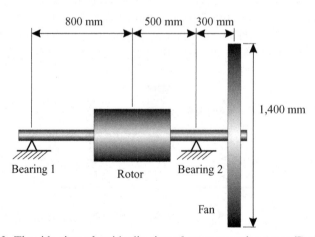

Figure 3.33. The side view of an idealization of an aero-engine rotor (Problem 3.13).

(a) Compute the diametral and polar moments of inertia of the two cylinders at their respective centers of gravity. Determine the center of gravity of the combined rotor and fan and calculate the combined diametral and polar moments of inertia.

(b) During takeoff conditions, the rotor rotates at 7,000 rev/min, and the aircraft is pitching upward with an angular velocity of 2 rad/s and an angular acceleration of 5 rad/s². Calculate the moments on the combined rotor and fan due to the polar and diametral inertias. Assume that the aircraft has zero angular velocity and angular acceleration in the roll and yaw directions.

(c) From (b), determine the net vertical-reaction loads at the bearings if the net vertical acceleration of bearing 1 of this engine is 0.4 g at the instant in question. Assume that g = 9.81 m/s².

4 Finite Element Modeling

4.1 Introduction

The finite element method (FEM) has developed into a sophisticated method for the analysis of stress, vibration, heat flow, and many other phenomena. Although the method is powerful, its derivation is simple and logical. It is undoubtedly the combination of mathematical versatility with a simple geometric interpretation that led to the immense popularity of the method across wide areas of engineering and science. The texts by Bickford (1994), Cook et al. (2001), Fagan (1992), Irons and Ahmad (1980), and Zienkiewicz et al. (2005) provide details of the formulation of element matrices for various structural element types (e.g., beams, plates, shells, and continua). The National Agency for Finite Element Methods and Standards (NAFEMS, 1986) produced *A Finite Element Primer*, which is an excellent introduction to finite element (FE) methodology. This chapter explains the principles of FEA as they relate to vibrating structures. The same principles apply to the FE modeling and analysis of rotating machines.

Two alternative methods produce the equations of motion of a system. The concept of *generalized coordinates* is explained in Section 4.2. The forces and moments produced by elastic deformation based on changes in these coordinates are calculated. For small deflections, these forces and moments, collectively called *generalized forces*, are linear functions of the generalized coordinates. Newton's second law is then used to equate the rate of change of momentum in the system to the forces on the system, both from the elastic deformation and externally applied forces, as in Chapter 2. The result is a set of second-order ODEs in the generalized coordinates. The alternative method is to express the kinetic energy, strain energy, and potential of the applied loads in terms of the generalized coordinates. The equations of motion are then derived from Lagrange's equations. Both routes generate the same equations of motion. Newton's second law is familiar but can become complex when dealing with large systems or systems with rotating frames of reference. Lagrange's equations are not particularly intuitive but they have the advantage of summing scalar quantities. This chapter derives the equations of motion using Newton's laws, although the derivation of element mass and stiffness matrices for continuous elements uses the energy approach. The initial derivation of the models does not consider damping.

The FEM may be separated into four steps, as follows:

1. *Define the finite element mesh.* The structure is divided into regions of simple geometry called *elements*. This is called *discretization*. The type of these regions is dependent on how the real system is to be modeled. Can the structure be modeled with bars and beams? Or, does the model require the modeling of plates or general three-dimensional regions? For the description of shafts, one-dimensional bar and beam models often are used. Restricting models to one-dimensional components substantially decreases the computational time. Symmetry, axisymmetry, and periodicity in systems also may be used to reduce the number of elements required in a model and, hence, the number of degrees of freedom, enabling the analysis to be performed more quickly.

   A beam is simply subdivided into smaller beams that may be represented by lines of the same length as the elements. Plates are divided into two-dimensional shapes, usually triangles and quadrilaterals. General three-dimensional shapes are split into a combination of tetrahedra, wedges, and cuboids. These elements are connected at points called *nodes*. Nodes are located at the ends of beam elements, the corners of triangular or quadrilateral elements, and the apexes of tetrahedra or cuboid elements. Nodes also may be located at intermediate points. The equations of motion of the system are written in terms of the translation and rotation at the nodes. These deformations represent the generalized coordinates in the FE problem.

2. *Express the elastic, inertia, and external forces on each element.* The forces and moments on each element must be expressed in terms of the local coordinates for the element. The local coordinates are the translation and rotations at the nodes of the element. For vibration analysis, the deformations usually are considered to be small, and forces and moments are linear functions of local coordinates and their derivatives. Essentially, the process approximates the forces and moments that are distributed throughout the element and produces equivalent forces at the nodes of the element. The power of the FEM partly arises because this must be done only once for each different description of an element.

3. *Assemble the elements.* The forces and moments from all of the elements must be assembled to produce the equivalent generalized forces on the complete system in terms of the generalized coordinates. Each force or moment produced by an element must be related to its equivalent generalized force, and each local coordinate must be matched to the corresponding generalized coordinate. Some elements may produce only inertia forces and some may produce only elastic forces.

4. *Perform the analysis.* Once the equations of motion have been generated, they may be solved using the techniques discussed in Chapter 2. The equations of motion also may be used in the analysis described in subsequent chapters.

Most analysis in rotordynamics is performed using a *shaft-line model*, in which the creation of the mesh is simply a process of choosing nodal locations along the shaft line. Nodes should be placed at key locations such as disks and bearings; other nodes are placed along the shaft to provide an adequate approximation to the deformation of the shaft. In Chapter 10, more complex three-dimensional models are discussed. In this chapter, we focus on element formulation, assembly, and analysis for simple bar and beam systems to introduce the process of FEA.

Before discussing the details of the FE approach, we emphasize that FEA can model only the abstract description of the physical system that we have used. For many simple systems, an *exact* solution can be found in closed form; hence, the solution can be determined to as many decimal places as required. For example, if we use the Euler-Bernoulli model (Rao, 1990) to determine the first natural frequency of a uniform pinned-pinned beam, the exact solution is $\omega_1 = \pi^2 \sqrt{\dfrac{EI}{\rho A L^4}}$.

An FE model of the system converges, to this solution as the number of degrees of freedom increases. However, this solution is exact only in the sense that the solution exactly satisfies the mathematical model used. If we model the beam using the Timoshenko model (Rao, 1990), then the exact solution is not simply proportional to π^2 but rather to a factor that depends on the geometric dimensions of the beam, the shear modulus of the beam material, and the shear coefficient of the beam cross-sectional shape. Han et al. (1999) provide an extensive discussion of four different beam theories.

4.2 Defining Generalized Coordinates

The modeling and analysis of any physical system requires that the configuration of the system be determined. In applied mechanics, including structural dynamics and rotordynamics, we must be able to define the position of a particle, rigid body, or structure. The position of a modeled system is given by coordinates based on a given frame of reference. The *independent generalized coordinates*, usually abbreviated to generalized coordinates, are a minimum set of independent coordinates required to specify the displacement of a modeled system. The coordinates are *generalized* because no distinction is made among linear displacements, rotations, or any other deflection quantity that may be used to define the system configuration. Using a minimum set of coordinates means that any one coordinate may be changed independently, and such changes produce a different configuration of the system. Usually, the symbols $q_1, q_2, \ldots, q_n$ are used to denote a set of generalized coordinates, but other letters may be used if it is necessary to distinguish among sets. When we want to refer to a set of coordinates, it is convenient to place $q_1, q_2, \ldots, q_n$ into a vector of generalized displacements, $\mathbf{q}$. The choice of generalized coordinates is not unique and, for a particular system, coordinates are chosen to make the subsequent analysis as simple as possible. Although the coordinate choice is not unique, the number of independent generalized coordinates for a particular model of a system is unique and is called the number of *degrees of freedom* of the system.

All material is continuously distributed; therefore, all real systems have distributed mass and stiffness. Also, the distortion of the system or structure is specified as a continuous function relative to its original position. For example, the transverse displacement of a beam in bending is a continuous function of the position along the undeformed beam, and continuous models of such systems have an infinite number of degrees of freedom. Often, the magnitudes of mass and stiffness are different in different components: Some components have a large mass and negligible flexibility; others have high flexibility and negligible mass. Thus, in some instances, we approximate the actual system by an assemblage of these discrete components.

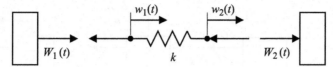

Figure 4.1. Hooke's law spring.

4.3 Finite Element Modeling of Discrete Components

Consider a discrete system of masses and springs. Figure 4.1 shows a simple Hooke's law spring element, in which the local coordinates are given by the displacement at the ends of the spring. The forces acting on the masses attached to the ends of the spring are shown as W_1 and W_2 and, for consistency, are positive in the direction of the displacements. From equilibrium, we have that $W_1 = -W_2$. Thus, if k is the spring stiffness

$$W_1 = -W_2 = -k(w_1 - w_2) \tag{4.1}$$

Therefore, we can express the force produced by changes in the local coordinates w_1 and w_2. Equation (4.1) can be expressed in matrix notation as

$$\begin{Bmatrix} W_1 \\ W_2 \end{Bmatrix} = -k \begin{bmatrix} 1 & -1 \\ -1 & 1 \end{bmatrix} \begin{Bmatrix} w_1 \\ w_2 \end{Bmatrix} = -\mathbf{K}_e \begin{Bmatrix} w_1 \\ w_2 \end{Bmatrix} \tag{4.2}$$

where $\mathbf{K}_e$ is called the *element stiffness matrix*. Equation (4.2) is the standard form for all expressions giving the force due to elastic strain in terms of the displacement. The deformation in the spring is completely specified by the relative displacement of its ends.

Using Newton's second law, the equation of motion of the rigid mass shown in Figure 4.2 is

$$m\ddot{w} = W \tag{4.3}$$

where W is the force applied to the mass, m, in the direction of w. This force arises from the springs as well as from external forces. In Equation (4.3), the mass may be interpreted as a 1×1 element mass matrix, $\mathbf{M}_e$; for other types of elements, the size of the matrix is larger.

These spring and mass elements are now assembled to give the complete equations of motion. Consider the system shown in Figure 4.3, consisting of three masses of mass m_1, m_2, and m_3 and four springs of stiffness k_1 through k_4. The system has three generalized coordinates, q_1, q_2, and q_3, which are the displacements of the

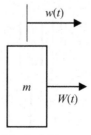

Figure 4.2. A rigid mass element.

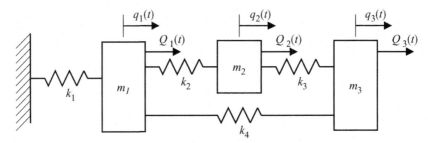

Figure 4.3. A discrete spring-mass example with three degrees of freedom.

three masses. The system also may be excited by external forces on the masses, which are shown by Q_i in Figure 4.3. The forces on the masses due to the spring k_2 may be expressed as

$$\mathbf{Q}_{k2} = \begin{Bmatrix} Q_{s1} \\ Q_{s2} \\ Q_{s3} \end{Bmatrix}_{k2} = -k_2 \begin{bmatrix} 1 & -1 & 0 \\ -1 & 1 & 0 \\ 0 & 0 & 0 \end{bmatrix} \begin{Bmatrix} q_1 \\ q_2 \\ q_3 \end{Bmatrix} \tag{4.4}$$

where Q_{si} denotes the force at generalized coordinate i, and the subscript ke indicates that force is produced by the spring element number e. This expression is obtained from Equation (4.2) by noting that $w_1 = q_1$, $w_2 = q_2$, $W_1 = Q_{s1}$, and $W_2 = Q_{s2}$. The force depends on q_1 and q_2 only because the third column of the matrix is zero, and the spring produces no force on mass 3 because the third row of the matrix is zero. This is as expected because the displacement of the third mass has no direct influence on the spring k_2. The force on the masses due to spring k_4 (noting that in this case, $w_1 = q_1$, $w_2 = q_3$, $W_1 = Q_{s1}$, and $W_2 = Q_{s3}$) is

$$\mathbf{Q}_{k4} = \begin{Bmatrix} Q_{s1} \\ Q_{s2} \\ Q_{s3} \end{Bmatrix}_{k4} = -k_4 \begin{bmatrix} 1 & 0 & -1 \\ 0 & 0 & 0 \\ -1 & 0 & 1 \end{bmatrix} \begin{Bmatrix} q_1 \\ q_2 \\ q_3 \end{Bmatrix} \tag{4.5}$$

which shows that the second generalized coordinate does not influence the force produced by this spring. The forces on the masses due to springs k_1 and k_3 can be derived in a similar fashion. Thus, the total force on the masses due to the springs is given by summing the contributions of all of the springs as

$$\mathbf{Q}_s = \mathbf{Q}_{k1} + \mathbf{Q}_{k2} + \mathbf{Q}_{k3} + \mathbf{Q}_{k4} \tag{4.6}$$

and, hence

$$\mathbf{Q}_s = \begin{Bmatrix} Q_{s1} \\ Q_{s2} \\ Q_{s3} \end{Bmatrix} = -k_1 \begin{bmatrix} 1 & 0 & 0 \\ 0 & 0 & 0 \\ 0 & 0 & 0 \end{bmatrix} \begin{Bmatrix} q_1 \\ q_2 \\ q_3 \end{Bmatrix} - k_2 \begin{bmatrix} 1 & -1 & 0 \\ -1 & 1 & 0 \\ 0 & 0 & 0 \end{bmatrix} \begin{Bmatrix} q_1 \\ q_2 \\ q_3 \end{Bmatrix}$$

$$- k_3 \begin{bmatrix} 0 & 0 & 0 \\ 0 & 1 & -1 \\ 0 & -1 & 1 \end{bmatrix} \begin{Bmatrix} q_1 \\ q_2 \\ q_3 \end{Bmatrix} - k_4 \begin{bmatrix} 1 & 0 & -1 \\ 0 & 0 & 0 \\ -1 & 0 & 1 \end{bmatrix} \begin{Bmatrix} q_1 \\ q_2 \\ q_3 \end{Bmatrix} \tag{4.7}$$

or

$$\begin{Bmatrix} Q_{s1} \\ Q_{s2} \\ Q_{s3} \end{Bmatrix} = - \begin{bmatrix} k_1 + k_2 + k_4 & -k_2 & -k_4 \\ -k_2 & k_2 + k_3 & -k_3 \\ -k_4 & -k_3 & k_3 + k_4 \end{bmatrix} \begin{Bmatrix} q_1 \\ q_2 \\ q_3 \end{Bmatrix} \qquad (4.8)$$

Applying Newton's second law to each mass in turn and assembling the resulting equations in matrix and vector notation give the equations of motion as

$$\begin{bmatrix} m_1 & 0 & 0 \\ 0 & m_2 & 0 \\ 0 & 0 & m_3 \end{bmatrix} \begin{Bmatrix} \ddot{q}_1 \\ \ddot{q}_2 \\ \ddot{q}_3 \end{Bmatrix} = \begin{Bmatrix} Q_{s1} \\ Q_{s2} \\ Q_{s3} \end{Bmatrix} + \begin{Bmatrix} Q_1 \\ Q_2 \\ Q_3 \end{Bmatrix} \qquad (4.9)$$

where Q_i is the external force applied to mass i. Thus,

$$\begin{bmatrix} m_1 & 0 & 0 \\ 0 & m_2 & 0 \\ 0 & 0 & m_3 \end{bmatrix} \begin{Bmatrix} \ddot{q}_1 \\ \ddot{q}_2 \\ \ddot{q}_3 \end{Bmatrix} + \begin{bmatrix} k_1 + k_2 + k_4 & -k_2 & -k_4 \\ -k_2 & k_2 + k_3 & -k_3 \\ -k_4 & -k_3 & k_3 + k_4 \end{bmatrix} \begin{Bmatrix} q_1 \\ q_2 \\ q_3 \end{Bmatrix} = \begin{Bmatrix} Q_1 \\ Q_2 \\ Q_3 \end{Bmatrix} \qquad (4.10)$$

or, more concisely

$$\mathbf{M\ddot{q}} + \mathbf{Kq} = \mathbf{Q} \qquad (4.11)$$

where

$$\mathbf{M} = \begin{bmatrix} m_1 & 0 & 0 \\ 0 & m_2 & 0 \\ 0 & 0 & m_3 \end{bmatrix}$$

is the mass matrix

$$\mathbf{K} = \begin{bmatrix} k_1 + k_2 + k_4 & -k_2 & -k_4 \\ -k_2 & k_2 + k_3 & -k_3 \\ -k_4 & -k_3 & k_3 + k_4 \end{bmatrix}$$

is the stiffness matrix, $\mathbf{q} = \{q_1 \ q_2 \ q_3\}^{\top}$ is the vector of generalized coordinates and $\mathbf{Q} = \{Q_1 \ Q_2 \ Q_3\}^{\top}$ is the corresponding vector of generalized forces.

The Hooke's law spring assumes negligible mass and the masses were assumed to be rigid in the discrete mass-spring system example. The principle is similar for the FE modeling of continuous systems. Systems containing continuous components are different only in the generation of the element mass and stiffness matrices. The principles involved in splitting the system into components or continuous regions and then assembling the full mass and stiffness matrices are similar in the FE modeling of continuous systems.

4.4 Axial Deflection in a Bar

Consider a bar of constant cross section subjected to motion in which the stresses are below the elastic limit. The bar is assumed to be elastic and also to have mass distributed evenly throughout the bar. The FEM provides a rational approach for this discretization that is vital for more complex systems. For a bar, the elements are one-dimensional with a node at each end, shown in Figure 4.4. At each node, there is

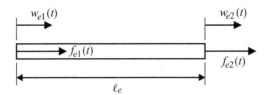

Figure 4.4. Uniform bar element.

only one degree of freedom, which is an axial displacement. The stiffness and inertia properties of an element are expressed in terms of these degrees of freedom.

The element mass and stiffness matrices are estimated by approximating the kinetic and strain energies and are derived in Section 4.6. For the axial motion of a bar, these matrices are

$$\mathbf{M}_e = \frac{\rho_e A_e \ell_e}{6} \begin{bmatrix} 2 & 1 \\ 1 & 2 \end{bmatrix}, \qquad \mathbf{K}_e = \frac{E_e A_e}{\ell_e} \begin{bmatrix} 1 & -1 \\ -1 & 1 \end{bmatrix} \qquad (4.12)$$

where E_e, A_e, and ρ_e are the Young's modulus, cross-sectional area, and mass density of the element and ℓ_e is its length. The equation of motion for this element in terms of the local displacements is then

$$\mathbf{M}_e \begin{Bmatrix} \ddot{w}_{e1} \\ \ddot{w}_{e2} \end{Bmatrix} + \mathbf{K}_e \begin{Bmatrix} w_{e1} \\ w_{e2} \end{Bmatrix} = \begin{Bmatrix} W_{e1} \\ W_{e2} \end{Bmatrix} \qquad (4.13)$$

where the forces W_{e1} and W_{e2} have components from the other elements as well as external sources.

The assembly of the full mass and stiffness matrices for bar problems is similar to that of discrete systems described previously. Figure 4.5 shows a bar split into three elements, together with the generalized coordinates for the discretized model. The elastic force at the four nodes due to the first element is obtained by noting that for this element, $w_{e1} = q_1$ and $w_{e2} = q_2$. Thus,

$$\mathbf{Q}_{b1} = \begin{Bmatrix} Q_{s1} \\ Q_{s2} \\ Q_{s3} \\ Q_{s4} \end{Bmatrix}_{b1} = -\frac{E_1 A_1}{\ell_1} \begin{bmatrix} 1 & -1 & 0 & 0 \\ -1 & 1 & 0 & 0 \\ 0 & 0 & 0 & 0 \\ 0 & 0 & 0 & 0 \end{bmatrix} \begin{Bmatrix} q_1 \\ q_2 \\ q_3 \\ q_4 \end{Bmatrix} \qquad (4.14)$$

where E_1, A_1, and ℓ_1 are Young's modulus, cross-sectional area, and length of the first element. Q_{si} denotes the strain force at generalized coordinate i and the subscript be denotes bar element number e. The element stiffness matrix has been inserted into a 2×2 submatrix of the full stiffness matrix. The position of this 2×2 matrix is determined by the generalized coordinates, which specify the displacement of the element: namely, q_1 and q_2.

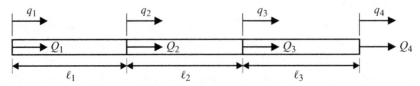

Figure 4.5. Three-element uniform bar example.

The forces due to the other elements are obtained in a similar manner. For example, the strain force produced by the second element is

$$\mathbf{Q}_{b2} = \begin{Bmatrix} Q_{s1} \\ Q_{s2} \\ Q_{s3} \\ Q_{s4} \end{Bmatrix}_{b2} = -\frac{E_2 A_2}{\ell_2} \begin{bmatrix} 0 & 0 & 0 & 0 \\ 0 & 1 & -1 & 0 \\ 0 & -1 & 1 & 0 \\ 0 & 0 & 0 & 0 \end{bmatrix} \begin{Bmatrix} q_1 \\ q_2 \\ q_3 \\ q_4 \end{Bmatrix} \tag{4.15}$$

For a uniform bar, $E_i = E$, $A_i = A$, and $\ell_i = \ell$ for all i, and the total strain force from all of the elements is

$$\mathbf{Q}_b = \mathbf{Q}_{b1} + \mathbf{Q}_{b2} + \mathbf{Q}_{b3}$$

$$= -\frac{EA}{\ell} \left(\begin{bmatrix} 1 & -1 & 0 & 0 \\ -1 & 1 & 0 & 0 \\ 0 & 0 & 0 & 0 \\ 0 & 0 & 0 & 0 \end{bmatrix} + \begin{bmatrix} 0 & 0 & 0 & 0 \\ 0 & 1 & -1 & 0 \\ 0 & -1 & 1 & 0 \\ 0 & 0 & 0 & 0 \end{bmatrix} + \begin{bmatrix} 0 & 0 & 0 & 0 \\ 0 & 0 & 0 & 0 \\ 0 & 0 & 1 & -1 \\ 0 & 0 & -1 & 1 \end{bmatrix} \right) \begin{Bmatrix} q_1 \\ q_2 \\ q_3 \\ q_4 \end{Bmatrix} \tag{4.16}$$

$$= -\mathbf{K}\mathbf{q}$$

where $\mathbf{q} = \{q_1 \; q_2 \; q_3 \; q_4\}^\top$ is the vector of generalized coordinates and the full stiffness matrix, $\mathbf{K}$, is

$$\mathbf{K} = \frac{EA}{\ell} \begin{bmatrix} 1 & -1 & 0 & 0 \\ -1 & 2 & -1 & 0 \\ 0 & -1 & 2 & -1 \\ 0 & 0 & -1 & 1 \end{bmatrix} \tag{4.17}$$

The inertia term produced by the first element is

$$\frac{\rho_1 A_1 \ell_1}{6} \begin{bmatrix} 2 & 1 & 0 & 0 \\ 1 & 2 & 0 & 0 \\ 0 & 0 & 0 & 0 \\ 0 & 0 & 0 & 0 \end{bmatrix} \begin{Bmatrix} \ddot{q}_1 \\ \ddot{q}_2 \\ \ddot{q}_3 \\ \ddot{q}_4 \end{Bmatrix} = \mathbf{M}_{b1} \ddot{\mathbf{q}} \tag{4.18}$$

where ρ_1 is the mass density of the first element.

The inertia terms for the other elements may be obtained similarly. Because $\rho_i = \rho$ for all i, the inertia term for the whole system is $\mathbf{M}\ddot{\mathbf{q}}$, where the mass matrix $\mathbf{M}$ is

$$\mathbf{M} = \mathbf{M}_{b1} + \mathbf{M}_{b2} + \mathbf{M}_{b3}$$

$$= \frac{\rho A \ell}{6} \left(\begin{bmatrix} 2 & 1 & 0 & 0 \\ 1 & 2 & 0 & 0 \\ 0 & 0 & 0 & 0 \\ 0 & 0 & 0 & 0 \end{bmatrix} + \begin{bmatrix} 0 & 0 & 0 & 0 \\ 0 & 2 & 1 & 0 \\ 0 & 1 & 2 & 0 \\ 0 & 0 & 0 & 0 \end{bmatrix} + \begin{bmatrix} 0 & 0 & 0 & 0 \\ 0 & 0 & 0 & 0 \\ 0 & 0 & 2 & 1 \\ 0 & 0 & 1 & 2 \end{bmatrix} \right) \tag{4.19}$$

$$= \frac{\rho A \ell}{6} \begin{bmatrix} 2 & 1 & 0 & 0 \\ 1 & 4 & 1 & 0 \\ 0 & 1 & 4 & 1 \\ 0 & 0 & 1 & 2 \end{bmatrix}$$

The equation of motion is obtained as

$$\mathbf{M\ddot{q}} = \mathbf{Q}_b + \mathbf{Q} \quad \text{or} \quad \mathbf{M\ddot{q}} + \mathbf{Kq} = \mathbf{Q} \tag{4.20}$$

where $\mathbf{Q} = \{Q_1 \ Q_2 \ Q_3 \ Q_4\}^\top$ is the vector of generalized (external) forces.

Thus far, we assume that the bar is free-free; hence, none of the coordinates is fixed. Suppose that the left end of the bar is fixed so that $q_1 = 0$. From Equation (4.20), the equations of motion become

$$\frac{\rho A\ell}{6} \begin{bmatrix} 2 & 1 & 0 & 0 \\ 1 & 4 & 1 & 0 \\ 0 & 1 & 4 & 1 \\ 0 & 0 & 1 & 2 \end{bmatrix} \begin{Bmatrix} 0 \\ \ddot{q}_2 \\ \ddot{q}_3 \\ \ddot{q}_4 \end{Bmatrix} + \frac{EA}{\ell} \begin{bmatrix} 1 & -1 & 0 & 0 \\ -1 & 2 & -1 & 0 \\ 0 & -1 & 2 & -1 \\ 0 & 0 & -1 & 1 \end{bmatrix} \begin{Bmatrix} 0 \\ q_2 \\ q_3 \\ q_4 \end{Bmatrix} = \begin{Bmatrix} Q_1 \\ Q_2 \\ Q_3 \\ Q_4 \end{Bmatrix} \tag{4.21}$$

The first row of Equation (4.21) gives the force at the fixed end as

$$Q_1 = \frac{\rho A\ell}{6}\ddot{q}_2 - \frac{EA}{\ell}q_2 \tag{4.22}$$

whereas the remaining rows give the equations of motion, in terms of the three unconstrained generalized coordinates, as

$$\frac{\rho A\ell}{6} \begin{bmatrix} 4 & 1 & 0 \\ 1 & 4 & 1 \\ 0 & 1 & 2 \end{bmatrix} \begin{Bmatrix} \ddot{q}_2 \\ \ddot{q}_3 \\ \ddot{q}_4 \end{Bmatrix} + \frac{EA}{\ell} \begin{bmatrix} 2 & -1 & 0 \\ -1 & 2 & -1 \\ 0 & -1 & 1 \end{bmatrix} \begin{Bmatrix} q_2 \\ q_3 \\ q_4 \end{Bmatrix} = \begin{Bmatrix} Q_2 \\ Q_3 \\ Q_4 \end{Bmatrix} \tag{4.23}$$

Essentially, we deleted the first row and the first column of the matrices in Equation (4.21).

The process may be repeated if another point on the beam is fixed – for example, the right end, $q_4 = 0$. The mass and stiffness matrices of the resulting two degrees of freedom system are obtained from the matrices in Equation (4.23) by deleting the last row and column. The reaction force Q_4 may be calculated from the last row of Equation (4.23), with $q_4 = \ddot{q}_4 = 0$.

EXAMPLE 4.4.1. Consider the axial vibration of a uniform bar or rod of length L, fixed at one end and free at the other, with no external forcing. Estimate the first two natural frequencies using three elements and compare with the exact results. How do the natural frequencies converge as the number of elements is increased?

Solution. The equations of motion using three elements, or three degrees of freedom, are obtained from Equation (4.23) using an element length of $\ell = L/3$ and are

$$\frac{\rho AL}{18} \begin{bmatrix} 4 & 1 & 0 \\ 1 & 4 & 1 \\ 0 & 1 & 2 \end{bmatrix} \begin{Bmatrix} \ddot{q}_2 \\ \ddot{q}_3 \\ \ddot{q}_4 \end{Bmatrix} + \frac{3EA}{L} \begin{bmatrix} 2 & -1 & 0 \\ -1 & 2 & -1 \\ 0 & -1 & 1 \end{bmatrix} \begin{Bmatrix} q_2 \\ q_3 \\ q_4 \end{Bmatrix} = \begin{Bmatrix} 0 \\ 0 \\ 0 \end{Bmatrix}.$$

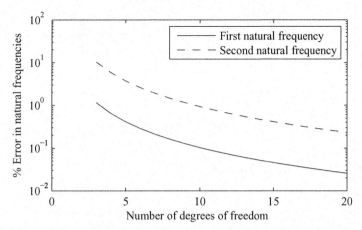

Figure 4.6. Convergence of the first two natural frequencies of a uniform bar clamped at one end (Example 4.4.1).

By formulating and solving an eigenvalue problem (see Section 2.4.2), these equations produce estimates of the first two natural frequencies of

$$\frac{1.5888}{L}\sqrt{\frac{E}{\rho}} \quad \text{and} \quad \frac{5.1962}{L}\sqrt{\frac{E}{\rho}}$$

Using four elements ($\ell = L/4$) gives the equations of motion

$$\frac{\rho A L}{24}\begin{bmatrix} 4 & 1 & 0 & 0 \\ 1 & 4 & 1 & 0 \\ 0 & 1 & 4 & 1 \\ 0 & 0 & 1 & 2 \end{bmatrix}\begin{Bmatrix} \ddot{q}_2 \\ \ddot{q}_3 \\ \ddot{q}_4 \\ \ddot{q}_5 \end{Bmatrix} + \frac{4EA}{L}\begin{bmatrix} 2 & -1 & 0 & 0 \\ -1 & 2 & -1 & 0 \\ 0 & -1 & 2 & -1 \\ 0 & 0 & -1 & 1 \end{bmatrix}\begin{Bmatrix} q_2 \\ q_3 \\ q_4 \\ q_5 \end{Bmatrix} = \begin{Bmatrix} 0 \\ 0 \\ 0 \\ 0 \end{Bmatrix}.$$

producing estimates of the first two natural frequencies of

$$\frac{1.5809}{L}\sqrt{\frac{E}{\rho}} \quad \text{and} \quad \frac{4.9872}{L}\sqrt{\frac{E}{\rho}}$$

These estimated natural frequencies compare with the exact frequencies (Inman, 2008) of

$$\frac{\pi}{2L}\sqrt{\frac{E}{\rho}} = \frac{1.5708}{L}\sqrt{\frac{E}{\rho}} \quad \text{and} \quad \frac{3\pi}{2L}\sqrt{\frac{E}{\rho}} = \frac{4.7124}{L}\sqrt{\frac{E}{\rho}}$$

By approximating the bar using increasingly more elements, the estimates of the natural frequencies converge to the exact values, although not necessarily monotonically. Figure 4.6 shows the convergence of the estimates of the first two natural frequencies as the number of elements are increased. Of particular significance are the facts that the lowest natural frequency converges most quickly and that the estimated natural frequencies are always higher than the exact natural frequencies. These two features are always present in beam structures modeled with consistent mass and stiffness matrices – that is, where the matrices are derived from the same displacement model.

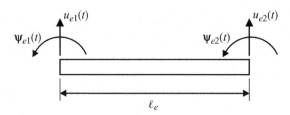

Figure 4.7. Uniform beam element.

4.5 Lateral Deflection of a Beam

The analysis of beam bending may be considered in a manner similar to bar extension. Figure 4.7 shows a typical element, together with the local coordinates. To ensure the continuity of deflection and slope across the element boundaries, the local coordinates consist of the translations and the rotations at the ends of the element.

Section 5.4.1 develops the element mass and stiffness matrices by approximating the kinetic and strain energies. For the bending motion of a slender beam, the Euler-Bernoulli assumptions hold and the element matrices are

$$\mathbf{M}_e = \frac{\rho_e A_e \ell_e}{420} \begin{bmatrix} 156 & 22\ell_e & 54 & -13\ell_e \\ 22\ell_e & 4\ell_e^2 & 13\ell_e & -3\ell_e^2 \\ 54 & 13\ell_e & 156 & -22\ell_e \\ -13\ell_e & -3\ell_e^2 & -22\ell_e & 4\ell_e^2 \end{bmatrix} \tag{4.24}$$

and

$$\mathbf{K}_e = \frac{E_e I_e}{\ell_e^3} \begin{bmatrix} 12 & 6\ell_e & -12 & 6\ell_e \\ 6\ell_e & 4\ell_e^2 & -6\ell_e & 2\ell_e^2 \\ -12 & -6\ell_e & 12 & -6\ell_e \\ 6\ell_e & 2\ell_e^2 & -6\ell_e & 4\ell_e^2 \end{bmatrix} \tag{4.25}$$

where E_e, I_e, A_e, and ρ_e are the Young's modulus, second moment of area, cross-sectional area, and mass density of the eth beam element.

The assembly of the full mass and stiffness matrices for beam problems is similar to that used for discrete systems and bar elements described previously. Figure 4.8 shows the simple example of a beam split into three elements, together with the generalized coordinates for the discretized model. The vector of forces at the four nodes due to the first element is obtained by noting that $q_1 = u_{e1}$, $q_2 = \psi_{e1}$, $q_3 = u_{e2}$,

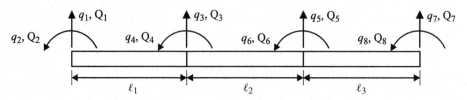

Figure 4.8. Three-element uniform beam example.

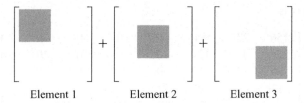

Element 1   Element 2   Element 3

Figure 4.9. The degrees of freedom affected during matrix assembly for the beam elements.

and $q_4 = \psi_{e2}$. Thus,

$$
\mathbf{Q}_{b1} = -\frac{E_1 I_1}{\ell_1^3}
\begin{bmatrix}
12 & 6\ell_1 & -12 & 6\ell_1 & 0 & 0 & 0 & 0 \\
6\ell_1 & 4\ell_1^2 & -6\ell_1 & 2\ell_1^2 & 0 & 0 & 0 & 0 \\
-12 & -6\ell_1 & 12 & -6\ell_1 & 0 & 0 & 0 & 0 \\
6\ell_1 & 2\ell_1^2 & -6\ell_1 & 4\ell_1^2 & 0 & 0 & 0 & 0 \\
0 & 0 & 0 & 0 & 0 & 0 & 0 & 0 \\
0 & 0 & 0 & 0 & 0 & 0 & 0 & 0 \\
0 & 0 & 0 & 0 & 0 & 0 & 0 & 0 \\
0 & 0 & 0 & 0 & 0 & 0 & 0 & 0
\end{bmatrix}
\begin{Bmatrix}
q_1 \\ q_2 \\ q_3 \\ q_4 \\ q_5 \\ q_6 \\ q_7 \\ q_8
\end{Bmatrix}
\tag{4.26}
$$

where $\mathbf{Q}_{be}$ denotes the strain force produced by the element number e at the eight generalized coordinates. The element stiffness matrix has been inserted into a 4×4 submatrix of the full stiffness matrix. The position of this 4×4 matrix is determined by the generalized coordinates specifying the displacement of the element: namely, q_1, q_2, q_3, and q_4. The elastic forces for the other elements and the inertia forces are produced in a similar way.

For a beam with three elements, shown in Figure 4.8, the total elastic force is

$$
\mathbf{Q}_b = \mathbf{Q}_{b1} + \mathbf{Q}_{b2} + \mathbf{Q}_{b3} = -\mathbf{Kq}
\tag{4.27}
$$

Equation (4.26) highlights that element 1 affects only the first four degrees of freedom, shown diagrammatically in Figure 4.9, in which the shaded areas represent the element stiffness matrices. Element 2 affects only degrees of freedom 3 to 6, and the location of the stiffness matrix for this element is also shown. Similarly, element 3 affects only degrees of freedom 5 to 8.

For a uniform slender beam, $E_i = E$, $I_i = I$, $\rho_i = \rho$, and $\ell_i = \ell$ for all i, and the global stiffness matrix becomes

$$
\mathbf{K} = \frac{EI}{\ell^3}
\begin{bmatrix}
12 & 6\ell & -12 & 6\ell & 0 & 0 & 0 & 0 \\
6\ell & 4\ell^2 & -6\ell & 2\ell^2 & 0 & 0 & 0 & 0 \\
-12 & -6\ell & 24 & 0 & -12 & 6\ell & 0 & 0 \\
6\ell & 2\ell^2 & 0 & 8\ell^2 & -6\ell & 2\ell^2 & 0 & 0 \\
0 & 0 & -12 & -6\ell & 24 & 0 & -12 & 6\ell \\
0 & 0 & 6\ell & 2\ell^2 & 0 & 8\ell^2 & -6\ell & 2\ell^2 \\
0 & 0 & 0 & 0 & -12 & -6\ell & 12 & -6\ell \\
0 & 0 & 0 & 0 & 6\ell & 2\ell^2 & -6\ell & 4\ell^2
\end{bmatrix}
\tag{4.28}
$$

The mass matrix is

$$\mathbf{M} = \frac{\rho A \ell}{420} \begin{bmatrix} 156 & 22\ell & 54 & -13\ell & 0 & 0 & 0 & 0 \\ 22\ell & 4\ell^2 & 13\ell & -3\ell^2 & 0 & 0 & 0 & 0 \\ 54 & 13\ell & 312 & 0 & 54 & -13\ell & 0 & 0 \\ -13\ell & -3\ell^2 & 0 & 8\ell^2 & 13\ell & -3\ell^2 & 0 & 0 \\ 0 & 0 & 54 & 13\ell & 312 & 0 & 54 & -13\ell \\ 0 & 0 & -13\ell & -3\ell^2 & 0 & 8\ell^2 & 13\ell & -3\ell^2 \\ 0 & 0 & 0 & 0 & 54 & 13\ell & 156 & -22\ell \\ 0 & 0 & 0 & 0 & -13\ell & -3\ell^2 & -22\ell & 4\ell^2 \end{bmatrix} \tag{4.29}$$

The equations of motion are then in the standard form

$$\mathbf{M\ddot{q}} = \mathbf{Q}_b + \mathbf{Q} \quad \text{or} \quad \mathbf{M\ddot{q}} + \mathbf{Kq} = \mathbf{Q} \tag{4.30}$$

where the lengths of the vector of generalized coordinates and the vector of generalized forces are now eight. Thus far, we have assumed that the beam is free-free; thus, none of the coordinates is fixed. Suppose that both ends of the beam are pinned, which implies that $q_1 = q_7 = 0$. The mass and stiffness matrices for this problem are obtained from Equations (4.28) and (4.29) by deleting the first and seventh rows and columns from the mass and stiffness matrices for the free-free case. Thus, for example

$$\mathbf{K} = \frac{EI}{\ell^3} \begin{bmatrix} 4\ell^2 & -6\ell & 2\ell^2 & 0 & 0 & 0 \\ -6\ell & 24 & 0 & -12 & 6\ell & 0 \\ 2\ell^2 & 0 & 8\ell^2 & -6\ell & 2\ell^2 & 0 \\ 0 & -12 & -6\ell & 24 & 0 & 6\ell \\ 0 & 6\ell & 2\ell^2 & 0 & 8\ell^2 & 2\ell^2 \\ 0 & 0 & 0 & 6\ell & 2\ell^2 & 4\ell^2 \end{bmatrix} \tag{4.31}$$

The rows deleted from the unconstrained equation of motion give the reaction force required on the pinned constraints to maintain the zero displacement. The process may be repeated if other boundary conditions are required. For example, if the left-hand end were fixed, giving the model of a cantilever beam, then $q_1 = q_2 = 0$.

EXAMPLE 4.5.1. Estimate the first two natural frequencies and mode shapes of a uniform cantilever beam of length L in lateral vibration.

Solution. The mass and stiffness matrices using three elements, or six degrees of freedom, are obtained from Equations (4.28) and (4.29) using an element length of $\ell = L/3$ and deleting the first two rows and columns. These matrices give estimates of the first two natural frequencies of

$$\frac{3.5164}{L^2} \sqrt{\frac{EI}{\rho A}} \quad \text{and} \quad \frac{22.1069}{L^2} \sqrt{\frac{EI}{\rho A}}$$

Increasing the number of elements produces better estimates of the natural frequencies, as shown in Figure 4.10. The exact values of natural frequency

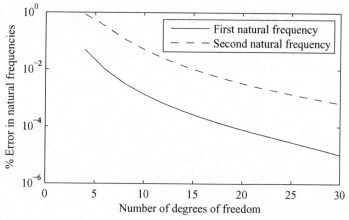

Figure 4.10. Convergence of the first two natural frequencies of a uniform cantilever beam (Example 4.5.1).

(Blevins, 1979) are

$$\frac{3.5160}{L^2}\sqrt{\frac{EI}{\rho A}} \quad \text{and} \quad \frac{22.0345}{L^2}\sqrt{\frac{EI}{\rho A}}$$

Using just three elements gives a surprisingly good result. In fact, the errors in the natural frequencies are proportional to the square of element length. The errors in the beam stress are directly proportional to the element length and are estimated with lower accuracy.

Figure 4.11 shows the first two modes of a uniform cantilever beam. The solid line shows the exact mode shapes and the crosses are the results from an FE model with six elements. The star on the left of the beam indicates the clamped end. Clearly, the results from the FE model are close to the exact modal displacements at the nodes. Figure 4.11 also shows that mode 2 has a location on the beam where the response is zero, shown by the circle. Such a point on a one-dimensional structure, including a beam, is called a *vibration node*.

4.6 Developing General Element Matrices

In the preceding sections, FE models of discrete systems, bar, and beam structures were assembled based on given element matrices. This section outlines how these

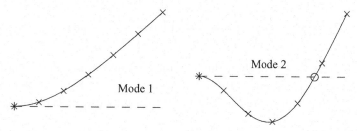

Figure 4.11. The first two mode shapes of a uniform cantilever beam (Example 4.5.1). The dashed line represents the centerline of the undeformed beam.

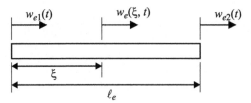

Figure 4.12. Uniform bar element.

element matrices are generated using energy methods. The approximation in the FEM arises from the specification of the displacement pattern within an element, using the local coordinates at the element nodes. The most convenient method to generate element matrices is to calculate the strain and kinetic energies as functions of the local coordinates. This is most easily demonstrated for the bar element in axial vibration and torsion.

4.6.1 Axial Bar Element

Consider the bar element, with local coordinates given in Figure 4.12. The simplest assumption for bar extension is that the displacement varies linearly along the element so that

$$w_e\left(\xi, t\right) = N_{e1}\left(\xi\right) w_{e1}\left(t\right) + N_{e2}\left(\xi\right) w_{e2}\left(t\right) = \begin{Bmatrix} N_{e1}(\xi) \\ N_{e2}(\xi) \end{Bmatrix}^\top \begin{Bmatrix} w_{e1}(t) \\ w_{e2}(t) \end{Bmatrix} \qquad (4.32)$$

where the *shape functions*, N_{ei}, are

$$N_{e1}\left(\xi\right) = \left(1 - \frac{\xi}{\ell_e}\right), \qquad N_{e2}\left(\xi\right) = \frac{\xi}{\ell_e} \qquad (4.33)$$

Equation (4.32) satisfies the conditions $w_e\left(0\right) = w_{e1}$ and $w_e\left(\ell_e\right) = w_{e2}$. For a constant cross-sectional element, loaded at its ends and in static equilibrium, this expression for deflection – called the *displacement model* – is exact. In general, for bars with a variable cross section or behaving dynamically, this displacement model is approximate but still may be used to produce accurate results if the element lengths are sufficiently small.

The strain energy within a bar due to axial extension, U_e, is given by the integral of the product of stress and strain as

$$U_e = \frac{1}{2} \int_{\text{bar}} \left(\frac{\partial w_e\left(\xi, t\right)}{\partial \xi}\right) \left(E_e \frac{\partial w_e\left(\xi, t\right)}{\partial \xi}\right) dV \qquad (4.34)$$

where the integration is over the volume of the bar and a linear elastic material that obeys Hooke's law is assumed. If the bar sections remain plane so that all points within a section have the same axial deformation, Equation (4.34) may be written as

$$U_e = \frac{1}{2} \int_0^{\ell_e} E_e A_e \left(\frac{\partial w_e\left(\xi, t\right)}{\partial \xi}\right)^2 d\xi \qquad (4.35)$$

Although both the Young's modulus, E_e, and the cross-sectional area, A_e, may vary within the element, for the purposes of illustration, we assume that they are constant. From Equation (4.32)

$$
\frac{\partial w_e(\xi, t)}{\partial \xi} = N'_{e1}(\xi)\, w_{e1}(t) + N'_{e2}(\xi)\, w_{e2}(t)
$$

$$
= \begin{Bmatrix} N'_{e1}(\xi) \\ N'_{e2}(\xi) \end{Bmatrix}^{\top} \begin{Bmatrix} w_{e1}(t) \\ w_{e2}(t) \end{Bmatrix} = \frac{1}{\ell_e} \begin{Bmatrix} -1 \\ 1 \end{Bmatrix}^{\top} \begin{Bmatrix} w_{e1}(t) \\ w_{e2}(t) \end{Bmatrix}
\tag{4.36}
$$

where the prime denotes differentiation with respect to ξ. The integration in Equation (4.35) is easy in this case because the shape function derivatives, N'_{ei}, are constant. Thus,

$$
U_e = \frac{1}{2} E_e A_e \begin{Bmatrix} w_{e1}(t) \\ w_{e2}(t) \end{Bmatrix}^{\top} \left(\int_0^{\ell_e} \begin{Bmatrix} N'_{e1}(\xi) \\ N'_{e2}(\xi) \end{Bmatrix} \begin{Bmatrix} N'_{e1}(\xi) \\ N'_{e2}(\xi) \end{Bmatrix}^{\top} d\xi \right) \begin{Bmatrix} w_{e1}(t) \\ w_{e2}(t) \end{Bmatrix}
$$

$$
= \frac{1}{2} \frac{E_e A_e}{\ell_e} \begin{Bmatrix} w_{e1}(t) \\ w_{e2}(t) \end{Bmatrix}^{\top} \begin{bmatrix} 1 & -1 \\ -1 & 1 \end{bmatrix} \begin{Bmatrix} w_{e1}(t) \\ w_{e2}(t) \end{Bmatrix}
\tag{4.37}
$$

$$
= \frac{1}{2} \begin{Bmatrix} w_{e1}(t) \\ w_{e2}(t) \end{Bmatrix}^{\top} \mathbf{K}_e \begin{Bmatrix} w_{e1}(t) \\ w_{e2}(t) \end{Bmatrix}
$$

where $\mathbf{K}_e$ is the element stiffness matrix quoted in Equation (4.12).

The element mass matrix is obtained by considering the approximation to the kinetic energy within the element, given by the integral of the product of the mass and the velocity squared. The kinetic energy of the element, T_e, is given by

$$
T_e = \frac{1}{2} \int_0^{\ell_e} \rho_e A_e \left(\frac{dw_e(\xi, t)}{dt} \right)^2 d\xi
\tag{4.38}
$$

Using the displacement model, Equation (4.32), and assuming that the mass density is constant, the approximate kinetic energy within the element is

$$
T_e = \frac{1}{2} \begin{Bmatrix} \dot{w}_{e1}(t) \\ \dot{w}_{e2}(t) \end{Bmatrix}^{\top} \left(\int_0^{\ell_e} \rho_e A_e \begin{Bmatrix} N_{e1}(\xi) \\ N_{e2}(\xi) \end{Bmatrix} \begin{Bmatrix} N_{e1}(\xi) \\ N_{e2}(\xi) \end{Bmatrix}^{\top} d\xi \right) \begin{Bmatrix} \dot{w}_{e1}(t) \\ \dot{w}_{e2}(t) \end{Bmatrix}
$$

$$
= \frac{1}{2} \frac{\rho_e A_e \ell_e}{6} \begin{Bmatrix} \dot{w}_{e1}(t) \\ \dot{w}_{e2}(t) \end{Bmatrix}^{\top} \begin{bmatrix} 2 & 1 \\ 1 & 2 \end{bmatrix} \begin{Bmatrix} \dot{w}_{e1}(t) \\ \dot{w}_{e2}(t) \end{Bmatrix}
\tag{4.39}
$$

$$
= \frac{1}{2} \begin{Bmatrix} \dot{w}_{e1}(t) \\ \dot{w}_{e2}(t) \end{Bmatrix}^{\top} \mathbf{M}_e \begin{Bmatrix} \dot{w}_{e1}(t) \\ \dot{w}_{e2}(t) \end{Bmatrix}
$$

where $\mathbf{M}_e$ is the element-mass matrix quoted in Equation (4.12). The integrands in Equation (4.39) are products of linear shape functions; therefore, the integrands are quadratic functions of ξ.

The equations of motion are obtained by summing the contributions to the strain and kinetic energies from all the elements, and by applying Lagrange's equations

$$
\frac{d}{dt} \left(\frac{\partial T}{\partial \dot{q}_k} \right) - \frac{\partial T}{\partial q_k} + \frac{\partial U}{\partial q_k} = Q_k, \quad \text{for } k = 1, \ldots, n
\tag{4.40}
$$

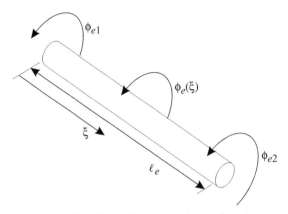

Figure 4.13. Local coordinates for the torsion element.

where T and U are the kinetic and strain energies for the assembled structure. In Equation (4.40), q_k is the kth generalized coordinate, Q_k is the applied generalized force at the kth coordinate, and n is the number of degrees of freedom in the system. The terms in Equation (4.40) give the force contributions to the equations of motion, and we may calculate these on an element-by-element basis. Thus, from Equation (4.37)

$$\left\{ \begin{array}{c} \dfrac{\partial U_e}{\partial w_{e1}} \\[2mm] \dfrac{\partial U_e}{\partial w_{e2}} \end{array} \right\} = \mathbf{K}_e \left\{ \begin{array}{c} w_{e1}(t) \\ w_{e2}(t) \end{array} \right\} \tag{4.41}$$

giving the strain-force expression for a bar used in Equation (4.13). Similarly, from Equation (4.39)

$$\frac{\mathrm{d}}{\mathrm{d}t} \left\{ \begin{array}{c} \dfrac{\partial T_e}{\partial \dot{w}_{e1}} \\[2mm] \dfrac{\partial T_e}{\partial \dot{w}_{e2}} \end{array} \right\} = \mathbf{M}_e \left\{ \begin{array}{c} \ddot{w}_{e1}(t) \\ \ddot{w}_{e2}(t) \end{array} \right\} \tag{4.42}$$

giving the inertia term for a bar used in Equation (4.13).

4.6.2 Torsion Element

The procedure for the FE modeling of the torsional motion of a rotor is similar to the analysis of the axial motion. For torsional problems, the deflection quantity of interest is the rotation of the bar about its axis, as shown in Figure 4.13, and the rotation angles at the two nodes constitute the generalized coordinates. Figure 4.13 shows a bar element with the angles of rotation at the two nodes given by ϕ_{e1} and ϕ_{e2}. The angle of rotation within the element is given by ϕ_e and may be approximated using linear shape functions (compare to Equation (4.32)) as

$$\phi_e(\xi, t) = \left(1 - \frac{\xi}{\ell_e}\right) \phi_{e1}(t) + \frac{\xi}{\ell_e} \phi_{e2}(t) \tag{4.43}$$

where ξ is the position within the element and ℓ_e is the length of the element.

Now that we have an approximation of the rotation throughout the element, the element mass and stiffness matrices may be generated via the approximate kinetic and strain energies. The strain energy (Petyt, 1990) is

$$U_e = \frac{1}{2} \int_0^{\ell_e} G_e C_e \left(\frac{\partial \phi_e (\xi, t)}{\partial \xi} \right)^2 d\xi \tag{4.44}$$

where G_e is the shear modulus and C_e is the torsional constant for the cross section.

The similarities with the axial motion in Section 4.6.1 are clear when Equation (4.44) is compared to Equation (4.35). It is shown that G_e is equivalent to E_e, C_e to A_e, and ϕ to w. Thus, from Equation (4.12), the element stiffness matrix is

$$\mathbf{K}_e = \frac{G_e C_e}{\ell_e} \begin{bmatrix} 1 & -1 \\ -1 & 1 \end{bmatrix} \tag{4.45}$$

If all cross sections of the bar remain plane, then C_e is equal to the polar moment of area of the cross section, J_e; thus, for a bar with a circular cross section

$$C_e = J_e = \frac{1}{32} \pi \left(d_o^4 - d_i^4 \right) \tag{4.46}$$

where d_i and d_o are the inside and outside diameters of the hollow bar. Expressions for the torsion constant, C_e, for various sections are tabulated in Blevins (1979) and a selection is provided in Appendix 3.

Similarly, the kinetic energy is obtained by integrating the product of the second moment of area by the angular velocity squared. Thus,

$$T_e = \frac{1}{2} \int_0^{\ell_e} \rho_e J_e \left(\frac{\partial \phi_e (\xi, t)}{\partial t} \right)^2 d\xi \tag{4.47}$$

Comparing Equation (4.47) to Equation (4.38), the element mass matrix is, from Equation (4.12)

$$\mathbf{M}_e = \frac{\rho_e J_e \ell_e}{6} \begin{bmatrix} 2 & 1 \\ 1 & 2 \end{bmatrix} \tag{4.48}$$

For a bar with a circular section, $J_e \ell_e = I_{pe}$, where I_{pe} is the total polar moment of inertia of the bar element.

In simple torsional models, disks are modeled as discrete elements and are thus assumed to be rigid. Only a single generalized coordinate is required to specify the angular displacement of the disk. The disk merely adds a discrete inertia – equal to the polar moment of inertia of the disk – to the FE model.

The element mass and stiffness matrices for beam bending and other elements are derived in Section 5.4, although the principle is exactly the same. The strain and kinetic energies of the element are approximated using the displacement model. Through Lagrange's equations, these energy expressions may be used to estimate the element forces and, hence, the contributions of the element to the global matrices.

4.7 Assembling Global Matrices

In Sections 4.4 and 4.5, the mass and stiffness matrices for a system are built up from the element matrices for bar-extension and beam-bending problems. For these simple examples, we can relate the local element coordinates directly to the global generalized coordinates, and directly generate the equations of motion. The previous section derived general mass and stiffness matrices in terms of local element coordinates; these element matrices must be assembled to give the global matrices. The advantage of the energy approach to the element-matrix derivation is that the tools for the assembly process are already in place. The total kinetic energy for the entire structure or system is simply the arithmetic sum of the contributions to the kinetic energy from all the individual elements.

Conceptually, the simplest method for element assembly is to determine the transformation from local to global coordinates and total the transformed matrices. Thus, if $\mathbf{w}_e$ is the local coordinate vector for the eth element and $\mathbf{q}$ is the vector of global generalized coordinates, there exists a mapping $\mathbf{H}_e$ such that

$$\mathbf{w}_e = \mathbf{H}_e \mathbf{q} \tag{4.49}$$

In the three-element bar extension example in Section 4.4 with four degrees of freedom

$$\mathbf{w}_e = \begin{Bmatrix} w_{e1} \\ w_{e2} \end{Bmatrix} \qquad \mathbf{H}_1 = \begin{bmatrix} 1 & 0 & 0 & 0 \\ 0 & 1 & 0 & 0 \end{bmatrix}$$

$$\mathbf{H}_2 = \begin{bmatrix} 0 & 1 & 0 & 0 \\ 0 & 0 & 1 & 0 \end{bmatrix} \qquad \mathbf{H}_3 = \begin{bmatrix} 0 & 0 & 1 & 0 \\ 0 & 0 & 0 & 1 \end{bmatrix} \tag{4.50}$$

This transformation then may be used to write the element strain energy in terms of the global coordinates to derive the stiffness matrix and the element kinetic energy in terms of the global coordinates to derive the mass matrix. Here, we derive the mass matrix by writing the kinetic energy as the sum of element energies. Thus,

$$T = \sum_e T_e = \frac{1}{2} \sum_e \mathbf{w}_e^\top \mathbf{M}_e \mathbf{w}_e = \frac{1}{2} \mathbf{q}^\top \mathbf{M} \mathbf{q} \tag{4.51}$$

where

$$\mathbf{M} = \sum_e \mathbf{H}_e^\top \mathbf{M}_e \mathbf{H}_e \tag{4.52}$$

For the bar example in Section 4.4, the element is given by Equation (4.12) and repeated here for convenience:

$$\mathbf{M}_e = \frac{\rho_e A_e \ell_e}{6} \begin{bmatrix} 2 & 1 \\ 1 & 2 \end{bmatrix}$$

Applying Equation (4.52), the system mass matrix is

$$\mathbf{M} = \frac{\rho_1 A_1 \ell_1}{6} \begin{bmatrix} 2 & 1 & 0 & 0 \\ 1 & 2 & 0 & 0 \\ 0 & 0 & 0 & 0 \\ 0 & 0 & 0 & 0 \end{bmatrix}$$

$$+ \frac{\rho_2 A_2 \ell_2}{6} \begin{bmatrix} 0 & 0 & 0 & 0 \\ 0 & 2 & 1 & 0 \\ 0 & 1 & 2 & 0 \\ 0 & 0 & 0 & 0 \end{bmatrix} + \frac{\rho_3 A_3 \ell_3}{6} \begin{bmatrix} 0 & 0 & 0 & 0 \\ 0 & 0 & 0 & 0 \\ 0 & 0 & 2 & 1 \\ 0 & 0 & 1 & 2 \end{bmatrix} \quad (4.53)$$

For a uniform bar, this is identical to Equation (4.19).

If a discrete mass, of mass m, is added to the second node, a term

$$\begin{bmatrix} 0 & 0 & 0 & 0 \\ 0 & m & 0 & 0 \\ 0 & 0 & 0 & 0 \\ 0 & 0 & 0 & 0 \end{bmatrix} \quad (4.54)$$

is simply added to the mass matrix.

Of course, performing the multiplication by the transformation matrix required by Equation (4.52) is computationally prohibitive in practice. However, Equation (4.53) shows that in the bar extension example, element 1 affects only the first two global coordinates. Therefore, for each element, only small submatrices of the global mass and stiffness matrices (corresponding to these coordinates) are affected. Thus, the element matrices are simply inserted into the correct parts of the global matrices, as illustrated in Section 4.4. Equation (4.53) shows this effectively, and it is readily seen how this could be automated in a computer software package.

4.8 General Finite Element Models

In the previous sections, we study in detail one-dimensional bar and beam elements because many rotors can be modeled with adequate accuracy using a one-dimension element (i.e., shaft-line models). For greater accuracy, it is sometimes desirable to model rotors using three-dimensional elements or axisymmetric elements that can represent a three-dimensional axisymmetric structure using only a two-dimensional model. Models of rotors using axisymmetric elements are described in Chapter 10. Detailed FE models of foundations often require two- and three-dimensional elements. These models are beyond the scope of this book but, for completeness, we briefly discuss the methods used to develop two- and three-dimensional elements. Readers are referred to Fagan (1992), Petyt (1990), and Zienkiewicz et al. (2005) for fuller discussion.

The general concept of FEA is the same regardless of the dimensions of the elements: namely, that a relatively complex geometry can be decomposed into a number of simpler geometries. For each of the simpler geometries, a reasonably accurate relationship can be calculated between the deformations of that shape and the forces required to cause those deformations.

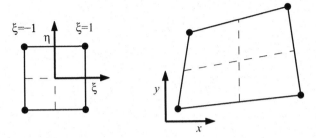

Figure 4.14. The natural coordinates of a four-node quadrilateral element.

A simple one-dimensional element can be varied in size (i.e., the length of a beam element can be varied), but it cannot be varied in shape. In contrast, it is desirable that two- and three-dimensional elements are capable of being varied in both size and shape so that they can represent parts of complex geometric shapes. Although some simple two-dimensional elements are restricted to triangular or rectangular shapes, elements can be developed that are capable of accommodating a wide range of shapes, including some with curved sides. The procedures for deriving the stiffness and mass matrices for such an element are outlined here. For simplicity, the outline is provided in terms of only in-plane loads and deformations of flat constant-thickness quadrilateral plate elements. Similar methods apply to almost all FE shapes.

To develop an element in two (or three) dimensions, we first derive a set of shape functions that cover the dimensions and satisfy the boundary conditions. These shape functions are conveniently expressed in terms of what are called *natural coordinates*. For example, in a square element, the natural coordinates are between $\xi = \pm 1$ in one direction and between $\eta = \pm 1$ in the perpendicular direction. This element is called a *parent* or *master element*. At first sight, this element is of limited value because it is a square element in the $O\eta\xi$ coordinates, with sides of length 2. We can map this four-node parent element onto any straight-sided quadrilateral shape, as shown in Figure 4.14. Similarly, we can map an eight-node square element onto a general quadrilateral shape, as shown in Figure 4.15.

The mapping between the parent element and the required shape can be expressed as

$$x = \Phi_x\left(\xi, \eta\right), \qquad y = \Phi_y\left(\xi, \eta\right) \tag{4.55}$$

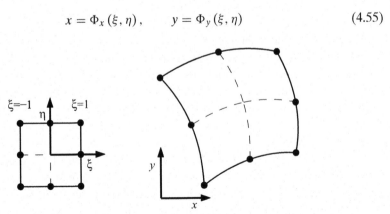

Figure 4.15. The natural coordinates of an eight-node quadrilateral element.

where the two functions, Φ_x and Φ_y, are continuous and differentiable with respect to the natural coordinates at all points (ξ, η) in the parent element. Also, for any point (ξ, η) in the parent element, there is only one corresponding point (x, y) in the actual element.

The deformation within the actual element also is expressed in terms of the natural coordinates using other functions that also must be continuous and differentiable. Thus, we can write

$$u = \Theta_x (\xi, \eta), \qquad v = \Theta_y (\xi, \eta) \tag{4.56}$$

where u and v are the deformations in the x and y directions.

Unfortunately, the functions Θ_x and Θ_y cannot be known in advance. The essence of the FEM is that these functions are approximated by linear combinations of known functions, called *shape functions*, that are chosen by the user. The coefficients in these linear combinations are simply the *nodal displacements*. The basic qualities that must be obeyed by the general ith shape function is that it has a value of unity at the ith node and a value of exactly zero at all other nodes. At points within the element other than the nodes, the shape function is generally non-zero, but it must be continuous and differentiable. The continuous displacement may be expressed mathematically as

$$\Theta_x (\xi, \eta) = \sum_i \beta_i (\xi, \eta) \, u_i = \begin{bmatrix} \beta_1 (\xi, \eta) & \beta_2 (\xi, \eta) & \ldots \end{bmatrix} \begin{Bmatrix} u_1 \\ u_2 \\ \vdots \end{Bmatrix} \tag{4.57}$$

$$= \mathbf{B} (\xi, \eta) \, \mathbf{u}$$

$$\Theta_y (\xi, \eta) = \sum_i \beta_i (\xi, \eta) \, v_i = \begin{bmatrix} \beta_1 (\xi, \eta) & \beta_2 (\xi, \eta) & \ldots \end{bmatrix} \begin{Bmatrix} v_1 \\ v_2 \\ \vdots \end{Bmatrix} \tag{4.58}$$

$$= \mathbf{B} (\xi, \eta) \, \mathbf{v}$$

where $\beta_i (\xi, \eta)$ is the ith shape function and u_i and v_i are the displacements of the ith node in the x and y directions. The same shape functions are used to define both Θ_x and Θ_y. Often, it is also convenient to use the same shape functions to define Φ_x and Φ_y; in such cases, the elements are described as being *isoparametric*.

Tables 4.1 through 4.4 show the nodal positions in natural coordinates and their associated shape functions for the most commonly used two-dimensional (in-plane) elements. The ordering of nodes in any element is completely arbitrary in one sense, but it is vital that corner nodes in the reference element correspond to corner nodes in each corresponding actual element and that midside nodes in the reference element correspond to midside nodes in each corresponding actual element. Figures 4.16 and 4.17 show example shape functions for the eight-node quadrilateral element.

The mass and stiffness matrices are obtained by approximating the kinetic energy and the potential (strain) energy within each actual element using the shape

Table 4.1. *Shape functions for a three-node triangle*

Node	ξ_i	η_i	Shape function, $\beta_i(\xi, \eta)$
1	0	0	$1 - (\xi + \eta)$
2	1	0	ξ
3	0	1	η

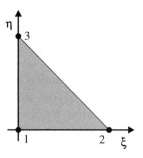

Table 4.2. *Shape functions for a four-node quadrilateral*

Node	ξ_i	η_i	Shape function, $\beta_i(\xi, \eta)$
1	-1	-1	$(1 - \xi)(1 - \eta)/4$
2	1	-1	$(1 + \xi)(1 - \eta)/4$
3	1	1	$(1 + \xi)(1 + \eta)/4$
4	-1	1	$(1 - \xi)(1 + \eta)/4$

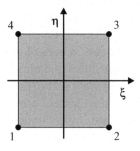

Table 4.3. *Shape functions for a six-node triangle*

Node	ξ_i	η_i	Shape function, $\beta_i(\xi, \eta)$
1	-1	-1	$(\xi + \eta)(\xi + \eta + 1)/2$
2	1	-1	$(1 + \xi) \times \xi/2$
3	-1	1	$(1 + \eta) \times \eta/2$
4	0	1	$-(1 + \xi)(\xi + \eta)$
5	0	0	$(1 + \xi)(1 + \eta)$
6	-1	0	$-(1 + \eta)(\xi + \eta)$

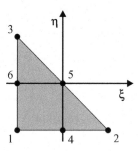

Table 4.4. *Shape functions for an eight-node quadrilateral*

Node	ξ_i	η_i	Shape function, $\beta_i(\xi, \eta)$
1	-1	-1	$-(1 - \xi)(1 - \eta)(1 + \xi + \eta)/4$
2	1	-1	$-(1 + \xi)(1 - \eta)(1 - \xi + \eta)/4$
3	1	1	$-(1 + \xi)(1 + \eta)(1 - \xi - \eta)/4$
4	-1	1	$-(1 - \xi)(1 + \eta)(1 + \xi - \eta)/4$
5	0	-1	$(1 - \xi)(1 + \xi)(1 - \eta)/2$
6	1	0	$(1 + \xi)(1 + \eta)(1 - \eta)/2$
7	0	1	$(1 - \xi)(1 + \xi)(1 + \eta)/2$
8	-1	0	$(1 - \xi)(1 + \eta)(1 - \eta)/2$

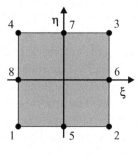

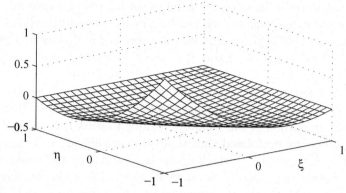

Figure 4.16. The shape function β_1 for the eight-node quadrilateral element.

functions. The kinetic energy of the element is

$$T_e = \frac{1}{2} \iint \rho h \left(\dot{u}^2 + \dot{v}^2 \right) \, dxdy = \frac{1}{2} \iint \rho h \begin{Bmatrix} \dot{u} \\ \dot{v} \end{Bmatrix}^\top \begin{Bmatrix} \dot{u} \\ \dot{v} \end{Bmatrix} \, dxdy \qquad (4.59)$$

where the double integral shows that the integration occurs over the complete area of the actual element, h is the element thickness, and ρ is the element density (both of which may vary within the element). In all practical FEA schemes, the integration is actually performed over the reference element because this has straight boundaries. Hence, we have

$$T_e = \frac{1}{2} \int\limits_{-1}^{1} \int\limits_{-1}^{1} \rho h \begin{Bmatrix} \dot{\Theta}_x \left(\xi, \eta \right) \\ \dot{\Theta}_y \left(\xi, \eta \right) \end{Bmatrix}^\top \begin{Bmatrix} \dot{\Theta}_x \left(\xi, \eta \right) \\ \dot{\Theta}_y \left(\xi, \eta \right) \end{Bmatrix} \det \left(\mathbf{J} \right) \, d\xi d\eta \qquad (4.60)$$

where the so-called *Jacobian matrix*, $\mathbf{J}$, is defined as

$$\mathbf{J} = \begin{bmatrix} \dfrac{\partial x}{\partial \xi} & \dfrac{\partial x}{\partial \eta} \\ \dfrac{\partial y}{\partial \xi} & \dfrac{\partial y}{\partial \eta} \end{bmatrix} \qquad (4.61)$$

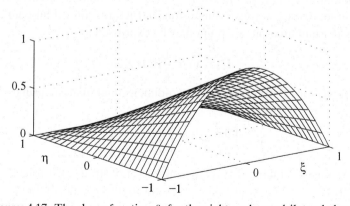

Figure 4.17. The shape function β_5 for the eight-node quadrilateral element.

The transformation between the natural and the physical coordinates should be such that the determinant of the Jacobian matrix is positive. The double integral in Equation (4.60) is identical to the one expressed in Equation (4.59) because $(\det(\mathbf{J})\, d\xi d\eta)$ is identical to $(dx dy)$. It is also clear that we can attempt to evaluate this integral – at least, approximately – by sampling the integrand (i.e., finding the value of the expression inside the integral sign) at an appropriate number of judiciously chosen points and finding a suitable weighted sum of these samples. A substantial proportion of the literature relating to FEA is concerned with how these integrations are performed numerically. In virtually all cases, the integration is achieved by sampling (i.e., evaluating) the integrand at a small number of *Gauss points*. Gauss points are specially chosen locations within a reference element where the smallest possible number of evaluations of the integrand results in the highest degree of accuracy. See Zienkiewicz et al. (2005) for a full discussion on the selection and use of Gauss points. The reason that so much attention is focused on numerical integration is that it often consumes the majority of all computer time associated with FEA.

The shape functions are independent of time, so that

$$\begin{Bmatrix} \dot{u} \\ \dot{v} \end{Bmatrix} = \begin{Bmatrix} \dot{\Theta}_x(\xi,\eta) \\ \dot{\Theta}_y(\xi,\eta) \end{Bmatrix} = \begin{Bmatrix} \mathbf{B}(\xi,\eta)\,\dot{\mathbf{u}} \\ \mathbf{B}(\xi,\eta)\,\dot{\mathbf{v}} \end{Bmatrix} = \begin{bmatrix} \mathbf{B}(\xi,\eta) & \mathbf{0} \\ \mathbf{0} & \mathbf{B}(\xi,\eta) \end{bmatrix} \begin{Bmatrix} \dot{\mathbf{u}} \\ \dot{\mathbf{v}} \end{Bmatrix} \tag{4.62}$$

Furthermore, the nodal displacements are independent of ξ and η, so that substituting Equation (4.62) into Equation (4.60) gives

$$T_e = \frac{1}{2} \begin{Bmatrix} \dot{\mathbf{u}} \\ \dot{\mathbf{v}} \end{Bmatrix}^\top \mathbf{M}_e \begin{Bmatrix} \dot{\mathbf{u}} \\ \dot{\mathbf{v}} \end{Bmatrix} \tag{4.63}$$

where the element mass matrix, $\mathbf{M}_e$, is

$$\mathbf{M}_e = \int_{-1}^{1} \int_{-1}^{1} \rho h \begin{bmatrix} \mathbf{B}^\top(\xi,\eta)\,\mathbf{B}(\xi,\eta) & \mathbf{0} \\ \mathbf{0} & \mathbf{B}^\top(\xi,\eta)\,\mathbf{B}(\xi,\eta) \end{bmatrix} \det(\mathbf{J})\, d\xi d\eta \tag{4.64}$$

We conclude the discussion of the element-mass matrix with the observation that this matrix is merged into the global (system) mass matrix in a fashion perfectly consistent with Section 4.7.

Similar methods may be applied to obtain the stiffness matrix by approximating the potential (strain) energy of an element. The instantaneous deformation determines the strain energy. For two-dimensional structures, the strain-energy density (i.e., energy per unit area) at any point may be expressed as

$$\hat{U}(x,y) = \frac{1}{2}\gamma^\top \mathbf{D}\gamma \tag{4.65}$$

where γ is a vector of strains that for two-dimensional analysis usually takes the form

$$\gamma = \begin{bmatrix} \dfrac{\partial u}{\partial x} & \dfrac{\partial v}{\partial y} & \left(\dfrac{\partial v}{\partial x} + \dfrac{\partial u}{\partial y} \right) \end{bmatrix}^\top \tag{4.66}$$

and where $\mathbf{D}$ is a 3×3 matrix that relates the vector of strains, γ, to a corresponding vector of stresses, σ. Different matrices $\mathbf{D}$ are applicable depending on whether the problem is plane stress (i.e., stresses normal to the surface of the plate element being zero) or plane strain (i.e., strains normal to the surface being constrained to zero). Plane-stress problems are, by far, the most common; for these problems, if the material is isotropic (i.e., has no particular directionality), then

$$\mathbf{D} = \frac{Eh}{(1 - v^2)} \begin{bmatrix} 1 & v & 0 \\ v & 1 & 0 \\ 0 & 0 & \frac{1}{2}(1 - v) \end{bmatrix} \tag{4.67}$$

To find the total strain energy in the actual element, we must integrate the strain-energy density over the area of the element to give

$$U_e = \frac{1}{2} \iint \boldsymbol{\gamma}^\top \mathbf{D} \boldsymbol{\gamma} \, dx dy \tag{4.68}$$

Again, we carry out this integration in the natural coordinates; thus,

$$U_e = \frac{1}{2} \int_{-1}^{1} \int_{-1}^{1} \boldsymbol{\gamma}^\top \mathbf{D} \boldsymbol{\gamma} \det(\mathbf{J}) \, d\xi d\eta \tag{4.69}$$

$\boldsymbol{\gamma}(\xi, \eta)$ is not immediately obtained from Equations (4.66) and (4.56) because the partial derivatives in Equation (4.66) involve x and y as independent variables rather than ξ and η. However, the strain vector, $\boldsymbol{\gamma}(\xi, \eta)$, can be found by observing that

$$\begin{bmatrix} \dfrac{\partial u}{\partial \xi} & \dfrac{\partial u}{\partial \eta} \end{bmatrix} = \begin{bmatrix} \dfrac{\partial u}{\partial x} & \dfrac{\partial u}{\partial y} \end{bmatrix} \begin{bmatrix} \dfrac{\partial x}{\partial \xi} & \dfrac{\partial x}{\partial \eta} \\ \dfrac{\partial y}{\partial \xi} & \dfrac{\partial y}{\partial \eta} \end{bmatrix} = \begin{bmatrix} \dfrac{\partial u}{\partial x} & \dfrac{\partial u}{\partial y} \end{bmatrix} \mathbf{J} \tag{4.70}$$

or

$$\begin{bmatrix} \dfrac{\partial u}{\partial x} & \dfrac{\partial u}{\partial y} \end{bmatrix} = \begin{bmatrix} \dfrac{\partial u}{\partial \xi} & \dfrac{\partial u}{\partial \eta} \end{bmatrix} \mathbf{J}^{-1} \tag{4.71}$$

A similar expression may be obtained for $\begin{bmatrix} \dfrac{\partial v}{\partial x} & \dfrac{\partial v}{\partial y} \end{bmatrix}$.

Thus, to obtain the element stiffness matrix, we need to calculate $\dfrac{\partial u}{\partial \xi}, \dfrac{\partial u}{\partial \eta}, \dfrac{\partial v}{\partial \xi}$, and $\dfrac{\partial v}{\partial \eta}$. The shape functions must be differentiable at every point within the element; therefore, we may differentiate both sides of Equations (4.57) and (4.58) to obtain

$$\frac{\partial u}{\partial \xi} = \begin{bmatrix} \dfrac{\partial \beta_1}{\partial \xi} & \dfrac{\partial \beta_2}{\partial \xi} & \dfrac{\partial \beta_3}{\partial \xi} & \cdots \end{bmatrix} \begin{Bmatrix} u_1 \\ u_2 \\ u_3 \\ \vdots \end{Bmatrix} = \mathbf{B}_\xi(\xi, \eta) \, \mathbf{u} \tag{4.72}$$

$$\frac{\partial u}{\partial \eta} = \begin{bmatrix} \dfrac{\partial \beta_1}{\partial \eta} & \dfrac{\partial \beta_2}{\partial \eta} & \dfrac{\partial \beta_3}{\partial \eta} & \cdots \end{bmatrix} \begin{Bmatrix} u_1 \\ u_2 \\ u_3 \\ \vdots \end{Bmatrix} = \mathbf{B}_\eta(\xi, \eta) \, \mathbf{u} \tag{4.73}$$

Similar expressions apply for $\dfrac{\partial v}{\partial \xi}$ and $\dfrac{\partial v}{\partial \eta}$. We provide these expressions here to emphasize that the same vector functions of ξ and η are employed in Equations (4.72) and (4.73). Thus,

$$\frac{\partial v}{\partial \xi} = \begin{bmatrix} \dfrac{\partial \beta_1}{\partial \xi} & \dfrac{\partial \beta_2}{\partial \xi} & \dfrac{\partial \beta_3}{\partial \xi} & \cdots \end{bmatrix} \begin{Bmatrix} v_1 \\ v_2 \\ v_3 \\ \vdots \end{Bmatrix} = \mathbf{B}_\xi (\xi, \eta) \, \mathbf{v} \qquad (4.74)$$

$$\frac{\partial v}{\partial \eta} = \begin{bmatrix} \dfrac{\partial \beta_1}{\partial \eta} & \dfrac{\partial \beta_2}{\partial \eta} & \dfrac{\partial \beta_3}{\partial \eta} & \cdots \end{bmatrix} \begin{Bmatrix} v_1 \\ v_2 \\ v_3 \\ \vdots \end{Bmatrix} = \mathbf{B}_\eta (\xi, \eta) \, \mathbf{v} \qquad (4.75)$$

The total potential (strain) energy of the element is then

$$U_e = \frac{1}{2} \begin{Bmatrix} \mathbf{u} \\ \mathbf{v} \end{Bmatrix}^\top \mathbf{K}_e \begin{Bmatrix} \mathbf{u} \\ \mathbf{v} \end{Bmatrix} \qquad (4.76)$$

where the element stiffness matrix, $\mathbf{K}_e$, is

$$\mathbf{K}_e = \int\limits_{-1}^{1} \int\limits_{-1}^{1} \mathbf{G} (\xi, \eta) \det (\mathbf{J}) \, \mathrm{d}\xi \mathrm{d}\eta \qquad (4.77)$$

where $\mathbf{G} (\xi, \eta)$ is a matrix that determines the local strain-energy density at any point (corresponding to point (ξ, η) in the reference element) given the nodal deflection vectors, $\mathbf{u}$ and $\mathbf{v}$. $\mathbf{G} (\xi, \eta)$ is derived numerically using Equations (4.72) through (4.75) to determine deflection gradients (with respect to the reference coordinates) in terms of $\mathbf{u}$ and $\mathbf{v}$, and then using Equation (4.70) with Equation (4.61) to convert these into deflection gradients with respect to the global coordinates, and finally using Equations (4.65) and (4.67) to express the strain-energy density at any one point in terms of $\mathbf{u}$ and $\mathbf{v}$.

The procedure by which the element-mass and stiffness matrices are obtained is outlined herein. Section 4.7 describes the method to assemble the complete system mass and stiffness matrices from element matrices. Exactly the same assembly procedure applies here. If the forcing applied to the model is known, then the (sometimes large) set of simultaneous equations may be solved to determine the deflections at every node within the model. The nodal deflections for each element then may be extracted in turn, which enables the computation of the stress and strain at any point within the element. In a continuum, all stresses should be continuous. In an FE model, the calculated stresses are continuous within each element but discontinuities occur at the element boundaries. When the object being analyzed has been discretized into sufficiently small elements, the discontinuities are small and most computer software allows the user to *smooth out* these discontinuities.

Extending these developments to three dimensions is conceptually straightforward. In three-dimensional models, each point has three translational degrees of freedom. There are three mapping functions – Φ_x, Φ_y, and Φ_z – relating the positions of points in the reference element to positions of the corresponding points in the actual element; each function is clearly a function of three variables.

The stress and strain vectors each have six components and the matrix, **D**, which describes the constitutive relationships for the element material, is 6×6. All integrations are triple integrals. The variety of elements available is increased greatly. The most common three-dimensional elements are four- and eight-node tetrahedra, six- and fifteen-node wedges, and eight- and twenty-node brick elements.

Three-dimensional models tend to employ vast numbers of degrees of freedom and the associated computation times can be extremely long. Model reduction is discussed in Chapter 2 as a means of reducing the dimension of large problems, and this is certainly useful for three-dimensional FE models. However, the model-reduction methods described in Chapter 2 do not avoid the expenditure of significant computational resources to derive the element mass and stiffness matrices for every single element in the model.

To make it feasible to run meaningful FE models of many structures and machines of interest, other classes of elements are available to the user. These include beam elements, plate/shell elements, axisymmetric elements (see Chapter 9), and others such as the two-dimensional in-plane elements mentioned in this section. In all cases, the bodies being examined are actually three-dimensional, but some additional assumptions are made to enable the deflections of the entire body to be expressed in terms of a smaller number of nodal deflections. In the case of the two-dimensional in-plane elements, for example, no mention is made of displacements normal to the plane. However, the implicit assumption is that these deflections can and do occur such that the stress normal to the plane is zero. In the case of shell elements, it is assumed that the deflections of any point within the shell can be determined from the nodal translations and rotations given (typically) the assumptions that normal stresses are considered to be zero, shear stresses in planes normal to the surface are also zero, and that displacements vary linearly through the thickness of the shell.

A simple and yet absolutely accurate way to regard the FEA using other than the three-dimensional elements is to consider that model reduction has been done analytically on a three-dimensional element to reduce the number of independent nodal deflections required to describe its deformed state and to reduce the computational burden of numerical integration.

4.9 Summary

In this chapter, it is shown that the FEM provides a rational way of discretizing a continuous system. We develop the method from Newton's laws of motion, beginning with discrete components. This force-equilibrium approach is satisfactory for simple problems; however, for the general development of elements and models, an energy approach is required. The energy method is used to develop elements specific to rotating machines in the following chapter. An attraction of the FEM is that mass and stiffness matrices for a specific type of element must be formulated only once.

4.10 Problems

4.1 Write the equations of motion for the axial motion of a uniform free-free bar using two elements of equal length. Estimate the first non-zero natural

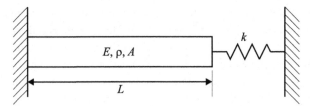

Figure 4.18. The bar clamped at one end with a grounded spring attached to the other end (Problem 4.3).

frequency of this system. Suppose that one end of the bar is now fixed; apply this constraint to the FE model and estimate the first natural frequency. An axial force $f(t)$ is applied to the free end of the bar; write the equations of motion of the bar including this force.

4.2 A uniform bar is clamped at both ends. Model the axial vibration using two, three, and four elements, and estimate the first natural frequency in each case. Compare these estimates to the first exact natural frequency of $\pi \sqrt{\dfrac{E}{\rho L^2}}$, where E is the Young's modulus, ρ is the density of the bar material, and L is the bar length.

4.3 Figure 4.18 shows a bar clamped at one end with a grounded spring of stiffness, k, attached to the other end. What are the equations of motion of the system, assuming the bar is modeled with two elements? If $k = \dfrac{EA}{2L}$, where E is the Young's modulus of the bar material and A is the cross-sectional area of the bar, estimate the first natural frequency of axial vibration.

4.4 Figure 4.19 shows a bar clamped at one end with a mass, m, attached at the other end. What are the equations of motion of the system, assuming the bar is modeled with two elements? If $m = 3\rho AL$, where ρ is the density of the bar material and A is the cross-sectional area of the bar, estimate the first natural frequency of axial vibration.

4.5 Consider the uniform bar of length L, clamped at one end and divided into two elements. If the element lengths are equal, as in Figure 4.20(a), estimate the first natural frequency of axial vibration. Now, split the bar into elements of length $L/4$ and $3L/4$, as in Figure 4.20(b), and estimate the first natural frequency. Compare these results to the exact first natural frequency and explain the differences.

4.6 Estimate the first natural frequency of a pinned-pinned beam in bending using one element. The pinned-boundary condition means that the translational deflection is zero, but the rotation is unconstrained. Compare this result to

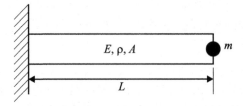

Figure 4.19. The bar clamped at one end with a discrete mass (Problem 4.4).

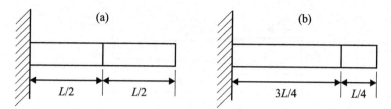

Figure 4.20. The bar clamped at one end discretized in two different ways (Problem 4.5).

the exact first natural frequency of $\dfrac{\pi^2}{L^2}\sqrt{\dfrac{EI}{\rho A}}$, where E and ρ are the material Young's modulus and density and I, A, and L are the second moment of area, cross-sectional area, and length of the beam.

4.7 Estimate the first natural frequency of a clamped-clamped beam in bending using two elements.

4.8 Estimate the first natural frequency of a clamped-pinned beam in bending using two elements.

4.9 Give the equations of motion of the three degrees of freedom system shown in Figure 4.21, where the beam in bending is modeled using a single element.

4.10 Suppose the cross section of a tapered bar varies linearly with the distance along the bar. Thus, for the element shown in Figure 4.22

$$A_e(\zeta) = A_{e1}(1 - \zeta/\ell_e) + A_{e2}\zeta/\ell_e$$

Calculate the element mass and stiffness matrices for such a tapered element using Equations (4.35) and (4.38), where now the cross-sectional area is not constant.

   The cross-sectional area of a 1m-long steel, clamped-free bar varies linearly from $0.2\,\mathrm{m}^2$ at the clamped end to $0.1\,\mathrm{m}^2$ at the free end. Model the beam using two tapered elements and estimate the first natural frequency of axial vibration. Suppose that the tapered bar is modeled using two uniform elements, as shown in Figure 4.23. Estimate the first natural frequency of the bar using this model and comment on the results.

4.11 Obtain the equations of motion for the axial motion of a uniform bar clamped at one end using three elements of equal length. Reduce the resulting model to a single degree of freedom using static reduction (see Section 2.5), retaining the displacement of the free end. Compare the estimated natural frequency from this reduced model to the estimated first natural frequency of the three

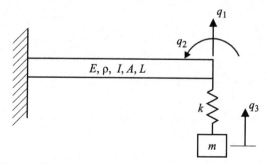

Figure 4.21. The cantilever beam with discrete spring-mass system (Problem 4.9).

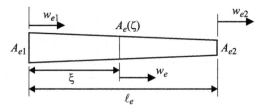

Figure 4.22. The tapered bar element (Problem 4.10).

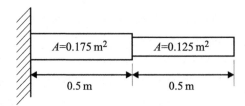

Figure 4.23. The tapered bar modeled using uniform elements (Problem 4.10).

degrees of freedom model and also to the natural frequency estimated from a model with a single element. Explain why, in this case, the result from the reduced model is exactly the same as that for the single element.

4.12 A uniform beam of length L is pinned at both ends so that its displacement is zero but the rotation is unconstrained. Calculate the first three natural frequencies using four and eight elements. Plot the convergence of these natural frequencies as the number of elements is increased further, and show that they converge to the exact natural frequencies given by $\omega_n = \dfrac{n^2\pi^2}{L^2}\sqrt{\dfrac{EI}{\rho A}}$ for mode n.

5 Free Lateral Response of Complex Systems

5.1 Introduction

In this chapter, we examine the characteristics of complex rotor–bearing systems in the absence of any applied forces. Chapter 3 shows how the characteristic dynamic behavior of simple rotors can be computed. As the rotor–bearing system becomes more complicated, the corresponding system model also becomes more complicated. The normal approach to handling this increased complexity is to discretize the system in a systematic way so that an approximate model can be created with a finite (but possibly large) number of coordinates. In the past, the transfer matrix method (or its predecessors: the Holtzer method for torsional analysis and the Myklestad-Prohl method for lateral analysis) were used for this purpose. With the increase in computing power in the last three decades, the FEM has become the de facto standard method for the static and dynamic analysis of structures as well as for the analysis of rotor–bearing systems. In this book, we choose to concentrate exclusively on the application of the FEM to model rotor–bearing systems. Example applications of the FEM to rotating machinery may be found in the works of Lalanne and Ferraris (1999), Nelson (1980), and Nelson and McVaugh (1976).

The problem of creating an adequate model of a rotating machine is considered and the model is then analyzed to determine its dynamic characteristics. By "dynamic characteristics," we mean the natural frequencies, the corresponding mode shapes, and the free response of the system. In Chapters 6, 7, and 8, we examine the effect of lateral forces acting on the rotor.

5.2 Coordinate Systems

It is shown in Chapter 4 that when applying the FEM to a structure, it is necessary first to break down the structure conceptually into relatively simple elements. The mathematical representations of these elements are then assembled to obtain a set of equations that model the system to an acceptable accuracy. In a shaft-line model of a rotor–bearing system, the rotor is divided into a number of shaft elements with nodes at both ends of each element, as shown in Figure 5.1. Each shaft element has two nodes, and disks and bearings are assumed to be attached to the shaft at these nodes where required. Usually, only degrees of freedom on the shaft of the rotor

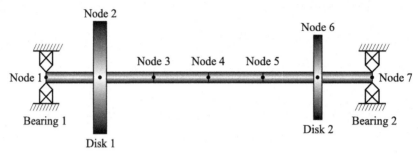

Figure 5.1. Side elevation of a typical rotor with the nodes indicated by solid dots.

are considered. Here, we consider only lateral or transverse vibrations, so each node has four generalized coordinates: transverse displacements in the x and y directions and rotations about the x- and y-axes. These are identical to the definitions given in Figure 3.2. The rotations are defined such that θ is a positive rotation about the x-axis and ψ is a positive rotation about the y-axis. The translations of the shaft from the equilibrium position are u and v in the x and y directions, respectively. We assume that these axes are fixed in space because doing so leads to simpler equations if the rotor is symmetric. We also assume that the rotations θ and ψ are small.

The equilibrium position of a rotor with a flexible shaft requires further discussion. Consider a rotor whose axis of rotation is vertical. If no dynamic forces are acting on it, then the rotor is in equilibrium due to static forces. In this equilibrium position, the rotor lies on a straight line passing through the bearing centers. If we now place the same rotor horizontally, due to the effect of gravity, the rotor then sags between the bearings and its equilibrium position lies along a curve between the bearing centers. For short and stiff rotors, the difference between a straight line through the bearings and the deflected shape is negligible. In contrast, long and flexible rotors deflect considerably due to gravity. The situation is further complicated if the rotor is supported by more than two bearings (see Section 10.11). Unless stated otherwise, the equations of motion are derived based on small displacements and rotations of the rotor from this equilibrium position.

5.3 Disk Elements

In Chapter 3, we determine the behavior of a rigid rotor and of a disk on a flexible rotor using Newton's laws of motion. In this section we develop the properties of a rigid disk from an energy viewpoint. Because the disk is assumed to be rigid, we can ignore the strain energy and only calculate its kinetic energy. The kinetic energy of the disk with respect to axes fixed in the disk is relatively easy to derive. Discounting any axial translation, the kinetic energy due to the translation of the disk is

$$\frac{1}{2}\,(\text{disk mass})\,(\text{linear velocity})^2 = \frac{1}{2}m_d\left(\dot{u}^2 + \dot{v}^2\right) \tag{5.1}$$

where m_d is the mass of the disk and $\dot{u}$ and $\dot{v}$ are the velocities in the x and y directions, respectively.

The kinetic energy due to the rotation of the disk is more difficult to calculate because the inertia properties are not identical in all directions. Here, we assume that the disk is symmetric so that the inertia properties may be calculated using the polar moment of inertia, I_p, about the shaft and the diametral moment of inertia, I_d, about any axis perpendicular to the shaft line. The kinetic energy due to the rotational motion of the disk is then

$$\frac{1}{2} I_d \left(\omega_{\tilde{x}}^2 + \omega_{\tilde{y}}^2 \right) + \frac{1}{2} I_p \omega_{\tilde{z}}^2 \tag{5.2}$$

where $\omega_{\tilde{x}}$, $\omega_{\tilde{y}}$, and $\omega_{\tilde{z}}$ are the instantaneous angular velocities about the $\tilde{x}$-, $\tilde{y}$-, and $\tilde{z}$-axes, which are fixed in the disk and rotate with it.

To use Equation (5.2), we need the angular velocity in axes fixed in the disk. We assume that the instantaneous angular velocity of the disk, with respect to axes fixed in space, is $\dot{\theta}$ about the x-axis and $\dot{\psi}$ about the y-axis. We must choose an order to apply the rotations to obtain the kinetic energy, although this choice is irrelevant once the equations of motions are generated. Suppose that the rotations are applied in the following order: ψ about y-axis, θ about the new x-axis, and then ϕ about the new z-axis (thus, ϕ is the angle of rotation about the shaft). The kinetic energy of rotation at this stage cannot be written directly in terms of $\dot{\theta}$ and $\dot{\psi}$ because the moments of inertia about the instantaneous axes are unknown. To make a valid expression for the kinetic-energy term, the rotations must be related to coordinate axes rotating with the disk; in order to do this, we use the transformation matrices (see Chapter 7).

Consider the angular velocity of the disk measured in a frame of reference that rotates with the disk. Making use of transformation matrices, this angular velocity is

$$\begin{Bmatrix} \omega_{\tilde{x}} \\ \omega_{\tilde{y}} \\ \omega_{\tilde{z}} \end{Bmatrix} = \begin{Bmatrix} 0 \\ 0 \\ \Omega \end{Bmatrix} + \begin{bmatrix} \cos\phi & \sin\phi & 0 \\ -\sin\phi & \cos\phi & 0 \\ 0 & 0 & 1 \end{bmatrix} \begin{Bmatrix} \dot{\theta} \\ 0 \\ 0 \end{Bmatrix}$$

$$+ \begin{bmatrix} \cos\phi & \sin\phi & 0 \\ -\sin\phi & \cos\phi & 0 \\ 0 & 0 & 1 \end{bmatrix} \begin{bmatrix} 1 & 0 & 0 \\ 0 & \cos\theta & \sin\theta \\ 0 & -\sin\theta & \cos\theta \end{bmatrix} \begin{Bmatrix} 0 \\ \dot{\psi} \\ 0 \end{Bmatrix} \tag{5.3}$$

Although Equation (5.3) appears complicated, the origin of each term is easy to describe. The instantaneous angular velocity about the z-axis is $\Omega = \dot{\phi}$, where Ω is the disk rotational speed (assumed constant). Because the rotation ϕ is applied last, this angular velocity is given directly in axes fixed in the disk (given as the first term on the right side of Equation (5.3)). The instantaneous angular velocity about the x-axis is $\dot{\theta}$, but this must be rotated about the z-axis by the angle ϕ. This is the second term on the right side of Equation (5.3), where the angular velocity is clearly in the local x-axis position of the vector. The matrix is the standard form of a rotation about the z-axis because the transformation does not affect the z direction. The last term transforms the instantaneous angular velocity about the y-axis $\dot{\psi}$, first by a rotation θ about the local x-axis and then by a rotation ϕ about the z-axis. Multiplying out Equation (5.3) produces

$$\begin{Bmatrix} \omega_{\tilde{x}} \\ \omega_{\tilde{y}} \\ \omega_{\tilde{z}} \end{Bmatrix} = \begin{Bmatrix} \dot{\theta}\cos\phi + \dot{\psi}\sin\phi\cos\theta \\ -\dot{\theta}\sin\phi + \dot{\psi}\cos\phi\cos\theta \\ \Omega - \dot{\psi}\sin\theta \end{Bmatrix} \tag{5.4}$$

The total kinetic energy is then

$$T_d = \frac{1}{2}m_d\left(\dot{u}^2 + \dot{v}^2\right) + \frac{1}{2}I_d\left(\omega_{\tilde{x}}^2 + \omega_{\tilde{y}}^2\right) + \frac{1}{2}I_p\omega_{\tilde{z}}^2$$

$$= \frac{1}{2}m_d\left(\dot{u}^2 + \dot{v}^2\right) + \frac{1}{2}I_d(\dot{\theta}^2 + \dot{\psi}^2\cos^2\theta) \tag{5.5}$$

$$+ \frac{1}{2}I_p\left(\Omega^2 - 2\Omega\dot{\psi}\sin\theta + \dot{\psi}^2\sin^2\theta\right)$$

Assuming the rotations θ and ψ are small, we can neglect terms higher than second order and their derivatives. Thus,

$$T_d = \frac{1}{2}m_d\left(\dot{u}^2 + \dot{v}^2\right) + \frac{1}{2}I_d\left(\dot{\theta}^2 + \dot{\psi}^2\right) + \frac{1}{2}I_p\left(\Omega^2 - 2\Omega\dot{\psi}\theta\right) \tag{5.6}$$

The last term arises from the gyroscopic effects of the disk. The reason that $\dot{\psi}$ appears but not $\dot{\theta}$ is because of the order in which the rotations are applied.

The element matrices are obtained by applying Lagrange's equations to Equation (5.6). If the local coordinates are arranged in the vector $[u, v, \theta, \psi]^\mathsf{T}$, then the inertia terms from Lagrange's equations are

$$\begin{Bmatrix} \dfrac{\mathrm{d}}{\mathrm{d}t}\left(\dfrac{\partial T_d}{\partial \dot{u}}\right) - \dfrac{\partial T_d}{\partial u} \\ \vdots \\ \dfrac{\mathrm{d}}{\mathrm{d}t}\left(\dfrac{\partial T_d}{\partial \dot{\psi}}\right) - \dfrac{\partial T_d}{\partial \psi} \end{Bmatrix} = \begin{bmatrix} m_d & 0 & 0 & 0 \\ 0 & m_d & 0 & 0 \\ 0 & 0 & I_d & 0 \\ 0 & 0 & 0 & I_d \end{bmatrix}\begin{Bmatrix} \ddot{u} \\ \ddot{v} \\ \ddot{\theta} \\ \ddot{\psi} \end{Bmatrix} + \Omega\begin{bmatrix} 0 & 0 & 0 & 0 \\ 0 & 0 & 0 & 0 \\ 0 & 0 & 0 & I_p \\ 0 & 0 & -I_p & 0 \end{bmatrix}\begin{Bmatrix} \dot{u} \\ \dot{v} \\ \dot{\theta} \\ \dot{\psi} \end{Bmatrix} \tag{5.7}$$

Thus, we have the element mass matrix for the disk

$$\mathbf{M}_e = \begin{bmatrix} m_d & 0 & 0 & 0 \\ 0 & m_d & 0 & 0 \\ 0 & 0 & I_d & 0 \\ 0 & 0 & 0 & I_d \end{bmatrix} \tag{5.8}$$

and the element gyroscopic matrix for the disk

$$\mathbf{G}_e = \begin{bmatrix} 0 & 0 & 0 & 0 \\ 0 & 0 & 0 & 0 \\ 0 & 0 & 0 & I_p \\ 0 & 0 & -I_p & 0 \end{bmatrix} \tag{5.9}$$

The same matrices are obtained if the rotations θ and ϕ are applied in the opposite order. Equation (5.9) gives the same gyroscopic matrix as derived in Section 3.3 using angular-momentum arguments.

5.4 Shaft Elements

The shaft contributes both mass and stiffness to the overall rotor model. It also may contribute gyroscopic effects and, although these are generally small, they are included at the end of this section for completeness. For a symmetric shaft, if the gyroscopic effects are neglected, the two bending planes are often uncoupled so that forces and moments in one plane cause displacements and rotations only in the same plane. Thus, the element mass and stiffness matrices for beam bending may be derived in a single plane, which are then used to generate the element matrix

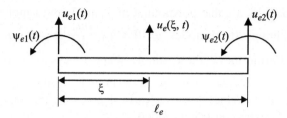

Figure 5.2. The local coordinates for the beam element.

for a flexible shaft that can bend in two planes. However, care must be taken with the signs of the rotations, as is demonstrated in Section 5.4.4. Other effects, such as axial torque, helical-shaft properties, and certain rotor–stator interactions, also can couple the two planes.

Every rotor has some degree of internal damping, which also leads to coupling between the planes. To model this requires the development of transformations between stationary and rotating frames; for this reason, we discuss internal damping of rotors in Chapter 7.

In discussing the bending of shafts, we frequently refer to the *neutral axis*, which is a line that experiences no longitudinal strain when the shaft bends. If the bending occurs only in one plane, then a surface containing the neutral axis and perpendicular to the bending plane experiences no longitudinal strain; it is called the *neutral plane*. We use the motion of the neutral axis to express the displacements of the entire shaft.

5.4.1 Euler-Bernoulli Beam Theory

We begin by considering the bending vibration in a single plane only. The element matrices are derived using energy methods, as described in Chapter 4 for bar elements. Beam-element matrices are obtained first by ignoring shear effects and rotary inertia, the so-called Euler-Bernoulli beam element. The process is to choose the degrees of freedom at the nodes of the element and approximate the displacement throughout the element in terms of these degrees of freedom using shape functions. Expressions for the kinetic and strain energy are used to compute the element matrices. The beam elements described here have two nodes per element and two degrees of freedom per node: the lateral displacement and slope at the node. Figure 5.2 shows these local coordinates. The deflection within the element is approximated by

$$u_e\left(\xi, t\right) = [N_{e1}\left(\xi\right) \ N_{e2}\left(\xi\right) \ N_{e3}\left(\xi\right) \ N_{e4}\left(\xi\right)] \begin{Bmatrix} u_{e1}(t) \\ \psi_{e1}(t) \\ u_{e2}(t) \\ \psi_{e2}(t) \end{Bmatrix} \tag{5.10}$$

where the shape functions, $N_{ei}\left(\xi\right)$, are

$$N_{e1}\left(\xi\right) = \left(1 - 3\frac{\xi^2}{\ell_e^2} + 2\frac{\xi^3}{\ell_e^3}\right), \quad N_{e2}\left(\xi\right) = \ell_e\left(\frac{\xi}{\ell_e} - 2\frac{\xi^2}{\ell_e^2} + \frac{\xi^3}{\ell_e^3}\right),$$

$$N_{e3}\left(\xi\right) = \left(3\frac{\xi^2}{\ell_e^2} - 2\frac{\xi^3}{\ell_e^3}\right), \qquad N_{e4}\left(\xi\right) = \ell_e\left(-\frac{\xi^2}{\ell_e^2} + \frac{\xi^3}{\ell_e^3}\right) \tag{5.11}$$

Equation (5.10) is a cubic polynomial in ξ and is the simplest function that satisfies the conditions at the nodes, $u_e(0) = u_{e1}$, $\dfrac{\partial u_e}{\partial \xi}(0) = \psi_{e1}$, $u_e(\ell_e) = u_{e2}$, and $\dfrac{\partial u_e}{\partial \xi}(\ell_e) = \psi_{e2}$. Furthermore, these simple shape functions reproduce the exact static deformation of a slender beam subject to point loads at its ends. More complex shape functions are possible; however, extra degrees of freedom for the element must be defined or extra constraints added.

The next requirement is to approximate the kinetic and strain energy within the beam element based on the lateral displacement, $u_e(\xi, t)$, of the beam neutral plane. The assumption is made that the cross sections remain planar and perpendicular to the beam centerline. The material is assumed to be linear elastic, obeying Hooke's law with Young's modulus E_e, which is assumed constant within the element. The strain energy, U_e, within the beam element (Inman, 2008) is

$$U_e = \frac{1}{2} \int_0^{\ell_e} E_e I_e(\xi) \left(\frac{\partial^2 u_e(\xi, t)}{\partial \xi^2} \right)^2 d\xi \qquad (5.12)$$

where I_e is the second moment of area of the cross section about the neutral plane. For a circular shaft of radius r, $I_e = \pi r^4/4$. For a rectangular beam, $I_e = bd^3/12$, where b is the width and d is the depth in the plane of bending.

It is possible to model tapered beams using the strain-energy expression Equation (5.12); this is discussed in Section 5.4.7. For now, assume that the cross section does not vary within the element. In this case, Equation (5.12) becomes

$$U_e = \frac{1}{2} E_e I_e \int_0^{\ell_e} \left(\frac{\partial^2 u_e(\xi, t)}{\partial \xi^2} \right)^2 d\xi \qquad (5.13)$$

This approximation to the strain energy may now be used – together with the approximation to the lateral displacement of the beam centerline given by Equation (5.10) – to obtain

$$U_e = \frac{1}{2} \begin{Bmatrix} u_{e1}(t) \\ \psi_{e1}(t) \\ u_{e2}(t) \\ \psi_{e2}(t) \end{Bmatrix}^{\mathsf{T}} \begin{bmatrix} k_{11} & k_{12} & k_{13} & k_{14} \\ k_{21} & k_{22} & k_{23} & k_{24} \\ k_{31} & k_{32} & k_{33} & k_{34} \\ k_{41} & k_{42} & k_{43} & k_{44} \end{bmatrix} \begin{Bmatrix} u_{e1}(t) \\ \psi_{e1}(t) \\ u_{e2}(t) \\ \psi_{e2}(t) \end{Bmatrix} \qquad (5.14)$$

where the elements of the stiffness matrix are

$$k_{ij} = E_e I_e \int_0^{\ell_e} N_{ei}''(\xi) N_{ej}''(\xi) d\xi \qquad (5.15)$$

and the prime represents differentiation with respect to ξ. The second derivatives of the shape functions are

$$N_{e1}''(\xi) = -\frac{6}{\ell_e^2} \left(1 - \frac{2\xi}{\ell_e} \right), \quad N_{e2}''(\xi) = \frac{2}{\ell_e} \left(-2 + 3\frac{\xi}{\ell_e} \right),$$

$$N_{e3}''(\xi) = \frac{6}{\ell_e^2} \left(1 - \frac{2\xi}{\ell_e} \right), \quad N_{e4}''(\xi) = \frac{2}{\ell_e} \left(-1 + \frac{3\xi}{\ell_e} \right) \qquad (5.16)$$

Obtaining explicit expressions for the k_{ij} is not difficult but is a little tedious, and only one example is developed here. Thus,

$$
\begin{aligned}
k_{12} &= E_e I_e \int_0^{\ell_e} N_{e1}''(\xi)\, N_{e2}''(\xi)\, \mathrm{d}\xi \\
&= E_e I_e \int_0^{\ell_e} -\frac{6}{\ell_e^2}\left(1 - \frac{2\xi}{\ell_e}\right)\frac{2}{\ell_e}\left(-2 + \frac{3\xi}{\ell_e}\right)\mathrm{d}\xi \\
&= \frac{12 E_e I_e}{\ell_e^3} \int_0^{\ell_e} \left(2 - 7\frac{\xi}{\ell_e} + 6\frac{\xi^2}{\ell_e^2}\right)\mathrm{d}\xi \\
&= \frac{12 E_e I_e}{\ell_e^3}\left[2\xi - \frac{7}{2}\frac{\xi^2}{\ell_e} + 2\frac{\xi^3}{\ell_e^2}\right]_0^{\ell_e} = \frac{12 E_e I_e}{\ell_e^2}\left[2 - \frac{7}{2} + 2\right] = \frac{6 E_e I_e}{\ell_e^2}
\end{aligned}
\tag{5.17}
$$

The integrand in Equation (5.15) is symmetric, thereby reducing the amount of computation because $k_{ij} = k_{ji}$. Calculating the other terms gives the element stiffness matrix as

$$
\mathbf{K}_e = \begin{bmatrix} k_{11} & k_{12} & k_{13} & k_{14} \\ k_{21} & k_{22} & k_{23} & k_{24} \\ k_{31} & k_{32} & k_{33} & k_{34} \\ k_{41} & k_{42} & k_{43} & k_{44} \end{bmatrix} = \frac{E_e I_e}{\ell_e^3}\begin{bmatrix} 12 & 6\ell_e & -12 & 6\ell_e \\ 6\ell_e & 4\ell_e^2 & -6\ell_e & 2\ell_e^2 \\ -12 & -6\ell_e & 12 & -6\ell_e \\ 6\ell_e & 2\ell_e^2 & -6\ell_e & 4\ell_e^2 \end{bmatrix}
\tag{5.18}
$$

The mass matrix is computed in a similar way but by using the kinetic energy. Neglecting the rotational effects, the kinetic energy of the beam is

$$
T_e = \frac{1}{2}\int_0^{\ell_e} \rho_e A_e(\xi)\,\dot{u}_e^2(\xi, t)\,\mathrm{d}\xi
\tag{5.19}
$$

where ρ_e is the density of the material, A_e is the cross-sectional area of the beam, and the dot denotes the derivative with respect to time. This assumes that, the entire cross section moves with the velocity of the beam neutral axis. Substituting the approximation to the lateral response, Equation (5.10), gives

$$
T_e = \frac{1}{2}\begin{Bmatrix} \dot{u}_{e1}(t) \\ \dot{\psi}_{e1}(t) \\ \dot{u}_{e2}(t) \\ \dot{\psi}_{e2}(t) \end{Bmatrix}^{\top}\begin{bmatrix} m_{11} & m_{12} & m_{13} & m_{14} \\ m_{21} & m_{22} & m_{23} & m_{24} \\ m_{31} & m_{32} & m_{33} & m_{34} \\ m_{41} & m_{42} & m_{43} & m_{44} \end{bmatrix}\begin{Bmatrix} \dot{u}_{e1}(t) \\ \dot{\psi}_{e1}(t) \\ \dot{u}_{e2}(t) \\ \dot{\psi}_{e2}(t) \end{Bmatrix}
\tag{5.20}
$$

where the elements of the mass matrix, for a uniform cross-sectional beam, are

$$
m_{ij} = \rho_e A_e \int_0^{\ell_e} N_{ei}(\xi)\, N_{ej}(\xi)\, \mathrm{d}\xi
\tag{5.21}
$$

As an example, the element m_{12} is computed. Thus,

$$
\begin{aligned}
m_{12} &= \rho_e A_e \int_0^{\ell_e} N_{e1}(\xi)\, N_{e2}(\xi)\, d\xi \\
&= \rho_e A_e \int_0^{\ell_e} \left(1 - \frac{3\xi^2}{\ell_e^2} + \frac{2\xi^3}{\ell_e^3}\right) \left(\xi - \frac{2\xi^2}{\ell_e} + \frac{\xi^3}{\ell_e^2}\right) d\xi \\
&= \rho_e A_e \int_0^{\ell_e} \left(\xi - \frac{2\xi^2}{\ell_e} - \frac{2\xi^3}{\ell_e^2} + \frac{8\xi^4}{\ell_e^3} - \frac{7\xi^5}{\ell_e^4} + \frac{2\xi^6}{\ell_e^5}\right) d\xi \qquad (5.22) \\
&= \rho_e A_e \left[\frac{\xi^2}{2} - \frac{2\xi^3}{3\ell_e} - \frac{\xi^4}{2\ell_e^2} + \frac{8\xi^5}{5\ell_e^3} - \frac{7\xi^6}{6\ell_e^4} + \frac{2\xi^7}{7\ell_e^5}\right]_0^{\ell_e} \\
&= \rho_e A_e \ell_e^2 \left[\frac{1}{2} - \frac{2}{3} - \frac{1}{2} + \frac{8}{5} - \frac{7}{6} + \frac{2}{7}\right] = \frac{11}{210}\rho_e A_e \ell_e^2
\end{aligned}
$$

Computing the other integrals gives the element mass matrix as

$$
\mathbf{M}_e =
\begin{bmatrix}
m_{11} & m_{12} & m_{13} & m_{14} \\
m_{21} & m_{22} & m_{23} & m_{24} \\
m_{31} & m_{32} & m_{33} & m_{34} \\
m_{41} & m_{42} & m_{43} & m_{44}
\end{bmatrix}
= \frac{\rho_e A_e \ell_e}{420}
\begin{bmatrix}
156 & 22\ell_e & 54 & -13\ell_e \\
22\ell_e & 4\ell_e^2 & 13\ell_e & -3\ell_e^2 \\
54 & 13\ell_e & 156 & -22\ell_e \\
-13\ell_e & -3\ell_e^2 & -22\ell_e & 4\ell_e^2
\end{bmatrix}
\qquad (5.23)
$$

Note that the mass matrix is symmetric.

5.4.2 Including Shear and Rotary Inertia Effects

The element mass and stiffness matrices for the Euler-Bernoulli beam are accurate approximations if the beam is slender. There are two important effects for beams and shafts that are relatively thick: shear effects and rotary inertia. Including rotary inertia produces a Rayleigh beam model; including both shear effects and rotary inertia produces a Timoshenko beam model (Astley, 1992; Davis et al., 1972; Inman, 2008; Petyt, 1990, Thomas et al., 1973). The effect of shear is to relax the assumption that the beam cross section always remains perpendicular to the beam centerline. However, Euler-Bernoulli, Rayleigh, and Timoshenko beam models all assume that plane sections remain plane. Figure 5.3 shows a small section of a beam and the effect of shear through an angle β_e, which is the difference between the plane of the beam cross section and the normal to the beam centerline. Thus, the angle of the beam cross section, ψ_e, is

$$
\psi_e(\xi, t) = \frac{\partial u_e(\xi, t)}{\partial \xi} + \beta_e(\xi, t) \qquad (5.24)
$$

Although extra degrees of freedom at the nodes may be included to account for the shear, the approach adopted here is to retain only two degrees of freedom per node. One possibility is to use the lateral displacement and the slope of the beam centerline as the degrees of freedom, as for the Euler-Bernoulli beam. However, a fixed-boundary condition requires the cross section to be fixed rather than the slope of the beam centerline; hence, the boundary conditions are difficult to apply. A similar problem arises with sudden changes in the cross section. Thus, the degrees of freedom used at the nodes are the lateral displacement, u_e, and angle of the beam

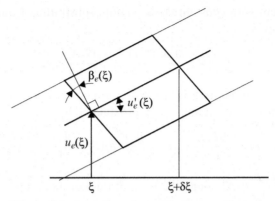

Figure 5.3. A small section of the beam with shear.

cross section, ψ_e; therefore, based on the nomenclature of Figure 5.2

$$\psi_{e1}(t) = \left.\frac{\partial u_e}{\partial \xi}\right|_{\xi=0} + \beta_e(0,t), \qquad \psi_{e2}(t) = \left.\frac{\partial u_e}{\partial \xi}\right|_{\xi=\ell_e} + \beta_e(\ell_e,t) \tag{5.25}$$

The shear angle, β_e, must now be related to the lateral displacement, u_e, which is assumed to be cubic and, thus,

$$u_e(\xi,t) = a_0(t) + a_1(t)\xi + a_2(t)\xi^2 + a_3(t)\xi^3 \tag{5.26}$$

for some constants a_i that depend on the nodal displacements. The relationship between shear angle and lateral displacement is obtained by considering the moment equilibrium of the beam. Neglecting inertia terms, the relationship (Inman, 2008) may be written as

$$\frac{\partial}{\partial \xi}\left(E_e I_e(\xi)\frac{\partial \psi_e(\xi,t)}{\partial \xi}\right) = \kappa_e G_e A_e \beta_e(\xi,t) \tag{5.27}$$

where κ_e is the shear constant that depends on the shape of the cross section of the beam and G_e is the shear modulus, with $G_e = E_e/2(1+\nu_e)$, where ν_e is Poisson's ratio. This shear constant compensates for the stiffening effect of the assumption that plane sections remain plane. Cowper (1966) gives values of the shear constant for typical cross sections and, for a hollow, circular shaft section, the shear constant is

$$\kappa_e = \frac{6(1+\nu_e)\left(1+\mu^2\right)^2}{(7+6\nu_e)\left(1+\mu^2\right)^2 + (20+12\nu_e)\mu^2} \tag{5.28}$$

where μ is the ratio of the inner shaft radius to the outer shaft radius, $\mu = r_i/r_o$. For a solid shaft, $\mu \to 0$ and $\kappa_e = 6(1+\nu_e)/(7+6\nu_e)$. Hutchinson (2001) gives shear constants that are slightly more accurate and, for a hollow circular shaft section

$$\kappa_e = \frac{6(1+\nu_e)^2\left(1+\mu^2\right)^2}{(7+12\nu_e+4\nu_e^2)\left(1+\mu^2\right)^2 + 4(5+6\nu_e+2\nu_e^2)\mu^2} \tag{5.29}$$

Hence, for a solid shaft, $\kappa_e = 6(1+\nu_e)^2/(7+12\nu_e+4\nu_e^2)$.

For an element with constant cross section, substituting Equation (5.24) into Equation (5.27) gives

$$\frac{\partial^2 \beta_e(\xi, t)}{\partial \xi^2} + \frac{\partial^3 u_e(\xi, t)}{\partial \xi^3} = \frac{\kappa_e G_e A_e}{E_e I_e} \beta_e(\xi, t) \tag{5.30}$$

The solution to Equation (5.30) for a uniform beam, when u_e is given by Equation (5.26), is a constant shear angle β_e approximated by

$$\beta_e(\xi, t) = \frac{6 E_e I_e}{\kappa_e G_e A_e} a_3(t) = \frac{\Phi_e \ell_e^2}{2} a_3(t) = \frac{\Phi_e \ell_e^2}{12} \frac{\partial^3 u_e(\xi, t)}{\partial \xi^3} \tag{5.31}$$

where $\Phi_e = \dfrac{12 E_e I_e}{\kappa_e G_e A_e \ell_e^2}$. Although the form of this constant appears unusual, it has the advantage that it is nondimensional.

Applying the lateral nodal conditions $u_e(0) = u_{e1}$ and $u_e(\ell_e) = u_{e2}$ gives

$$a_0 = u_{e1} \quad \text{and} \quad a_0 + a_1 \ell_e + a_2 \ell_e^2 + a_3 \ell_e^3 = u_{e2} \tag{5.32}$$

Similarly, applying the rotational nodal conditions, Equation (5.25) gives

$$a_1 + a_3 \frac{\Phi_e \ell_e^2}{2} = \psi_{e1} \quad \text{and} \quad a_1 + 2a_2 \ell_e + a_3 \left(3\ell_e^2 + \frac{\Phi_e \ell_e^2}{2} \right) = \psi_{e2} \tag{5.33}$$

Solving for a_i; substituting back into Equation (5.26); and grouping terms in u_{e1}, u_{e2}, ψ_{e1}, and ψ_{e2} gives

$$u_e(\xi, t) = [N_{e1}(\xi) \ N_{e2}(\xi) \ N_{e3}(\xi) \ N_{e4}(\xi)] \begin{Bmatrix} u_{e1}(t) \\ \psi_{e1}(t) \\ u_{e2}(t) \\ \psi_{e2}(t) \end{Bmatrix} \tag{5.34}$$

where

$$
\begin{aligned}
N_{e1}(\xi) &= \frac{1}{1 + \Phi_e} \left(1 + \Phi_e - \Phi_e \frac{\xi}{\ell_e} - 3\frac{\xi^2}{\ell_e^2} + 2\frac{\xi^3}{\ell_e^3} \right) \\
N_{e2}(\xi) &= \frac{\ell_e}{1 + \Phi_e} \left(\frac{2 + \Phi_e}{2} \frac{\xi}{\ell_e} - \frac{4 + \Phi_e}{2} \frac{\xi^2}{\ell_e^2} + \frac{\xi^3}{\ell_e^3} \right) \\
N_{e3}(\xi) &= \frac{1}{1 + \Phi_e} \left(\Phi_e \frac{\xi}{\ell_e} + 3\frac{\xi^2}{\ell_e^2} - 2\frac{\xi^3}{\ell_e^3} \right) \\
N_{e4}(\xi) &= \frac{\ell_e}{1 + \Phi_e} \left(-\frac{\Phi_e}{2} \frac{\xi}{\ell_e} - \frac{2 - \Phi_e}{2} \frac{\xi^2}{\ell_e^2} + \frac{\xi^3}{\ell_e^3} \right)
\end{aligned} \tag{5.35}
$$

If $\Phi_e = 0$, shear effects are neglected and Equation (5.35) is identical to Equation (5.11).

The strain energy for the element, including the shear, is

$$U_e = \frac{1}{2} \int_0^{\ell_e} E_e I_e(\xi) \left(\frac{\partial \psi_e(\xi, t)}{\partial \xi} \right)^2 d\xi + \frac{1}{2} \int_0^{\ell_e} \kappa_e^2 G_e A_e(\xi) \beta_e^2(\xi, t) d\xi \tag{5.36}$$

Note that for a uniform element cross section

$$\frac{\partial \psi_e(\xi, t)}{\partial \xi} = \frac{\partial^2 u_e(\xi, t)}{\partial \xi^2} + \frac{\partial \beta_e(\xi, t)}{\partial \xi} = \frac{\partial^2 u_e(\xi, t)}{\partial \xi^2} \tag{5.37}$$

because β_e is constant along the length of the element. Thus,

$$U_e = \frac{1}{2} \begin{Bmatrix} u_{e1}(t) \\ \psi_{e1}(t) \\ u_{e2}(t) \\ \psi_{e2}(t) \end{Bmatrix}^{\top} \begin{bmatrix} k_{11} & k_{12} & k_{13} & k_{14} \\ k_{21} & k_{22} & k_{23} & k_{24} \\ k_{31} & k_{32} & k_{33} & k_{34} \\ k_{41} & k_{42} & k_{43} & k_{44} \end{bmatrix} \begin{Bmatrix} u_{e1}(t) \\ \psi_{e1}(t) \\ u_{e2}(t) \\ \psi_{e2}(t) \end{Bmatrix} \tag{5.38}$$

where the elements of the stiffness matrix are

$$k_{ij} = E_e I_e \int_0^{\ell_e} N_{ei}''(\xi) N_{ej}''(\xi) \, d\xi + \frac{E_e I_e \Phi_e \ell_e^2}{12} \int_0^{\ell_e} N_{ei}'''(\xi) N_{ej}'''(\xi) \, d\xi \tag{5.39}$$

The definition of Φ_e is used to express the second integral in terms of Young's modulus rather than shear modulus. The calculations involved in Equation (5.39) are demonstrated on a typical matrix element. Thus,

$$\begin{aligned} k_{12} &= E_e I_e \int_0^{\ell_e} N_{e1}''(\xi) N_{e2}''(\xi) \, d\xi + \frac{E_e I_e \Phi_e \ell_e^2}{12} \int_0^{\ell_e} N_{e1}'''(\xi) N_{e2}'''(\xi) \, d\xi \\ &= E_e I_e \int_0^{\ell_e} \frac{6}{(1+\Phi_e)\ell_e^2} \left(-1 + \frac{2\xi}{\ell_e} \right) \frac{1}{(1+\Phi_e)\ell_e} \left(-4 - \Phi_e + \frac{6\xi}{\ell_e} \right) d\xi \\ &\quad + \frac{E_e I_e \Phi_e \ell_e^2}{12} \int_0^{\ell_e} \frac{12}{(1+\Phi_e)\ell_e^3} \frac{6}{(1+\Phi_e)\ell_e^2} \, d\xi \\ &= \frac{6 E_e I_e}{(1+\Phi_e)^2 \ell_e^3} \int_0^{\ell_e} \left(4 + \Phi_e - (14 + \Phi_e)\frac{\xi}{\ell_e} + 12\frac{\xi^2}{\ell_e^2} \right) d\xi + \frac{6 E_e I_e \Phi_e}{(1+\Phi_e)^2 \ell_e^3} \int_0^{\ell_e} d\xi \\ &= \frac{6 E_e I_e}{(1+\Phi_e)^2 \ell_e^3} \left[(4+\Phi_e)\xi - (7+\Phi_e)\frac{\xi^2}{\ell_e} + 4\frac{\xi^3}{\ell_e^2} \right]_0^{\ell_e} + \frac{6 E_e I_e \Phi_e}{(1+\Phi_e)^2 \ell_e^2} \tag{5.40} \\ &= \frac{6 E_e I_e}{(1+\Phi_e)^2 \ell_e^2} \left[(4+\Phi_e) - (7+\Phi_e) + 4 + \Phi_e \right] \\ &= \frac{6 E_e I_e}{(1+\Phi_e)^2 \ell_e^2} [1 + \Phi_e] = \frac{6 E_e I_e}{(1+\Phi_e)\ell_e^2} \end{aligned}$$

The result is surprisingly simple given the complexity of the intermediate integration. The calculation of the remaining matrix elements are left as an exercise for readers. The resulting stiffness matrix is

$$\mathbf{K}_e = \frac{E_e I_e}{(1+\Phi_e)\ell_e^3} \begin{bmatrix} 12 & 6\ell_e & -12 & 6\ell_e \\ 6\ell_e & \ell_e^2(4+\Phi_e) & -6\ell_e & \ell_e^2(2-\Phi_e) \\ -12 & -6\ell_e & 12 & -6\ell_e \\ 6\ell_e & \ell_e^2(2-\Phi_e) & -6\ell_e & \ell_e^2(4+\Phi_e) \end{bmatrix} \tag{5.41}$$

The kinetic energy is computed using the same shape functions – that is, including the shear effect. The effect of shear on the mass matrices is likely to be small and is often neglected, but we include it here for completeness. The expression for the kinetic energy is extended to account for rotary inertia. The kinetic energy for a shaft element is

$$T_e = \frac{1}{2} \int_0^{\ell_e} \left(\rho_e A_e \dot{u}_e^2 + \rho_e I_e \dot{\psi}_e^2 \right) d\xi$$

$$= \frac{1}{2} \int_0^{\ell_e} \rho_e A_e \dot{u}_e^2 + \rho_e I_e \left(\dot{\beta}_e + \frac{\partial \dot{u}_e}{\partial \xi} \right)^2 d\xi \tag{5.42}$$

The kinetic energy then may be written in terms of the mass matrix as

$$T_e = \frac{1}{2} \begin{Bmatrix} \dot{u}_{e1}(t) \\ \dot{\psi}_{e1}(t) \\ \dot{u}_{e2}(t) \\ \dot{\psi}_{e2}(t) \end{Bmatrix}^\top \begin{bmatrix} m_{11} & m_{12} & m_{13} & m_{14} \\ m_{21} & m_{22} & m_{23} & m_{24} \\ m_{31} & m_{32} & m_{33} & m_{34} \\ m_{41} & m_{42} & m_{43} & m_{44} \end{bmatrix} \begin{Bmatrix} \dot{u}_{e1}(t) \\ \dot{\psi}_{e1}(t) \\ \dot{u}_{e2}(t) \\ \dot{\psi}_{e2}(t) \end{Bmatrix} \tag{5.43}$$

where, for a uniform cross-sectional beam, the m_{ij} terms are

$$m_{ij} = \rho_e A_e \int_0^{\ell_e} N_{ei}(\xi) N_{ej}(\xi) \, d\xi$$

$$+ \rho_e I_e \int_0^{\ell_e} \left(\frac{\Phi_e \ell_e^2}{12} N_{ei}'''(\xi) + N_{ei}'(\xi) \right) \left(\frac{\Phi_e \ell_e^2}{12} N_{ej}'''(\xi) + N_{ej}'(\xi) \right) d\xi \tag{5.44}$$

Although the integration is tedious, the process is exactly as before. The result is the following symmetric mass matrix:

$$\mathbf{M}_e = \frac{\rho_e A_e \ell_e}{840 \left(1 + \Phi_e\right)^2} \begin{bmatrix} m_1 & m_2 & m_3 & m_4 \\ m_2 & m_5 & -m_4 & m_6 \\ m_3 & -m_4 & m_1 & -m_2 \\ m_4 & m_6 & -m_2 & m_5 \end{bmatrix}$$

$$+ \frac{\rho_e I_e}{30 \left(1 + \Phi_e\right)^2 \ell_e} \begin{bmatrix} m_7 & m_8 & -m_7 & m_8 \\ m_8 & m_9 & -m_8 & m_{10} \\ -m_7 & -m_8 & m_7 & -m_8 \\ m_8 & m_{10} & -m_8 & m_9 \end{bmatrix} \tag{5.45}$$

where

$$m_1 = 312 + 588\Phi_e + 280\Phi_e^2, \qquad m_6 = -\left(6 + 14\Phi_e + 7\Phi_e^2\right) \ell_e^2,$$

$$m_2 = \left(44 + 77\Phi_e + 35\Phi_e^2\right) \ell_e, \qquad m_7 = 36,$$

$$m_3 = 108 + 252\Phi_e + 140\Phi_e^2, \qquad m_8 = \left(3 - 15\Phi_e\right) \ell_e,$$

$$m_4 = -\left(26 + 63\Phi_e + 35\Phi_e^2\right) \ell_e, \qquad m_9 = \left(4 + 5\Phi_e + 10\Phi_e^2\right) \ell_e^2,$$

$$m_5 = \left(8 + 14\Phi_e + 7\Phi_e^2\right) \ell_e^2, \qquad m_{10} = \left(-1 - 5\Phi_e + 5\Phi_e^2\right) \ell_e^2$$

The second matrix represents the effect of rotary inertia. If shear and rotary inertia effects are ignored, Equation (5.45) gives the Euler-Bernoulli mass matrix.

5.4.3 The Effect of Axial Loading

Forces along the undeformed axis of the rotor often occur – for example, in a fan as a reaction to the air flow. The result is a stiffening effect if the shaft is in tension or a softening effect if the force is compressive (and the ultimate result is buckling). Suppose that a force acts along the shaft. We now consider the deformations in a single plane. The strain energy (Dawe, 1984) is

$$U_{Fe} = \int_0^{\ell_e} f_e(\xi) \frac{1}{2} \left(\frac{\partial u_e(\xi, t)}{\partial \xi} \right)^2 d\xi \tag{5.46}$$

where f_e is the axial tensile force within the element. If this force is constant within the element, then

$$U_{Fe} = \frac{f_e}{2} \int_0^{\ell_e} \left(\frac{\partial u_e(\xi, t)}{\partial \xi} \right)^2 d\xi \tag{5.47}$$

The displacement within the element is approximated by the shape functions, given by Equation (5.11) for the Euler-Bernoulli beam or by Equation (5.35) including shear effects. Here, we derive the additional stiffness matrix including shear effects. Setting $\Phi_e = 0$ neglects the shear effects and recovers the Euler-Bernoulli beam equations. Because the energy in Equation (5.47) is an additional scalar strain energy, the resulting stiffness matrix is simply added to that given in Equation (5.41). From Equations (5.34) and (5.47), the k_{Fij} term of this additional stiffness matrix is

$$k_{Fij} = f_e \int_0^{\ell_e} N'_{ei}(\xi) N'_{ej}(\xi) d\xi \tag{5.48}$$

Substituting for the shape functions and integrating as before gives the additional stiffness matrix due to the axial force as

$$\mathbf{K}_{Fe} = \frac{f_e}{60(1 + \Phi_e)^2 \ell_e} \begin{bmatrix} k_1 & 6\ell_e & -k_1 & 6\ell_e \\ 6\ell_e & k_2 & -6\ell_e & k_3 \\ -k_1 & -6\ell_e & k_1 & -6\ell_e \\ 6\ell_e & k_3 & -6\ell_e & k_2 \end{bmatrix} \tag{5.49}$$

where

$$k_1 = 72 + 120\Phi_e + 60\Phi_e^2, \quad k_3 = -\ell_e^2 \left(2 + 10\Phi_e + 5\Phi_e^2 \right),$$

$$k_2 = \ell_e^2 \left(8 + 10\Phi_e + 5\Phi_e^2 \right)$$

The corresponding stiffness matrix for a shaft element, neglecting shear effects, is given in Equation (5.52).

5.4.4 Mass and Stiffness Matrices for Shaft Elements

Based on the local coordinates described in Section 3.2, the coordinates defining bending in the two planes are shown in Figure 5.4. Comparing this to Figure 5.2 shows that the angles θ_1 and θ_2 for beam bending in the yz plane have the opposite sense to the angles ψ_1 and ψ_2, relative to the positive transverse translation and the positive z direction.

Thus, the element mass and stiffness matrices for the Euler-Bernoulli beam may be obtained directly from Equations (5.18) and (5.23), based on the local coordinate vector $[u_1, v_1, \theta_1, \psi_1, u_2, v_2, \theta_2, \psi_2]^\top$. The assumption is made that the two bending planes do not couple; therefore, the element matrices for the two planes are merely inserted into the correct location in the 8×8 shaft element matrices (noting the

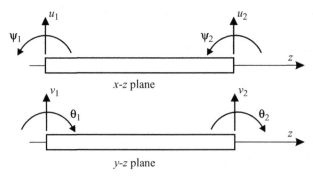

Figure 5.4. The local coordinates in the two bending planes.

change in sign for matrix elements involving θ_1 and θ_2 and the corresponding moments). If the shaft is symmetric so that the second moment of area in both directions is the same, then the mass and stiffness matrices are

$$\mathbf{M}_e = \frac{\rho_e A_e \ell_e}{420} \begin{bmatrix} 156 & 0 & 0 & 22\ell_e & 54 & 0 & 0 & -13\ell_e \\ 0 & 156 & -22\ell_e & 0 & 0 & 54 & 13\ell_e & 0 \\ 0 & -22\ell_e & 4\ell_e^2 & 0 & 0 & -13\ell_e & -3\ell_e^2 & 0 \\ 22\ell_e & 0 & 0 & 4\ell_e^2 & 13\ell_e & 0 & 0 & -3\ell_e^2 \\ 54 & 0 & 0 & 13\ell_e & 156 & 0 & 0 & -22\ell_e \\ 0 & 54 & -13\ell_e & 0 & 0 & 156 & 22\ell_e & 0 \\ 0 & 13\ell_e & -3\ell_e^2 & 0 & 0 & 22\ell_e & 4\ell_e^2 & 0 \\ -13\ell_e & 0 & 0 & -3\ell_e^2 & -22\ell_e & 0 & 0 & 4\ell_e^2 \end{bmatrix} \quad (5.50)$$

and

$$\mathbf{K}_e = \frac{E_e I_e}{\ell_e^3} \begin{bmatrix} 12 & 0 & 0 & 6\ell_e & -12 & 0 & 0 & 6\ell_e \\ 0 & 12 & -6\ell_e & 0 & 0 & -12 & -6\ell_e & 0 \\ 0 & -6\ell_e & 4\ell_e^2 & 0 & 0 & 6\ell_e & 2\ell_e^2 & 0 \\ 6\ell_e & 0 & 0 & 4\ell_e^2 & -6\ell_e & 0 & 0 & 2\ell_e^2 \\ -12 & 0 & 0 & -6\ell_e & 12 & 0 & 0 & -6\ell_e \\ 0 & -12 & 6\ell_e & 0 & 0 & 12 & 6\ell_e & 0 \\ 0 & -6\ell_e & 2\ell_e^2 & 0 & 0 & 6\ell_e & 4\ell_e^2 & 0 \\ 6\ell_e & 0 & 0 & 2\ell_e^2 & -6\ell_e & 0 & 0 & 4\ell_e^2 \end{bmatrix} \quad (5.51)$$

where the constants were defined previously. The change in stiffness matrix due to the application of an axial force is obtained in a similar way, from Equation (5.49), using the Euler-Bernoulli shape functions, as

$$\mathbf{K}_{Fe} = \frac{F_e}{30\ell_e} \begin{bmatrix} 36 & 0 & 0 & 3\ell_e & -36 & 0 & 0 & 3\ell_e \\ 0 & 36 & -3\ell_e & 0 & 0 & -36 & -3\ell_e & 0 \\ 0 & -3\ell_e & 4\ell_e^2 & 0 & 0 & 3\ell_e & -\ell_e^2 & 0 \\ 3\ell_e & 0 & 0 & 4\ell_e^2 & -3\ell_e & 0 & 0 & -\ell_e^2 \\ -36 & 0 & 0 & -3\ell_e & 36 & 0 & 0 & -3\ell_e \\ 0 & -36 & 3\ell_e & 0 & 0 & 36 & 3\ell_e & 0 \\ 0 & -3\ell_e & -\ell_e^2 & 0 & 0 & 3\ell_e & 4\ell_e^2 & 0 \\ 3\ell_e & 0 & 0 & -\ell_e^2 & -3\ell_e & 0 & 0 & 4\ell_e^2 \end{bmatrix} \quad (5.52)$$

The shaft-element matrices generated thus far have neglected shear and rotary inertia effects. Including these effects is straight forward and merely requires that the element matrices from Section 5.4.3 are used rather than the Euler-Bernoulli beam-element matrices. The positions of the terms and the negative signs generated by the change in sign of the θ angles are readily determined from this example.

5.4.5 Gyroscopic Effects

The shaft also produces gyroscopic effects, in much the same way as the rigid disks in Section 5.3. These effects are generally small unless the shaft has a large diameter. The gyroscopic effects arise from the kinetic energy, particularly the $\Omega\dot\psi\theta$ term in Equation (5.6) for the disk. For a thin disk of thickness $d\xi$ taken from the shaft, the polar moment of inertia, I_p, may be written in terms of the second moment of area of the shaft about the neutral plane, I_e, as

$$I_p = 2\rho_e I_e d\xi \tag{5.53}$$

The factor of 2 arises because for a very thin disk, the polar moment of inertia is twice the diametral moment of inertia, which is equal to the second moment of area multiplied by the disk mass per unit area. The increment in kinetic energy of this thin disk due to the rotation about a diameter is

$$dT = -I_p \Omega \dot\psi_e(\xi, t)\,\theta_e(\xi, t) = -2\rho_e I_e \Omega \dot\psi_e(\xi, t)\,\theta_e(\xi, t)\,d\xi \tag{5.54}$$

Integrating over the length of a uniform shaft element gives the contribution to the kinetic energy as

$$T_{Ge} = -2\rho_e I_e \Omega \int_0^{\ell_e} \dot\psi_e(\xi, t)\,\theta_e(\xi, t)\,d\xi \tag{5.55}$$

Notice that the gyroscopic effects couple the two planes. The shape functions now must be used to describe the rotations ψ_e and θ_e in terms of the local coordinates at the nodes. Care must be exercised concerning the signs of the θ terms. Shear is neglected initially to demonstrate the calculation of the element gyroscopic matrix. The rotations may be written in terms of the shape functions, by noting

$$\theta_e(\xi, t) = -\frac{dv_e}{d\xi}, \qquad \psi_e(\xi, t) = \frac{du_e}{d\xi} \tag{5.56}$$

and, thus,

$$\begin{Bmatrix} \theta_e(\xi, t) \\ \psi_e(\xi, t) \end{Bmatrix} = \begin{bmatrix} 0 & -N_1' & N_2' & 0 & 0 & -N_3' & N_4' & 0 \\ N_1' & 0 & 0 & N_2' & N_3' & 0 & 0 & N_4' \end{bmatrix} \mathbf{q}_e$$

$$= \begin{bmatrix} B_{11} & B_{12} & B_{13} & B_{14} & B_{15} & B_{16} & B_{17} & B_{18} \\ B_{21} & B_{22} & B_{23} & B_{24} & B_{25} & B_{26} & B_{27} & B_{28} \end{bmatrix} \mathbf{q}_e \tag{5.57}$$

where $\mathbf{q}_e = [u_1, v_1, \theta_1, \psi_1, u_2, v_2, \theta_2, \psi_2]^\top$ and the B terms are defined by Equation (5.57) and introduced as a notational convenience. The negative signs in the shape-function matrix arise from the orientation of the angles shown in Figure 5.4. Substituting Equation (5.57) into Equation (5.55) gives

$$T_{Ge} = \dot{\mathbf{q}}_e^\top \mathbf{A}\mathbf{q}_e \tag{5.58}$$

where

$$A_{ij} = -2\rho_e I_e \Omega \int_0^{\ell_e} B_{2i}(\xi) B_{1j}(\xi) \, d\xi \tag{5.59}$$

Then, from Lagrange's equations

$$\left\{ \begin{array}{c} \dfrac{d}{dt}\left(\dfrac{\partial T_{Ge}}{\partial \dot{q}_1}\right) - \dfrac{\partial T_{Ge}}{\partial q_1} \\ \vdots \\ \dfrac{d}{dt}\left(\dfrac{\partial T_{Ge}}{\partial \dot{q}_8}\right) - \dfrac{\partial T_{Ge}}{\partial q_8} \end{array} \right\} = \left[\mathbf{A} - \mathbf{A}^\top \right] \dot{\mathbf{q}} = \Omega \mathbf{G}_e \dot{\mathbf{q}} \tag{5.60}$$

where $\mathbf{G}_e$ is the element gyroscopic matrix. Clearly, from the definition given by Equation (5.60), this matrix is skew-symmetric. Equation (5.59) gives the elements of the gyroscopic matrix as

$$G_{ij} = -2\rho_e I_e \int_0^{\ell_e} \left(B_{2i}(\xi) B_{1j}(\xi) - B_{2j}(\xi) B_{1i}(\xi) \right) d\xi \tag{5.61}$$

Thus, the full matrix may be obtained by integrating Equation (5.61), a task that is facilitated by the large number of zero terms. For example,

$$G_{12} = -2\rho_e I_e \int_0^{\ell_e} \left(B_{21}(\xi) B_{12}(\xi) - B_{22}(\xi) B_{11}(\xi) \right) d\xi$$

$$= 2\rho_e I_e \int_0^{\ell_e} N_1'(\xi) N_1'(\xi) \, d\xi = \frac{12\rho_e I_e}{5\ell_e} \tag{5.62}$$

Calculating the remaining terms gives

$$\mathbf{G}_e = \frac{\rho_e I_e}{15\ell_e} \begin{bmatrix} 0 & 36 & -3\ell_e & 0 & 0 & -36 & -3\ell_e & 0 \\ -36 & 0 & 0 & -3\ell_e & 36 & 0 & 0 & -3\ell_e \\ 3\ell_e & 0 & 0 & 4\ell_e^2 & -3\ell_e & 0 & 0 & -\ell_e^2 \\ 0 & 3\ell_e & -4\ell_e^2 & 0 & 0 & -3\ell_e & \ell_e^2 & 0 \\ 0 & -36 & 3\ell_e & 0 & 0 & 36 & 3\ell_e & 0 \\ 36 & 0 & 0 & 3\ell_e & -36 & 0 & 0 & 3\ell_e \\ 3\ell_e & 0 & 0 & -\ell_e^2 & -3\ell_e & 0 & 0 & 4\ell_e^2 \\ 0 & 3\ell_e & \ell_e^2 & 0 & 0 & -3\ell_e & -4\ell_e^2 & 0 \end{bmatrix} \tag{5.63}$$

If shear were included, then the angles in Equation (5.55) would represent rotations of cross sections, given by Equation (5.24). The shape-functions would also include the shear effects, as shown in Equation (5.35). The resulting integrations are similar to those for the inclusion of rotary inertia, Equation (5.44), and result in the following element gyroscopic matrix:

$$\mathbf{G}_e = \frac{\rho_e I_e}{15\left(1+\Phi_e\right)^2 \ell_e} \begin{bmatrix} 0 & g_1 & -g_2 & 0 & 0 & -g_1 & -g_2 & 0 \\ -g_1 & 0 & 0 & -g_2 & g_1 & 0 & 0 & -g_2 \\ g_2 & 0 & 0 & g_3 & -g_2 & 0 & 0 & g_4 \\ 0 & g_2 & -g_3 & 0 & 0 & -g_2 & -g_4 & 0 \\ 0 & -g_1 & g_2 & 0 & 0 & g_1 & g_2 & 0 \\ g_1 & 0 & 0 & g_2 & -g_1 & 0 & 0 & g_2 \\ g_2 & 0 & 0 & g_4 & -g_2 & 0 & 0 & g_3 \\ 0 & g_2 & -g_4 & 0 & 0 & -g_2 & -g_3 & 0 \end{bmatrix} \tag{5.64}$$

where

$$g_1 = 36, \qquad\qquad g_3 = \left(4 + 5\Phi_e + 10\Phi_e^2\right)\ell_e^2,$$

$$g_2 = \left(3 - 15\Phi_e\right)\ell_e, \quad g_4 = \left(-1 - 5\Phi_e + 5\Phi_e^2\right)\ell_e^2$$

Note that the gyroscopic matrix is skew-symmetric.

5.4.6 The Effect of Torque

Most rotors carry some torque about the axis of rotation. Indeed, the purpose of the rotating machine usually depends fundamentally on this transmission of torque. In the case of a turbo-alternator, for example, the rotor carries mechanical power from the turbine stages to the alternator. In the case of a gas turbine, a proportion of the mechanical power developed in the turbine is transmitted through rotors up the middle of the engine to the compressor side, where air is pressurized prior to feeding into the turbine.

Torque within a rotor may have a significant effect on the lateral behavior, which is similar to the effect caused by the axial forces in the rotor discussed in Section 5.4.3. However, whereas axial force in a shaft or rotor does not cause any coupling between orthogonal planes containing the axis of rotation, torque does. In this sense, the effect of torque has something in common with gyroscopic couples discussed in Section 5.4.5. Relatively few papers or texts on rotordynamics consider this coupling, although Zorzi and Nelson (1980) derive the contribution to the stiffness for the lateral motion of an Euler-Bernoulli beam.

Torque exists at any position along the rotor in the direction of the neutral axis, which results in a *follower torque*. The sign convention for torque within the rotor requires careful definition. At any plane in the rotor, if the rotational speed of the rotor is positive, a positive torque within the shaft and rotor indicates that mechanical power is flowing in the positive z-axis. In FE modeling of rotors, it is natural to consider that torques add or subtract discretely at nodes of the FE model. For example, a rotor blade-stage that is instantaneously at some angle to the z-axis will naturally produces an increment of torque on the rotor in the direction of the normal to the blade-stage.

Follower forces and moments are not conservative; hence, the equations of motion must be derived via Newton's law or via a virtual work argument. Zorzi and Nelson (1980) derived the contribution to the stiffness matrix for an Euler-Bernoulli beam element as

$$\mathbf{K}_{Te} = \frac{\tau_e}{2\ell_e}
\begin{bmatrix}
0 & 0 & 2 & 0 & 0 & 0 & -2 & 0 \\
0 & 0 & 0 & 2 & 0 & 0 & 0 & -2 \\
2 & 0 & 0 & -\ell_e & -2 & 0 & 0 & \ell_e \\
0 & 2 & \ell_e & 0 & 0 & -2 & -\ell_e & 0 \\
0 & 0 & -2 & 0 & 0 & 0 & 2 & 0 \\
0 & 0 & 0 & -2 & 0 & 0 & 0 & 2 \\
-2 & 0 & 0 & -\ell_e & 2 & 0 & 0 & \ell_e \\
0 & -2 & \ell_e & 0 & 0 & 2 & -\ell_e & 0
\end{bmatrix}
\tag{5.65}$$

where τ_e is the transmitted torque within the element. This stiffness matrix is not symmetric, which highlights that the follower torque is nonconservative.

5.4.7 Tapered-Shaft Elements

Tapered shafts often occur in rotating machinery, and one option is to model these shafts with a relatively large number of cylindrical elements of different diameters. The alternative is to incorporate the changes in the cross section of the shaft into the element definition, producing a tapered or conical element. The procedure is exactly the same as in the previous sections that considered shafts of constant cross section. The element matrices based on Euler-Bernoulli beam theory are derived in this section, but the extension to Timoshenko beam models is straightforward. Bickford and Nelson (1985), Edney et al. (1990), and Genta and Gugliotta (1988) may be consulted for further details.

The shape functions are exactly the same as the constant-sectional case, and the differences arise in the integrations to obtain the kinetic and strain energies, where the cross-sectional area and second moment of area become functions of position, ξ. Thus, Equation (5.11) gives the shape functions for Euler-Bernoulli beam theory and Equations (5.12) and (5.19) give the expressions for strain and kinetic energy. Suppose that the shaft has a linear taper; that is, the shaft radius, r_e, is a linear function of the axial position within the element, ξ. Section 5.4.1 describes the notation in detail. Assume that the shaft is hollow and that the inside radii at the two ends of the element are $r_{ie}(0) = r_{ie1}$ and $r_{ie}(\ell_e) = r_{ie2}$. Thus,

$$r_{ie}(\xi) = \left(1 - \frac{\xi}{\ell_e}\right) r_{ie1} + \frac{\xi}{\ell_e} r_{ie2} \tag{5.66}$$

Similarly, the outside radius of the shaft element varies linearly from $r_{oe}(0) = r_{oe1}$ and $r_{oe}(\ell_e) = r_{oe2}$ so that

$$r_{oe}(\xi) = \left(1 - \frac{\xi}{\ell_e}\right) r_{oe1} + \frac{\xi}{\ell_e} r_{oe2} \tag{5.67}$$

The cross-sectional area and second moment of area are then

$$A_e(\xi) = \pi \left(r_{oe}(\xi)^2 - r_{ie}(\xi)^2\right), \qquad I_e(\xi) = \pi \left(r_{oe}(\xi)^4 - r_{ie}(\xi)^4\right)/4 \tag{5.68}$$

Comparing Equations (5.12), (5.13), and (5.15) and assuming that the Young's modulus is constant within the element show that the elements of the stiffness matrix are

$$k_{ij} = E_e \int_0^{\ell_e} I_e(\xi)\, N_{ei}''(\xi)\, N_{ej}''(\xi)\, \mathrm{d}\xi \tag{5.69}$$

where the second derivative of the shape functions are given in Equation (5.16). The integrands in Equation (5.69) are polynomials of degree 6 in ξ, and an explicit expression for the stiffness matrix may be obtained. The result is

$$\mathbf{K}_e = \frac{E\pi}{70\ell_e^3} \begin{bmatrix} k_1 & k_2 & -k_1 & k_3 \\ k_2 & k_4 & -k_2 & k_5 \\ -k_1 & -k_2 & k_1 & -k_3 \\ k_3 & k_5 & -k_3 & k_6 \end{bmatrix} \tag{5.70}$$

where

$$k_1 = 210p_1 + 420p_2 + 504p_3 + 294p_4 + 66p_5,$$

$$k_2 = (105p_1 + 140p_2 + 147p_3 + 84p_4 + 19p_5)\,\ell_e,$$

$$k_3 = (105p_1 + 280p_2 + 357p_3 + 210p_4 + 47p_5)\,\ell_e,$$

$$k_4 = (70p_1 + 70p_2 + 56p_3 + 28p_4 + 6p_5)\,\ell_e^2,$$

$$k_5 = (35p_1 + 70p_2 + 91p_3 + 56p_4 + 13p_5)\,\ell_e^2,$$

$$k_6 = (70p_1 + 210p_2 + 266p_3 + 154p_4 + 34p_5)\,\ell_e^2,$$

and

$$p_1 = r_{oe1}^4 - r_{ie1}^4,$$

$$p_2 = r_{oe1}^3 (r_{oe2} - r_{oe1}) - r_{ie1}^3 (r_{ie2} - r_{ie1}),$$

$$p_3 = r_{oe1}^2 (r_{oe2} - r_{oe1})^2 - r_{ie1}^2 (r_{ie2} - r_{ie1})^2,$$

$$p_4 = r_{oe1} (r_{oe2} - r_{oe1})^3 - r_{ie1} (r_{ie2} - r_{ie1})^3,$$

$$p_5 = (r_{oe2} - r_{oe1})^4 - (r_{ie2} - r_{ie1})^4$$

Exactly the same procedure may be followed to obtain the mass matrix from the approximation to the strain energy given by Equation (5.19) as

$$\mathbf{M}_e = \frac{\rho \ell_e \pi}{2520} \begin{bmatrix} m_1 & m_2 & m_3 & -m_4 \\ m_2 & m_5 & m_6 & -m_7 \\ m_3 & m_6 & m_8 & -m_9 \\ -m_4 & -m_7 & -m_9 & m_{10} \end{bmatrix} \tag{5.71}$$

where

$$m_1 = 936p_6 + 432p_7 + 76p_8, \qquad m_6 = (78p_6 + 84p_7 + 25p_8)\,\ell_e,$$

$$m_2 = (132p_6 + 84p_7 + 17p_8)\,\ell_e, \qquad m_7 = (18p_6 + 18p_7 + 5p_8)\,\ell_e^2,$$

$$m_3 = 324p_6 + 324p_7 + 92p_8, \qquad m_8 = 936p_6 + 1440p_7 + 580p_8,$$

$$m_4 = (78p_6 + 72p_7 + 19p_8)\,\ell_e, \qquad m_9 = (132p_6 + 180p_7 + 65p_8)\,\ell_e,$$

$$m_5 = (24p_6 + 18p_7 + 4p_8)\,\ell_e^2, \qquad m_{10} = (24p_6 + 30p_7 + 10p_8)\,\ell_e^2,$$

and

$$p_6 = r_{oe1}^2 - r_{ie1}^2,$$

$$p_7 = r_{oe1} (r_{oe2} - r_{oe1}) - r_{ie1} (r_{ie2} - r_{ie1}),$$

$$p_8 = (r_{oe2} - r_{oe1})^2 - (r_{ie2} - r_{ie1})^2$$

Rotary inertia and shear effects may be included by using the shape-function and energy expressions in Section 5.4.2. Other changes in cross section that are not described by a linear taper may be modeled by inserting the corresponding expressions for cross-sectional area and second moment of area into the energy integrals. Alternatively, other changes in cross section can be represented by a small number of linearly tapered elements. The expressions for the mass and stiffness matrices, given by Equations (5.70) and (5.71), are for beam elements rather than shaft elements. Section 5.4.4 outlined the procedure to obtain the shaft elements, and the process is exactly the same for tapered elements.

The gyroscopic matrix for the tapered element uses an expression for kinetic energy similar to Equation (5.55), except that now the second moment of area is a function of position in the element. Thus,

$$T_{Ge} = -2\rho_e \Omega \int_0^{\ell_e} I_e(\xi) \dot{\psi}_e(\xi, t) \theta_e(\xi, t) \, d\xi \tag{5.72}$$

and (compare to Equation (5.61))

$$G_{ij} = -2\rho_e \int_0^{\ell_e} I_e(\xi) \left(B_{2i}(\xi) B_{1j}(\xi) - B_{2j}(\xi) B_{1i}(\xi) \right) d\xi \tag{5.73}$$

Performing the integrations gives

$$\mathbf{G}_e = \frac{\rho \pi}{840\ell_e} \begin{bmatrix} 0 & g_1 & -g_2 & 0 & 0 & -g_1 & -g_3 & 0 \\ -g_1 & 0 & 0 & -g_2 & g_1 & 0 & 0 & -g_3 \\ g_2 & 0 & 0 & g_4 & -g_2 & 0 & 0 & -g_5 \\ 0 & g_2 & -g_4 & 0 & 0 & -g_2 & g_5 & 0 \\ 0 & -g_1 & g_2 & 0 & 0 & g_1 & g_3 & 0 \\ g_1 & 0 & 0 & g_2 & -g_1 & 0 & 0 & g_3 \\ g_3 & 0 & 0 & -g_5 & -g_3 & 0 & 0 & g_6 \\ 0 & g_3 & g_5 & 0 & 0 & -g_3 & -g_6 & 0 \end{bmatrix} \tag{5.74}$$

where

$$g_1 = 504 p_1 + 1008 p_2 + 864 p_3 + 360 p_4 + 60 p_5,$$

$$g_2 = (42 p_1 + 168 p_2 + 180 p_3 + 84 p_4 + 15 p_5) \ell_e,$$

$$g_3 = (42 p_1 - 72 p_3 - 60 p_4 - 15 p_5) \ell_e,$$

$$g_4 = (56 p_1 + 56 p_2 + 48 p_3 + 22 p_4 + 4 p_5) \ell_e^2,$$

$$g_5 = (14 p_1 + 28 p_2 + 36 p_3 + 22 p_4 + 5 p_5) \ell_e^2,$$

$$g_6 = (56 p_1 + 168 p_2 + 216 p_3 + 130 p_4 + 30 p_5) \ell_e^2,$$

and $p_1, \ldots, p_5$ are given previously.

5.4.8 Rotor Couplings

The individual rotors comprising a large turbo-generator unit are rigidly coupled by being bolted together and therefore may be treated as a single composite rotor. On other types of machines, however – such as an electric motor driving a fan – it is often desirable to introduce a flexible component to allow for such influences as changes in alignment and differential expansion. This is achieved by the use of flexible couplings that may decouple one or more degrees of freedom. For example, flexible couplings are often used to allow relative lateral motion yet transmit the full torsional moments. The couplings develop neither significant forces in lateral or axial directions nor moments about the transverse axes. In this case, the flexible coupling in effect decouples both the lateral and axial motions of the two rotors in question while giving a high torsional stiffness between the two shafts. These couplings take several forms: gear, flexible membrane, helical, bellows, and rubber-block couplings are just a few of the many types in use. All types have their own distinct dynamic properties and may have a significant influence on the overall dynamics of the machine.

Couplings often may be modeled simply as either constant stiffness or represented as rigid connections. The stiffness values for the translational or rotational degrees of freedom are different and may be incorporated into the model as linear or rotary springs. If the coupling is flexible in rotation about the Ox and Oy axes but stiff in translation, then it may be modeled as a hinge and degrees of freedom eliminated (see Chapter 2).

5.5 Bearings, Seals, and Rotor–Stator Interactions

To a greater or lesser extent, all bearings are flexible and all bearings absorb energy. For most types of bearing, the load-deflection relationship is nonlinear, which makes dynamic analysis more difficult than for a linear system. Furthermore, load-deflection relationships are often a function of shaft speed. To simplify dynamic analysis, one widely used approach is to assume that the bearing has a linear load-deflection relationship. This assumption is reasonably valid provided that the dynamic displacements are small and the degree of error inherent in this approach may be illustrated graphically or formally by the use of a Taylor series. Thus, the relationship between the forces acting on the shaft due to the bearing and the resultant velocities and displacements of the shaft may be approximated by

$$\begin{Bmatrix} f_x \\ f_y \end{Bmatrix} = - \begin{bmatrix} k_{uu} & k_{uv} \\ k_{vu} & k_{vv} \end{bmatrix} \begin{Bmatrix} u \\ v \end{Bmatrix} - \begin{bmatrix} c_{uu} & c_{uv} \\ c_{vu} & c_{vv} \end{bmatrix} \begin{Bmatrix} \dot{u} \\ \dot{v} \end{Bmatrix} \tag{5.75}$$

where f_x and f_y are the dynamic forces in the x and y directions, and u and v are the dynamic displacements of the shaft journal relative to the bearing housing in the x and y directions. Inertia terms also may be included in Equation (5.75) if they are significant (Smith, 1969). In vector notation, with $\mathbf{Q}_s = \begin{Bmatrix} f_x \\ f_y \end{Bmatrix}$ and $\mathbf{q} = \begin{Bmatrix} u \\ v \end{Bmatrix}$

$$\mathbf{Q}_s = -\mathbf{K}\left(\Omega\right)\mathbf{q} - \mathbf{C}\left(\Omega\right)\dot{\mathbf{q}} \tag{5.76}$$

If the housing is fixed, $\mathbf{q}$ is the absolute dynamic displacement of the shaft journal. If the housing can displace due to lack of rigidity in the bearing supports, then

$$\mathbf{q} = \mathbf{q}_s - \mathbf{q}_f \tag{5.77}$$

where $\mathbf{q}_s$ are the coordinates at the shaft and $\mathbf{q}_f$ are the coordinates at the foundation or housing. Thus,

$$\mathbf{Q}_s = -\mathbf{K}\left(\Omega\right)\left(\mathbf{q}_s - \mathbf{q}_f\right) - \mathbf{C}\left(\Omega\right)\left(\dot{\mathbf{q}}_s - \dot{\mathbf{q}}_f\right) \tag{5.78}$$

The force on the foundation due to the bearing is

$$\mathbf{Q}_f = -\mathbf{Q}_s \tag{5.79}$$

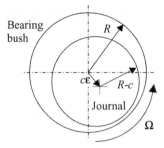

Figure 5.5. Fluid-bearing geometry. R is the bearing radius ($D/2$), c is the bearing clearance, and ε is the eccentricity.

Thus, combining these two equations, we have

$$\begin{Bmatrix} \mathbf{Q}_s \\ \mathbf{Q}_f \end{Bmatrix} = - \begin{bmatrix} \mathbf{K}(\Omega) & -\mathbf{K}(\Omega) \\ -\mathbf{K}(\Omega) & \mathbf{K}(\Omega) \end{bmatrix} \begin{Bmatrix} \mathbf{q}_s \\ \mathbf{q}_f \end{Bmatrix} - \begin{bmatrix} \mathbf{C}(\Omega) & -\mathbf{C}(\Omega) \\ -\mathbf{C}(\Omega) & \mathbf{C}(\Omega) \end{bmatrix} \begin{Bmatrix} \dot{\mathbf{q}}_s \\ \dot{\mathbf{q}}_f \end{Bmatrix} \tag{5.80}$$

We now discuss specific types of bearings and other rotor–stator interactions.

5.5.1 Hydrodynamic Journal Bearings

The *hydrodynamic bearing*, often called the oil or fluid film bearing, is extensively used in large rotating machines because of its high load-carrying capacity. It consists of a bearing bush in which the shaft or journal rotates, as shown in Figure 5.5. The bush has an internal diameter that is slightly greater than the diameter of the journal, thereby providing a clearance space between the bush and the journal, which is typically between 0.1 and 0.2 percent of the journal diameter. Oil is fed into the clearance space through one or more holes or grooves and, due to its viscosity and the rotation of the journal, it is swept circumferentially to create an oil film between the journal and bush. The oil eventually leaves the bearing by leakage from the ends.

The dynamics of the bearing at startup are complex and are not considered further. However, once a sufficient speed of rotation is reached, the journal and bearing bush surfaces separate and the journal is supported by an oil film, greatly reducing friction. This thin oil film is called a *hydrodynamic oil film* because its existence depends mainly on the relative motion between two surfaces. The journal center is now displaced horizontally as well as vertically from the center of the bush. If the shaft rotates at a high speed and the bearing carries a relatively light load, the journal runs almost concentrically in the bush. Conversely, if the shaft rotates at a low speed and/or the bearing carries a relatively large load, the minimum clearance between the journal and bush is reduced. When a bearing is designed to carry a steady load, the designer must ensure that there is a sufficient thickness of oil film to take up the effect of surface undulations and possible changes in load during the machine's lifetime. Most bearings are designed to operate with the ratio between the journal displacement and the radial clearance of about 0.6–0.7. This ratio is called the *eccentricity*, ε. When $\varepsilon = 1$, the oil film has zero thickness at one point and the bearing surfaces are in contact.

Although the cylindrical or plain bearing – consisting of a cylindrical bush completely or partially surrounding the journal – is the simplest and widely used hydrodynamic bearing, other more complex designs exist. Hydrodynamic bearings sometimes incorporate circumferential grooves, may have a slightly noncylindrical bush (i.e., lemon bore), offset halves, or the bush may consist of tilting pads. A full description of these bearings are outside the scope of this book, and the work of Hamrock et al. (2004) should be consulted for more information.

To determine the load-carrying capacity of a hydrodynamic bearing, we must model its fluid film. The film may be modeled using Reynolds's equation (Hamrock et al., 2004; Smith, 1969), where it is assumed that the viscosity and density of the fluid are constant throughout the film. For most fluid film geometries, Reynolds's equation can be solved only numerically, typically by the methods of finite elements or finite differences. However, an approximate solution in closed form can be determined for the fluid film that exists in a short bearing. Because of the assumptions made, this solution has a limited range of applicability, but it does illustrate many of the characteristics of hydrodynamic bearings in a relatively simple manner; for this reason, it is developed here. Let us assume that

- the flow is laminar and Reynolds's equation applies
- the bearing is very short, so that $L/D \ll 1$, where L is the bearing length and D is the bearing diameter, which means that the pressure gradients are much larger in the axial than in the circumferential direction
- the lubricant pressure is zero at the edges of the bearing
- the bearing is operating under steady running conditions
- the lubricant properties do not vary substantially throughout the oil film
- the shaft does not tilt in the bearing

The pressure distribution then may be obtained as an analytical expression; for example, the variation of pressure with axial position is parabolic. The force exerted on the journal by the bearing (i.e., the load that the bearing carries) is obtained by integrating this pressure over the surface area. Any areas of negative pressure are susceptible to cavitation; for the purposes of integration, we assume zero pressure in these areas. The radial and tangential forces, f_r and f_t (Cameron, 1976; Childs, 1993) are

$$f_r = -\frac{D\Omega\eta L^3 \varepsilon^2}{2c^2 \left(1 - \varepsilon^2\right)^2} \quad \text{and} \quad f_t = -\frac{\pi D\Omega\eta L^3 \varepsilon}{8c^2 \left(1 - \varepsilon^2\right)^{3/2}} \tag{5.81}$$

where η is the absolute viscosity of the oil, Ω is the speed of rotation, c is the bearing radial clearance, and L, D, and ε are defined previously. These forces are applied to both the bearing bush and the journal. The tangential force f_t opposes the sliding motion so that the power dissipation is $f_t \Omega D/2$. The resultant force on the stator (through the bush) must be equal and opposite to the load applied to the rotor (through the journal). From Equation (5.81), the magnitude of the resultant force f is

$$f = \sqrt{f_r^2 + f_t^2} = \frac{\pi D\Omega\eta L^3 \varepsilon}{8c^2 \left(1 - \varepsilon^2\right)^2} \left(\left(\frac{16}{\pi^2} - 1\right)\varepsilon^2 + 1\right)^{1/2} \tag{5.82}$$

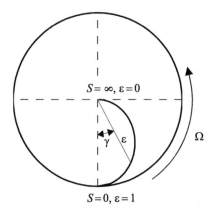

Figure 5.6. Locus of the equilibrium position of the journal, assuming short-bearing theory.

A vertical resultant load is very common – for example, where the load is due to the rotor weight; in this case, the position the journal takes in the bearing ensures that this load is indeed vertical. If the magnitude of this load is known, then the bearing eccentricity may be obtained by rearranging Equation (5.82) to give the following quartic equation in ε^2:

$$\varepsilon^8 - 4\varepsilon^6 + \left(6 - S_s^2\left(16 - \pi^2\right)\right)\varepsilon^4 - \left(4 + \pi^2 S_s^2\right)\varepsilon^2 + 1 = 0 \qquad (5.83)$$

where

$$S_s = \frac{D\Omega\eta L^3}{8fc^2} \qquad (5.84)$$

is called the *modified Sommerfeld number* or the *Ocvirk number* and is known for a particular speed, load, and oil viscosity. The smallest root of Equation (5.83) is taken and is always between 0 and 1.

A more general nondimensional parameter is called the *Sommerfeld number* or *duty parameter*, S, which is typically defined as

$$S = \frac{S_s}{\pi}\left(\frac{D}{L}\right)^2 = \frac{D^3\Omega\eta L}{8\pi fc^2} \qquad (5.85)$$

We should note that the rotor spin speed is given here in units of rad/s, but some definitions of the Sommerfeld number in the literature give this speed in units of rev/s or rev/min. Also, the load is sometimes given as a pressure over a projected area – that is, in terms of f/DL – rather than the force.

The force f must be in the direction of the force applied to the bearing. This direction is given by $\tan\gamma = f_t/f_r$. Thus,

$$\tan\gamma = \frac{\pi\sqrt{1 - \varepsilon^2}}{4\varepsilon} \qquad (5.86)$$

Assuming a vertical load, γ is the angle between this vertical load and the direction of the displacement of the journal and is called the *attitude angle*. Figure 5.6 shows the angle as a function of eccentricity. The locus of the journal center is close to but not exactly a semicircle.

From Equation (5.82), the load-carrying capacity of a particular short bearing for any rotational speed and eccentricity may be determined. Conversely, if the applied load and rotational speed are known, the eccentricity may be determined from Equation (5.83). There are other issues that must be addressed in the design of bearings, such as oil flow and thermal properties, but they are not considered here.

We now consider the effect of dynamic forces acting on the bearing. In general, the force-displacement relationship is nonlinear but, provided that the amplitude of the resultant displacements is small, we can assume a linear force-displacement relationship. When a linear bearing model is used in a machine analysis, the displacement predicted at the bearings should be checked to ensure that it is indeed small because a linear analysis does not include any constraints on the displacement of the rotor in a bearing journal. Here, we consider only short bearings, where the matrices are 2×2 and the shaft rotations about x and y are unconstrained. The stiffness and damping matrices may be written in closed form in terms of the eccentricity and load as

$$\mathbf{K}_e = \frac{f}{c} \begin{bmatrix} a_{uu} & a_{uv} \\ a_{vu} & a_{vv} \end{bmatrix}, \qquad \mathbf{C}_e = \frac{f}{c\Omega} \begin{bmatrix} b_{uu} & b_{uv} \\ b_{vu} & b_{vv} \end{bmatrix} \tag{5.87}$$

where

$$a_{uu} = h_0 \times 4 \left(\pi^2 \left(2 - \varepsilon^2 \right) + 16\varepsilon^2 \right),$$

$$a_{uv} = h_0 \times \frac{\pi \left(\pi^2 \left(1 - \varepsilon^2 \right)^2 - 16\varepsilon^4 \right)}{\varepsilon \sqrt{1 - \varepsilon^2}},$$

$$a_{vu} = -h_0 \times \frac{\pi \left(\pi^2 \left(1 - \varepsilon^2 \right) \left(1 + 2\varepsilon^2 \right) + 32\varepsilon^2 \left(1 + \varepsilon^2 \right) \right)}{\varepsilon \sqrt{1 - \varepsilon^2}},$$

$$a_{vv} = h_0 \times 4 \left(\pi^2 \left(1 + 2\varepsilon^2 \right) + \frac{32\varepsilon^2 \left(1 + \varepsilon^2 \right)}{\left(1 - \varepsilon^2 \right)} \right),$$

$$b_{uu} = h_0 \times \frac{2\pi \sqrt{1 - \varepsilon^2} \left(\pi^2 \left(1 + 2\varepsilon^2 \right) - 16\varepsilon^2 \right)}{\varepsilon},$$

$$b_{uv} = b_{vu} = -h_0 \times 8 \left(\pi^2 \left(1 + 2\varepsilon^2 \right) - 16\varepsilon^2 \right),$$

$$b_{vv} = h_0 \times \frac{2\pi \left(\pi^2 \left(1 - \varepsilon^2 \right)^2 + 48\varepsilon^2 \right)}{\varepsilon \sqrt{1 - \varepsilon^2}},$$

and

$$h_0 = \frac{1}{\left(\pi^2 \left(1 - \varepsilon^2 \right) + 16\varepsilon^2 \right)^{3/2}}$$

Clearly, the stiffness matrix is not symmetric; therefore, hydrodynamic bearings introduce anisotropic supports into the machine model.

Hamrock et al. (2004) and Smith (1969) give full details. Figures 5.7, 5.8, and 5.9 show the variation of eccentricity, nondimensional stiffness ($\mathbf{K}_e c / f$), and damping ($\mathbf{C}_e c\Omega / f$) coefficients with the modified Sommerfeld number. The dynamic performance of the simple fluid film bearing has been improved by methods such as the addition of grooves, the use of tilting pads, and noncircular bores (Goodwin, 1989, Kramer, 1993).

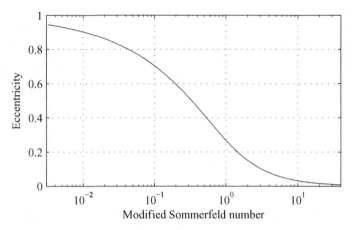

Figure 5.7. Bearing eccentricity as a function of the modified Sommerfeld number.

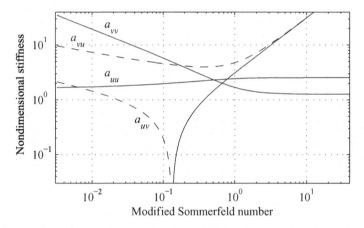

Figure 5.8. Nondimensional stiffness of a short fluid bearing as a function of the modified Sommerfeld number (negative coefficients are shown as dashed lines) from Equation (5.87).

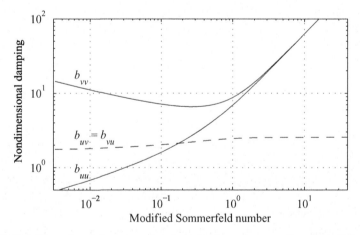

Figure 5.9. Nondimensional damping of a short fluid bearing as a function of the modified Sommerfeld number (negative coefficients are shown as dashed line), from Equation (5.87).

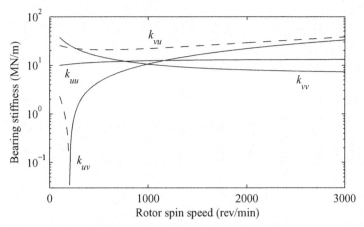

Figure 5.10. Fluid-bearing stiffness as a function of rotor spin speed (Example 5.5.1) (negative coefficients are shown as dashed lines).

EXAMPLE 5.5.1. An oil film bearing, 100 mm diameter and 30 mm long, supports a static load of 525 N. The radial clearance in the bearing is 0.1 mm and the oil film has a viscosity of 0.1 Pa s. Calculate the bearing eccentricity when running at 1,500 rev/min. Determine the bearing stiffness and damping matrices under these conditions.

Solution. The shaft speed $\Omega = 1,500 \times 2\pi/60 = 157.1$ rad/s. Assuming that the short-bearing theory is sufficiently accurate for the task at hand, then S_s is

$$S_s = \frac{D\Omega\eta L^3}{8fc^2} = \frac{0.1 \text{ m} \times 157.1 \text{ rad/s} \times 0.1 \text{ Pa s} \times 0.03^3 \text{ m}^3}{8 \times 525 \text{ N} \times (0.1 \times 10^{-3} \text{ m})^2} = 1.010$$

The Sommerfeld number is $S = 3.571$. Equation (5.83) becomes

$$\varepsilon^8 - 4\varepsilon^6 - 0.2511\varepsilon^4 - 14.06\varepsilon^2 + 1 = 0$$

Solving this quartic equation provides the roots for ε^2. There are only two real roots and only one between 0 and 1: namely, 0.07091. Thus, $\varepsilon = 0.2663$ and, from Equation (5.87)

$$\mathbf{K}_e = \begin{bmatrix} 12.81 & 16.39 \\ -25.06 & 8.815 \end{bmatrix} \text{MN/m} \quad \text{and} \quad \mathbf{C}_e = \begin{bmatrix} 232.9 & -81.92 \\ -81.92 & 294.9 \end{bmatrix} \text{kNs/m}$$

Figures 5.10 and 5.11 show the stiffness and damping coefficients for the bearing analyzed in this example. The variation of eccentricity and stiffness with shaft speed are scaled versions of Figures 5.7 and 5.8. The shaft speed of 3,000 rev/min corresponds to a modified Sommerfeld number of 2.020. The variation of damping coefficients is different than that shown in Figure 5.9 because the shaft speed is required for the nondimensionalization.

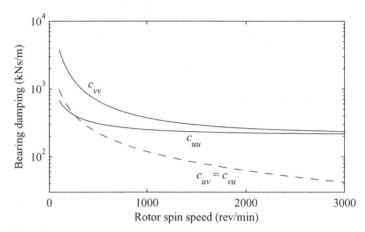

Figure 5.11. Fluid-bearing damping as a function of rotor speed (Example 5.5.1) (negative coefficients are shown as dashed line).

5.5.2 Hydrostatic Journal Bearings

The hydrostatic, or externally pressurized, journal bearing again consists of a journal and bush separated by a fluid film. Here, the fluid film is maintained by an external supply of fluid at a high pressure. The most commonly used fluid is oil that is supplied to the bearing under pressure and enters recesses located in the bearing bush. Here, the pressure drops slightly and the oil then flows out of the recesses over the bearing and into the drain holes. Thus, unlike the hydrodynamic journal bearing, the fluid film does not depend on the relative motion of the journal and bush. A key advantage of hydrostatic bearings is that the oil film is present when the shaft is just commencing to rotate, because the film can exist when the shaft is stationary. Furthermore, the stiffness of the bearing is high and can even be varied in use (see Goodwin et al., 1983). The disadvantage of the hydrostatic bearing is that an external high-pressure supply of fluid is required; if this fails, the bearing could fail catastrophically. Hydrostatic bearings can support large loads even at zero rotational speed because there is always full film lubrication. The frictional losses of this type of bearing are proportional to rotating speed; therefore, it has a low startup resistance.

   In theory, the load-carrying capacity of a hydrostatic bearing is directly proportional to the supply pressure, irrespective of the film thickness or flow rates. However, in practice, because the pump pressurizing the oil has a limited performance, increasing the fluid flow (by increasing the film thickness) causes the supply pressure to fall.

5.5.3 Rolling-Element Bearings

The key feature of the rolling-element bearing is that the journal and bearing housing are separated by a cage (sometimes called a train) containing rolling elements. As the journal rotates, the rolling elements are forced to rotate between the inner raceway attached to the rotating journal and the outer raceway, which is usually stationary and attached to the bearing housing. To allow the rolling elements to rotate, the cage in which they are carried also must be free to rotate. Because surfaces are essentially in rolling contact rather than in sliding contact, the friction and wear are greatly reduced, and the starting friction torque is only slightly greater than the moving friction torque. For lubrication, the bearings are packed with grease or a continuous supply of oil is provided. The bore of a rolling-element bearing can be as small as 1.5 mm (e.g., for instruments) or as large as 2 m or more (e.g., in wind turbines). Their life is limited by fatigue of the races and rolling elements, but this is predictable.

Rolling-element bearings can be designed to carry radial loads only, axial loads only, or a combination of both. For example, deep-groove ball bearings can be designed to carry radial and/or axial loads. Furthermore, a double-row arrangement can carry higher radial loads. Still higher axial loads can be carried by angular contact bearings. In contrast, self-aligning bearings have a lower radial and axial load-carrying capacity than deep-groove ball bearings but have the advantage of accommodating large shaft misalignment.

In the roller-bearings family, cylindrical, needle, and spherical roller bearings can carry high radial loads but only low axial loads. Generally, spherical roller bearings operate at lower speeds than cylindrical roller bearings. Relative to the shaft diameter, needle roller bearings have a small outer diameter. Tapered roller bearings support high radial and axial loads. However, compared to angular contact ball bearings, they have a lower operating speed.

We now consider the stiffness of a rolling-element bearing. It has been shown (Harris, 2001; Kramer, 1993) that the deflection δ (in meters) of a single steel ball or roller being compressed between two flat plates by a force f (in Newtons) is given by

$$\delta = 4.36 \times 10^{-8} \, d^{-1/3} f^{2/3} \quad \text{(for a ball of diameter } d \text{ in meters)} \tag{5.88}$$

and

$$\delta = 3.06 \times 10^{-10} \, l^{-0.8} f^{0.9} \quad \text{(for a roller of length } l \text{ in meters)} \tag{5.89}$$

These approximations are sufficient for rolling-element bearings in which the rolling-element radius is small compared to the radii of the inner and outer races.

The deflection of a complete bearing consisting of many rolling elements can be derived by distributing the applied force over all the elements of the bearing that are carrying load. Having derived the nonlinear force-deflection relationship for the bearing, the stiffness then can be determined as the slope of this function. Often, the static load is due to the weight of the rotor; hence, the static force and deflection are in the negative y direction. The vertical stiffness of a ball bearing, df/dy, is then approximated (Kramer, 1993) by

$$k_{vv} = \kappa_b \, n_b^{2/3} d^{1/3} f_s^{1/3} \cos^{5/3} \alpha \tag{5.90}$$

where n_b is the number of steel balls of diameter d in the race, α is the contact angle (Figure 5.12), f_s is the vertical static load acting on the bearing, and the constant is $\kappa_b = 13 \times 10^6 \, \text{N}^{2/3} \, \text{m}^{-4/3}$. Similarly, for rolling-element bearings

$$k_{vv} = \kappa_r \, n_r^{0.9} l^{0.8} f_s^{0.1} \cos^{1.9} \alpha \tag{5.91}$$

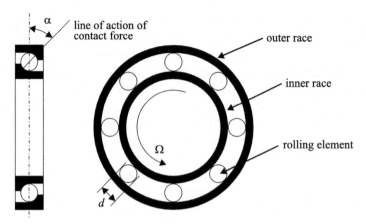

Figure 5.12. A typical rolling-element cross section.

where n_r is the number of steel rolling elements of length l in the race, f_s is the static load due to gravity acting on the bearing in the vertical direction, α is again the contact angle, and the constant is $\kappa_r = 1.0 \times 10^9 \, \text{N}^{0.9} \, \text{m}^{-1.8}$. These expressions are given in SI units, whereas some authors define component dimensions and displacements in mm or μm. Also note that the bearing stiffness increases with increasing load; that is, it behaves as a hardening spring. As the inner race rotates, at certain instances, there is a rolling element directly on the line of the load; at other instances, the point of application of the load is midway between rolling elements. The fluctuation in the stiffness is typically less that 1 percent, however, in some circumstances, the effect may be noticeable.

Assuming the load is vertical, the ratio of the horizontal stiffness, k_{uu}, to the vertical stiffness, k_{vv}, can be calculated. Kramer (1993) gives the following data for ball and rolling-element bearings:

Ball bearing: $k_{uu}/k_{vv} = 0.46, 0.64, 0.73$ for 8, 12, 16 balls, respectively.
Roller bearing: $k_{uu}/k_{vv} = 0.49, 0.66, 0.74$ for 8, 12, 16 rollers, respectively.

This ratio is independent of the static load, f_s. Damping is low in rolling-element bearings and the damping coefficient c is typically in the range of $(0.25 \sim 2.5) \times 10^{-5} \, \text{s} \times k$, where k is a typical bearing stiffness.

Lim and Singh (1990, 1994) developed 5×5 mean-bearing stiffness matrices for ball bearings and roller bearings. Their mean stiffness matrices relate generalized displacements in u, v, w, θ, and ψ and the corresponding forces. The displacement w along the axis of the shaft is significant only for axial thrust bearings. The mean bearing stiffness coefficients and the mean bearing forces are determined from the bearing nonlinear load-deflection relationships by varying the mean bearing displacements systematically. White (1979) considered the effect of bearing clearance on the stiffness.

Rolling-element bearings, in common with all other flexible bearings, influence the overall dynamics of the rotor–bearing–foundation system. However, rolling-element bearings also can be a source of vibration, producing time-varying forces that cause the system to vibrate. These forces are inherent in the design of the bearings but they also can be caused by defects in the bearings such as a damaged rolling-element (Harris, 2001). Assuming no gross skidding, we can develop expressions for the speed of rotation of the cage and the rolling elements from the geometry of the bearing. The angular velocity of the cage (or train) is

$$\Omega_c = \frac{\Omega}{2} \left\{ 1 - \frac{d}{D} \cos \alpha \right\} \tag{5.92}$$

and the angular velocity of the balls or rollers is

$$\Omega_r = \frac{\Omega}{2} \left(\frac{D}{d} \right) \left\{ 1 - \left(\frac{d}{D} \right)^2 \cos^2 \alpha \right\} \tag{5.93}$$

where Ω is the angular velocity of the rotor (or inner raceway), D is the pitch-circle diameter of the balls (or rollers), d is the diameter of the balls (or rollers), and α is the contact angle. The angular velocities Ω_c and Ω_r are often referred to as the *fundamental train frequency* (FTF) and the *ball spin frequency* (BSF), respectively.

A damaged rolling-element bearing can generate many frequencies, although one of four predictable frequencies often dominates. A defect in the inner or outer raceways causes vibration at the ball (or roller) passing frequencies (BPFI or BPFO, respectively). These frequencies are

$$\Omega_i = n_b \frac{\Omega}{2} \left\{ 1 + \frac{d}{D} \cos \alpha \right\} \tag{5.94}$$

and

$$\Omega_o = n_b \frac{\Omega}{2} \left\{ 1 - \frac{d}{D} \cos \alpha \right\} \tag{5.95}$$

A defect in the rolling element impacts on the inner and outer raceways at *twice* the BSF. A defect in the cage or train causes a vibration at the FTF.

5.5.4 Magnetic Bearings

Magnetic bearings are becoming increasingly popular in rotating machines because of low losses, an ability to operate without lubricant, and an extremely long life. In broad terms, they divide into two types: *active* and *passive* magnetic bearings. Magnetic bearings for rotating machines also may be divided according to whether they are *radial* or *axial*. Radial magnetic bearings exert lateral forces. They might be described better as lateral magnetic bearings, but the terminology is now established. The descriptions here are restricted to radial magnetic bearings.

Passive magnetic bearings produce restoring forces that are determined completely by the position of the bearing journal relative to the bearing stator, the rate of change of that position, and the rotational speed. Most passive magnetic bearings have permanent magnet material on both the journal and the stator, and all of them have permanent magnet material on at least one side. Consider first the passive bearings with permanent magnet material on both rotor and stator and with no deliberate paths for electrical currents to be induced. For these bearings, the dependence on linear and angular speeds is usually minor, and the bearing forces in the x and y directions are functions of the deflections in these directions only, u and v. The

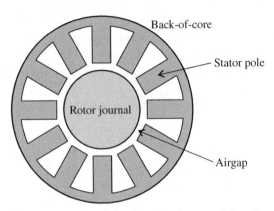

Figure 5.13. Schematic of a typical magnetic bearing.

magnetic fields in a passive magnetic bearing may be predominantly axial or radial in nature. Delamare et al. (1995) discuss various configurations. By changing the shape of the airgap, we can obtain hardening or softening stiffness characteristics of the bearing.

Irrespective of the shape of the airgap, passive magnetic bearings share the following three characteristics: (a) the maximum bearing force achievable is small compared to the forces that can be reacted by mechanical bearings of similar size; (b) these bearings behave like flexible linear springs; and (c) passive magnetic bearings have negative stiffness characteristics in the axial direction. Earnshaw (1842) proved that for any stationary body in a passive magnetic field, the sum of the stiffnesses in any three orthogonal directions must be zero.

Active magnetic bearings provide complete control over the force exerted between the bearing rotor and stator. The principle of operation normally employed is essentially the same as that used in controlled electromagnets. Figure 5.13 illustrates schematically a conventional magnetic bearing. The bearing stator contains *poles* that protrude inward from a cylindrical *back-of-core* with spaces between these poles. The stator poles come close to the rotating part of the bearing (usually laminated), but there is a space between the rotor and the tips of the stator poles that is usually called the *airgap*. Coils are fitted in the spaces between poles and when an electrical current flows in these coils, magnetic flux travels in loops that cross the airgap twice (normal to the rotor surface). A tensile stress of up to 1 MPa is created at the stator pole ends; by adjusting the currents such that this stress is not the same beneath each pole, a net transverse force can be exerted on the rotor. If the outer diameter of the bearing rotor is D and the active length of the rotor is L, then a maximum net force on the rotor on the order of $DL \times 10^6$ N can be realized in the bearing (where D and L are measured in meters).

Active magnetic bearings can provide excellent performance. These bearings exhibit a strongly unstable behavior if the currents within the bearings are constant. Small displacements of the rotor from the equilibrium position causes the magnetic flux to redistribute and produce a force that tends to deflect the rotor still farther. In effect, the bearing has a negative stiffness. Conversely if the voltage across the coils is constant, then a short-term displacement of the bearing rotor does not substantially change the flux passing through any one of the stator poles and initially does not change the net force. Magnetic FEA is usually employed in the modeling of active magnetic bearings, and this analysis can provide the complete prediction of forces from rotor position and currents. This relationship may be *linearized* for small perturbations of currents about a mean and for small bearing rotor movements, as discussed in Section 2.2.

In operation, the forces required to be exerted by the active magnetic bearing are determined from measurements of the position of the bearing rotor relative to the stator using noncontact probes of some type. The position signals are inputs to a controller that, in effect, determines what voltages should be applied to the bearing by a power amplifier to achieve the appropriate currents in the bearing coils. In high-performance active magnetic bearings systems, this dynamic behavior is modeled and taken into account in the bearing controller. It is typical for magnetic bearings to be capable of exerting significant forces on the rotor at frequencies up to several kHz.

When the complete transfer function between rotor positions and bearing forces is analyzed, it is common for it to be approximately linear because the controller can compensate for the nonlinear behaviors of the bearing itself arising from magnetic saturation and significant deflections. The controllers for magnetic bearings are often developed using modern control theory (i.e., state–space form). Typically, one or two state variables are devoted to representing the dynamics of each channel of position sensing and, typically, an additional two or three state variables are devoted to each independent current in the bearing. Thus, for each magnetic bearing in a machine, between 10 and 20 state variables may be employed. A state–space model of the complete rotating machine (after appropriate model reduction) might have between 20 and 40 state variables.

5.5.5 Rigid Bearings

Some bearings are much stiffer than the shaft they support. Common examples are ball and roller bearings used to support a flexible shaft. Although these bearings may be modeled using a high value of stiffness, given in Equations (5.90) and (5.91), using terms that differ greatly in magnitude (in this case, bearing stiffness versus shaft stiffness) can lead to numerical problems. An alternative is to model the bearings as being rigid. Two alternative forms of rigid bearing are possible. A short rigid bearing constrains the shaft deflections in the Ox and Oy directions to zero but does not constrain shaft rotations about the Ox and Oy axes. This is equivalent to the pinned-beam constraint considered in Chapter 4 for nonrotating structures. A long rigid bearing constrains both the displacement along and rotation about the Ox and Oy axes. These descriptions of short and long bearings assume that support for the stationary part of the bearing is rigid; if the support is flexible, then the zero constraints are interpreted as enforcing the corresponding degrees of freedom on the rotating and stationary parts of the bearing to be equal.

In the FE model of a machine with short bearings, the zero displacements (in both the x and y directions) are obtained by eliminating the corresponding columns of the mass, damping, gyroscopic, and stiffness matrices. The corresponding rows are also eliminated because they give the reaction forces required to enforce the zero constraints. These forces may be recovered from the rotor response if required.

The result is a reduction of two degrees of freedom for each short rigid bearing. Long rigid bearings are treated similarly, except that both the displacements and the rotations at the bearing location are set to zero. This is equivalent to the clamped constraint for beam structures. The number of degrees of freedom is now reduced by four for each long rigid bearing. Section 2.5 considers the formal imposition of constraints.

5.5.6 Seals

The role of seals is to prevent the flow of working fluid to an inappropriate part of the machine or between the inside and outside of the machine. They invariably involve fine clearances between the rotor and stator, within which a pressure field develops that can exert a force on the rotor. In essence, each seal acts as an extra bearing, in the sense that it introduces a rotor–stator interaction. Seals are crucially important in virtually all turbo-machinery, such as steam turbines, pumps, and compressors. These seals take several different forms and can significantly modify the dynamic behavior of a machine.

Extensive discussions of seal problems are in Adams (2001), Childs (1993), and Kramer (1993). Here, we only give an overview of the analysis and the effect of liquid seals, which are important in multistage centrifugal pumps. From the dynamics viewpoint, the high-power density within a pump generates high forces on a relatively light rotor. Black (1969) reported the first detailed study of this important problem. His basic model is based on an empirical study of turbulent flow down an annulus and estimates the pressure across the seal, which is then integrated to give the force. Black included shaft rotation in his analysis by formulating the equations in a frame of reference rotating at half rotor speed (i.e., the mean fluid rotation rate). The force is approximated using Equation (5.76) but also includes inertia terms. The 2×2 mass, damping, and stiffness matrices for the translational degrees of freedom (i.e., neglecting moments and rotations) in the stationary frame may be written as

$$
\mathbf{M}_e = \begin{bmatrix} m_d & 0 \\ 0 & m_d \end{bmatrix}, \qquad \mathbf{C}_e = \begin{bmatrix} c_d & m_d\Omega \\ -m_d\Omega & c_d \end{bmatrix},
$$

$$
\mathbf{K}_e = \begin{bmatrix} k_d - \frac{1}{4}m_d\Omega^2 & \frac{1}{2}c_d\Omega \\ -\frac{1}{2}c_d\Omega & k_d - \frac{1}{4}m_d\Omega^2 \end{bmatrix},
\tag{5.96}
$$

where

$$
m_d = \gamma\tau^2 \frac{19\sigma + 18\sigma^2 + 8\sigma^3}{(1.5 + 2\sigma)^3}, \qquad c_d = \gamma\tau \frac{(3 + 2\sigma)^2(1.5 + 2\sigma) - 9\sigma}{(1.5 + 2\sigma)^2},
$$

$$
k_d = \gamma \frac{9\sigma}{1.5 + 2\sigma}, \qquad \gamma = \frac{\pi}{6\lambda}\left(\frac{\sigma}{1.5 + 2\sigma}\right)RP
$$

In these definitions, $\sigma = \lambda L/c$, λ is the dimensionless seal friction coefficient, L is the axial length of the seal, R is the seal radius, P is the pressure difference across the seal, c is the radial clearance, $\tau = L/V$ where V is the average axial stream velocity, and Ω is the shaft speed. This approximation relies on $\tau \ll 2\pi/\Omega$. The asymmetric stiffness matrix may seem surprising at first sight; however, the system

is nonconservative. The net effect of the stiffness asymmetry is often to reduce the stability of the system. Sections 7.8 and 7.9 consider in more detail the instabilities due to the skew-symmetric component of the stiffness matrix.

This is probably the simplest model to represent a liquid seal and it involves a number of assumptions and approximations. The approach, however, does provide a good indication of the influences of seals; for a real example, the study can become complicated. Because the flow patterns and pressure distributions within a pump depend not only on the running speed but also on the volumetric flow rate, the seal characteristics are also functions of both parameters. The relationship between flow and speed is not fixed but rather is dependent on the external system that is variable due to, for instance, valve settings. Other features may influence the analysis; for example, annular grooving is frequently used to reduce leakage flows for a given clearance. Such grooves modify the mass, damping, and stiffness matrices of the seal (Childs et al., 2007).

5.5.7 Alford's Force

In high-pressure steam and gas turbines, an instability arises that is often referred to as *steam whirl* (Adams, 2001; Alford, 1965; Ding et al., 2006; Thomas et al., 1976). The effect is similar to oil whirl (see Section 10.12), but the stability threshold depends on the machine output rather than the rotor spin speed. The phenomenon arises because the lateral motion of the rotor causes an asymmetry in the forces on the blades due to the nonuniform gap between the blades and the casing. Adams (2001) gives a simple model of Alford's force as

$$\left\{ \begin{matrix} f_x \\ f_y \end{matrix} \right\} = - \begin{bmatrix} 0 & k_{sw} \\ -k_{sw} & 0 \end{bmatrix} \left\{ \begin{matrix} u \\ v \end{matrix} \right\} \tag{5.97}$$

where f_x and f_y are the forces in the x and y directions acting on the rotor, and u and v are the rotor displacements. The coefficient k_{sw} is given by

$$k_{sw} = \frac{\beta T}{DL} \tag{5.98}$$

where T is the turbine torque, D is the mean diameter of the blades, L is the radial length of the turbine blades, and β is a dimensionless factor related to the force reduction due to radial-tip clearance. Although β has an interpretation in terms of the blade force, this often is an empirical factor obtained from experience or laboratory tests. For unshrouded turbines, experience has shown that $2 < \beta < 5$.

5.5.8 Squeeze-Film Dampers

Squeeze-film dampers are essentially hydrodynamic bearings in which the two bearing surfaces do not (usually) rotate relative to one another. Because of the lack of rotation, squeeze-film dampers have no stiffness and no ability to carry static loads. They are incorporated into rotating machines to provide damping in a bearing support (Cookson and Kossa, 1979, 1980). Invariably, the bearing concerned is a rolling-element bearing that has high stiffness and little capability to absorb any vibration energy. Typically, squeeze-film dampers are formed between the outer race

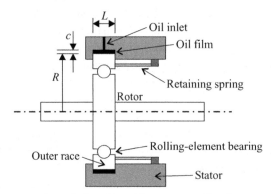

Figure 5.14. The cross section of a bearing with a squeeze-film damper and retaining spring.

of a ball bearing and the slightly larger inner diameter of the bearing seating. Figure 5.14 illustrates this situation with a cross section of a bearing and its housing parallel to the axis of bearing. The radial gap in the squeeze-film is generally of the order of one-thousandth of the mean radius of the gap, and the gap, is filled with oil.

Two different provisions are common for preventing the two surfaces of the device from rotating relative to one another. One or more radial pegs may be fixed into one of the damper surfaces, and these pegs are trapped within small slots in the other surface. Alternatively, the outer bearing case is supported on a retaining spring, as Figure 5.14 illustrates. The retaining spring takes the form of a set of flexible parallel bars in a circular arrangement; for this reason, it is often referred to as a *squirrel cage*. This spring is usually set so that at equilibrium, the squeeze-film is an annulus.

The oil present in the squeeze-film damper is invariably the same as the oil used to lubricate the bearing, and the axial length of the squeeze-film is the same as the axial dimension of the outer race of the bearing – which is often short in comparison to the circumferential direction. In order that a useful degree of damping can be achieved, it is sometimes necessary to impede the flow of oil in the axial direction. To this end, seals of one type or another are commonly included in squeeze-film dampers.

Adams (2001) shows that the linear dynamic properties of a short squeeze-film damper may be approximated by diagonal damping and stiffness matrices. The stiffness is determined by the centering spring (if present) and the damping value is

$$c_{SFD} = \frac{\pi \eta R L^3}{2c^3} \qquad (5.99)$$

where η is the oil viscosity, R is the damper radius, L is the damper length, and c is the damper radial clearance. Goodwin (1989) and Proctor and Gunter (2005) consider the case when the squeeze-film damper precesses and therefore has both damping and stiffness. Often, the stiffness and damping characteristics of squeeze-film dampers are highly nonlinear (Mohan and Hahn, 1974), but a detailed analysis is outside the scope of this book.

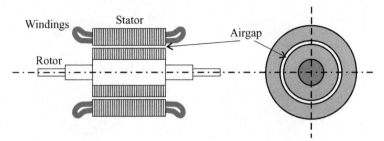

Figure 5.15. Cross section of an electrical machine.

5.5.9 Unbalanced Magnetic Pull

Unbalanced magnetic pull (UMP) arises in electrical machines when the magnetic field linking the rotor and stator of a machine is not perfectly symmetric. More specifically, if the magnetic flux lines coupling the rotor and stator are rotated by 180° about the axis of the machine and they do not reproduce the same pattern, there almost certainly is a net magnetic force between the rotor and the stator. For all practical machines, there are essentially two ways that such a lack of symmetry can occur in the machine: (1) the distribution of magneto-motive force (MMF) on either the rotor or stator may lack this symmetry; or (2) as a result of transverse movement of the rotor relative to the stator.

We focus on the latter root cause first because this has obvious direct linkage to rotordynamics. Figure 5.15 shows a cross section and end section of an electrical machine. The airgap is shown exaggerated in the figure. Typically, airgaps are of the order of 0.4 mm for small machines and they increase for larger machines (e.g., 75 mm for turbo-alternators) but not in proportion to the diameter.

All electromagnetic machines contain a source of MMF on at least one side of the airgap, which acts to push the magnetic flux across the airgap (twice) and around the rest of a loop that is partly in the rotor and partly in the stator. MMF arises as a result of either permanent magnet material or currents being present. In the design of most electrical machines, a substantial fraction of the available MMF is absorbed in pushing the magnetic flux across the airgap. If the total length of airgap traversed by any magnetic flux line is reduced when the rotor moves transversely relative to the stator, then flux density is increased by that movement.

Figure 5.16 shows the same cross section (normal to the axis of rotation) that is shown in Figure 5.15. In Figure 5.16(a), a symmetric four-pole pattern of MMF is present and the rotor is centered within the stator. The result is a symmetric pattern of magnetic flux. In Figure 5.16(b), the same MMF pattern may be present but because the rotor is offset relative to the stator, the airgap on the right-hand side is much shorter than the airgap on the left-hand side. It is clear that more flux will circulate on the right-hand side of the airgap than on the left. The result of this concentration of flux on one side of the machine is a pull on the rotor toward the right-hand side.

Clearly, there is a positive-feedback effect in UMP. The more that the rotor deflects, the more magnetic force comes to exist, encouraging it to deflect further. In effect, the magnetic field acts as a negative spring for static deflections of the rotor

(a)   (b)

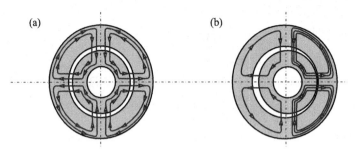

Figure 5.16. The flux distribution in a four-pole machine: (a) symmetric flux distribution; (b) unsymmetric flux distribution due to eccentricity.

(i.e., if the rotor is neither rotating nor translating with any velocity relative to the stator).

It is not difficult to quantify approximately this static stiffness. If the flux field in the airgap is known completely at any instant, then the net force on the rotor can be found by integrating the contributions due to stresses called *Maxwell stresses*, σ_{rr} and $\sigma_{r\theta}$, about a circle passing through the center of the airgap between the rotor and the stator. The resulting forces are

$$f_x = \int_{-\pi}^{\pi} RL \left(\sigma_{rr} \cos \theta - \sigma_{r\theta} \sin \theta \right) d\theta, \tag{5.100}$$

$$f_y = \int_{-\pi}^{\pi} RL \left(\sigma_{rr} \sin \theta + \sigma_{r\theta} \cos \theta \right) d\theta \tag{5.101}$$

where R is the radius to the middle of the airgap, L is the length of the machine, and

$$\sigma_{rr} = \frac{B_r^2 - B_\theta^2}{2\mu_0} \quad \text{and} \quad \sigma_{r\theta} = \frac{2 B_r B_\theta}{2\mu_0} \tag{5.102}$$

where B_r and B_θ are the flux densities in the radial and circumferential directions, respectively, and μ_0 is the permeability of free space ($4\pi \times 10^{-7} \text{N/A}^2$).

For a given rotor eccentricity and a given set of currents in a machine, an electromagnetic FEA package could be used to compute the Maxwell stresses, and these integrations could be performed numerically. The integrations clearly have two components: one due to normal stress, σ_{rr}, and one due to shear stress, $\sigma_{r\theta}$. In virtually all practical cases, the former term is substantially greater than the latter and, for approximation purposes, the latter may be ignored. An upper limit to the UMP force in any iron-bearing machine can be obtained by recognizing that over any nontrivial area of airgap in a machine, the root mean square flux density does not usually exceed 1.2 Tesla. This limit is absolute and it occurs because of the saturation of ferromagnetic materials at around 2 Tesla. The locally averaged value of σ_{rr} is usually less than 600 kPa. Neglecting the lower value of σ_{rr} over half of the circumference of the airgap and assuming that $\sigma_{rr} = 600$ kPa on the other side, an overestimate of the maximum UMP force is computed as $2RL \times 600 \times 10^3$ N, where R and L are the mean radius of the airgap and the axial length of the machine, respectively, in meters. A pessimistic upper limit for the UMP negative stiffness then can be obtained by dividing this force by the nominal thickness of the airgap, δ. This is consistent with the expression obtained by Merrill (1994), where the stiffness is

$$k_{UMP} = -\frac{\pi}{2} \frac{B_p^2}{2\mu_0} \frac{2RL}{\delta} \qquad (5.103)$$

where B_p represents the amplitude of the fundamental component of the magnetic flux density and is approximately 1 Tesla.

For some machines, the effect of UMP can be modeled reasonably well as a simple negative stiffness. The negative stiffness has the effect of reducing the lower critical speeds of the machine; in extreme cases, it may be sufficient to cause pullover in the machine. In addition to the negative stiffness effect, there is also a pulsating force in the direction of the smallest gap. The pulsating force peaks each time a pole of the machine magnetic field passes the location of the smallest gap.

In other machines, the unbalanced magnetic field induces currents in closed circuits within the machine and the dynamics are more complicated; induction machines are a case in point. Früchtenicht et al. (1982) derive equations for the net magnetic force between rotor and stator when the rotor has a circular orbit at a fixed frequency within the stator. Arkkio et al. (2000) observe that the UMP behavior of most induction machines can be represented well by introducing extra state-variables relating to the magnetic field, thereby extending the analysis of Früchtenicht et al. to the general case of noncircular and nonperiodic orbits. These additional state-variables represent forces due to patterns of current in the rotor having $(p-1)$ and $(p+1)$ complete waves of current around the circumference. The differential equations for these extra state-variables, denoted by q_{mx}, q_{my}, q_{px}, and q_{py}, are

$$\begin{Bmatrix} \dot{q}_{mx} \\ \dot{q}_{my} \\ \dot{q}_{px} \\ \dot{q}_{py} \end{Bmatrix} = \begin{bmatrix} z_{mr} & -z_{mi} & 0 & 0 \\ z_{mi} & z_{mr} & 0 & 0 \\ 0 & 0 & z_{pr} & -z_{pi} \\ 0 & 0 & z_{pi} & z_{pr} \end{bmatrix} \begin{Bmatrix} q_{mx} \\ q_{my} \\ q_{px} \\ q_{py} \end{Bmatrix} + \begin{bmatrix} k_{mr} & -k_{mi} \\ k_{mi} & k_{mr} \\ k_{pr} & -k_{pi} \\ k_{pi} & k_{pr} \end{bmatrix} \begin{Bmatrix} u \\ v \end{Bmatrix} \qquad (5.104)$$

where u and v are the instantaneous deflections of the rotor center relative to the center of the stator in the x and y directions. The coefficients z_{mr}, z_{mi}, z_{pr}, z_{pi}, k_{mr}, k_{mi}, k_{pr}, and k_{pi} are dependent on the operating conditions of the machine (i.e., voltage and rotational speed). Quantities $\{z_{mr}, z_{pr}\}$ are invariably negative numbers determined only by the operating voltage, whereas $\{z_{mi}, z_{pi}\}$ are given by

$$z_{mi} = \frac{\omega_{supply}}{p}(1 - sl(1 - p)) \qquad (5.105)$$

$$z_{pi} = \frac{\omega_{supply}}{p}(1 - sl(1 + p)) \qquad (5.106)$$

where p represents the number of pole pairs in the machine, ω_{supply} represents the angular frequency of the supplied electricity, and sl represents the *slip* of the machine through the formula

$$\Omega = \frac{\omega_{supply}}{p}(1 - sl) \qquad (5.107)$$

The net magnetic force acting on the rotor is given by

$$
\begin{Bmatrix} f_x \\ f_y \end{Bmatrix} = \begin{bmatrix} 1 & 0 & 1 & 0 \\ 0 & 1 & 0 & 1 \end{bmatrix} \begin{Bmatrix} q_{mx} \\ q_{my} \\ q_{px} \\ q_{py} \end{Bmatrix} + \begin{bmatrix} k_{0r} & -k_{0i} \\ k_{0i} & k_{0r} \end{bmatrix} \begin{Bmatrix} u \\ v \end{Bmatrix}
\tag{5.108}
$$

where f_x and f_y are the forces in the x and y directions and the coefficients k_{0r} and k_{0i} depend on the operating conditions of the machine. The coefficients in Equations (5.104) and (5.108) may be determined either experimentally or from an electromagnetic FEA of the machine. Tenhunen et al. (2003), for example, presented an elegant method of analysis to extract all of the coefficients from a single time domain simulation in which the rotor center is forced to undergo a step in displacement. Remarkably, although modern machine designs usually involve significant saturation of the magnetic iron in the rotor and stator (i.e., magnetic nonlinearity), the previous linear equations model the machine behavior extremely well.

Equation (5.104) is invariably stable in its own right in the sense that all eigenvalues of the 4×4 matrix have negative real parts. Equation (5.108) introduces negative stiffness but this is rarely sufficiently large to overcome the positive mechanical stiffness. Nevertheless, the dynamic coupling between the rotor displacements $\{u, v\}$ and the rotor forces $\{f_x, f_y\}$ through the mechanical behavior of the machine can lead to instabilities. Even the simplest mechanical model of a rotor (i.e., Jeffcott) can be used with Equations (5.104) and (5.108) to discover instabilities. The likelihood of rotordynamic problems is especially high for machines running close to or above the first critical speed and with relatively low values of slip.

5.6 Modeling Foundations and Stators

The dynamic characteristics of the bearing supports are important, particularly for very large machines. These supports consist of the case of the machine, often called the *stator*, and the remaining structure, referred to as the *foundation*. Although they are physically distinct, dynamically they appear as a single entity. Modeling foundations and stators is difficult, and structures that are nominally identical can have differing dynamic characteristics. This is due to the fact that the ground conditions to which the foundations are anchored can vary, reinforced concrete properties can vary depending on the exact position of the reinforcing bars, the quality of the concrete and even plain steel constructions can vary due to the details of welds, and so on. Detailed foundation and stator modeling is beyond the scope of this book but the following discussion gives the reader some insight into the problems.

A straightforward modeling approach assumes that the combined stator and foundation can be represented as an assembly of springs, masses, and dampers. Engineers must use their judgment and experience to decide how many masses and springs are required, how they are interconnected, and which parameter values should be used. The number of degrees of freedom required in this model increases the size of the rotor–bearing system mass and stiffness matrices; the foundation model parameters are inserted in these matrices as appropriate. This approach has the advantage of introducing a relatively small number of extra degrees of freedom into the rotor–bearing model.

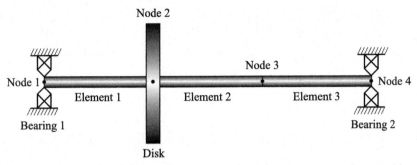

Figure 5.17. The simple three-element example to demonstrate matrix assembly.

For more geometrically complex foundations and/or stators, the FE modeling approach may be used, but this often requires many extra degrees of freedom, more than those of the rotor–bearing model. Under these circumstances, it might be necessary to use a model-reduction method to reduce the size of the foundation dynamic-stiffness matrix. One approach (if there is no coupling between bearing pedestals) incorporates the dynamic-stiffness matrix of the foundation into the speed-dependent bearing parameters.

Because some nominally identical foundations have differing dynamic properties, it is difficult to model them accurately; all models, including FE models, are limited by the accuracy with which parameter values are known. An alternative approach measures the bearing-support dynamics experimentally. Provided the experiment is conducted with care, the measured dynamic characteristics are more accurate than any model. However, the dynamic characteristics are not necessarily in a form that easily can be incorporated into the rotor–bearing model. FE models can be improved and experimental data and the techniques of model updating (Friswell and Mottershead, 1995).

5.7 Assembly of the Full Equations of Motion

The general form of the equation for an n degree of freedom system after assembly becomes

$$\mathbf{M\ddot{q}} + \mathbf{C\dot{q}} + \Omega\mathbf{G\dot{q}} + \mathbf{Kq} = \mathbf{0} \tag{5.109}$$

where $\mathbf{q}$ consists of the displacement and rotations at the nodes. For certain types of bearings and seals, the damping and stiffness matrices are dependent on rotor spin speed. Internal damping also causes the stiffness matrix to depend on rotor spin speed (see Chapter 7). Assuming the foundation is rigid and the bearings do not contain any internal degrees of freedom, then all of the degrees of freedom lie on the shaft (i.e., a shaft-line model). Consider the example shown in Figure 5.17, with three elements, four nodes, a disk at the second node, and short rigid bearings at each end.

The matrices for the shaft are assembled and then the constraints are applied. Before applying the support constraints, the shaft model has the following 16 degrees of freedom

$$\mathbf{q} = [u_1, v_1, \theta_1, \psi_1, u_2, v_2, \theta_2, \psi_2, u_3, v_3, \theta_3, \psi_3, u_4, v_4, \theta_4, \psi_4]^\top \tag{5.110}$$

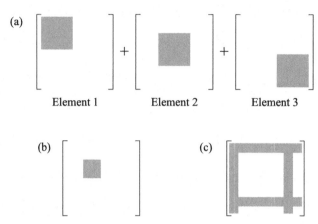

Figure 5.18. The degrees of freedom affected during matrix assembly: (a) shaft elements, (b) disk, and (c) constraints (bearings).

Detailed assembly of the matrices is not shown because it is explained fully in Chapter 4. However, how the element matrices contribute to the full system matrices is indicated. The shaft-element matrices are 8×8 and the full-system matrices are 16×16. The first element corresponds to global degrees of freedom 1 to 8, the second element to degrees of freedom 5 to 12, and the third element to degrees of freedom 9 to 16. Thus, the full matrices consist of the sum of contributions from the elements shown in Figure 5.18(a), where the gray blocks indicate the degrees of freedom corresponding to each element. In a similar way, the mass and gyroscopic matrices due to the disk involve degrees of freedom 5 to 8, which is shown in Figure 5.18(b). Once the shaft matrices have been assembled, the effect of the bearings must be included. For flexible bearings, additional terms are added in degrees of freedom 1 to 4 for bearing 1 and degrees of freedom 13 to 16 for bearing 2 (irrespective of whether the stiffness and damping varies with rotor speed). In Figure 5.17, the bearings are short and rigid so that the displacements are fixed at the bearings, but the rotations are not constrained. This means that the rows and columns corresponding to degrees of freedom 1, 2, 13, and 14 must be removed, which is shown in Figure 5.18(c). This is achieved formally by the transformations described in Section 2.5.

Thus far, damping has not been mentioned other than in the bearing models. Often most of the energy dissipates in the bearings; however, in some cases (e.g., laminated rotors), significant damping occurs in the rotor. Rotor damping can cause instabilities, which are considered in Chapter 7.

5.7.1 Speed Dependence of the System Matrices

In general, rotating machines have several effects that can cause system matrices to be a function of speed. These effects include the following:

- Gyroscopic effects make a (skew-symmetric) contribution to the imaginary part of the dynamic-stiffness matrix. This is proportional to speed and the frequency of oscillation.

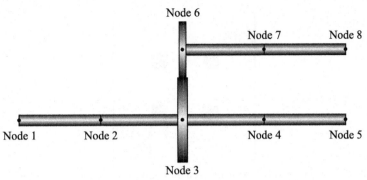

Figure 5.19. Rotors coupled by a gear.

- Stiffness and damping properties of bearings depend on speed. This is particularly true for fluid bearings.
- Internal damping in the rotor results in a skew-symmetric contribution to the stiffness matrix, which is proportional to speed and independent of the frequency of oscillation. This may cause instabilities and is considered in detail in Chapter 7.
- At high rotational speeds, the stiffness of the rotor can be reduced due to loss of tightness-of-shrink fits. Internal damping also can be affected.
- There are interactions between the rotor and the stator other than at bearings, which can depend on speed. These include dynamic UMP and rotor–stator forces at seals.

When these effects are taken together, it is clear that the mass, damping, and stiffness properties can be strong functions of rotational speed.

5.7.2 Branching

Thus far, we have considered the dynamics of a single rotor supported in bearings. However, more complex systems of rotors can occur and these are generally called *branched systems*. Figure 5.19 shows such a system. The nodes on each gear are connected by a stiffness matrix, although this condition could be changed and, for example, a rigid connection applied. This complexity does not present any difficulty in an FEA, and the elements are assembled in a similar way as before. The assembly process is slightly different from that shown in Figure 5.18. In a branched system, once the global coordinates are defined, the element matrices are inserted into the system matrices in the positions determined by the positions of the local element coordinates in the global vector. This is demonstrated diagrammatically in Figure 5.20 for the example of Figure 5.19, in which the non-zero degrees of freedom are shaded. The black squares denote the stiffness matrix for the connection at the gears, which is split into four 4×4 blocks that slot into the system matrix at positions corresponding to degrees of freedom 9–12 and 21–24. The blocks in two different shades of gray denote the positions corresponding to each of the two shaft lines.

If two or more rotors are connected by gears so that each shaft rotates at a different speed, the gyroscopic terms and the element matrices for the fluid bearings

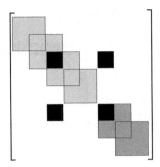

Figure 5.20. The degrees of freedom affected during matrix assembly for a branched system.

must account for this. Similarly, the gearing affects the forcing frequencies. Connecting shafts through gears often causes coupling between the lateral and torsion degrees of freedom (see Section 10.9). Another example of branching occurs when a rotor carries a flexible disk (see Section 10.4).

5.8 Free Response of Complex Systems

To determine the system eigenvalues, we must rewrite the equation of motion, Equation (5.109), in state–space form (compare to Equation (2.71)), giving

$$\begin{bmatrix} \mathbf{C}+\Omega\mathbf{G} & \mathbf{M} \\ \mathbf{M} & \mathbf{0} \end{bmatrix} \frac{\mathrm{d}}{\mathrm{d}t} \begin{Bmatrix} \mathbf{q} \\ \dot{\mathbf{q}} \end{Bmatrix} + \begin{bmatrix} \mathbf{K} & \mathbf{0} \\ \mathbf{0} & -\mathbf{M} \end{bmatrix} \begin{Bmatrix} \mathbf{q} \\ \dot{\mathbf{q}} \end{Bmatrix} = \begin{Bmatrix} \mathbf{0} \\ \mathbf{0} \end{Bmatrix} \tag{5.111}$$

This may be expressed conveniently in terms of the state vector $\mathbf{x} = \begin{Bmatrix} \mathbf{q} \\ \dot{\mathbf{q}} \end{Bmatrix}$ as

$$\mathbf{A}\dot{\mathbf{x}} + \mathbf{B}\mathbf{x} = \mathbf{0} \tag{5.112}$$

where the definition of the matrices follows from Equation (5.111). Of course, $\mathbf{A}$ and $\mathbf{B}$ depend on the shaft speed, Ω, because the bearing properties and gyroscopic forces vary with shaft speed. Equation (5.112) has the same form as Equation (2.72); the only differences arise from the inclusion of the gyroscopic matrix and the fact that the matrices may depend on rotor speed, as indicated previously. The process of calculating the eigenvalues and eigenvectors of Equation (5.112) is the same as described in Chapter 2; however, the eigenvalue problem must be solved for each shaft speed of interest. Even if the rotor is undamped but the gyroscopic effects are significant, the solution cannot be obtained directly from the second-order form. The dependence of the natural frequency on shaft speed is shown by the natural frequency map, where the natural frequencies are plotted as a function of shaft speed. Many of the features follow from the discussion of natural frequency maps for simple systems in Chapter 3. However, for complex systems, many more natural frequencies may be significant.

A feature of rotor-system models is that the second-order system matrices in Equation (5.109) are no longer symmetric, in general. The eigenvalue problem in second-order form is

$$\left[\mathbf{M}s^2 + \mathbf{C}s + \Omega\mathbf{G}s + \mathbf{K} \right] \mathbf{u}_R = \mathbf{0} \tag{5.113}$$

where the damping and stiffness matrices may be speed-dependent and non-symmetric and the gyroscopic matrix is skew-symmetric. The eigenvector, $\mathbf{u}_R$, has the subscript R to denote the right eigenvector. There is also a left eigenvector defined by

$$\mathbf{u}_L^H \left[\mathbf{M}s^2 + \mathbf{C}s + \Omega \mathbf{G}s + \mathbf{K} \right] = \mathbf{0} \tag{5.114}$$

where the superscript H denotes the *Hermitian transpose* or, equivalently

$$\left[\mathbf{M}^\top \bar{s}^2 + \mathbf{C}^\top \bar{s} + \Omega \mathbf{G}^\top \bar{s} + \mathbf{K}^\top \right] \mathbf{u}_L = \mathbf{0}$$

where the overbar denotes the complex conjugate. For systems with symmetric matrices, these eigenvectors are equal. The right eigenvectors are associated with the response, whereas the left eigenvectors are associated with the force input to the modes.

5.8.1 Features of Eigenvalues and Eigenvectors

The eigenvalues and natural frequencies of rotating systems have a number of important features:

- With a cyclically symmetrical rotor and isotropic support characteristics, eigenvalues occur in groups of four, instead of the normal groupings of two associated with static structures. These eigenvalues occur in complex conjugate pairs, each corresponding to a natural frequency and damping ratio.
- Anisotropy in the bearings and supports causes a splitting of a group of four eigenvalues into two pairs.
- Gyroscopic couples (associated with a steady shaft speed) also may contribute to separation of the eigenvalues with rotational speed. These couples also cause the mode shapes to become associated with particular whirl directions – namely, forward and backward.
- Damping in the stationary part of the system causes the real parts in the computed characteristic roots to become more negative (implying stability). Chapter 7 considers the effect of damping in the rotor.
- *Overdamped* roots may occur that have a zero imaginary part and a substantial negative real part.

When a rotating machine or a static structure is vibrating in one of its natural modes, a transfer of energy occurs between strain energy and kinetic energy. This transfer of energy from one form into another occurs at twice the frequency of the oscillations; the distribution of strain energy in the structure or machine is often quite different from the distribution of kinetic energy. The natural modes of any rotating machine can be divided roughly into those in which the kinetic energy is held predominantly in the rotor and the others. Broadly speaking, the modes that are significantly excited by unbalance forces are those in the former category, which is why a relatively crude representation of the bearings and bearing-support dynamics is often sufficient in models of rotating machines. In general, rotor modes do occur in groups of two, associated with four eigenvalues. These groups are tightly bunched if the characteristics of the bearings and bearing supports are close to isotropic, and if gyroscopic effects are not very significant. Often, when engineers speak about the

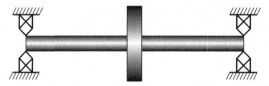

Figure 5.21. The rotor system used to demonstrate discretization errors (Example 5.8.1).

first natural frequency of a machine, they are referring to a group of two natural frequencies and four roots of the characteristic equation.

For each mode of vibration, a particular node can precess either forward or backward relative to the direction of rotation. Section 3.6.1 determines the direction of rotation for the mode of a rigid rotor, and the same approach may be adopted for the modal displacement and rotation at each node of a complex rotor. The displacement and the rotation are treated separately. Sometimes, for a particular mode, some nodes may rotate in a forward direction and some in a backward direction.

When using FEMs to model a rotor–bearing system – or, indeed, any structure – three factors that affect the accuracy of the eigenvalues must considered, as follows

- How many degrees of freedom should be used to describe the system?
- What type of elements should be used to model the system?
- Are there any *special* modeling features that should be considered?

These effects are now considered in turn.

5.8.2 Number of Degrees of Freedom Required in a Model

The minimum number of coordinates – and, hence, the number of degrees of freedom required in a model – is fixed by the geometric complexity of the system. For example, a uniform shaft supported by a pair of short bearings and carrying two rigid disks requires at least 12 degrees of freedom. However, the more degrees of freedom used in a model, the greater the accuracy with which the natural frequencies are computed. An analyst should perform a convergence test on any model produced. A convenient procedure is to double the number of elements in the model by subdividing each existing element into two, and then check that the natural frequencies of interest do not change significantly.

Examples 4.4.1 and 4.5.1 show that as the number of elements in a model is increased, the accuracy of the model is also increased and predicted natural frequencies and other dynamic characteristics become more accurate. Of course, this accuracy is only within the limitations of the assumed model (see Section 4.1). The effect of the number of elements is now demonstrated for a simple rotor example.

EXAMPLE 5.8.1. Consider the symmetric rotor with a single central disk, shown in Figure 5.21. The shaft is hollow with an outside diameter of 80 mm, an inside diameter of 30 mm, and a length of 1.2 m. The shaft is modeled using Euler-Bernoulli elements, which neglect the shear and rotary inertia effects. There is no internal shaft damping. The disk has a diameter of 400 mm and a thickness of 80 mm. The shaft and disk are both made of steel, with material properties

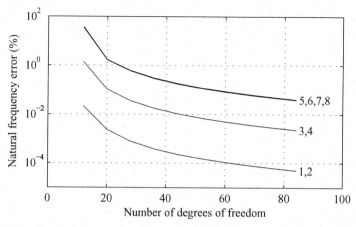

Figure 5.22. The convergence of the natural frequencies at rest for Example 5.8.1. The numbers denote the modes.

$E = 200\,\text{GN/m}^2$, $\rho = 7{,}800\,\text{kg/m}^3$, and Poisson's ratio $\nu = 0.27$. The bearings at the ends of the shaft are both assumed to be rigid and short. Investigate the effect of increasing the number of elements used to model the shaft.

Solution. Because the bearings are assumed to be short and stiff, the translational degrees of freedom at the corresponding nodes are constrained to be zero. Because there is a central disk, the smallest number of elements possible is two, which produces a model with eight degrees of freedom. More accurate models are generated by splitting the shaft into elements of shorter length. A model with 80 elements (i.e., 324 degrees of freedom) is used as the reference model. Figures 5.22 and 5.23 show the convergence of the lower natural frequencies and demonstrate that they are estimated more accurately than the higher natural frequencies, for a given number of degrees of freedom.

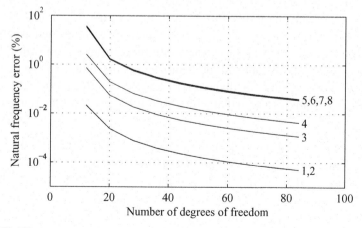

Figure 5.23. The convergence of the natural frequencies at 5,000 rev/min for Example 5.8.1. The numbers denote the modes.

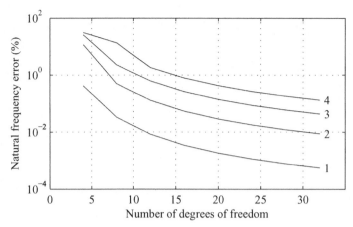

Figure 5.24. The convergence of the first four natural frequencies of the pinned beam, $R = 0.04$ (Example 5.8.2). The numbers denote the modes.

5.8.3 The Effect of Shear and Rotary Inertia

It is common practice to model the rotor using shaft elements. An engineer must decide whether to use Euler-Bernoulli or Timoshenko shaft elements, with or without the inclusion of the beam gyroscopic effects. Generally, the use of Timoshenko shaft elements is recommended, although in many instances, Euler-Bernoulli elements give satisfactory results. To illustrate the accuracy of a model composed of Timoshenko beam elements, first consider a stationary, simply supported, uniform circular shaft. When shear deflection and rotary inertia effects are ignored (i.e., we use the Euler-Bernoulli beam theory), the *exact* natural frequencies may be computed directly from the partial differential equations (Inman, 2008). Goodman and Sutherland (1951) have showed that it is possible to determine the exact natural frequencies of a simply supported beam or shaft – including the effect of shear deformation and rotary inertia – by introducing a modification to the Euler-Bernoulli beam theory. Blevins (1979) gives the expressions for the natural frequencies.

> **EXAMPLE 5.8.2.** Consider a beam of length 1 m with pinned ends. The material properties for the shaft are steel, with Poisson's ratio taken as 0.27. Demonstrate the convergence of the natural frequencies with an increasing number of elements by comparing the results with those from the analytical solution. Also, vary the shaft diameter to demonstrate the effect of the slenderness ratio, R (i.e., the ratio of diameter to length), on the model accuracy. Although $R = 0$ implies a zero diameter, this represents the Euler-Bernoulli beam theory.

Solution. Figure 5.24 shows the convergence of the first four natural frequencies as the number of elements is increased for the case when $R = 0.04$. There are two degrees of freedom per node and the boundary conditions at the ends of the beam are pinned, implying that the displacement is zero and the rotation unconstrained. Thus, the number of degrees of freedom is twice the number of elements. Clearly, the model converges to the exact solution with increasing model order, and the resonances associated with the lower modes are more accurate.

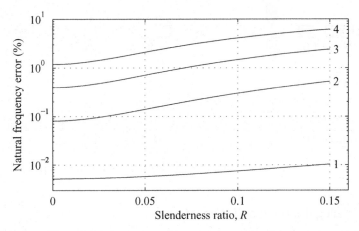

Figure 5.25. The effect of slenderness on the accuracy of the first four natural frequencies of the pinned beam with 12 degrees of freedom (Example 5.8.2). The numbers denote the modes.

The number of elements in the FE model is now fixed at six (i.e., 12 degrees of freedom) and the slenderness ratio R varied. Note that $R = 0$ is a Euler-Bernoulli beam-theory model. Figure 5.25 shows the error in the natural frequencies between the FE model and the exact solution as R varies. It is clear that as the beam diameter increases, the accuracy decreases, and the lower modes are more accurate.

Table 5.1 shows the modeling accuracy for different values of the slenderness ratio as well as when the shear and rotary inertia are neglected in different combinations. A model with six finite elements is used. From the table, we see that the lower natural frequencies are accurately predicted for a slenderness ratio as high as 0.1 using Timoshenko beam elements. However, as the slenderness ratio increases, the relative error increases. Also, from Table 5.1, it is more important to include shear effects than rotary inertia in this example.

Table 5.1. *The percentage error in the first two natural frequencies of the simply supported stationary shaft (Example 5.8.2)*

Shear effects		Included		Included
Rotary inertia			Included	Included
Mode number	Slenderness, R			
	0	0.0052	0.0052	0.0052	0.0052
	0.02	0.0529	0.0183	0.0406	0.0060
	0.04	0.1960	0.0574	0.1466	0.0084
1	0.06	0.4337	0.1219	0.3223	0.0123
	0.08	0.7647	0.2107	0.5664	0.0177
	0.10	1.1875	0.3223	0.8769	0.0246
	0	0.0810	0.0810	0.0810	0.0810
	0.02	0.2720	0.1424	0.2225	0.0932
	0.04	0.8411	0.3228	0.6426	0.1294
2	0.06	1.7773	0.6115	1.3282	0.1877
	0.08	3.0635	0.9921	2.2592	0.2654
	0.10	4.6778	1.4447	3.4097	0.3594

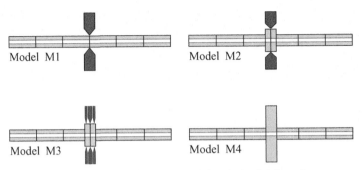

Figure 5.26. The four models of the shaft–disk interface.

5.8.4 Modeling the Shaft and Disk Interface

Although some shafts consist of a single forged or machined rotor, disks often are manufactured separately and then mounted on a shaft. In such cases, the disk may be keyed to or shrunk onto the shaft. If the disk is shrunk onto the shaft, the disk increases the stiffness of the shaft locally. In contrast, if it is merely keyed to the shaft, it provides little if any local shaft stiffening. We can accommodate these conditions using one of the four models shown in Figure 5.26. Model M1 attaches the disk to one node of a shaft. There is no local stiffening, only an increase in mass and inertia at the node due to the disk. Models M2 and M3 provide a hub on the shaft onto which the disk is attached. Let h denote the width of the disk and d denote the outside diameter of the shaft. One approach makes the hub of length h, with a diameter $h + d$. In model M2, the disk is attached to the node in the middle of the hub. In model M3 the disk is divided into three parts with thicknesses and, hence, inertias in the proportion to 1:2:1. This allows for the fact that the mass and inertia are actually distributed over the disk thickness h. Model M4 uses a large-diameter shaft element to represent the disk. This model has an even higher stiffness. In all models, the total mass of the shaft and any attached disk is the same.

Another technique used to improve the accuracy of models is to introduce a *stiffness diameter* into a shaft element. In this model, separate diameters are used for the mass and stiffness, thus allowing the shaft to be artificially stiffened. Engineering insight is required to estimate the level of stiffening required, and this approach is not considered further.

EXAMPLE 5.8.3. Consider a hollow shaft carrying a central disk and mounted on short, rigid bearings at each end. The shaft is 1.2 m long and it has an outside diameter of 80 mm and an inside diameter of 30 mm. The central disk has a diameter of 400 mm and a thickness of 80 mm. The shaft and disk have a modulus of elasticity of 200 GN/m^2 and a density of 7,800 kg/m^3. Poisson's ratio is 0.27. This is the same as the rotor used in Example 5.8.1. The rotor, consisting of the shaft and central disk, spins at 3,000 rev/min. Investigate the effect of the shaft–disk interface models on the natural frequencies of the machine.

Solution. The FE model of the shaft uses six Timoshenko shaft elements, but the disk–shaft interface is modeled in the four ways shown in Figure 5.26. The resulting system natural frequencies are given in Table 5.2. The natural

Table 5.2. *Natural frequencies for the four models of the shaft–disk interface*

Model	Degrees of freedom in model	Natural frequency (Hz)			
		Pair 1	Pair 2	Pair 3	Pair 4
M1	24	53.64	261.7	736.6	792.8
		53.70	326.3	737.9	821.1
M2	32	58.62	278.2	826.6	919.8
		58.69	337.1	828.0	950.4
M3	32	58.64	276.9	827.9	925.2
		58.70	334.5	829.3	954.6
M4	28	58.99	279.3	835.2	931.4
		59.06	337.8	836.7	961.9
M5	1,950	56.63	265.6	789.7	830.6
		56.69	328.3	791.0	859.3

frequencies for model M1 are significantly lower than for models M2, M3, and M4. This can be explained by the fact that model M1 does not increase the local shaft stiffness at the disk and approximates a disk keyed onto the shaft. The other three models give frequencies that are much closer to one another. These models approximate a disk shrunk on the shaft. Of course, none of the models is an exact representation of reality. For example, model M3 attempts to allow for the distributed nature of the disk, but still only allows the disk to join to the shaft hub at three points.

Mode pairs 1 and 3 separate slightly due to the gyroscopic effects. The machine is symmetric; hence, the disk has zero rotation in these modes, resulting in zero gyroscopic effects due to the disk. However, gyroscopic effects are included in the shaft elements, which gives rise to the slight separation in mode pairs 1 and 3. Of course, the separation in mode pairs 2 and 4 is significantly larger because of the gyroscopic effects arising from the disk.

Also shown in Table 5.2 (denoted model M5) are the results obtained from a detailed FE model using solid elements. These elements are discussed in more detail in Chapter 10 but may be considered as a more accurate representation of the shaft–disk interface. The natural frequencies fall somewhere between model M1 and the other three models. These results show the need for care when modeling the shaft–disk interface.

5.9 Modeling Examples

We now consider a range of examples that demonstrate the features highlighted in Section 5.8.1, using many of the models described in this chapter.

EXAMPLE 5.9.1. *Isotropic Bearings.* A 1.5-m-long shaft, shown in Figure 5.27, has a diameter of 0.05 m. The disks are keyed to the shaft at 0.5 and 1 m from one end. The left disk is 0.07 m thick with a diameter of 0.28 m; the right disk is 0.07 m thick with a diameter of 0.35 m. For the shaft, $E = 211\,\text{GN/m}^2$ and $G = 81.2\,\text{GN/m}^2$. There is no internal shaft damping. For both the shaft and the disks, $\rho = 7,810\,\text{kg/m}^3$. The shaft is supported by identical bearings at its ends.

Figure 5.27. The layout of the two-disk, two-bearing rotor (Example 5.9.1).

These bearings are isotropic and have a stiffness of 1 MN/m in both the x and y directions. The bearings contribute no additional stiffness to the rotational degrees of freedom and there is no damping or cross-coupling in the bearings. Create an FE model of the shaft using six Timoshenko beam elements and investigate the dynamics of the machine at 0 and 4,000 rev/min.

Solution. The shaft is divided into elements of equal length, the FE model assembled using the techniques described in this chapter, and the roots of the characteristic equation determined. These roots are of the form $s_1, s_2 = -\zeta\omega_n \pm j\omega_n\sqrt{1-\zeta^2}$ and they occur in complex conjugate pairs. From Table 5.3, we see that at 0 rev/min, the roots are purely imaginary; hence, the damping ratio is zero, $\zeta = 0$. This is to be expected because no damping has been specified in the system. From the imaginary part of the root, we can deduce the natural frequencies, which are shown in Table 5.3.

At 0 rev/min, the natural frequencies occur in pairs. This is because in the x and y directions, the rotor–bearing system is uncoupled and the inertia and stiffness properties of the rotor in the x and y directions are identical. When the shaft is spinning at 4,000 rev/min, each pair of natural frequencies separate due to gyroscopic effects (see Table 5.3). However, the eigenvalues remain purely imaginary and therefore undamped. The separation of the natural frequencies is more clearly illustrated by the natural frequency map, shown in Figure 5.28. As the shaft speed increases, each natural frequency pair diverges: one frequency increases and one decreases. This also happens to the first and second natural frequencies, although the divergence is so small that it is barely detectable on the plot.

Table 5.3. *Eigenvalues and natural frequencies for a rotor supported by isotropic bearings (Example 5.9.1)*

0 rev/min		4,000 rev/min	
Root s (rad/s)	ω_n (Hz)	Root s (rad/s)	ω_n (Hz)
$0 \pm 86.66j$	13.79	$0 \pm 85.39j$	13.59
$0 \pm 86.66j$	13.79	$0 \pm 87.80j$	13.97
$0 \pm 274.31j$	43.66	$0 \pm 251.78j$	40.07
$0 \pm 274.31j$	43.66	$0 \pm 294.71j$	46.90
$0 \pm 716.78j$	114.08	$0 \pm 600.18j$	95.52
$0 \pm 716.78j$	114.08	$0 \pm 827.08j$	131.63

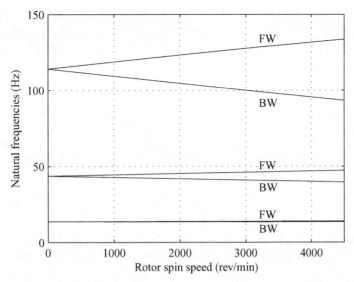

Figure 5.28. The natural frequency map for a rotor supported by isotropic bearings (Example 5.9.1). FW indicates forward whirl and BW indicates backward whirl.

The mode shapes of the system are also important. Consider first when the shaft is stationary. It has already been stated that the model is actually two uncoupled and identical systems vibrating in the x and y directions. Thus, the modes of vibration in the x and y directions are identical. These are shown for the x direction in Figure 5.29. Mode 3 has one *vibration node* where the displacement is zero, shown by a circle. Mode 5 has two vibration nodes, also shown by circles in Figure 5.29.

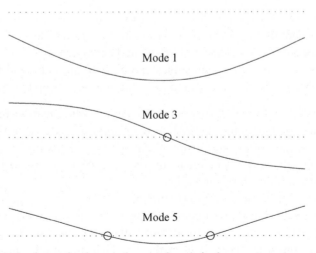

Figure 5.29. Mode shapes in the xz plane at 0 rev/min for a rotor supported by isotropic bearings (Example 5.9.1). The dotted lines represent the shaft centerline at rest and the circles are vibration nodes. Modes 2, 4, and 6 in the yz plane are identical to modes 1, 3, and 5, respectively.

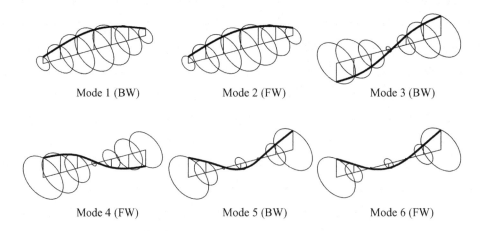

Figure 5.30. Mode shapes at 4,000 rev/min for a rotor supported by isotropic bearings (Example 5.9.1). FW indicates forward whirl and BW indicates backward whirl.

If a system has a pair of identical natural frequencies, then a linear combination of the modes is also a mode. Thus, the modes shown in Figure 5.29 are not unique. Because of the symmetry, the same modes can occur in any direction. We have arbitrarily chosen to separate the mode shapes into motion in the xz plane and motion in the yz plane.

At 4,000 rev/min, there are no repeated frequencies; therefore, the modes have a unique shape (but their amplitude is still arbitrary). When we plot the modes, we see that they form circles at points along the shaft (Figure 5.30), and the relative amplitude of the mode varies along the shaft. Furthermore, in this case, we find that the odd-numbered modes precess backward at all nodes, whereas the even-numbered modes precess forward at all nodes. This is shown in Figure 5.31.

EXAMPLE 5.9.2. *Anisotropic Bearings.* This system is the same as that of Example 5.9.1 except that the isotropic bearings are replaced by anisotropic bearings. Both bearings have a stiffness of 1 MN/m in the x direction and 0.8 MN/m in the y direction. Calculate the eigenvalues and mode shapes at 0 and 4,000 rev/min and plot the natural frequency map for rotational speeds up to 4,500 rev/min.

Solution. Table 5.4 shows the roots of the characteristic equation and the natural frequencies. At 0 rev/min, the system is uncoupled in the x and y directions; however, because of the anisotropy of the bearings, the two uncoupled systems are no longer dynamically identical. In consequence, they have different natural frequencies in the x and y directions.

The natural frequency map (Figure 5.32) shows the divergence of the natural frequencies due to gyroscopic effects, but it also shows that at 0 rev/min, the frequencies are not in identical pairs. The mode shapes at 0 rev/min appear identical to those in Figure 5.29 in the xz plane and similar in the yz plane. The bearing anisotropy means that these mode shapes are now unique. Figure 5.33 shows the mode shapes at 4,000 rev/min; the main difference between

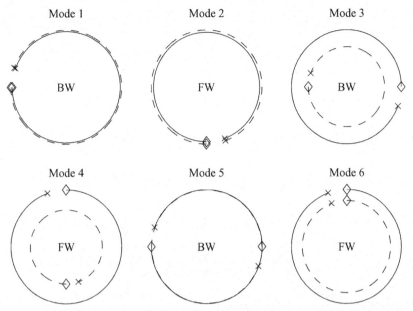

Mode 1   Mode 2   Mode 3

BW   FW   BW

Mode 4   Mode 5   Mode 6

FW   BW   FW

Figure 5.31. The axial view of the mode shapes at the left disk (node 3, solid) and right disk (node 5, dashed) for a rotor supported by isotropic bearings (Example 5.9.1) at 4,000 rev/min. The cross denotes the start of the orbit and the diamond denotes the end. FW indicates forward whirl and BW indicates backward whirl.

these mode shapes and those for the isotropic bearings (see Figure 5.30) is that the orbit of any point on the shaft is now elliptical rather than circular. This is highlighted in Figure 5.34, which shows an axial view of the mode shapes.

EXAMPLE 5.9.3. *System with Mixed Modes.* Analyze the system of Example 5.9.2 with bearing stiffnesses of 1 MN/m in the x direction and 0.2 MN/m in the y direction. Calculate the eigenvalues and mode shapes at 4,000 rev/min and show that some modes contain both forward and backward whirling components.

Solution. Table 5.5 shows the natural frequencies and orbit properties for the given anisotropic support stiffnesses. The orbit properties are given using the

Table 5.4. *Eigenvalues and natural frequencies for a rotor supported by anisotropic bearings (Example 5.9.2)*

0 rev/min		4,000 rev/min	
Root s (rad/s)	ω_n (Hz)	Root s (rad/s)	ω_n (Hz)
$0 \pm 82.65_J$	13.15	$0 \pm 82.33_J$	13.10
$0 \pm 86.66_J$	13.79	$0 \pm 86.86_J$	13.82
$0 \pm 254.52_J$	40.51	$0 \pm 239.64_J$	38.14
$0 \pm 274.31_J$	43.66	$0 \pm 287.25_J$	45.72
$0 \pm 679.49_J$	108.14	$0 \pm 583.49_J$	92.86
$0 \pm 716.79_J$	114.08	$0 \pm 806.89_J$	128.42

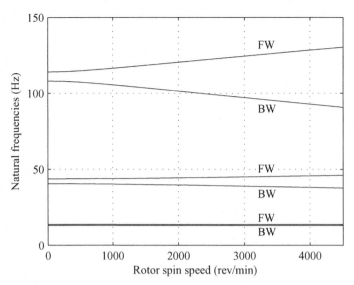

Figure 5.32. The natural frequency map for a rotor supported by anisotropic bearings (Example 5.9.2). FW indicates forward whirl and BW indicates backward whirl.

parameter κ defined in Section 3.6.1, based on the modal displacements in the two translational directions at each node of the FE model. If κ is negative, then the mode at that location is backward-whirling; if κ is positive, then the mode is forward-whirling. The magnitude of κ gives the aspect ratio of the ellipse. Mode 1 is backward-whirling and mode 6 is forward-whirling. However, modes 2 to 5 are mixed modes, where the rotor is forward-whirling at some locations along the shaft and backward-whirling at others. Figure 5.35 shows mode 5 and highlights that the mode is forward-whirling at the left disk (node 3) and backward-whirling at the right disk (node 5). Only node 3 is forward-whirling in mode 5, and the relative modal amplitude at this node is small.

EXAMPLE 5.9.4. *Cross-Coupling in the Bearings.* This system is the same as that of Example 5.9.1 except that some coupling is introduced in the bearings between

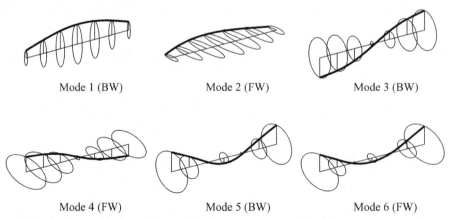

Figure 5.33. Mode shapes at 4,000 rev/min for a rotor supported by anisotropic bearings (Example 5.9.2). FW indicates forward whirl and BW indicates backward whirl.

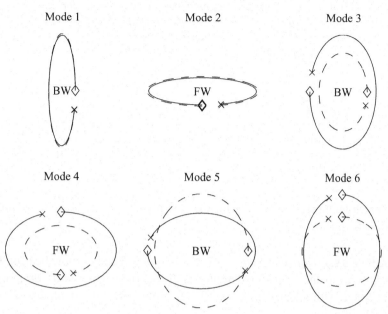

Figure 5.34. The axial view of the mode shapes at the left disk (node 3, solid) and right disk (node 5, dashed) for a rotor supported by anisotropic bearings (Example 5.9.2) at 4,000 rev/min. The cross denotes the start of the orbit and the diamond denotes the end. FW indicates forward whirl and BW indicates backward whirl.

the x and y directions. The bearings have direct stiffnesses of 1 MN/m and cross-coupling stiffnesses of 0.5 MN/m.

Solution. Table 5.6 shows the eigenvalues and natural frequencies for this example. Although there is symmetry in the x and y directions, the coupling terms mean that even at zero rotor speed, the natural frequencies do not occur in pairs. This is in contrast to the case of isotropic bearings with no coupling, in which the natural frequencies always occur in identical pairs at zero shaft speed. Figure 5.36 highlights that the orbits at the two disks are elliptical. The direction of the whirl is not marked because most of the modes are mixed; modes 1, 3, and 5

Table 5.5. *Natural frequencies and orbit directions (κ) at the nodes of the finite element model (Example 5.9.3), at 4,000 rev/min*

	Mode 1	Mode 2	Mode 3	Mode 4	Mode 5	Mode 6
Freq. (Hz)	8.545	13.77	22.35	44.06	78.76	120.4
Node 1	−0.0030	−0.081	−0.063	0.426	−0.371	0.685
Node 2	−0.0076	−0.030	−0.075	0.306	−0.510	0.509
Node 3	−0.0106	−0.012	−0.073	0.211	0.410	0.254
Node 4	−0.0116	0.004	0.075	−0.071	−0.357	0.481
Node 5	−0.0096	0.032	−0.192	0.222	−0.445	0.479
Node 6	−0.0058	0.083	−0.151	0.157	−0.345	0.546
Node 7	−0.0010	0.221	−0.117	0.156	−0.294	0.662

Table 5.6. *Eigenvalues and natural frequencies for a rotor supported by cross coupled bearings (Example 5.9.4)*

0 rev/min		4,000 rev/min	
Root s (rad/s)	ω_n (Hz)	Root s (rad/s)	ω_n (Hz)
$0 \pm 73.28_J$	11.66	$0 \pm 73.21_J$	11.65
$0 \pm 93.00_J$	14.80	$0 \pm 92.92_J$	14.79
$0 \pm 213.4_J$	33.97	$0 \pm 208.3_J$	33.16
$0 \pm 309.1_J$	49.19	$0 \pm 312.2_J$	49.69
$0 \pm 615.5_J$	97.97	$0 \pm 561.8_J$	89.41
$0 \pm 795.5_J$	126.61	$0 \pm 840.6_J$	133.79

are predominantly backward-whirling and modes 2, 4, and 6 are predominantly forward-whirling.

EXAMPLE 5.9.5. *Isotropic Bearings with Damping.* The isotropic bearing Example 5.9.1 is repeated but with damping in the bearings. The x and y directions are uncoupled, with a translational stiffness of 1 MN/m and a damping of 3 kNs/m in each direction.

Solution. Table 5.7 shows the eigenvalues, natural frequencies, and damping ratios for this example. At zero speed, there is no coupling between the x and y directions and the symmetry means that the modes occur in pairs. Figure 5.37 shows the natural frequency map and Figure 5.38 shows the first six mode shapes. The eigenvalues in Table 5.7 are ordered by increasing natural frequency; however, because the damping ratio for mode 8 is higher than that for mode 7, the damped natural frequency for mode 8 is lower than that for mode 7. The orbits (not shown) indicate that mode 7 is forward and mode 8 is backward.

EXAMPLE 5.9.6. *Hydrodynamic Bearings.* Repeat the analysis of Example 5.9.1 when the bearings are replaced with hydrodynamic bearings. The oil-film bearings have a diameter of 100 mm, are 30 mm long, and each supports a static load of 525 N, which represents half of the weight of the rotor. The radial clearance in the bearings is 0.1 mm and the oil film has a viscosity of 0.1 Pa s. These bearings have the same characteristics as Example 5.5.1.

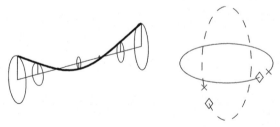

Figure 5.35. Mode shape 5 at 4,000 rev/min (Example 5.9.3). On the left side is the mode shape. On the right side is the axial view of the mode shape at the left disk (node 3, solid) and right disk (node 5, dashed). The cross denotes the start of the orbit and the diamond denotes the end.

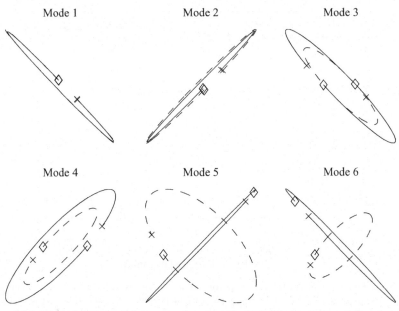

Figure 5.36. The axial view of the mode shapes at the left disk (node 3, solid) and right disk (node 5, dashed) for a rotor supported by bearings with coupling in the x and y directions (Example 5.9.4) at 4,000 rev/min. The cross denotes the start of the orbit and the diamond denotes the end.

Solution. Table 5.8 shows the natural frequencies, and damping ratios at 200 and 4,000 rev/min. There are also four negative real eigenvalues that are not shown. It is clear that the system is unstable at 4,000 rev/min because the real part of the second eigenvalue has become positive. This instability arises from

Table 5.7. *Eigenvalues (s), natural frequencies (ω_n), damped natural frequencies (ω_d), and damping ratios (ζ) for a rotor supported by isotropic bearings with damping (Example 5.9.5)*

Speed (rev/min)	Root s (rad/s)	ω_n (Hz)	ω_d (Hz)	ζ
	$-4.424 \pm 87.26j$	13.91	13.89	0.051
	$-4.424 \pm 87.26j$	13.91	13.89	0.051
	$-78.24 \pm 292.4j$	48.18	46.54	0.258
0	$-78.24 \pm 292.4j$	48.18	46.54	0.258
	$-566.5 \pm 648.6j$	137.06	103.22	0.658
	$-566.5 \pm 648.6j$	137.06	103.22	0.658
	$-657.3 \pm 834.8j$	169.10	132.86	0.619
	$-657.3 \pm 834.8j$	169.10	132.86	0.619
	$-4.083 \pm 85.97j$	13.70	13.68	0.048
	$-4.742 \pm 88.41j$	14.09	14.07	0.054
	$-74.10 \pm 263.8j$	43.61	41.98	0.270
4,000	$-78.81 \pm 318.2j$	52.18	50.65	0.240
	$-402.6 \pm 655.0j$	122.37	104.25	0.524
	$-667.2 \pm 663.9j$	149.81	105.66	0.709
	$-609.5 \pm 868.5j$	168.87	138.23	0.574
	$-694.7 \pm 818.2j$	170.82	130.22	0.647

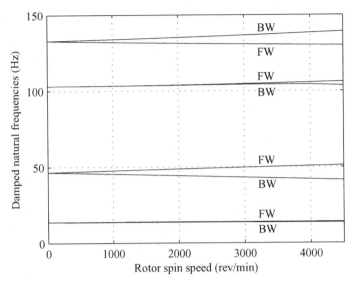

Figure 5.37. The natural frequency map for a rotor supported by isotropic bearings with damping (Example 5.9.5). FW indicates forward whirl and BW indicates backward whirl.

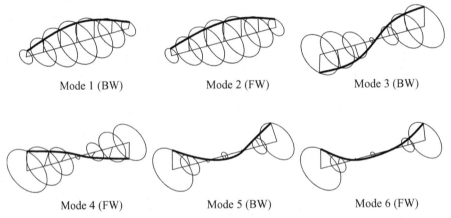

Figure 5.38. Mode shapes at 4,000 rev/min for a rotor supported by isotropic bearings with damping (Example 5.9.5).

Table 5.8. *Eigenvalues, natural frequencies, and damping ratios for a rotor supported by hydrodynamic bearings (Example 5.9.6)*

200 rev/min			4,000 rev/min		
Root s (rad/s)	ω_n (Hz)	ζ	Root s (rad/s)	ω_n (Hz)	ζ
$-17.36 \pm 14.74j$	3.62	0.762	$-0.4277 \pm 107.50j$	17.11	0.004
$-17.37 \pm 14.79j$	3.63	0.761	$1.476 \pm 113.63j$	18.09	-0.013
$-1.108 \pm 110.90j$	17.65	0.010	$-44.62 \pm 212.09j$	34.49	0.206
$-0.1877 \pm 111.04j$	17.67	0.002	$-49.92 \pm 212.00j$	34.66	0.229
$-2.110 \pm 436.39j$	69.45	0.005	$-2.245 \pm 421.40j$	67.07	0.005
$-0.5857 \pm 436.46j$	69.46	0.001	$-4.987 \pm 447.40j$	71.21	0.011

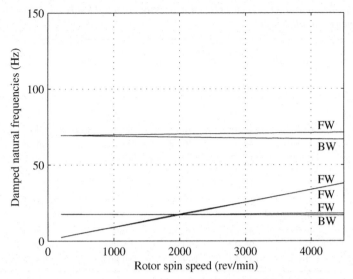

Figure 5.39. The natural frequency map for a rotor supported by hydrodynamic bearings (Example 5.9.6). FW indicates forward whirl and BW indicates backward whirl.

the asymmetry in the bearing-stiffness matrix (see Chapter 7 for more details). Figure 5.39 shows the natural frequency map and Figure 5.40 shows the first six mode shapes for this example. The mode shapes split into two types: (1) those in which the displacement at the bearings is relatively small and the damping is low; and (2) those that involve substantial motion at the bearings and therefore have high damping. Type (2) modes are sensitive to rotor speed, causing these modes to swap order. The orbits (not shown) also highlight that both the third and fourth modes are forward-whirling modes.

EXAMPLE 5.9.7. *The Effect of Axial Load and Follower Torque.* Investigate the effect on the model of Example 5.9.1 of axial loads of −10, 10, and 100 kN and torques of 50 and 100 kNm.

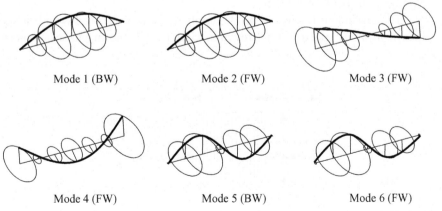

Figure 5.40. Mode shapes at 4,000 rev/min for a rotor supported by hydrodynamic bearings with damping (Example 5.9.6). FW indicates forward whirl and BW indicates backward whirl.

Table 5.9. *The natural frequencies (Hz) due to different axial loads (positive is tensile) and torque (Example 5.9.7)*

Speed (rev/min)	Axial Load (kN)				Torque (kNm)	
	0	10	100	−10	50	100
	13.79	13.94	15.02	13.64	13.66	13.23
	13.79	13.94	15.02	13.64	13.66	13.23
0	43.66	43.91	46.13	43.40	43.31	42.24
	43.66	43.91	46.13	43.40	43.31	42.24
	114.08	114.76	120.62	113.39	113.83	113.06
	114.08	114.76	120.62	113.39	113.83	113.06
	13.59	13.74	14.87	13.43	13.44	12.94
	13.97	14.11	15.16	13.83	13.86	13.49
4,000	40.07	40.33	42.57	39.81	39.72	38.65
	46.90	47.15	49.33	46.65	46.57	45.54
	95.52	96.09	100.97	94.95	95.18	94.14
	131.63	132.40	139.06	130.86	131.62	131.62

Solution. Table 5.9 shows the effect of these axial loads and torques on the natural frequencies. Clearly, an axial compressive load reduces and an axial tensile load increases the natural frequencies. The torque reduces the natural frequencies. If the natural frequency reduces to zero, then the shaft buckles.

EXAMPLE 5.9.8. *UMP.* The rotor of an experimental four-pole induction machine ($p = 2$) is mounted centrally between two bearings 500 mm apart, as shown in Figure 5.41. The rotor shaft has a diameter of 50 mm throughout. The stiffening effect of the machine rotor core on the shaft is negligible so that the mass of the machine rotor can be treated as being concentrated at the central x-y plane. The properties of the shaft are $E = 210$ GPa, $v = 0.285$, and $\rho = 7,800$ kg/m³.

The bearings have isotropic properties. In every transverse direction, the stiffness of each bearing is 12 MN/m and the damping constant for each bearing is 500 Ns/m. The rotor core has a diameter of 200 mm, length of 200 mm, and an average density of 7,500 kg/m².

The machine is fed from a 50 Hz supply ($\omega_{supply} = 100\pi$ rad/s) and, as such, its synchronous speed is 1,500 rev/min. Under load, the machine is expected

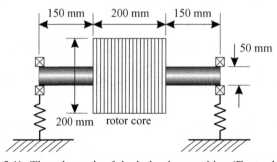

Figure 5.41. The schematic of the induction machine (Example 5.9.8).

to operate between 1,425 and 1,500 rev/min. The parameters representing the UMP behavior for this machine are as follows:

Negative UMP stiffness: $k_{0r} = 4\,\text{MN/m}$, $k_{0i} = 0$

Dynamics of the "$(p-1)$" magnetic forces: $k_{mr} = 16\,\text{MN/m}$, $k_{mi} = 0$, $z_{mr} = -4.2\,\text{s}^{-1}$

Dynamics of the "$(p+1)$" magnetic forces: $k_{pr} = 84\,\text{MN/m}$, $k_{pi} = 0$, $z_{pr} = -12.5\,\text{s}^{-1}$

Prepare a reduced-dimension model of the rotating machine (without any magnetic effects) retaining only two displacement coordinates: x and y translations at the center of the rotor. Use this to compute the first critical speed of the machine in the absence of magnetic effects.

Then, use the reduced model to determine the range of values of slip, sl, over which the machine is stable.

Solution. The shaft is initially divided into 10 distinct sections each of length 0.05 m. The results are not sensitive to how many divisions are chosen. The complete stiffness, gyroscopic, damping, and mass matrices are formed for the rotor. With 10 distinct shaft sections, there are 11 shaft nodes and, hence, 44 degrees of freedom. Guyan reduction (see Section 2.5.1) is applied such that only two degrees of freedom are retained as masters (i.e., x and y translation at the center of the rotor core) and all others are eliminated. The resulting matrices are

$$\mathbf{M} = \begin{bmatrix} 49.9731 & 0 \\ 0 & 49.9731 \end{bmatrix}\text{kg}, \qquad \mathbf{K} = \begin{bmatrix} 12.0547 & 0 \\ 0 & 12.0547 \end{bmatrix}\text{MN/m},$$

$$\mathbf{D} = \begin{bmatrix} 252.285 & 0 \\ 0 & 252.285 \end{bmatrix}\text{Ns/m}, \qquad \mathbf{G} = \begin{bmatrix} 0 & -0.010908 \\ 0.010908 & 0 \end{bmatrix}\text{Ns/m}$$

Solving for the critical speed yields $\Omega_{crit} = 491.0859\,\text{rad/s}$ (4,689.5 rev/min). In general, to compute the eigenvalues of the system at any given value of slip, sl, we form the state–space matrix

$$\mathbf{A} = \begin{bmatrix} \mathbf{0} & \mathbf{I} & \mathbf{0} & \mathbf{0} \\ -\mathbf{M}^{-1}\mathbf{K} & -\mathbf{M}^{-1}\mathbf{D} & \mathbf{M}^{-1}\mathbf{S} & \mathbf{M}^{-1}\mathbf{S} \\ \mathbf{K}_m\mathbf{S}^\top & \mathbf{0} & \mathbf{Z}_m & \mathbf{0} \\ \mathbf{K}_p\mathbf{S}^\top & \mathbf{0} & \mathbf{0} & \mathbf{Z}_p \end{bmatrix}$$

and calculate its eigenvalues. Here, the matrix $\mathbf{S}$ is a $N \times 2$ selection matrix that selects the lateral translations of the center of the rotor core relative to the full-length vector of N displacement coordinates. In the present case, the model is reduced to the point where only the lateral translations of the rotor center are retained; therefore, $\mathbf{S}$ is the 2×2 identity matrix.

By performing the substitutions

$$\mathbf{K}_m = \begin{bmatrix} 16 \times 10^6 & 0 \\ 0 & 16 \times 10^6 \end{bmatrix}\text{N/m}, \qquad \mathbf{K}_p = \begin{bmatrix} 84 \times 10^6 & 0 \\ 0 & 84 \times 10^6 \end{bmatrix}\text{N/m},$$

$$\mathbf{Z}_m = \begin{bmatrix} -4.2 & 0 \\ 0 & -4.2 \end{bmatrix} + (1 - sl\,(1 - p))\frac{\omega_{supply}}{p}\begin{bmatrix} 0 & -1 \\ 1 & 0 \end{bmatrix},$$

$$\mathbf{Z}_p = \begin{bmatrix} -12.5 & 0 \\ 0 & -12.5 \end{bmatrix} + (1 - sl\,(1 + p))\frac{\omega_{supply}}{p}\begin{bmatrix} 0 & -1 \\ 1 & 0 \end{bmatrix}$$

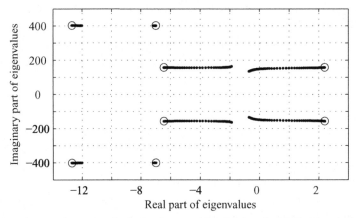

Figure 5.42. The root locus for the induction machine (Example 5.9.8). The circles denote a slip angle of 0.

the eigenvalues of **A** may be calculated for any given value of slip. When this is done for all values of slip between 0 and 5 percent, corresponding to speeds between 1,500 and 1,425 rev/min, the roots may be calculated. The machine is found to be unstable for values of slip less than 2.215 percent. Figure 5.42 shows the root locus for the range of slips considered and Figure 5.43 show the maximum real part of any eigenvalue of this system as a function of the slip.

The same computation can be done with the full model in place of the reduced model. In this case, the matrix **A** is 52×52, but the prediction for the stable range of rotational speeds is imperceptibly altered.

EXAMPLE 5.9.9. *An Overhung Rotor.* Consider the overhung rotor shown in Figure 5.44. The shaft is 1.5 m-long and the diameter is 50 mm with a disk of diameter 350 mm and thickness 70 mm. The two bearings, with positions given in Figure 5.44, have a stiffness of 10 MN/m in each direction. The shaft and disk are made of steel, with material properties $E = 211 \text{ GN/m}^2$, $G = 81.2 \text{ GN/m}^2$,

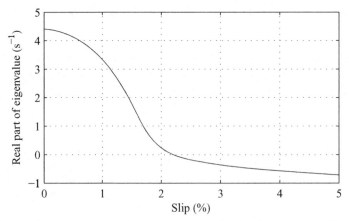

Figure 5.43. The largest real part of the eigenvalues for the induction machine (Example 5.9.8).

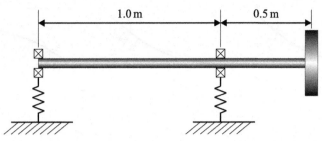

Figure 5.44. The layout for an overhung rotor (Example 5.9.9).

and $\rho = 7{,}810\,\mathrm{kg/m^3}$. Damping is neglected. Estimate the first six natural frequencies and mode shapes between 0 and 4,000 rev/min.

Solution. An FE model with six Timoshenko shaft elements generated the estimated natural frequencies shown in Table 5.10, and the mode shapes are shown in Figure 5.45. The corresponding natural frequency map, Figure 5.46, shows that all of the pairs of modes separate significantly, highlighting that the gyroscopic effects are more pronounced than in the previous examples. At first glance, the shape of mode 3 at 4,000 rev/min appears slightly odd and out of order. However, the natural frequency map shows that modes 3 to 5 interact significantly around 3,000 rev/min, which accounts for the reordering of the mode shapes.

EXAMPLE 5.9.10. *A Tapered Shaft.* Consider a tapered shaft of length 1.5 m and a diameter that changes linearly from 25 to 40 mm. A disk of diameter 250 mm and thickness 40 mm is placed at the center of the shaft, and short bearings of stiffness 10 MN/m and damping 1 kNs/m are attached at the ends of the shaft. The Young's modulus and mass density are 211 GN/m² and 7,810 kg/m³, respectively. Estimate the first pair of natural frequencies of this machine at 3,000 rev/min using a stepped shaft diameter and elements of uniform diameter and by using tapered elements.

Solution. Shear and rotary inertia effects are ignored but gyroscopic effects are included. The shaft is split into a number of elements of equal length and the

Table 5.10. *Eigenvalues and natural frequencies for an overhung rotor (Example 5.9.9)*

0 rev/min		4,000 rev/min	
Root s (rad/s)	ω_n (Hz)	Root s (rad/s)	ω_n (Hz)
$0 \pm 90.14 j$	14.35	$0 \pm 76.19 j$	12.13
$0 \pm 90.14 j$	14.35	$0 \pm 103.91 j$	16.54
$0 \pm 630.73 j$	100.38	$0 \pm 565.99 j$	90.08
$0 \pm 630.73 j$	100.38	$0 \pm 634.23 j$	100.94
$0 \pm 830.43 j$	132.17	$0 \pm 647.75 j$	103.09
$0 \pm 830.43 j$	132.17	$0 \pm 1174.2 j$	186.88

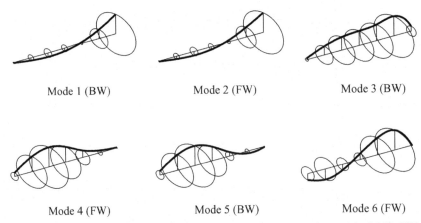

Mode 1 (BW)	Mode 2 (FW)	Mode 3 (BW)
Mode 4 (FW)	Mode 5 (BW)	Mode 6 (FW)

Figure 5.45. Mode shapes at 4,000 rev/min for an overhung rotor (Example 5.9.9). FW indicates forward whirl and BW indicates backward whirl.

diameter at each end is determined. For the uniform element, the average diameter is used; this average is not optimal for either the inertia or the stiffness properties of the shaft and is used only for illustration. Figure 5.47 shows how the first pair of natural frequencies changes as the number of elements increases. It is clear that the natural frequencies of the tapered element model converge much more quickly than those for the model with uniform elements. Thus, for a given accuracy, fewer tapered elements are required. The tapered elements are consistent and the natural frequencies converge from above. Although the matrices for a model with uniform elements are derived using a consistent formulation, the model geometry changes with the number of elements. This explains why, in this case, the natural frequencies converge from below.

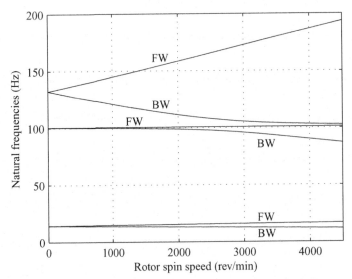

Figure 5.46. The natural frequency map for an overhung rotor (Example 5.9.9). FW indicates forward whirl and BW indicates backward whirl.

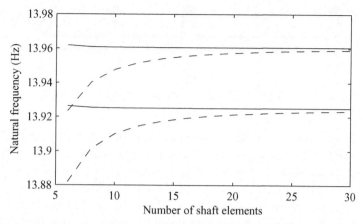

Figure 5.47. The natural frequencies for the tapered shaft at 3,000 rev/min, using tapered elements (solid) and uniform elements (dashed) (Example 5.9.10).

5.10 Summary

In this chapter, the FEM of Chapter 4 is extended to include some of the features required for the analysis of rotating systems. Models of shafts, disks, bearings, and other rotor–stator interactions are developed. Using these models, the FEM is applied to a selection of rotating-system problems, and the eigenvalues and eigenvectors are calculated. The effects of gyroscopic couples, isotropic and anisotropic bearings, and damping in these bearings are illustrated. Asymmetric rotors and the effect of rotor damping are discussed in Chapter 7.

5.11 Problems

5.1 A uniform steel shaft, shown in Figure 5.48, is 1.6 m long and supported by plain, self-aligning bearings at its ends. It carries two identical disks, each with a diameter of 400 mm and a thickness of 80 mm. The disks are located 0.5 and 1 m from the left bearing. For steel, $E = 200$ GPa, $v = 0.27$, and $\rho = 7,800$ kg/m^3. Using both Timoshenko and Euler-Bernoulli beam models with 16 elements, determine the first five natural frequencies for the systems with the following shafts, spinning at 3,000 rev/min:

(a) a solid shaft of 60 mm diameter

(b) a hollow shaft with 80 mm outside diameter and 73 mm inside diameter

Show that the percentage difference between the Timoshenko and Euler-Bernoulli results for the rotor with the hollow shaft is about five times greater

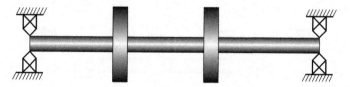

Figure 5.48. A schematic of the machine described in Problem 5.1.

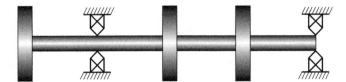

Figure 5.49. A schematic of the machine described in Problem 5.2.

than for the rotor with the smaller diameter solid shaft. The shaft diameters are chosen to give similar shaft stiffness in the two cases.

5.2 A uniform solid-steel shaft, shown in Figure 5.49, is 1.6 m long and 75 mm in diameter. It is supported in two self-aligning bearings located 0.4 and 1.6 m from its left-hand end. It carries three identical disks, 400 mm in diameter and 80 mm thick, located at the left end and at 0.8 and 1.2 m from the left end.

   Determine the first five natural frequencies for this system when the rotor spins at 3,000 rev/min. Use 4, 8, and 16 Euler-Bernoulli shaft elements in the analysis; in each case, let the shaft elements be equal in length. For steel, $E = 200$ GPa and $\rho = 7,800$ kg/m^3.

5.3 The solid-steel rotor shown in Figure 5.50 consists of a shaft 1.4 m long that is supported by simple self-aligning bearings 0.4 and 1 m from the left end of the shaft. Two disks, each with a diameter of 400 mm and a thickness of 80 mm, are carried at the ends of the shaft. A third disk of 320 mm diameter and 80 mm thickness is carried between the bearings, 0.8 m from the left end of the shaft. Between the bearings, the shaft is uniform with a diameter of 80 mm; between the disks and the bearings, the shaft tapers so that its diameter increases linearly from 40 mm at the disks to 80 mm at the bearings, as shown in Figure 5.50.

   (a) Model the system with seven elements of equal length, using tapered and uniform elements as appropriate.
   (b) Model each tapered section with four uniform elements of equal length, where the diameter is taken as the mean diameter of the tapered shaft at the element location. Model the section between the bearings with three uniform elements of equal length.

For both models, determine the first five natural frequencies of the system when the rotor spins at 3,000 rev/min. Neglect the shear and rotary inertia effects. Assume that the inner diameter of the disks is 40 mm for all cases.

   The natural frequencies of the system, computed using 12 uniform elements and 16 tapered elements, are 27.692, 32.528, 40.775, 50.762, and 90.877 Hz. Which of the two models developed gives more accurate natural frequencies? For steel, $E = 200$ GPa and $\rho = 7,800$ kg/m^3. In Example 5.8.1, the natural frequencies converged from above; that is, the natural frequencies

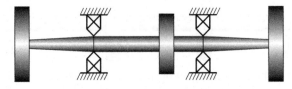

Figure 5.50. A schematic of the machine with a tapered shaft (Problem 5.3).

decreased as the number of elements increased. Why are lowest estimates of the lowest natural frequencies not the most accurate in this case?

5.4 The rigid rotor shown in Figure 3.26 is 0.5 m long. It is carried by a rigidly supported, short, self-aligning bearing at its No. 1 end and by a short hydrodynamic bearing at its No. 2 end. The rotor has a polar moment of inertia of $0.6 \, kg \, m^2$ and a diametral moment of inertia about the No. 1 bearing of $10 \, kg \, m^2$. The weight of the rotor is 1,200 N and the center of mass of the rotor is 0.25 m from each bearing.

The hydrodynamic bearing has a diameter of 100 mm, a length of 20 mm, and a radial clearance of 0.2 mm, and the oil in the bearing has a viscosity of 0.03 Pa s. Develop the equations of motion for this system. Then:

(a) Calculate the Sommerfeld number for the bearing when the rotor spins at 3,000 rev/min.

(b) When the rotor spins at 3,000 rev/min, determine the eccentricity of the bearing, the radial force, the tangential force, and γ, the angle between the vertical load and the direction of maximum displacement of the journal in the bearing.

(c) At 3,000 rev/min, calculate the eigenvalues of the system as well as the damped natural frequencies and damping factors for any underdamped modes.

(d) Using the same approach, calculate the damped natural frequencies and corresponding damping factors, at a rotor spin speed of 6,000 rev/min.

5.5 A Jeffcott rotor has a lateral stiffness k, a lateral damping coefficient c, and a mass m at its midspan. Forces $f_y = k_{sw}u$ and $f_x = -k_{sw}v$ also act at the midspan of the rotor due to the effect of steam whirl. By considering the eigenvalues, determine when the system is at the limit of stable operation and obtain the corresponding whirl frequency. (Hint: The stability boundary is obtained when the real part of the eigenvalues is zero.) For a particular running speed, also determine an expression for the maximum power output for stable running.

A small experimental turbine has blades that are 50 mm long with a mean diameter of 150 mm. The turbine develops 30 kW at 9,600 rev/min. Assuming for this system that $\beta = 3$, determine the value of k_{sw}, defined by Equation (5.98). The turbine is modeled as a Jeffcott rotor. The length of the steel shaft between the bearings is 300 mm and the shaft diameter is 15 mm. The bearings are the short, self-aligning type and provide negligible damping. A disk that carries the turbine blades is located at the midspan; the disk and blades have a mass of 3 kg. To increase the damping force in the lateral direction, an auxiliary bearing attached to a damping device is fitted close to the central disk. Determine the minimum damping that must be added to this system for stable operation. Also determine the natural frequencies, damped natural frequencies, and damping ratios (ζ) when the effective damping coefficient provided via the auxiliary bearing is 0, 20, and 40 Ns/m. Determine the directions of the stable and unstable whirl orbits. Assume that $E = 200 \, GPa$.

5.6 A computer programmer is asked to write some computer code to solve Problem 5.5. He accidentally reverses the sign of the stiffness coefficient k_{sw} in Equation (5.98) (i.e., the matrix in this equation is transposed). What is the

Figure 5.51. A schematic of the machine with angular contact ball bearings (Problem 5.8).

effect on the frequencies and mode shapes predicted by the program? To illustrate the effect of the programmer's error, determine the natural frequencies, damped natural frequencies, damping ratios (ζ), and directions of the stable and unstable whirl orbits using the data in Problem 5.5 with an effective damping coefficient of 20 Ns/m. Compare your results with those from Problem 5.5.

5.7 Show that for a Jeffcott rotor with a seal close to the central disk, the limit of stability is given when $\Omega = 2\sqrt{k/m_0}$. (Hint: The stability boundary is obtained when the real part of the eigenvalues is zero.) In this relationship, Ω is the spin speed of the rotor, k is the total system stiffness (including the contribution from the seal), and m_0 is the mass of the disk at midspan.

A Jeffcott rotor consists of the steel shaft of length 0.6 m between bearings and the diameter is 50 mm. The bearings are the short, self-aligning type and provide negligible damping. The central disk has a mass of 600 kg and a seal close to it has the parameters $m_d = 120$ kg, $c_d = 200$ Ns/m, and $k_d = 20$ MN/m. Determine the shaft speed and the whirl frequency at the limit of stability. For steel, $E = 200$ GPa.

A damping unit is attached to the system via a bearing located close to the midspan disk. The unit provides damping forces at the center of the shaft, $f_x = c\dot{u}$ and $f_y = c\dot{v}$ in the x and y directions, respectively. Determine the shaft speeds and whirl frequencies at the limit of stability when the viscous damping coefficient, c, is 80 and 160 Ns/m.

5.8 The two-bearing machine shown in Figure 5.51 has a shaft of diameter 40 mm and a length of 1.2 m. Disks of thickness 50 mm are present on the rotor at mean axial positions 0.5 and 0.9 m from the left end, and these disks have outside diameters of 400 and 200 mm, respectively. Their inner diameters are each 40 mm, but they are considered to not add stiffness to the shaft. The nominal position of the bearings is 0.1 m from each end of the shaft. The rotor is manufactured from steel with Young's modulus 200 GPa, mass density 7,800 kg/m³, and Poisson's ratio of 0.285. The shaft is modeled using 12 elements of equal length.

Assume initially that the bearings are rigid and that the bearing reactions act through a plane containing the ball centers. Calculate the lowest five natural frequencies of this system at 3,000 rev/min under this assumption.

This case is appropriate for ball bearings with a small radial clearance. In practice, it is common to use angular contact ball bearings that have the potential to allow for different radial growth of the inner and outer races of the

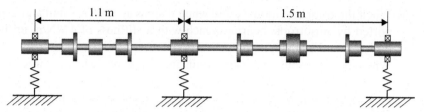

Figure 5.52. A schematic of the machine with three bearings (Problem 5.9).

bearing. Allowing for a contact angle of 20°, a mean diameter of the ball centers of 70 mm, and a tension in the shaft of 500 N, recalculate the natural frequencies for this rotor. (Hint: The effective locations of the bearings are 35 mm × tan(20°) away from the planes containing the ball centers.) There are two competing effects on the natural frequencies caused by the shaft tension: the movement of the reaction centers lowers the resonance frequencies, and the shaft tension raises the frequencies. Calculate the shaft tension so that the first natural frequency is identical to that for the simple model with the negligible radial clearance given previously.

5.9 The rotor shown in Figure 5.52 consists of a steel shaft 2.8 m long, supported on three bearings, located at 0.1, 1.2, and 2.7 m from the left end of the shaft. Table 5.11 shows the diameters of the shaft sections. The rotor carries five disks, located at 0.4, 0.8, 1.6, 2.0, and 2.4 m from the left bearings. Each disk is 200 mm in diameter and 25 mm thick, except for the disk located at 2.0 m from the left end, which is 100 mm thick. The inside diameter of each disk is 110 mm. Assume $E = 200\,\text{GPa}$, $\nu = 0.27$, and $\rho = 7{,}800\,\text{kg/m}^3$.

Using appropriate software, develop a shaft-line model of the system. Model the shaft with 28 Timoshenko elements, each 100 mm long. Determine the six lowest damped natural frequencies (and damping factors where appropriate) for the machine (neglect real eigenvalues) and the shape of the orbit of the corresponding modes at the disk 2 m from the left end of the rotor for the following conditions:

(a) When the rotor is stationary and when it spins at 3,000 rev/min and is supported by isotropic bearings, each with a stiffness of 5 MN/m.

Table 5.11. *The diameters of the shaft sections of the machine in Figure 5.52 (Problem 5.9)*

Position of shaft section		Shaft	Position of shaft section		Shaft
Left end (m)	Right end (m)	diameter (mm)	Left end (m)	Right end (m)	diameter (mm)
0.0	0.2	100	1.3	1.6	38
0.2	0.3	38	1.6	1.7	110
0.3	0.4	110	1.7	1.9	38
0.4	0.5	38	1.9	2.1	110
0.5	0.6	110	2.1	2.4	38
0.6	0.7	38	2.4	2.5	110
0.7	0.8	110	2.5	2.6	38
0.8	1.1	38	2.6	2.8	100
1.1	1.3	100			

(b) When the rotor is stationary and when it spins at 3,000 rev/min and is supported by anisotropic bearings, each with a stiffness of 4 MN/m in the x direction and 5 MN/m in the y direction.

(c) When the rotor spins at 3,000 rev/min and is supported by the three identical oil-film bearings, indicated schematically in Figure 5.52 by springs. Each bearing has a length of 30 mm and a journal diameter of 100 mm. The radial clearance in each bearing is 0.2 percent of the diameter and the viscosity of the oil is $\eta = 0.030$ Pa s. The loads on the bearings result from the weight of the rotor (whose mass is 136.0976 kg) and are assumed to have three possible distributions: (1) the load on all three bearings is equal; (2) the load on the inner bearing is three times that of the outer bearings; and (3) the load on the inner bearing is half that of the outer bearings. Comment on the stability of the rotor on oil-film bearings.

5.10 Before the widespread availability of computers, the natural frequencies of rotating machines were often estimated using *Rayleigh's method*. The method is based on an energy approach and is similar to the FEM except that the shaft displacements are approximated using functions defined over the whole shaft. Consider the kinetic and strain energy of the shaft when it vibrates in its first mode of vibration. Suppose the transverse displacement is approximated as $u(z, t) = u_1(z) \sin(\omega_1 t)$, where $u_1(z)$ is the first mode shape and ω_1 is the first natural frequency of the rotor. The shaft kinetic and strain energies are $T = \frac{1}{2} \int_0^L \rho A \dot{u}^2 dz$ and $U = \frac{1}{2} \int_0^L EI(u'')^2 dz$, where $\dot{u}$ and u'' are the derivative of u with respect to time and the second derivative with respect to z, respectively; L is the length of the shaft; and shear and rotary inertia effects are ignored. Thus, the maximum kinetic and strain energies are $T_{max} = \frac{1}{2}\omega_1^2 \int_0^L \rho A u_1^2 dz$ and $U_{max} = \frac{1}{2} \int_0^L EI(u_1'')^2 dz$. Equating these maximum energies provides an expression for the square of the first natural frequency. In practice, we do not know the first mode shape. However, if $u_1(z)$ is a reasonable estimate for the mode shape (i.e., it at least satisfies the geometric boundary conditions for the system), then equating the energies provides a good (over) estimate of the first natural frequency. If the shaft carries an added concentrated mass, M_d, at $z = z_d$, then the approximate kinetic energy of the mass is $T_{max} = \frac{1}{2}\omega_1^2 M_d \{u_1(z_d)\}^2$. This term must be added to T_{max} for the shaft to determine the kinetic energy of the system. The effect of the angular displacement of the disk may be included using the diametral moment of inertia of the disk and the slope of the approximate mode shape at the disk location.

Use Rayleigh's method to estimate the first natural frequency for the following cases:

(a) For the rotor with the hollow shaft described in Problem 5.1, by approximating the first mode of the rotor by $u_1(z) = \sin(\pi z/L)$.

(b) For the rotor described in Problem 5.2, by approximating the first mode of the rotor by $u_1(z) = 0.2148 - 0.8815\tilde{z} + 1.6667\tilde{z}^4 - \tilde{z}^5$, where $\tilde{z} = z/L$. Also show that this polynomial satisfies the geometric boundary conditions.

5.11 A rotor consists of a shaft, 1 m long and 40 mm diameter, supported at each end by short, rigid bearings. It can carry one of several possible disks at a distance

Table 5.12. *The dimensions of the disks (Problem 5.11)*

Diameter (mm)	300	400	500	600	700	800
Thickness (mm)	30	40	50	60	70	80

of 0.6 m from one bearing. Table 5.12 gives the dimensions of the disks. The shaft and disk are both made of steel with $E = 210$ GPa and $\rho = 7,800$ kg/m^3.

Determine the first and second natural frequencies of the stationary shaft carrying each disk in turn, as follows:

(a) Model the rotor with two degrees of freedom. Use the stiffness coefficients given in Appendix 2 and assume the shaft is mass-less and the disk does *not* have a central hole.

(b) Model the rotor with two degrees of freedom. Use the stiffness and mass coefficients given in Appendix 2 – that is, account for the shaft mass and model the disk *with* a central hole.

(c) Model the system using FEA. Use 20 Euler-Bernoulli elements to model the system.

For the models, determine the percentage difference between the natural frequencies obtained from the two degrees of freedom models (with and without shaft mass) and the natural frequencies obtained from the detailed FE model.

6 Forced Lateral Response and Critical Speeds

6.1 Introduction

In Chapters 3 and 5, methods are presented to determine the dynamic character-istics of a rotor–bearing system, such as the natural frequencies, damping factors, and mode shapes. In this chapter, we examine how rotor–bearing systems respond to forces and moments. The most common forces acting in rotating machines are lateral forces and moments whose frequencies are locked to the rotor speed or mul-tiples of rotor speed. A force whose frequency is identical to rotor speed is said to be a *synchronous force*. We also examine how rotor–bearing systems respond to forces the frequency of which is unrelated to rotor speed, called *asynchronous forces* – for example, external forces acting on the rotor via the bearings and foundation.

The most significant lateral forces and moments are usually caused by an im-perfect distribution of mass in the rotor. As the rotor spins about its equilibrium position, forces and moments are generated that are called *out-of-balance forces* and *moments*. The direction of these forces and moments is fixed relative to the ro-tor; therefore, their direction rotates with the rotor. Thus, the excitation frequency in any plane, lateral to the axis of the rotor, is locked to the speed of rotation; for this reason, they are synchronous forces and moments. Due to manufacturing toler-ances and other factors, it is not possible to ensure that rotors are perfectly balanced. Although when new or recently commissioned, a rotor is balanced so that the resid-ual out-of-balance is minimal, this out-of-balance may increase with the passage of time. For example, in a gas turbine, blades may become unevenly fouled or unevenly eroded, which may lead to an increase in residual out-of balance.

Often, there is concern about frequency components of forcing other than syn-chronous unbalance. In some cases, this is because nonlinearity is known to be in the system (particularly in the bearings), which causes the displacement response to contain harmonics of the synchronous frequency. There are other sources of ex-citation, such as those due to the imperfect meshing of gear teeth, misalignment of shafts, slight asymmetries in the stiffness properties of the shaft, ball- (or roller-) passing in rolling-element bearings, and the unsymmetrical intake or exhaust of work-ing fluid in radial-flow turbo-machines interacting with rotating blades or vanes.

In some cases, the rotor of a machine experiences forcing from an outside source that has no direct connection with its own spin speed. This happens most

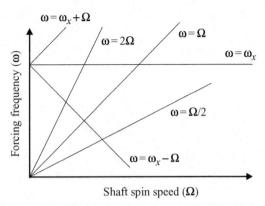

Figure 6.1. A plot of forcing frequencies as a function of shaft speed.

commonly through base excitation, in which one machine is mounted on a fixture that vibrates caused by the action of another machine. It is occasionally found that fixed-frequency lateral forcing exists on an electrical machine that is modulated by a multiple of rotational speed because of some type of asymmetry on the rotor. Certain electrical machines exhibit this forcing when there is unsymmetrical residual magnetism on their rotors. Other forces acting on rotors include foundation forces that are transmitted to the rotor via the bearings; gravity forces on large horizontal rotors; and effects of rotor bow, misaligned couplings, and cracks. Some of these forces rotate at the rotor speed or multiples of it; others act in a fixed direction with excitation frequencies unrelated to rotor speed.

The forcing frequencies of rotating machines are often dependent on operating speed. Figure 6.1 shows typical variations of forcing frequency with rotor spin speed. We also note that the mass, damping, and stiffness properties of a machine can be strongly dependent on shaft speed, and it follows that the natural frequencies computed at any one shaft speed may be unique to that speed. For this reason, it is appropriate to plot a graph showing the variation (i.e., the absolute values) of natural frequencies with shaft speed, as described in Section 3.7. This is called a *natural frequency map*. When such a graph is plotted, it is possible to superimpose on it a number of lines or curves representing the variation (i.e., the absolute values) of forcing frequency with shaft speed. The result is called a *Campbell diagram*. In practical situations, it is usual to express the natural frequencies in Hz and the speed of rotation in rev/min. Common nomenclature for forces and response at the rotational speed is 1X. The response at the rotational speed is called the *synchronous response*. Forces with frequencies of twice the rotor speed are denoted by 2X and so forth.

The response of a rotor–bearing system to various types of excitation can be large. The speeds at which such large responses occur are called *critical speeds* and locating them is of the utmost importance to designers. Simply stated, a critical speed is a rotational speed of a machine or shaft line at which the machine behaves poorly in that large vibrations or shaft whirl occurs. If a machine is run continuously at a critical speed, then damage can occur very quickly. Often, this happens when an excitation frequency coincides with a resonance frequency of the machine. Either the maximum response criterion or the coincidence of excitation and

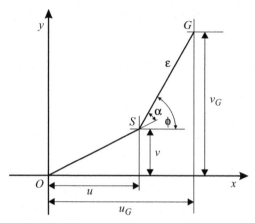

Figure 6.2. Instantaneous position of bearing centerline S and rotor mass center G.

resonance frequency may be used to estimate critical speeds for lightly damped, simple systems. A more formal definition of critical speeds is provided in Section 6.7. Methods to calculate critical speeds and expose their significance in engineering terms are described in Section 6.8.

6.2 Simple Models of Rotors

Chapter 3 discusses the free response of rigid rotors on simple flexible supports and simple flexible rotors on rigid supports. This section extends these models to investigate the response of these simple machines to out-of-balance forces and moments and due to bent rotors.

6.2.1 Modeling Out-of-Balance Forces and Moments

We begin by examining the synchronous response of a rotor to out-of-balance forces and moments. The analysis is developed in terms of a rigid rotor but extends readily to flexible rotors. To determine the effect of out-of-balance mass on the rigid circular rotor shown in Figure 3.6, we initially assume that the center of mass of the rotor is displaced a distance ε from the shaft centerline at equilibrium.

Consider the displacement of the rotor center of mass along axes Ox and Oy. Figure 6.2 shows the equilibrium position, O; the instantaneous position of the disturbed rotor centerline, S; and the position of the mass center of the rotor, G. Note that $|SG| = \varepsilon$. The instantaneous angle between the line SG (which represents a line on the rotor) and the Ox axis is ϕ, and the instantaneous angle between the line OS and the Ox axis is then $(\phi - \alpha)$. The distance $|OS|$, which is calculated during the analysis, is the amplitude of whirl of the rotor. The center of mass of the rotor moves u_G and v_G in the x and y directions, respectively, whereas the centerline of the rotor deflects u and v in the corresponding directions at the flexible bearings.

Now, from Figure 6.2, it is shown that

$$u_G = u + \varepsilon \cos \phi$$
$$v_G = v + \varepsilon \sin \phi$$

(6.1)

Differentiating these equations twice with respect to time and noting that ε is constant gives

$$\ddot{u}_G = \ddot{u} + \varepsilon \left(-\dot{\phi}^2 \cos\phi - \ddot{\phi} \sin\phi \right)$$
$$\ddot{v}_G = \ddot{v} + \varepsilon \left(-\dot{\phi}^2 \sin\phi + \ddot{\phi} \cos\phi \right)$$

(6.2)

In deriving the previous equation, we have not restricted the analysis to the case of a rotor spinning with a constant angular velocity. If we now introduce this simplification, then at a constant speed of rotation, Ω, $\dot{\phi} = \Omega$ and $\ddot{\phi} = 0$. Thus,

$$\ddot{u}_G = \ddot{u} - \varepsilon\Omega^2 \cos\Omega t$$
$$\ddot{v}_G = \ddot{v} - \varepsilon\Omega^2 \sin\Omega t$$

(6.3)

The equations of motion for the free vibration of this rotor, including damping at the supports and gyroscopic effects, are given in Equation (3.63) and are repeated here for convenience:

$$m\ddot{u} + c_{xT}\dot{u} + c_{xC}\dot{\psi} + k_{xT}u + k_{xC}\psi = 0$$
$$m\ddot{v} + c_{yT}\dot{v} - c_{yC}\dot{\theta} + k_{yT}v - k_{yC}\theta = 0$$
$$I_d\ddot{\theta} + I_p\Omega\dot{\psi} - c_{yC}\dot{v} + c_{yR}\dot{\theta} - k_{yC}v + k_{yR}\theta = 0$$
$$I_d\ddot{\psi} - I_p\Omega\dot{\theta} + c_{xC}\dot{u} + c_{xR}\dot{\psi} + k_{xC}u + k_{xR}\psi = 0$$

For the rotor being considered, the center of mass is offset from the shaft centerline at equilibrium by a small quantity ε, and the displacement of the center of mass is given by u_G and v_G. However, the displacements of the springs and dampers (at the bearings) are still in terms of u and v. Thus, replacing $\ddot{u}$ by $\ddot{u}_G$ and $\ddot{v}$ by $\ddot{v}_G$ in Equation (3.63), we have

$$m\ddot{u}_G + c_{xT}\dot{u} + c_{xC}\dot{\psi} + k_{xT}u + k_{xC}\psi = 0$$
$$m\ddot{v}_G + c_{yT}\dot{v} - c_{yC}\dot{\theta} + k_{yT}v - k_{yC}\theta = 0$$
$$I_d\ddot{\theta} + I_p\Omega\dot{\psi} - c_{yC}\dot{v} + c_{yR}\dot{\theta} - k_{yC}v + k_{yR}\theta = 0$$
$$I_d\ddot{\psi} - I_p\Omega\dot{\theta} + c_{xC}\dot{u} + c_{xR}\dot{\psi} + k_{xC}u + k_{xR}\psi = 0$$

(6.4)

Substituting for $\ddot{u}_G$ and $\ddot{v}_G$ from Equation (6.3) and rearranging gives

$$m\ddot{u} + c_{xT}\dot{u} + c_{xC}\dot{\psi} + k_{xT}u + k_{xC}\psi = m\varepsilon\Omega^2 \cos\Omega t$$
$$m\ddot{v} + c_{yT}\dot{v} - c_{yC}\dot{\theta} + k_{yT}v - k_{yC}\theta = m\varepsilon\Omega^2 \sin\Omega t$$
$$I_d\ddot{\theta} + I_p\Omega\dot{\psi} - c_{yC}\dot{v} + c_{yR}\dot{\theta} - k_{yC}v + k_{yR}\theta = 0$$
$$I_d\ddot{\psi} - I_p\Omega\dot{\theta} + c_{xC}\dot{u} + c_{xR}\dot{\psi} + k_{xC}u + k_{xR}\psi = 0$$

(6.5)

Equation (6.5) shows that the lateral offset of the mass center from the equilibrium position causes out-of-balance forces to act on the system. Thus, we can develop the equations of motion for a system with a disk or rotor with an offset either by modifying the position of the center of mass or more directly by adding forces on

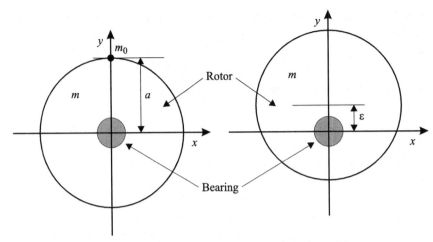

Figure 6.3. Models of out-of-balance causing a lateral force.

the right-hand side of the equations of motion, Equation (3.63), whichever is more convenient.

In the previous discussion, we imagine that the out-of-balance force arose because the total mass of the rotor is offset from the line joining the bearings by a small amount ε. This is shown in the right-hand diagram in Figure 6.3. In such a case, the out-of-balance force is given by

$$f = m\varepsilon\Omega^2 \qquad (6.6)$$

where m is the total mass of the rotor. This force appears on the right-hand side of Equation (6.5), resolved into two components. The alternative approach assumes that the original rotor mass center is coincident with the bearing centerline but that an extra (insignificant) mass, m_0, is fixed to the rotor a distance a from the bearing centerline. The added mass, m_0, is negligible compared to the rotor mass. This model is shown in the left-hand diagram in Figure 6.3. The lateral acceleration of mass m_0 is $a\Omega^2$ and the resultant force is

$$f = m_0 a\Omega^2 \qquad (6.7)$$

From Equations (6.6) and (6.7), it is seen that the out-of-balance forces are identical if

$$m_0 a = m\varepsilon$$

Because m_0 is small compared to m, for equal out-of-balance forces, ε is small compared to a. It must be stressed that these models are equivalent to one another; they are simply different ways of visualizing or representing the same phenomenon. In practice, we are normally unaware of whether the out-of-balance force arose from a small offset from the bearing centerline of a significant mass such as a disk, or an extra but small mass such as a bolt attached to the rotor at a large radius from the equilibrium position, or a combination of both.

Out-of-balance moments also can exist on a rotor. Consider again a rigid rotor and suppose that the rotor is supported in bearings in such a way that at very low speed, the centerline of the rotor is skewed relative to the axis of rotation by a small angle β, as shown in Figure 6.4. Thus, for a constant rotational speed, when the

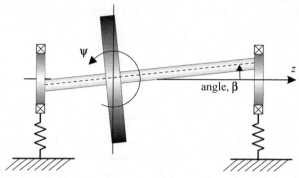

Figure 6.4. Rigid rotor with skewed principal axis of inertia.

rotor axis rotates by θ and ψ clockwise about the axes Ox and Oy, respectively, the angular position of the rotor (θ_A, ψ_A) is

$$\theta_A = \theta - \beta \sin \Omega t$$
$$\psi_A = \psi + \beta \cos \Omega t \tag{6.8}$$

Thus,

$$\dot{\theta}_A = \dot{\theta} - \beta \Omega \cos \Omega t$$
$$\dot{\psi}_A = \dot{\psi} - \beta \Omega \sin \Omega t \tag{6.9}$$

and

$$\ddot{\theta}_A = \ddot{\theta} + \beta \Omega^2 \sin \Omega t$$
$$\ddot{\psi}_A = \ddot{\psi} - \beta \Omega^2 \cos \Omega t \tag{6.10}$$

We again begin by considering the equations of motion for free vibrations, Equation (3.63), repeated here for convenience:

$$m\ddot{u} + c_{xT}\dot{u} + c_{xC}\dot{\psi} + k_{xT}u + k_{xC}\psi = 0$$
$$m\ddot{v} + c_{yT}\dot{v} - c_{yC}\dot{\theta} + k_{yT}v - k_{yC}\theta = 0$$
$$I_d\ddot{\theta} + I_p\Omega\dot{\psi} - c_{yC}\dot{v} + c_{yR}\dot{\theta} - k_{yC}v + k_{yR}\theta = 0$$
$$I_d\ddot{\psi} - I_p\Omega\dot{\theta} + c_{xC}\dot{u} + c_{xR}\dot{\psi} + k_{xC}u + k_{xR}\psi = 0$$

When the rigid rotor is skewed from the axis of rotation by a small angle, β, the angular displacements of the rotor are given by θ_A and ψ_A; however, the displacements of the springs and dampers at the bearings, caused by the angular displacements of the rotor about Ox and Oy, are still in terms of θ and ψ. Thus, replacing the angular accelerations of the rotor, $\ddot{\theta}$ and $\ddot{\psi}$, by $\ddot{\theta}_A$ and $\ddot{\psi}_A$ in Equation (3.63), and the velocities in the gyroscopic terms, $\dot{\theta}$ and $\dot{\psi}$, by $\dot{\theta}_A$ and $\dot{\psi}_A$, respectively, we have

$$m\ddot{u} + c_{xT}\dot{u} + c_{xC}\dot{\psi} + k_{xT}u + k_{xC}\psi = 0$$
$$m\ddot{v} + c_{yT}\dot{v} - c_{yC}\dot{\theta} + k_{yT}v - k_{yC}\theta = 0$$
$$I_d\ddot{\theta}_A + I_p\Omega\dot{\psi}_A - c_{yC}\dot{v} + c_{yR}\dot{\theta} - k_{yC}v + k_{yR}\theta = 0$$
$$I_d\ddot{\psi}_A - I_p\Omega\dot{\theta}_A + c_{xC}\dot{u} + c_{xR}\dot{\psi} + k_{xC}u + k_{xR}\psi = 0$$
$$\tag{6.11}$$

Substituting for $\dot{\theta}_A, \ddot{\theta}_A, \dot{\psi}_A$, and $\ddot{\psi}_A$ from Equations (6.9) and (6.10) leads to

$$m\ddot{u} + c_{xT}\dot{u} + c_{xC}\dot{\psi} + k_{xT}u + k_{xC}\psi = 0$$

$$m\ddot{v} + c_{yT}\dot{v} - c_{yC}\dot{\theta} + k_{yT}v - k_{yC}\theta = 0$$

$$I_d\ddot{\theta} + I_p\Omega\dot{\psi} - c_{yC}\dot{v} + c_{yR}\dot{\theta} - k_{yC}v + k_{yR}\theta = -(I_d - I_p)\beta\Omega^2\sin\Omega t$$

$$I_d\ddot{\psi} - I_p\Omega\dot{\theta} + c_{xC}\dot{u} + c_{xR}\dot{\psi} + k_{xC}u + k_{xR}\psi = (I_d - I_p)\beta\Omega^2\cos\Omega t$$

$$(6.12)$$

Thus, the effect of a skewed principal axis of inertia relative to the bearing centerline is to cause out-of-balance moments to act on the system. The term *swash* is often used by engineers to describe an angle between the shaft and the normal to a disk.

6.2.2 Response of a Rigid Rotor on Isotropic Supports to Out-of-Balance Forces

We now examine the solution of Equation (6.5) when the rigid rotor is supported in isotropic bearings. Then, we have $k_{xT} = k_{yT} = k_T, k_{xR} = k_{yR} = k_R, k_{xC} = k_{yC} = k_C,$ $c_{xT} = c_{yT} = c_T, c_{xR} = c_{yR} = c_R,$ and $c_{xC} = c_{yC} = c_C$. Thus, Equation (6.5) becomes

$$m\ddot{u} + c_T\dot{u} + c_C\dot{\psi} + k_T u + k_C\psi = m\varepsilon\Omega^2\cos\Omega t$$

$$m\ddot{v} + c_T\dot{v} - c_C\dot{\theta} + k_T v - k_C\theta = m\varepsilon\Omega^2\sin\Omega t$$

$$I_d\ddot{\theta} + I_p\Omega\dot{\psi} - c_C\dot{v} + c_R\dot{\theta} - k_C v + k_R\theta = 0$$

$$I_d\ddot{\psi} - I_p\Omega\dot{\theta} + c_C\dot{u} + c_R\dot{\psi} + k_C u + k_R\psi = 0$$

$$(6.13)$$

The bearing supports are isotropic and, hence, the two pairs of coupled equations can be reduced to a single pair by using complex coordinates, as described in Section 3.5.4. Combining the first and second and the third and fourth equations of Equation (6.13) by letting $r = u + jv$ and $\varphi = \psi - j\theta$, we have

$$m\ddot{r} + c_T\dot{r} + c_C\dot{\varphi} + k_T r + k_C\varphi = m\varepsilon\Omega^2 e^{j\Omega t}$$

$$I_d\ddot{\varphi} - jI_p\Omega\dot{\varphi} + c_C\dot{r} + c_R\dot{\varphi} + k_C r + k_R\varphi = 0$$

$$(6.14)$$

Note that r represents the vector OS in Figure 6.2. We begin by considering the solution of Equation (6.14) for the case when $k_C = 0$ and $c_C = 0$. Thus, we have

$$m\ddot{r} + c_T\dot{r} + k_T r = m\varepsilon\Omega^2 e^{j\Omega t}$$

$$I_d\ddot{\varphi} - jI_p\Omega\dot{\varphi} + c_R\dot{\varphi} + k_R\varphi = 0$$

$$(6.15)$$

This pair of equations is uncoupled and because there is no excitation on the second equation, there is no steady-state response in φ. Thus, we only need to solve the first equation of Equation (6.15). Dividing through by m, we have

$$\ddot{r} + 2\zeta\omega_n\dot{r} + \omega_n^2 r = \varepsilon\Omega^2 e^{j\Omega t} \qquad (6.16)$$

where $\omega_n = \sqrt{k_T/m}$ and $\zeta = c_T/(2m\omega_n)$. Note that ω_n is an undamped system natural frequency. The other system natural frequency is derived from the second equation of Equation (6.15) but is of no concern in this analysis because the rotational motion is decoupled and the resonance is not excited. We can solve Equation (6.16) only for the steady-state solution by assuming a solution $r(t) = r_0 e^{J\Omega t}$, where r_0 is complex. Thus, we have $\dot{r} = J\Omega r_0 e^{J\Omega t}$ and $\ddot{r} = -\Omega^2 r_0 e^{J\Omega t}$. Substituting into Equation (6.16) gives

$$\left(-\Omega^2 + J2\Omega\zeta\omega_n + \omega_n^2\right) r_0 e^{J\Omega t} = \varepsilon\Omega^2 e^{J\Omega t}$$

and, hence

$$r_0 = \frac{\varepsilon\Omega^2}{\left((\omega_n^2 - \Omega^2) + J2\Omega\zeta\omega_n\right)} = \frac{\left((\omega_n^2 - \Omega^2) - J2\Omega\zeta\omega_n\right)\varepsilon\Omega^2}{\left((\omega_n^2 - \Omega^2)^2 + (2\Omega\zeta\omega_n)^2\right)} \qquad (6.17)$$

From Equation (6.17), we see that r_0 is complex and can be expressed in the form

$$r_0 = (a + Jb) = |r_0| e^{-J\alpha}$$

where $|r_0| = \sqrt{a^2 + b^2}$ and $\alpha = \tan^{-1}(b/a)$. Thus, the response r can be written

$$r(t) = r_0 e^{J\Omega t} = |r_0| e^{-J\alpha} e^{J\Omega t} = |r_0| e^{J(\Omega t - \alpha)} \qquad (6.18)$$

This definition of α is consistent with that in Figure 6.2 because $\varphi(t) = \Omega t$; therefore, $r(t) = |r_0| e^{J(\varphi - \alpha)}$. Equation (6.18) defines a circular orbit in the same direction as the rotor spin, called *forward whirl*. One period of the whirl orbit corresponds to one revolution of the shaft. From Equation (6.17), we have

$$|r_0| = \frac{\varepsilon\Omega^2}{\sqrt{(\omega_n^2 - \Omega^2)^2 + (2\Omega\zeta\omega_n)^2}} \quad \text{and} \quad \alpha = \tan^{-1}\left(\frac{2\Omega\zeta\omega_n}{\omega_n^2 - \Omega^2}\right) \qquad (6.19)$$

These expressions can be nondimensionalized by letting $R = \Omega/\omega_n$. Hence,

$$|r_0| = \frac{\varepsilon R^2}{\sqrt{(1 - R^2)^2 + (2R\zeta)^2}} \quad \text{and} \quad \alpha = \tan^{-1}\left(\frac{2R\zeta}{1 - R^2}\right) \qquad (6.20)$$

R is a nondimensional parameter and is the ratio of the angular velocity of the rotor to the natural frequency of the system. For the case in which the damping in the bearings is zero, the response becomes infinite when the angular velocity of the rotor is equal to the natural frequency of the system. The maximum response may be obtained by differentiating $|r_0|^2$ with respect to R^2 and setting the result to zero. The maximum response occurs at

$$R^2 = \frac{1}{1 - 2\zeta^2} \quad \text{or} \quad \Omega^2 = \frac{\omega_n^2}{1 - 2\zeta^2}$$

From Equation (6.19) or (6.20), we see that at a steady speed Ω, the center of the rotor moves in a circle in the Oxy plane with angular velocity Ω and radius $|r_0|$. This whirl motion is equivalent to simultaneous vibration along both the x- and y-axes at frequency Ω. Figure 6.5 shows a nondimensional plot of $|r_0|/\varepsilon$ against the frequency ratio R and also the phase angle α against R. It is interesting to note the similarity between Figures 6.5 and 2.10. The latter figure shows the response of a static structure to a harmonic force, the amplitude of which is constant at all frequencies. However, although similar, there are significant differences between

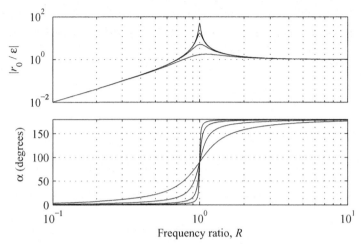

Figure 6.5. Nondimensional plots of $|r_0|/\varepsilon$ and the phase angle α against the frequency ratio R, for $\zeta = 0.01, 0.03, 0.1, 0.3$.

these two nondimensional plots. In particular, because the amplitude of the out-of-balance force is proportional to Ω^2, Figure 6.5 shows that at low rotation speeds, $|r_0|/\varepsilon$ tends to zero, and at high rotational speeds, $|r_0|/\varepsilon$ tends to unity. In these features, the response curves for a rotor excited by an out-of-balance force differ from the response curves shown in Figure 2.10.

When the amplitude of whirl is a maximum, the corresponding rotor speed is called the *critical speed*. A comprehensive discussion of critical speeds is provided in Section 6.7. In a simple model such as this, the critical speed is identical to the resonant frequency of the system in lateral vibrations. When ζ is small and Ω is close to ω_n (i.e., R is close to unity), the amplitude of rotor orbit can be large, and it is unlikely that the rotor can be safely operated at a steady speed in this region. Examining the phase response plot of Figure 6.5, we see that at low speeds, the phase angle α is close to zero. In Figure 6.6, it is shown that under these conditions, the undisturbed rotor center, O, the disturbed rotor center, S, and the center of mass of the rotor, G, lie almost in a straight line. Thus, as the rotor spins about O, the center of mass, G, rotates outside the attachment point, S. Near the critical speed, α is close to $90°$ so that the lines OS and SG form a right angle. At this speed, the out-of-balance force is in the same direction as the velocity, and power is flowing into the lateral vibrations. The whirl amplitude, OS, becomes large. The center of mass, G, leads the rotor center, S, as both points rotate about O. At high rotor speeds, well above the critical speed, α is close to $180°$ and the center of mass of the rotor, G, has moved so that the line SG rotates around and lies close to the line OS. Furthermore, because $(r_0/\varepsilon) \approx 1$, points O and G become almost coincident. The rotor center of mass is now virtually stationary at the center of rotation. These three conditions are illustrated in Figure 6.6. Figure 6.6(a) shows the whirl orbit (greatly enlarged for clarity) at a very low speed, well below the critical speed; Figure 6.6(b) shows the whirl orbit close to the critical speed; Figure 6.6(c) shows the whirl orbit at a high speed, well above the critical speed.

The phase angle α shown in Figure 6.5 is a physical (i.e., spatial) angle that might be measured if appropriate instrumentation were in place. In Section 2.3.3, we

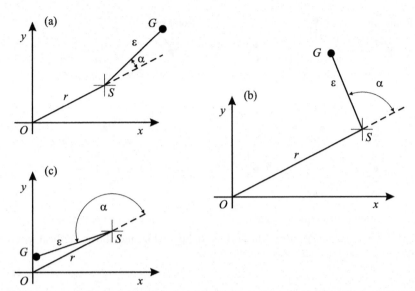

Figure 6.6. The relative positions of the rotor center of mass G and the bearing centerline O at three different rotor speeds: (a) low speed, r and α are small; (b) speed close to critical, r is large, $\alpha \approx 90°$; and (c) speed well above critical, $r \approx \epsilon$, $\alpha \approx 180°$, G and O almost coincident.

discuss the meaning of phase in terms of the relationship between two sinusoidal functions of time; we can describe this as a *temporal angle*. In the dynamics of rotating machines in which the stator properties are isotropic and the machine is vibrating with forward synchronous motion (i.e., 1X), there is a perfect equivalence between spatial and temporal angles. A comparison of Figure 6.5 with Figure 2.10 reinforces this point.

The phase relationship between the response in the two transverse directions, $u(t)$ and $v(t)$, is also important. The synchronous steady-state responses are

$$u(t) = \Re\left(r_0 e^{j\Omega t}\right) \quad \text{and} \quad v(t) = \Im\left(r_0 e^{j\Omega t}\right) \tag{6.21}$$

where r_0 is given by Equation (6.20). Because

$$v(t + \pi/(2\Omega)) = \Im\left(r_0 e^{j(\Omega t + \pi/2)}\right) = \Im\left(r_0 j e^{j\Omega t}\right) = \Re\left(r_0 e^{j\Omega t}\right) = u(t) \tag{6.22}$$

we see that the response $v(t)$ is identical to $u(t)$ but with a lag of one quarter of the synchronous time period. Furthermore, if the steady-state responses are written as

$$u(t) = \Re\left(u_0 e^{j\Omega t}\right) \quad \text{and} \quad v(t) = \Re\left(v_0 e^{j\Omega t}\right) \tag{6.23}$$

then

$$\Re\left(u_0 e^{j\Omega t}\right) = u(t) = v(t + \pi/(2\Omega)) = \Re\left(v_0 e^{j(\Omega t + \pi/2)}\right) = \Re\left(v_0 j e^{j\Omega t}\right) \tag{6.24}$$

Thus,

$$u_0 = j v_0 \quad \text{and, hence,} \quad u_0 - j v_0 = 0 \tag{6.25}$$

Note that this is true only for systems with istropic supports.

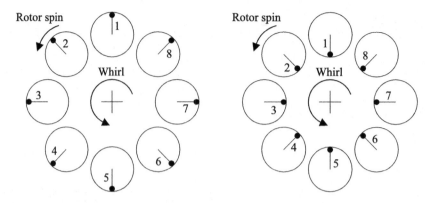

Figure 6.7. Forward synchronous whirl orbits slightly below and above a critical speed. The dark spot indicates the "heavy" side of the rotor.

Figure 6.7 also shows the position of the rotor geometric center relative to the equilibrium position at speeds just below and just above the critical speed. The displacements of the rotor are exaggerated for clarity. The so-called *heavy spot* is the point on the rotor at which the out-of-balance force acts. Figure 6.7 shows that below the critical speed, the rotor whirls with the heavy spot located outside of the orbit of the centerline of the bearings. Above the critical speed, the rotor has moved so that the heavy spot is now located inside the orbit of the centerline of the bearings. As the rotor speed increases, the whirl orbit decreases in size and the location of the heavy spot is closer to the equilibrium position of the rotor. At a very high speed above the critical speed, the rotor rotates about the heavy spot.

The fact that an out-of-balance mass causes the rotor to whirl and that this whirl becomes very large at the critical speed is a major reason why it is necessary to study the dynamics of rotors.

We now consider the situation when coupling is present between the equations of motion (i.e., $k_C \neq 0$ and $c_C \neq 0$). Equation (6.14) is repeated here for convenience:

$$m\ddot{r} + c_T\dot{r} + c_C\dot{\varphi} + k_Tr + k_C\varphi = m\varepsilon\Omega^2 e^{J\Omega t}$$

$$I_d\ddot{\varphi} - JI_p\Omega\dot{\varphi} + c_C\dot{r} + c_R\dot{\varphi} + k_Cr + k_R\varphi = 0$$

For the steady-state solution, $r(t) = r_0e^{J\Omega t}$ and $\varphi(t) = \varphi_0e^{J\Omega t}$, where r_0 and φ_0 are complex, then $\dot{r} = Jr_0\Omega e^{J\Omega t}$ and so forth. Substituting in these equations leads to

$$\left(-m\Omega^2 + Jc_T\Omega + k_T\right)r_0 + \left(Jc_C\Omega + k_C\right)\varphi_0 = m\varepsilon\Omega^2$$

$$\left(Jc_C\Omega + k_C\right)r_0 + \left(-\left(I_d - I_p\right)\Omega^2 + Jc_R\Omega + k_R\right)\varphi_0 = 0 \tag{6.26}$$

From the second equation, we can express φ_0 in terms of r_0. Substituting for φ_0 in terms of r_0 in the first equation of Equation (6.26) and rearranging gives

$$r_0 = \frac{m\varepsilon\Omega^2}{D}\left\{-\left(I_d - I_p\right)\Omega^2 + Jc_R\Omega + k_R\right\} \tag{6.27}$$

where

$$D = \left(-\left(I_d - I_p\right)\Omega^2 + jc_R\Omega + k_R\right)\left(-m\Omega^2 + jc_T\Omega + k_T\right) - \left(jc_C\Omega + k_C\right)^2 \quad (6.28)$$

Substituting in the second equation of Equation (6.26) the expression for r_0 given by Equation (6.27) and rearranging gives

$$\varphi_0 = \frac{m\varepsilon\Omega^2}{D}\{-jc_C\Omega - k_C\} \quad (6.29)$$

where D is defined in Equation (6.28).

When the damping is zero, there are values of Ω for which $D = 0$. An infinite response occurs at these speeds, which are determined from

$$m\left(I_d - I_p\right)\Omega^4 - \{k_R m + k_T \left(I_d - I_p\right)\}\Omega^2 + \left(k_R k_T - k_C^2\right) = 0 \quad (6.30)$$

If we set $s = j\Omega$ in Equation (3.42) with positive imaginary linear and cubic coefficients, we obtain (with some rearrangement) Equation (6.30). Equation (3.42) is the characteristic equation for a simple rotor on isotropic supports. By forcing $s = j\Omega$, we are determining the values of Ω that correspond to system roots and, hence, natural frequencies. For an undamped system, the condition for maximum response is governed by Equation (6.30). We can deduce that this condition occurs when the rotational speed equals a natural frequency with an orbit in the same direction of rotation (i.e., a forward orbit). Section 6.8 describes in detail methods to calculate critical speeds.

EXAMPLE 6.2.1. A uniform rigid rotor is shown in Figure 3.8. The rotor has a length of 0.5 m and a diameter of 0.2 m, and it is made from steel with a density of 7,810 kg/m^3. The rotor is supported at the ends by bearings. The horizontal and vertical support stiffnesses are 1 MN/m at bearing 1 and 1.3 MN/m at bearing 2. The damping values in the horizontal and vertical supports are 10 Ns/m at bearing 1 and 13 Ns/m at bearing 2. This rotor is the same as that described in Example 3.5.3(b). Find the response to a mass eccentricity of 0.1 mm and plot the Campbell diagram.

Solution. From Appendix 1, the mass and inertia properties of the rotor are

$$m = \rho\pi D^2 L/4 = 7,810 \times \pi \times 0.2^2 \times 0.5/4 = 122.68\,\text{kg}$$

$$I_p = mD^2/8 = 122.68 \times 0.2^2/8 = 0.6134\,\text{kg m}^2$$

$$I_d = I_p/2 + mL^2/12 = 0.6134/2 + 122.68 \times 0.5^2/12 = 2.8625\,\text{kg m}^2$$

The stiffness coefficients are

$$k_T = 1,000 + 1,300 = 2,300\,\text{kN/m}$$

$$k_R = 0.25^2 \times 1,000 + 0.25^2 \times 1,300 = 143.75\,\text{kNm}$$

$$k_C = -0.25 \times 1,000 + 0.25 \times 1,300 = 75\,\text{kN}$$

Similarly,

$$c_T = 23\,\text{Ns/m}, \quad c_R = 1.4375\,\text{Nms}, \quad c_C = 0.75\,\text{Ns}$$

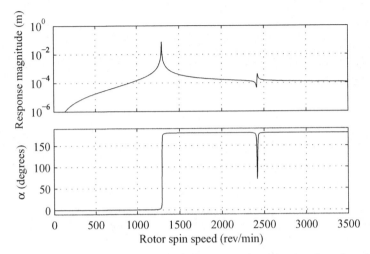

Figure 6.8. Response of a rigid rotor (Example 6.2.1), $|r_0|$ and α plotted again Ω.

Using Equation (6.27), the response can be obtained as a function of rotational speed, which is shown in Figure 6.8. The figure shows that there are two peaks in the response; the speeds of rotation at which these peaks occur are called *critical speeds*. They occur when the frequency of the out-of-balance force is close to a natural frequency of the rotor–bearing system. The rotor–bearing system has four degrees of freedom and, hence, four natural frequencies; yet, in this example, there are only two critical speeds. This condition is best explained by examining the Campbell diagram for the system shown in Figure 6.9. In this case, the unbalance produces a synchronous force; hence, the line where the rotational speed (in rev/min) is equal to the natural frequency (in Hz) is plotted on the Campbell diagram. At the points where this line intersects a system natural frequency line, the rotor speed and the natural frequency are equal and we might expect a critical speed at that rotational speed. Because there are four frequencies, there are four intersections. Looking first at the higher pair of natural frequencies, it is shown (by reference to the peaks in the response

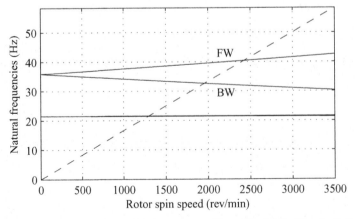

Figure 6.9. The Campbell diagram for a rigid rotor (Example 6.2.1). The dashed line represents $\omega = \Omega$. FW indicates forward whirl and BW indicates backward whirl.

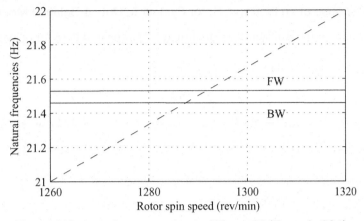

Figure 6.10. An enlargement of part of Figure 6.9 (Example 6.2.1).

shown in Figure 6.8) that the natural frequency that is excited is the higher of the pair. In Example 3.5.3(b), it is shown that the mode shape of this natural frequency rotates in the direction of the rotor spin. Due to the scale of Figure 6.9, when looking at the lower pair of natural frequencies, it is difficult to separate the point where the line of equal frequency and rotational speed intersects these frequencies. A larger-scale plot of this region of the Campbell diagram is shown in Figure 6.10. Close inspection of the response shows that the higher frequency of the pair is again excited. This is the frequency with a mode that rotates in the same direction as the rotor spin. Thus, we may deduce that for this problem, at least, peaks in the frequency-response curves occur only when the speed of rotation and the natural frequency of a forward-rotating mode of vibration are coincident or – in the case of damped systems – close.

6.2.3 Response of a Jeffcott Rotor to Out-of-Balance Forces

A simple model for the free lateral vibration of a flexible rotor is developed in Section 3.9. The equations of motion for a simple flexible rotor on rigid bearings are shown to be essentially the same as the equations of motion for a rigid rotor on flexible bearings; only the derivation and origin of the stiffness coefficients are different. In the case of the rigid rotor, deflections occur at the bearings; whereas, in a flexible rotor on rigid bearings, it is the rotor that deforms. Because of the essential similarity of the two systems, we can state immediately that Equation (6.16) describes the motion of a Jeffcott rotor excited by out-of-balance forces. The solution of this equation is given by Equation (6.19) and Figure 6.4 shows the response of the rotor to the out-of-balance force.

If the disk carried by the flexible, massless shaft is not at the midspan, there is coupling between the displacements along and rotation about the Ox and Oy axes. Then, the governing equations are given by Equation (6.14) and the responses to the out-of-balance force are given by Equations (6.27) and (6.29). Other simple rotor systems, such as an overhung rotor, also have coupling between the displacements and rotations of the disk.

6.2.4 Response of an Isotropic Rotor System to Out-of-Balance Moments

We now examine the solutions of Equation (6.12), repeated here for convenience:

$$m\ddot{u} + c_{xT}\dot{u} + c_{xC}\dot{\psi} + k_{xT}u + k_{xC}\psi = 0$$

$$m\ddot{v} + c_{yT}\dot{v} - c_{yC}\dot{\theta} + k_{yT}v - k_{yC}\theta = 0$$

$$I_d\ddot{\theta} + I_p\Omega\dot{\psi} - c_{yC}\dot{v} + c_{yR}\dot{\theta} - k_{yC}v + k_{yR}\theta = -(I_d - I_p)\beta\Omega^2\sin\Omega t$$

$$I_d\ddot{\psi} - I_p\Omega\dot{\theta} + c_{xC}\dot{u} + c_{xR}\dot{\psi} + k_{xC}u + k_{xR}\psi = (I_d - I_p)\beta\Omega^2\cos\Omega t$$

Considering the case of isotropic bearings, $k_{xT} = k_{yT} = k_T, k_{xR} = k_{yR} = k_R$, and so on, Equation (6.12) becomes

$$m\ddot{u} + c_T\dot{u} + c_C\dot{\psi} + k_Tu + k_C\psi = 0$$

$$m\ddot{v} + c_T\dot{v} - c_C\dot{\theta} + k_Tv - k_C\theta = 0$$

$$I_d\ddot{\theta} + I_p\Omega\dot{\psi} - c_C\dot{v} + c_R\dot{\theta} - k_Cv + k_R\theta = -(I_d - I_p)\beta\Omega^2\sin\Omega t \qquad (6.31)$$

$$I_d\ddot{\psi} - I_p\Omega\dot{\theta} + c_C\dot{u} + c_R\dot{\psi} + k_Cu + k_R\psi = (I_d - I_p)\beta\Omega^2\cos\Omega t$$

Because the supports are isotropic, the two pairs of coupled equations can be reduced to a single pair. Combining the first and second equations and the third and fourth equations of Equation (6.31) by letting $r = u + \jmath v$ and $\varphi = \psi - \jmath\theta$, we have

$$m\ddot{r} + c_T\dot{r} + c_C\dot{\varphi} + k_Tr + k_C\varphi = 0$$

$$I_d\ddot{\varphi} - \jmath I_p\Omega\dot{\varphi} + c_C\dot{r} + c_R\dot{\varphi} + k_Cr + k_R\varphi = (I_d - I_p)\beta\Omega^2 e^{\jmath\Omega t} \qquad (6.32)$$

We begin by considering the solution of Equation (6.32) for the case when $k_C = 0$ and $c_C = 0$. Thus, we have

$$m\ddot{r} + c_T\dot{r} + k_Tr = 0$$

$$I_d\ddot{\varphi} - \jmath I_p\Omega\dot{\varphi} + c_R\dot{\varphi} + k_R\varphi = (I_d - I_p)\beta\Omega^2 e^{\jmath\Omega t} \qquad (6.33)$$

These equations are uncoupled and because there is no force in the steady-state lateral direction, the lateral displacement, r, is zero. The second equation of Equation (6.33) can be solved for the steady state by assuming a solution of the form $\varphi(t) = \varphi_0 e^{\jmath\Omega t}$, where φ_0 is complex. Thus,

$$\left(-I_d\Omega^2 + I_p\Omega^2 + \jmath c_R\Omega + k_R\right)\varphi_0 e^{\jmath\Omega t} = (I_d - I_p)\beta\Omega^2 e^{\jmath\Omega t} \qquad (6.34)$$

Rearranging, and because $e^{\jmath\Omega t}$ is non-zero, gives

$$\left(-(I_d - I_p)\Omega^2 + \jmath c_R\Omega + k_R\right)\varphi_0 = (I_d - I_p)\beta\Omega^2 \qquad (6.35)$$

Although this equation may be nondimensionalized, these terms cannot be related to physical quantities. The two natural frequencies for this system are cumbersome expressions given by Equation (3.31) and cannot be used to simplify

Equation (6.35). Instead, we divide the previous equation by I_d and let $\gamma = I_p/I_d$, $\chi = c_R/\sqrt{I_d k_R}$, and $R = \Omega/\sqrt{(k_R/I_d)}$ to give

$$\left(1 - (1 - \gamma)\, R^2 + J\chi R\right) \varphi_0 = (1 - \gamma)\, \beta R^2 \tag{6.36}$$

Thus,

$$\varphi_0 = \frac{(1 - \gamma)\, \beta R^2}{(1 - (1 - \gamma)\, R^2) + J\chi R} = \frac{\left\{(1 - (1 - \gamma)\, R^2) - J\chi R\right\}(1 - \gamma)\, \beta R^2}{(1 - (1 - \gamma)\, R^2)^2 + (\chi R)^2} \tag{6.37}$$

From Equation (6.37), we see that φ_0 is of the form

$$\varphi_0 = (a + Jb) = |\varphi_0|\, e^{-J\alpha}$$

where $|\varphi_0| = \sqrt{a^2 + b^2}$ and $\alpha = -\tan^{-1}(b/a)$. Thus, the response φ can be written

$$\varphi(t) = \varphi_0 e^{J\Omega t} = |\varphi_0|\, e^{-J\alpha} e^{J\Omega t} = |\varphi_0|\, e^{J(\Omega t - \alpha)}$$

From Equation (6.37), we have

$$|\varphi_0| = \frac{|1 - \gamma|\, \beta R^2}{\sqrt{(1 - (1 - \gamma)\, R^2)^2 + (\chi R)^2}} \quad \text{and} \quad \alpha = \tan^{-1}\left(\frac{\chi R}{1 - (1 - \gamma)\, R^2}\right) \tag{6.38}$$

When $\Omega = 0$, then $R = 0$ and the response is zero. As Ω tends to infinity, R tends to infinity, and the response tends to β. For a system with no damping, $\chi = 0$ and then the maximum response occurs when $(1 - \gamma)\, R^2 = 1$. This occurs at a rotational speed given by

$$\Omega^2 = \frac{k_R}{I_d(1 - \gamma)} = \frac{k_R}{I_d - I_p} \tag{6.39}$$

For many rigid rotors, $I_d > I_p$ and, hence, a maximum response exists. However, for a thin disk, generally $I_d < I_p$ and there is no real value of Ω that satisfies Equation (6.39), so there is no peak in the response curve. An explanation for this is that the resonance frequency remains above the rotational speed. Figure 6.11 shows the response of a system described by Equation (6.38) for various values of γ. If $\chi \neq 0$, then the maximum response is obtained by differentiating $|\varphi_0|^2$, given in Equation (6.38), by R^2 and setting the result to zero. Thus, the maximum response occurs at

$$R^2 = \frac{2}{2(1 - \gamma) - \chi^2} \quad \text{and, thus,} \quad \Omega^2 = \frac{k_R}{I_d - I_p - c_R^2/(2k_R)}$$

If k_C and c_C are not zero, then for the steady-state solution $r(t) = r_0 e^{J\Omega t}$ and $\varphi(t) = \varphi_0 e^{J\Omega t}$, where r_0 and φ_0 are complex. Thus, $\dot{r} = Jr_0\Omega e^{J\Omega t}$ and so on, and substituting into Equation (6.32) leads to

$$\left(-m\Omega^2 + Jc_T\Omega + k_T\right) r_0 + (Jc_C\Omega + k_C)\, \varphi_0 = 0$$

$$(Jc_C\Omega + k_C)\, r_0 + \left(-(I_d - I_p)\,\Omega^2 + Jc_R\Omega + k_R\right)\varphi_0 = (I_d - I_p)\,\beta\Omega^2 \tag{6.40}$$

The solution of these equations is

$$r_0 = -\frac{\beta}{D}(Jc_C\Omega + k_C)(I_d - I_p)\,\Omega^2$$

$$\varphi_0 = \frac{\beta}{D}\left(-(I_d - I_p)\,\Omega^2 + Jc_R\Omega + k_R\right)(I_d - I_p)\,\beta\Omega^4 \tag{6.41}$$

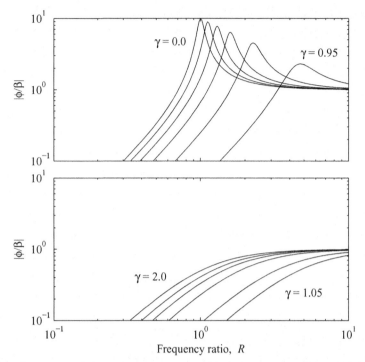

Figure 6.11. Response of the system to an out-of-balance moment for $\chi = 0.1$, and $\gamma = 0.0, 0.2, 0.4, 0.6, 0.8, 0.95$ (top plot), $\gamma = 1.05, 1.1, 1.3, 1.5, 1.7, 2$ (bottom plot).

where

$$D = \left(-\left(I_d - I_p\right)\Omega^2 + jc_R\Omega + k_R\right)\left(-m\Omega^2 + jc_T\Omega + k_T\right) - \left(jc_C\Omega + k_C\right)^2$$

Note that D here is identical to D given in Equation (6.28).

EXAMPLE 6.2.2. A 38 mm-diameter solid shaft is supported in self-aligning bearings 1.1 m apart. The shaft carries a disk 650 mm in diameter and 100 mm thick. The shaft and disk have a mass density of $\rho = 7{,}810\,\text{kg/m}^3$ and a modulus of elasticity of $E = 211\,\text{GPa}$. Determine the response of the system at 900 rev/min if the disk is misaligned with the axis of the shaft by $1°$, if

(a) The disk is shrunk onto the shaft at the midspan.
(b) The disk is shrunk onto the shaft 0.8 m from one end.

Solution. From Example 3.9.1, we have

disk mass and inertia, $m = 259.2\,\text{kg}$, $I_p = 13.69\,\text{kg m}^2$, $I_d = 7.06\,\text{kg m}^2$, shaft stiffness, $EI = 21597\,\text{Nm}^2$

(a) Given that $a = b = 0.55$ m, then from Appendix 2

$k_T = 3 \times 21{,}579\left(0.55^2 + 0.55^2\right) / \left(0.55^2 \times 0.55^2\right) = 0.4280 \times 10^6\,\text{N/m}$,

$k_C = 0$

$k_R = 3 \times 21{,}597 \times 1.1\left(0.55 \times 0.55\right) = 0.2354 \times 10^6\,\text{N m}$

Because $k_C = 0$ and $r_0 = 0$, we can use Equation (6.38) to calculate φ_0, which is the solution of Equation (6.33). Thus,

$$\beta = 1 \times \frac{\pi}{180} = 0.01745,\ \gamma = \frac{13.69}{7.06} = 1.9391,\ \chi = 0,$$

$$\sqrt{k_R/I_d} = \sqrt{\frac{0.2354 \times 10^6}{7.06}} = 182.6\,\text{rad/s},\ \Omega = 900 \times \frac{2\pi}{60} = 94.25\,\text{rad/s},$$

$$R = \frac{94.25}{182.6} = 0.5161.$$

Hence, $\varphi_0 = -0.003492\,\text{rad}$ or $\varphi_0 = -0.003492 \times \dfrac{180}{\pi} = -0.20°$.

(b) Given $a = 0.8\,\text{m}$ and $b = 0.3\,\text{m}$, using Appendix 2 gives

$$k_T = 3 \times 21{,}597 \times (0.83 + 0.33)/(0.33 \times 0.83) = 2.5262 \times 10^6\,\text{N/m}$$

$$k_C = 3 \times 21{,}597 \times 1.1 \times (0.8 - 0.3)/(0.32 \times 0.82) = 6.1866 \times 10^5\,\text{N}$$

$$k_R = 3 \times 21{,}597 \times 1.1/(0.3 \times 0.8) = 2.9696 \times 10^5\,\text{Nm}$$

Because $k_C \neq 0$, we must use Equation (6.41), which is the solution of Equation (6.40). Thus,

$$D = \left(k_R - (I_d - I_p)\,\Omega^2\right)\left(k_T - m\Omega^2\right) - k_C^2 = 1.0198 \times 10^{11}\,\text{N}^2$$

$$r_0 = -k_C\,(I_d - I_p)\,\beta\Omega^2/D = 0.006236\,\text{m} = 6.236\,\text{mm}$$

$$\varphi_0 = \left(k_R - \Omega^2\,(I_d - I_p)\right)(I_d - I_p)\,\beta\Omega^2/D = -0.003587\,\text{rad} = -0.21°$$

Although the steady-state rotational response is similar to (a), in this case, we also have a steady-state translational response.

6.2.5 Response of a Rigid Rotor on Anisotropic Supports to Out-of-Balance Forces and Moments

In many practical situations, the stiffnesses of the rotor supports are different in each plane. For example, if a bearing is mounted on a supporting pedestal, then the horizontal stiffness of the combined bearing and pedestal is typically lower than the vertical stiffness. In this section, we examine the effect of such anisotropic stiffness on a rigid rotor.

In Sections 6.2.1 and 6.2.3, the equations of motion for a rigid rotor excited by an out-of-balance force and an out-of-balance moment are developed. Combining these two sets of equations, we can consider the general case of a rigid rotor excited simultaneously by out-of-balance forces and moments. Combining Equations (6.5) and (6.31) gives

$$m\ddot{u} + c_{xT}\dot{u} + c_{xC}\dot{\psi} + k_{xT}u + k_{xC}\psi = m\varepsilon\Omega^2\cos\left(\Omega t + \delta\right)$$

$$m\ddot{v} + c_{yT}\dot{v} - c_{yC}\dot{\theta} + k_{yT}v - k_{yC}\theta = m\varepsilon\Omega^2\sin\left(\Omega t + \delta\right)$$

$$I_d\ddot{\theta} + I_p\Omega\dot{\psi} - c_{yC}\dot{v} + c_{yR}\dot{\theta} - k_{yC}v + k_{yR}\theta = -\left(I_d - I_p\right)\beta\Omega^2\sin\left(\Omega t + \gamma\right) \tag{6.42}$$

$$I_d\ddot{\psi} - I_p\Omega\dot{\theta} + c_{xC}\dot{u} + c_{xR}\dot{\psi} + k_{xC}u + k_{xR}\psi = \left(I_d - I_p\right)\beta\Omega^2\cos\left(\Omega t + \gamma\right)$$

We include angles δ and γ that determine the directions, when $t = 0$, of the out-of-balance force and moment vectors relative to the Oxy axes. When only one unbalance is present, we can calculate the response relative to that unbalance.

However, when more than one unbalance is present, we need to define the angular positions of the unbalance relative to a reference fixed in the shaft.

In matrix notation, Equation (6.42) can be written as

$$\mathbf{M\ddot{q}} + (\Omega\mathbf{G} + \mathbf{C})\,\dot{\mathbf{q}} + \mathbf{Kq} = \mathbf{Q}$$

where $\mathbf{q}^\top = \begin{bmatrix} u & v & \theta & \psi \end{bmatrix}$,

$$\mathbf{M} = \begin{bmatrix} m & 0 & 0 & 0 \\ 0 & m & 0 & 0 \\ 0 & 0 & I_d & 0 \\ 0 & 0 & 0 & I_d \end{bmatrix}, \quad \mathbf{G} = \begin{bmatrix} 0 & 0 & 0 & 0 \\ 0 & 0 & 0 & 0 \\ 0 & 0 & 0 & I_p \\ 0 & 0 & -I_p & 0 \end{bmatrix},$$

$$\mathbf{C} = \begin{bmatrix} c_{xT} & 0 & 0 & c_{xC} \\ 0 & c_{yT} & -c_{yC} & 0 \\ 0 & -c_{yC} & c_{yR} & 0 \\ c_{xC} & 0 & 0 & c_{xR} \end{bmatrix}, \quad \mathbf{K} = \begin{bmatrix} k_{xT} & 0 & 0 & k_{xC} \\ 0 & k_{yT} & -k_{yC} & 0 \\ 0 & -k_{yC} & k_{yR} & 0 \\ k_{xC} & 0 & 0 & k_{xR} \end{bmatrix}$$

and

$$\mathbf{Q} = \begin{Bmatrix} m\varepsilon\Omega^2\cos(\Omega t + \delta) \\ m\varepsilon\Omega^2\sin(\Omega t + \delta) \\ -(I_d - I_p)\beta\Omega^2\sin(\Omega t + \gamma) \\ (I_d - I_p)\beta\Omega^2\cos(\Omega t + \gamma) \end{Bmatrix} = \Re \begin{Bmatrix} m\varepsilon e^{J\delta} \\ -Jm\varepsilon e^{J\delta} \\ J(I_d - I_p)\beta e^{J\gamma} \\ (I_d - I_p)\beta e^{J\gamma} \end{Bmatrix} \Omega^2 e^{J\Omega t} \qquad (6.43)$$

The steady-state solution is found by assuming a response of the form $\mathbf{q}(t) = \Re\left(\mathbf{q}_0 e^{J\Omega t}\right)$. Thus,

$$\left[-\Omega^2\mathbf{M} + J\Omega(\Omega\mathbf{G} + \mathbf{C}) + \mathbf{K}\right]\mathbf{q}_0 e^{J\Omega t} = \Omega^2\mathbf{b}_0 e^{J\Omega t}$$

where $\mathbf{b}_0 = \begin{Bmatrix} m\varepsilon e^{J\delta} \\ -Jm\varepsilon e^{J\delta} \\ J(I_d - I_p)\beta e^{J\gamma} \\ (I_d - I_p)\beta e^{J\gamma} \end{Bmatrix}$ and, hence

$$\mathbf{q}_0 = \left[-\Omega^2\mathbf{M} + J\Omega(\Omega\mathbf{G} + \mathbf{C}) + \mathbf{K}\right]^{-1}\Omega^2\mathbf{b}_0 \qquad (6.44)$$

Even for a simple four degrees of freedom model with anisotropic properties, it is still necessary to invert a 4×4 matrix at every rotor speed for which the response is required. A more efficient approach for very large systems computes the modal decomposition in state–space form and calculates the response from this. The response is obtained in a way similar to Section 2.4.6 as

$$\mathbf{q}_0 = \sum_{i=1}^{2n} \frac{\mathbf{u}_{Ri}\mathbf{u}_{Li}^\top}{J\omega - s_i} \Omega^2\mathbf{b}_0 \qquad (6.45)$$

where s_i is the ith eigenvalue and $\mathbf{u}_{Li}$ and $\mathbf{u}_{Ri}$ are the corresponding left and right eigenvectors discussed in Section 5.8.

6.2.6 Forward- and Backward-Whirl Orbits

Section 3.6.1 demonstrates that for a particular mode, the orbit of the nodes on a rigid rotor on anisotropic supports is elliptical and gives a method to calculate the

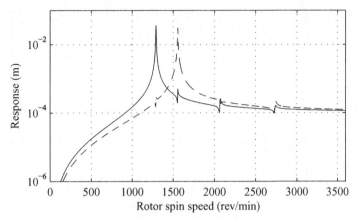

Figure 6.12. Response in the x (solid) and y (dashed) directions at bearing 1 of a rigid rotor on anisotropic bearings to an out-of-balance force (Example 6.2.3).

dimensions and orientation of the ellipse. It is now shown that the unbalanced response of a rigid rotor also whirls either forward or backward in an elliptical orbit. The direction of the orbit is calculated using the methods in Section 3.6.1. Elliptical orbits, unlike forward circular orbits, cause alternating stresses and, hence, the potential to cause fatigue, particularly in flexible shafts. The practical implications of these alternating stresses are greater for backward-whirling orbits.

We now calculate the unbalance response and orbits of a rigid rotor at different rotational speeds.

EXAMPLE 6.2.3. A uniform rigid rotor is shown in Figure 3.7. The rotor has a length of 0.5 m, a diameter of 0.2 m, and is made from steel having a density of 7,810 kg/m^3. The rotor is supported at the ends by bearings. The horizontal and vertical support stiffnesses are 1 and 1.5 MN/m, respectively, at bearing 1 and 1.3 and 1.8 MN/m, respectively, at bearing 2. The damping values in the horizontal and vertical supports at both bearings are proportional to the stiffnesses and are 20 and 30 Ns/m at bearing 1 and 26 and 36 Ns/m at bearing 2. This rotor is essentially the same as that described in Example 3.6.1(b). Determine the unbalance response over the speed range 0 to 4,000 rev/min.

Solution. Figure 6.12 shows the response in the x and y directions. It shows that the maximum response in the x and y directions occur at different frequencies because the stiffnesses in the two directions are different from one another. The plot of the radius of the major axis of the whirl ellipse, Figure 6.13, shows that there are four speeds at which the orbit has a maximum value, but two of them are rather insignificant. Figure 6.13 also shows the rotor speed range in which the whirl is backward. Superficially, it appears to be in the speed range between the pair of critical speeds. However, this is not strictly the case, and this point is discussed in more detail in Example 6.3.1. Suffice it to say that there are definite speed ranges in which backward whirl can occur in response to unbalance forcing. Figure 6.14 shows the orbits of the rotor at spin speeds just below and just above the first boundary at approximately 1,288 rev/min.

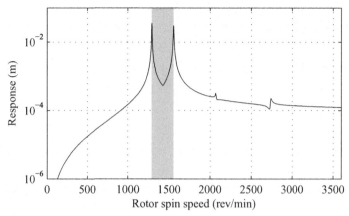

Figure 6.13. Maximum response at bearing 1 of a rigid rotor on anisotropic bearings to an out-of-balance force (Example 6.2.3). The shaded band shows the rotor speeds where the rotor whirl is backward.

Figure 6.15 shows the response for the same system but with zero damping in the bearings. Apart from the increase in amplitude at the peaks in the response (in this case, the response is infinite at these rotor speeds), it is seen that backward whirl exists in two additional but small speed ranges. In Figure 6.15, these ranges are so small that they appear as lines. Thus, damping in the bearings can sometimes prevent backward whirl from occurring.

It is clear from the Campbell diagram, shown in Figure 6.16, that when a rotor resonance frequency is coincident with the rotational speed, a maximum response to out-of-balance occurs. These maximum responses are the system critical speeds; thus, there are four critical speeds in this four degrees of freedom model of a system.

6.2.7 Response of Bent Rotors

We now examine the influence of a bent rotor on a rotating system. For simplicity, we consider first a Jeffcott rotor carrying a central disk. If the rotor is permanently bent, plastic deformation has occurred and the bent shape is now the equilibrium

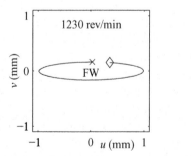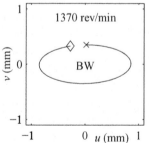

Figure 6.14. The orbits of the rigid rotor on anisotropic bearings at bearing 1 due to an out-of-balance force (Example 6.2.3), at rotor spin speeds of 1,230 and 1,370 rev/min. FW indicates forward whirl and BW indicates backward whirl.

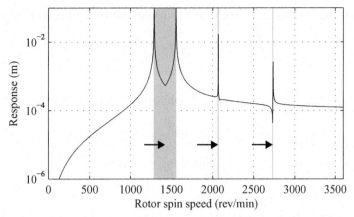

Figure 6.15. Maximum response at bearing 1 of a rigid rotor on anisotropic bearings to an out-of-balance force (Example 6.2.3). There is no damping in the bearings. The three shaded bands (indicated by arrows) show the rotor speeds in which the rotor whirl is backward.

state when no external forces are applied. We assume that the rotor system is perfectly balanced; that is, if we imagine the rotor straightened, then the center of mass of the disk is coincident with the line joining the bearings. Because the rotor is bent, the disk mass center is now offset from the straight line joining the bearing centers by r_b. We also assume that when $t = 0$, the displacement r_b lies along the x-axis. As the rotor spins, the deflection due to the permanent bend in the rotor is $r_b \cos \Omega t$ in the x direction and $r_b \sin \Omega t$ in the y direction. As the rotor spins, there is an additional elastic deformation of the rotor equal to u_e and v_e in the x and y directions, respectively. Hence, the total displacements of the disk from the centerline of the bearings are

$$u = u_e + r_b \cos \Omega t \quad \text{and} \quad v = v_e + r_b \sin \Omega t \tag{6.46}$$

in the x and y directions, respectively. Thus,

$$u_e = u - r_b \cos \Omega t \quad \text{and} \quad v_e = v - r_b \sin \Omega t$$

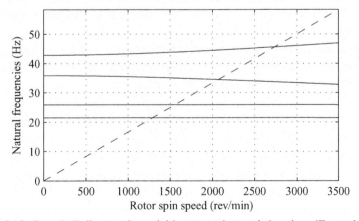

Figure 6.16. Campbell diagram for a rigid rotor on isotropic bearings (Example 6.2.3).

The elastic force restoring the rotor disk to its equilibrium position is

$$f_x = ku_e = k(u - r_b \cos \Omega t)$$
$$f_y = kv_e = k(v - r_b \sin \Omega t)$$

where k is the stiffness of the rotor at the midspan. Applying Newton's second law,

$$f_x = -m\ddot{u} \quad \text{and} \quad f_y = -m\ddot{v}$$

Equating the forces in the x and y directions, we have

$$m\ddot{u} + k(u - r_b \cos \Omega t) = 0$$
$$m\ddot{v} + k(v - r_b \sin \Omega t) = 0$$

Hence,

$$m\ddot{u} + ku = kr_b \cos \Omega t$$
$$m\ddot{v} + kv = kr_b \sin \Omega t \tag{6.47}$$

Let $r = u + jv$. Then, combining these equations, we have

$$m\ddot{r} + kr = kr_b e^{j\Omega t}$$

Dividing by m

$$\ddot{r} + \omega_n^2 r = \omega_n^2 r_b e^{j\Omega t} \tag{6.48}$$

where $\omega_n^2 = k/m$. Letting $r(t) = r_0 e^{j\Omega t}$, where r_0 is complex, leads to

$$|r_0| = \frac{r_b \omega_n^2}{\omega_n^2 - \Omega^2} = \frac{r_b}{1 - R^2} \tag{6.49}$$

where $R = \Omega/\omega_n$.

From this equation, we see that when Ω tends to zero, r tends to r_b. Conversely, at high speed, r tends to zero because the elastic deformation equals the deformation due to the bent rotor but is 180° out-of-phase with this deformation; this is shown in Figure 6.17.

We can compare the deformation of a Jeffcott rotor with a bent rotor, described by the previous equation, with the deformation due to out-of-balance (see Equation (6.12)). This equation is repeated here with $\zeta = 0$:

$$|r_0| = \frac{\varepsilon R^2}{1 - R^2}$$

Although similar, these equations are not identical. If a rotor is spinning at a constant speed, then it exhibits whirl and it is impossible to deduce whether that whirl is caused by a bent or an out-of-balance rotor. If the cause is a bent rotor, it is possible to balance the rotor and remove the whirl at only one speed. If the rotational speed changes, the whirl again arises (see Problem 6.6).

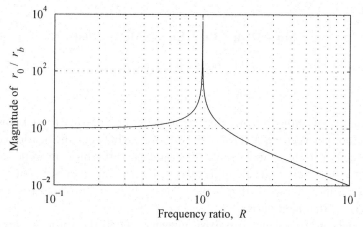

Figure 6.17. The response of a bent rotor.

6.3 Complex Rotor Models

Chapter 5 discusses the FE modeling of the rotor and bearing and determines the free response. We now consider the forced response of complex rotor models to various forms of excitation using FEA. Here, we focus on the derivation of the vector of forces acting on the system and an interpretation of the resulting responses.

Once the system response is determined at each node in terms of the displacements in the x and y directions, then the operating deflection shape and the orbit of the rotor at a particular node can be constructed. We also can determine the size of the major and minor axes of the elliptical orbit and the direction of rotor whirl around the orbit at each node using the methods described in Section 3.6.1. Within a response, all of the nodes are often whirling either forward or backward; however, part of a rotor may be executing forward whirl and part may be executing backward whirl. We now calculate the unbalance response and orbits of a flexible rotor at a number of rotational speeds.

6.3.1 Response of Rotors to Out-of-Balance Forces and Moments

In Section 5.7, the equation of motion for the free vibration of a multiple degrees of freedom rotor–bearing system is in Equation (5.109); it is repeated here for convenience:

$$\mathbf{M}\ddot{\mathbf{q}} + \Omega\mathbf{G}\dot{\mathbf{q}} + \mathbf{C}\dot{\mathbf{q}} + \mathbf{K}\mathbf{q} = \mathbf{0}$$

If there is a difference between the equilibrium position and the mass centerline of the rotor in any plane of rotation along the rotor, then out-of-balance forces and moments arise in that plane when the rotor spins. If we let the differences between the equilibrium position and the rotor center of mass at various locations along the rotor be $\mathbf{q}_\varepsilon$, and then replace $\mathbf{q}$ by $\mathbf{q} + \mathbf{q}_\varepsilon$ in the inertia and gyroscopic terms of Equation (5.109), we have

$$\mathbf{M}\left(\ddot{\mathbf{q}} + \ddot{\mathbf{q}}_\varepsilon\right) + \Omega\mathbf{G}\left(\dot{\mathbf{q}} + \dot{\mathbf{q}}_\varepsilon\right) + \mathbf{C}\dot{\mathbf{q}} + \mathbf{K}\mathbf{q} = \mathbf{0} \qquad (6.50)$$

Thus,

$$\mathbf{M\ddot{q}} + \Omega\mathbf{G\dot{q}} + \mathbf{C\dot{q}} + \mathbf{Kq} = -\mathbf{M\ddot{q}}_\varepsilon - \Omega\mathbf{G\dot{q}}_\varepsilon \tag{6.51}$$

Because the excitation and response is harmonic, $\ddot{\mathbf{q}}_\varepsilon = -\Omega^2\mathbf{q}_\varepsilon$; hence,

$$\mathbf{M\ddot{q}} + \Omega\mathbf{G\dot{q}} + \mathbf{C\dot{q}} + \mathbf{Kq} = \Omega^2\mathbf{Mq}_\varepsilon - \Omega\mathbf{G\dot{q}}_\varepsilon \tag{6.52}$$

In general, it is difficult to use this equation because to solve for $\mathbf{q}$, knowledge of $\mathbf{q}_\varepsilon$ is required. The vector $\mathbf{q}_\varepsilon$ is a list of the differences between the bearing centerline and mass center in the x and y directions (and the differences in slopes about the x- and y-axes) at every node along the rotor – information that is difficult to obtain. The value of this equation is that it allows us to study and predict the effect of any out-of-balance that might arise.

Although out-of-balance can exist anywhere along the rotor, we assume that it exists only at nodes of the rotor model. This assumption is not as limiting as it might appear because the important factor is how the unbalance excites the modes within the frequency range of the machine. Thus, a generally distributed unbalance is equivalent (in the sense of the excitation provided to a limited set of modes) to unbalance at a discrete set of nodes. This is why machines may be balanced using balance masses applied to a rotor at a small number of planes. Chapter 8 provides more detail about balancing. Thus, for a particular node k, if a disk at this node is offset by a displacement ε and an angle β, the vector $\mathbf{q}_k$ associated with the disk element is

$$\mathbf{q}_k = \begin{Bmatrix} \varepsilon\cos(\Omega t + \delta) \\ \varepsilon\sin(\Omega t + \delta) \\ -\beta\sin(\Omega t + \gamma) \\ \beta\cos(\Omega t + \gamma) \end{Bmatrix} \quad \text{and} \quad \dot{\mathbf{q}}_k = \begin{Bmatrix} -\varepsilon\Omega\sin(\Omega t + \delta) \\ \varepsilon\Omega\cos(\Omega t + \delta) \\ -\beta\Omega\cos(\Omega t + \gamma) \\ -\beta\Omega\sin(\Omega t + \gamma) \end{Bmatrix}$$

where in these vectors, δ and γ are the angles (when $t = 0$) of the out-of-balance force and moment vectors relative to the Oxy axes, and the four degrees of freedom are $4k - 3$, $4k - 2$, $4k - 1$, and $4k$. These two vectors can be expressed alternatively in the form

$$\mathbf{q}_k = \Re\left(\begin{Bmatrix} \varepsilon e^{J\delta} \\ -J\varepsilon e^{J\delta} \\ J\beta e^{J\gamma} \\ \beta e^{J\gamma} \end{Bmatrix} e^{J\Omega t} \right) \quad \text{and} \quad \dot{\mathbf{q}}_k = \Re\left(\begin{Bmatrix} J\Omega\varepsilon e^{J\delta} \\ \Omega\varepsilon e^{J\delta} \\ -\Omega\beta e^{J\gamma} \\ J\Omega\beta e^{J\gamma} \end{Bmatrix} e^{J\Omega t} \right)$$

Thus, at the particular node k

$$\Omega^2\mathbf{M}_k\mathbf{q}_\varepsilon - \Omega\mathbf{G}_k\dot{\mathbf{q}}_\varepsilon = \Re\left(\Omega^2 \begin{bmatrix} m_k & 0 & 0 & 0 \\ 0 & m_k & 0 & 0 \\ 0 & 0 & I_{dk} & 0 \\ 0 & 0 & 0 & I_{dk} \end{bmatrix} \begin{Bmatrix} \varepsilon e^{J\delta} \\ -J\varepsilon e^{J\delta} \\ J\beta e^{J\alpha} \\ \beta e^{J\alpha} \end{Bmatrix} e^{J\Omega t} \right.$$

$$\left. - \Omega \begin{bmatrix} 0 & 0 & 0 & 0 \\ 0 & 0 & 0 & 0 \\ 0 & 0 & 0 & I_{pk} \\ 0 & 0 & -I_{pk} & 0 \end{bmatrix} \begin{Bmatrix} J\Omega\varepsilon e^{J\delta} \\ \Omega\varepsilon e^{J\delta} \\ -\Omega\beta e^{J\gamma} \\ J\Omega\beta e^{J\gamma} \end{Bmatrix} e^{J\Omega t} \right)$$

Figure 6.18. The two-disk and two-bearing machine.

Hence,

$$\Omega^2 \mathbf{M}_k \mathbf{q}_\varepsilon - \Omega \mathbf{G}_k \dot{\mathbf{q}}_\varepsilon = \Re \left(\Omega^2 \begin{Bmatrix} m_k \varepsilon e^{J\delta} \\ -J m_k \varepsilon e^{J\delta} \\ J I_{dk} \beta e^{J\gamma} \\ I_{dk} \beta e^{J\gamma} \end{Bmatrix} e^{J\Omega t} - \Omega^2 \begin{Bmatrix} 0 \\ 0 \\ J I_{pk} \beta e^{J\gamma} \\ I_{pk} \beta e^{J\gamma} \end{Bmatrix} e^{J\Omega t} \right)$$

and, thus,

$$\Omega^2 \mathbf{M}_k \mathbf{q}_\varepsilon - \Omega \mathbf{G}_k \dot{\mathbf{q}}_\varepsilon = \Re \left(\Omega^2 \begin{Bmatrix} m_k \varepsilon e^{J\delta} \\ -J m_k \varepsilon e^{J\delta} \\ J \left(I_{dk} - I_{pk} \right) \beta e^{J\gamma} \\ \left(I_{dk} - I_{pk} \right) \beta e^{J\gamma} \end{Bmatrix} e^{J\Omega t} \right) = \Re \left(\Omega^2 \mathbf{b}_{0e} e^{J\Omega t} \right) \quad (6.53)$$

Note that $\Omega^2 \mathbf{b}_{0e}$ is the vector of forces acting at node k due to a disk offset and tilt and the similarity between Equations (6.53) and (6.43). Assuming no other out-of-balance is present in the rotor, the rest of the force vector, denoted by $\Omega^2 \mathbf{b}_0$, is zero. Thus, we have

$$\mathbf{M}\ddot{\mathbf{q}} + \Omega \mathbf{G}\dot{\mathbf{q}} + \mathbf{C}\dot{\mathbf{q}} + \mathbf{K}\mathbf{q} = \Re \left(\Omega^2 \mathbf{b}_0 e^{J\Omega t} \right) \quad (6.54)$$

We must solve Equation (6.54) to determine the steady-state response to the unbalance forces. Letting $\mathbf{q}(t) = \Re \left(\mathbf{q}_0 e^{J\Omega t} \right)$ (where $\mathbf{q}_0$ is complex) gives

$$\mathbf{q}_0 = \left[\left(\mathbf{K} - \Omega^2 \mathbf{M} \right) + J\Omega \left(\Omega \mathbf{G} + \mathbf{C} \right) \right]^{-1} \Omega^2 \mathbf{b}_0 \quad (6.55)$$

The steady-state response of the rotor at any node is an orbit. If the supports are isotropic, then the orbit is a circle and its radius is the magnitude of the response in both the x and y directions. For anisotropic bearings, the orbit is an ellipse and the magnitude is best represented by the length of the semimajor axis. This may be calculated using the method described in Section 3.6.1.

EXAMPLE 6.3.1. The rotor system shown in Figure 6.18 consists of a shaft 1.5 m long with a 50 mm diameter supported by bearings at each end. Disks are mounted on the shaft at one-third and two-third spans. Each disk is 70 mm thick and the left and right disks are 280 and 350 mm in diameter, respectively. The shaft and disks have material properties $E = 211$ GPa, $G = 81.1$ GPa, and $\rho = 7,810$ kg/m^3. Determine the response of the system at the disks due to an out-of-balance on the left disk of 0.001 kg m, if each bearing has a stiffness of (a) 1 MN/m and a damping of 100 Ns/m in both the x and y directions; and (b) 1 MN/m in the x direction and 0.8 MN/m in the y direction and a damping of 100 Ns/m in both directions. The natural frequencies and mode shapes for these rotor systems are calculated in Examples 5.9.1 and 5.9.2.

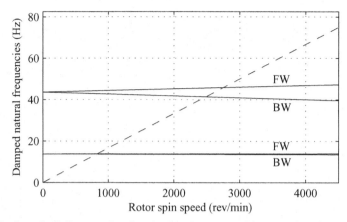

Figure 6.19. Campbell diagram showing the first four natural frequencies of the rotor system (Example 6.3.1(a)). The dashed line represents $\omega = \Omega$. FW indicates forward whirl and BW indicates backward whirl.

Solution

(a) *Isotropic system:* From FEA using six shaft elements (giving 28 degrees of freedom), we can obtain the Campbell diagram shown in Figure 6.19. At 3,000 rev/min, the first six natural frequencies are 13.64, 13.93, 41.00, 46.13, 100.0, and 127.6 Hz, although only the first four are shown on the Campbell diagram. The corresponding damping factors range from $\zeta = 0.0016$ to 0.02.

The response of the system to the out-of-balance force at node 3 is shown in Figure 6.20 for the left disk (node 3) and the right disk (node 5). The response to the out-of-balance force is a maximum at 830 and 2,755 rev/min. Comparing these speeds with the rotor speeds that are coincident with natural frequencies in the Campbell diagram, we see that only the forward modes are excited. This is particularly clear in the case of the third (backward) and fourth (forward) frequencies, which are coincident with the rotor speed at 2,489 and 2,755 rev/min,

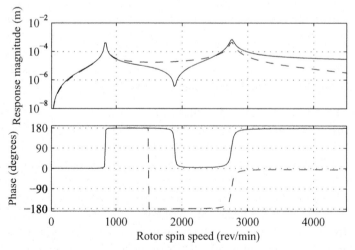

Figure 6.20. Steady-state response of the isotropic system (Example 6.3.1(a)) at nodes 3 (solid) and 5 (dashed) to an out-of-balance force at node 3. Note that the response at node 5 has no anti-resonance between the response peaks.

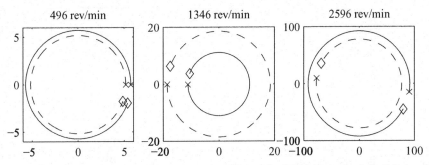

Figure 6.21. Forward-whirl orbits for the rotor (Example 6.3.1(a)) at the left disk (node 3, solid) and right disk (node 5, dashed). The cross denotes the start of the orbit and the diamond denotes the end. The dimensions are μm.

respectively. In the out-of-balance response plot, it is shown that there is a peak in the response at 2,755 rev/min but not at 2,489 rev/min. The response of the rotor at node 3 has a zero value in the region of 1,887 rev/min. At this frequency, the phase of the response of the disk at node 3 changes by 180°. There is no change of phase at this rotor speed at node 5.

The effects of these phase changes are shown in Figure 6.21. The figure shows the whirl orbits at selected rotor speeds. The whirl-orbit plots are circular and it is shown that both disks whirl in a forward direction (i.e., in the same direction as the rotation). This is because the stiffnesses of the system in the x and y directions are identical. At any particular node, if the phase of the whirl component in the x direction changes, the component in the y direction also changes; therefore, although the phase of an orbit changes, the direction does not. At 496 and 1,346 rev/min, both disks whirl in-phase; however, at 1,346 rev/min, both disks have changed phase due to the 180° phase change that occurs at the first critical speed. At 2,596 rev/min, there is a phase difference between the two disks of approximately 180° because at 1,888 rev/min, the disk at node 3 passes through a phase change, whereas the disk at node 5 does not.

Figure 6.22 shows the operating deflection shape of the rotor at 2,596 rev/min. This diagram helps to clarify the phase relationship among the nodes at that speed.

(b) *Anisotropic system*: The FE analysis again provides the Campbell diagram (Figure 6.23). The lowest six natural frequencies of the system at 3,000 rev/min are 13.12, 13.81, 38.96, 45.03, 97.14, and 124.5 Hz, although only the first four are shown on the Campbell diagram. For these frequencies, the damping factor, ζ, ranges from 0.002 to 0.02.

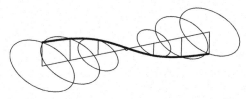

Figure 6.22. Operating deflection shape at 2,596 rev/min (Example 6.3.1(a)).

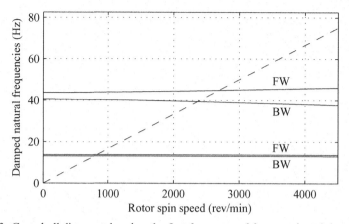

Figure 6.23. Campbell diagram showing the first four natural frequencies of the system (Example 6.3.1(b)). The dashed line represents $\omega = \Omega$. FW indicates forward whirl and BW indicates backward whirl.

The stiffness of the system is different in the x and y directions; hence, the amplitudes of whirl in these directions are different and the rotor speed at which the response to the out-of-balance force in the x and y directions is zero is also different. This is illustrated in Figures 6.24 and 6.25 for nodes 3 and 5, respectively. Here, the peaks in the out-of-balance responses occur at the same rotor speeds in the x and y directions. The phase of the responses in the x and y directions changes at a peak in the response and when the response is zero. Because peaks in the responses in the x and y directions occur at the same rotor speed, the responses in both the x and y directions change phase. This has the effect of changing the phase of the orbit relative to the location of the out-of-balance force. However, because zero out-of-balance responses in the x and y direction occur at different rotor speeds, the phase of the response in the x direction changes at a different rotor speed from that of the y direction. If the phase of the response in either the x or the y direction changes substantially, the direction

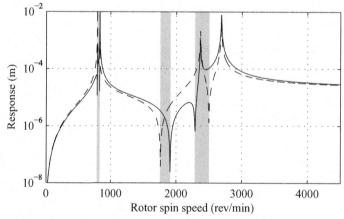

Figure 6.24. Response in the x (solid) and y (dashed) directions at node 3 for Example 6.3.1(b). The shaded regions indicate backward whirl at this node.

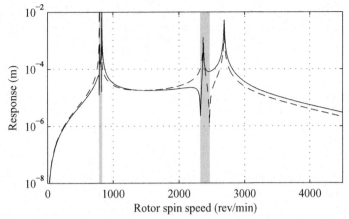

Figure 6.25. Response in the x (solid) and y (dashed) directions at node 5 for Example 6.3.1(b). The shaded regions indicate backward whirl at this node.

of whirl reverses. Regions of backward whirl are indicated by shaded regions in Figures 6.24 and 6.25, where the direction of whirl for each node is computed using Equation (3.60). Comparing the regions of backward whirl in Figures 6.24 and 6.25, we reach the counterintuitive result that at some rotor speeds, one disk is in backward whirl and the other is in forward whirl. Figure 6.26 shows the orbits at nodes 3 and 5 at various rotor speeds, illustrating that the orbits are elliptical – not circular – and can be forward or backward.

The amplitude of the response of the system (given by the semimajor axis of the orbital ellipse) at nodes 3 and 5 to the out-of-balance force at node 3 is shown in Figure 6.27. Comparing the rotor speed when the response to the out-of-balance force is a maximum with the rotor speeds that are coincident with natural frequencies in the Campbell diagram, we see that both forward and backward modes are excited. This is particularly clear in the case of the third (backward) and fourth (forward) frequencies, which are coincident with the rotor speed at 2,368 and 2,688 rev/min, respectively. In the out-of-balance response plot, it is shown that there are maxima in the response at both of these speeds.

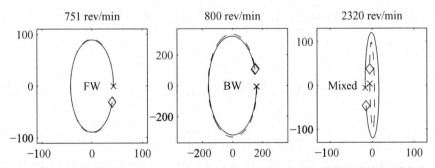

Figure 6.26. Whirl orbits for the rotor at the left disk (node 3, solid) and right disk (node 5, dashed), for Example 6.3.1(b). The cross denotes the start of the orbit and the diamond denotes the end. The dimensions are μm.

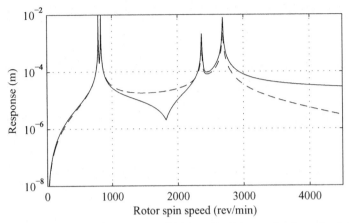

Figure 6.27. The amplitude of the response (the length of the semimajor axis of the orbit) of the system with anisotropic supports at nodes 3 (solid) and 5 (dashed) to an out-of-balance force at node 3, for Example 6.3.1(b).

6.3.2 Harmonic or Sub-Harmonic Response of Rotors to Sinusoidal Forces

Lateral forces can arise that rotate at multiples or submultiples of rotor speed. These forces are not necessarily proportional to the square of the rotational speed because they are not out-of-balance forces. Four-stroke reciprocating engines generate strong forcing components at multiples of half the rotor speed. A four-bladed helicopter generates significant 4X forcing.

As an example, we consider the case of lateral forces and/or moments rotating at frequency $n\Omega$ and acting on a rotor supported by isotropic bearings. We allow either $n = m$ or $n = 1/m$, where m is an integer. Ignoring rotor internal damping,

$$\mathbf{M}\ddot{\mathbf{q}} + \Omega\mathbf{G}\dot{\mathbf{q}} + \mathbf{C}\dot{\mathbf{q}} + \mathbf{K}\mathbf{q} = \Re\left(\mathbf{Q}_0 e^{jn\Omega t}\right) \qquad (6.56)$$

where $\Re\left(\mathbf{Q}_0 e^{jn\Omega t}\right)$ is a vector of forces and moments to account for the forces and moments acting at particular degrees of freedom. Then, letting $\mathbf{q}(t) = \Re\left(\mathbf{q}_0 e^{jn\Omega t}\right)$, we have

$$\left(-n^2\Omega^2\mathbf{M} + jn\Omega^2\mathbf{G} + jn\Omega\mathbf{C} + \mathbf{K}\right)\mathbf{q}_0 e^{jn\Omega t} = \mathbf{Q}_0 e^{jn\Omega t}$$

and, hence,

$$\mathbf{q}_0 = \left[\left(\mathbf{K} - n^2\Omega^2\mathbf{M}\right) + jn\Omega\left(\Omega\mathbf{G} + \mathbf{C}\right)\right]^{-1}\mathbf{Q}_0 \qquad (6.57)$$

Thus, we can determine the response of the system. Of course, when the rotor speed is Ω, the response occurs at a frequency corresponding to $n\Omega$.

The significance of this type of excitation can be assessed easily by plotting a $\omega = n\Omega$ line on the Campbell diagram. For Example 6.3.1, the Campbell diagram in Figure 6.28 shows the $\omega = 2\Omega$ as well as a $\omega = \Omega$ relationship. Thus, a force rotating at twice the rotor speed excites modes at 13.15, 13.8, 40.2, 44.0, 97.5, and 127.5Hz.

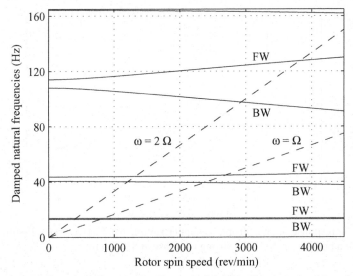

Figure 6.28. Campbell diagram for Example 6.3.1(b), showing $\omega = 2\Omega$ and $\omega = \Omega$ lines.

6.3.3 Response of Bent Rotors

We now consider the general case of a bent rotor consisting of a shaft with several disks. From an FEA, we obtain an equation of the form

$$\mathbf{M}\ddot{\mathbf{q}} + (\Omega\mathbf{G} + \mathbf{C})\,\dot{\mathbf{q}} + \mathbf{K}\mathbf{q}_e = \mathbf{0} \tag{6.58}$$

where $\mathbf{q}_e$ is a vector of the elastic deflections at the nodes and $\mathbf{q}$ is a vector of the total deflections at the nodes. Thus, $\mathbf{q} = \mathbf{q}_e + \mathbf{q}_b$, where $\mathbf{q}_b$ denotes the vector of deflections of the stationary rotor due to the permanent bend. Hence, $\mathbf{q}_e = \mathbf{q} - \mathbf{q}_b$. Thus, Equation (6.50) can be written

$$\mathbf{M}\ddot{\mathbf{q}} + (\Omega\mathbf{G} + \mathbf{C})\,\dot{\mathbf{q}} + \mathbf{K}\mathbf{q} = \mathbf{K}\mathbf{q}_b \tag{6.59}$$

Because the rotor is spinning, the bend in the shaft in a particular direction is a function of time. Thus, knowing the system properties and therefore the system matrices, as well as the displacements and slopes due the bent shaft at each node, we can determine the total displacements of the system by solving Equation (6.59). In practice, it is unlikely that the shape of a bent shaft is known with any degree of precision. A more practical but difficult problem is to determine from measurements of the shaft overall response whether the shaft is bent and, if it is, what is the size and general shape of the bend. Such analysis is beyond the scope of this book; here, we simply determine the response due to a bent rotor the geometry of which is known. Because the bend rotates with the shaft, $\mathbf{q}_b(t) = \Re\left(\mathbf{q}_{b0}e^{J\Omega t}\right)$, where the elements of $\mathbf{q}_{b0}$ are complex and are defined in a way similar to the unbalance forcing. Equation (6.59) becomes

$$\mathbf{M}\ddot{\mathbf{q}} + (\Omega\mathbf{G} + \mathbf{C})\,\dot{\mathbf{q}} + \mathbf{K}\mathbf{q} = \Re\left(\mathbf{K}\mathbf{q}_{b0}e^{J\Omega t}\right)$$

Assuming the solution of the form $\mathbf{q}(t) = \Re\left(\mathbf{q}_0e^{J\Omega t}\right)$, where $\mathbf{q}_0$ is complex, gives

$$\mathbf{q}_0 = \left[(\mathbf{K} - \Omega^2\mathbf{M}) + J\Omega(\Omega\mathbf{G} + \mathbf{C})\right]^{-1}\mathbf{K}\mathbf{q}_{b0} \tag{6.60}$$

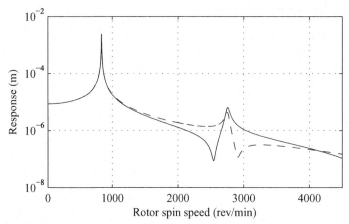

Figure 6.29. The response of an isotropic system (Example 6.3.2(a)) at nodes 3 (solid) and 5 (dashed) due to a bent rotor. Note the strong response at 830 rev/min.

The vector $\mathbf{q}_{b0}$ defines not only the displacement at each node due to the bend but also the slopes at each node. Even in an idealized situation, in which it is assumed that the deflections at each node due to the bend are known, it is unlikely that the slopes are also known. However, a stiffness matrix of the shaft exists and expansion based on the static (i.e., Guyan) reduction transformation is possible. The transformation is given by Equation (2.115), where the master degrees of freedom are those that are specified and the slave degrees of freedom are then estimated.

EXAMPLE 6.3.2. For the rotor–bearing system described in Example 6.3.1 (and shown in Figure 6.18), determine the response of the rotor at the disks due to a bent shaft if each bearing has a stiffness of (a) 1 MN/m and a damping of 100 Ns/m in both the x and y directions; and (b) 1 MN/m in the x direction and 0.8 MN/m in the y direction and a damping of 10 Ns/m in both directions. The system is modeled using six elements, giving 28 degrees of freedom. The lateral displacements of the stationary shaft are 0 at nodes 1 and 7, 0.00500 mm at nodes 2 and 6, 0.00866 mm at nodes 3 and 5, and 0.0100 mm at node 4.

Solution

(a) *Isotropic system*: The natural frequencies of the system are shown in the Campbell diagram in Figure 6.19. In this example, the lateral displacements associated with the static bend follow a half-sine wave. The response at nodes 3 and 5 (i.e., the disks) to this permanent bend is shown in Figure 6.29. Peak responses occur at 830 and 2,750 rev/min (i.e., disk at node 5) and 2,757 rev/min (i.e., disk at node 3). This slight difference in the shaft speed at which the peak response occurs arises because of the slightly different effects of damping on the response at the two disks. Because the bent shape is similar to that of the first mode of vibration, we expect the first mode to be excited strongly, which is the case. The system is isotropic; therefore, only the forward modes are excited and the whirl orbits are circular, as shown in Figure 6.30.

(b) *Anisotropic system*: The Campbell diagram for this system is shown in Figure 6.23. Figure 6.31 shows the amplitude of the response of the system to a permanent bend in the rotor. It is shown that the response has a maximum when

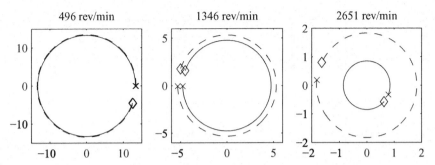

Figure 6.30. Forward-whirl orbits for the bent rotor on isotropic supports (Example 6.3.2(a)) at the left disk (node 3, solid) and right disk (node 5, dashed). The cross denotes the start of the orbit and the diamond denotes the end. The dimensions are μm.

the rotor speed is 789, 828, 2,368, and 2,688 rev/min and the bent rotor is exciting both forward and backward modes. Furthermore, because the stiffness of the system is different in the x and y directions, the zero responses in those directions occur at different rotor speeds; thus, the phase of the response in those directions changes at different speeds. When the phases of the responses in the x and y directions are different, the direction of whirl reverses. Figures 6.32 and 6.33 show the response magnitude at the two disks in the two directions. The regions of backward whirl are indicated by the shaded areas.

The behavior of a rotor excited by a bent shaft is almost indistinguishable from one excited by out-of-balance. Figure 6.34 is the response of the anisotropic rotor–bearing system to out-of-balances of 0.001 kg m on both disks (i.e., nodes 3 and 5) and at the same angular position. Comparing this figure to Figure 6.32, it is shown that the two responses are very similar; however, there are differences. The responses at very low speeds and very high speeds are different, but in a practical situation on a real and complex rotor–bearing system, these differences can be easily overlooked. For the rotor under discussion, suppose that the measured response is available only in the speed range of 500 to 1,500 rev/min; deciding whether this response is due to a bent rotor or out-of-balance is difficult.

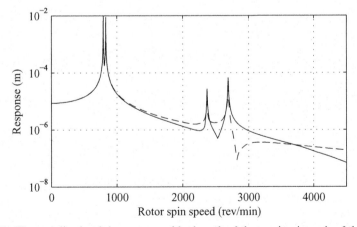

Figure 6.31. The amplitude of the response (the length of the semimajor axis of the orbit) of the bent rotor with anisotropic supports (Example 6.3.2(b)) at nodes 3 (solid) and 5 (dashed).

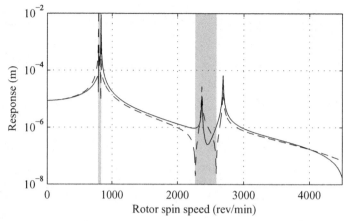

Figure 6.32. Response of the bent rotor in anisotropic supports (Example 6.3.2(b)) in the x (solid) and y (dashed) directions at node 3. The shaded regions indicate backward whirl at this node.

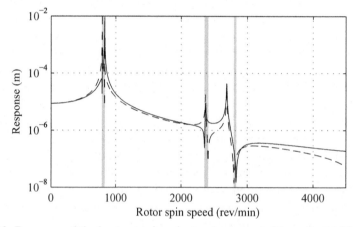

Figure 6.33. Response of the bent rotor in anisotropic supports (Example 6.3.2(b)) in the x (solid) and y (dashed) directions at node 5. The shaded regions indicate backward whirl at this node.

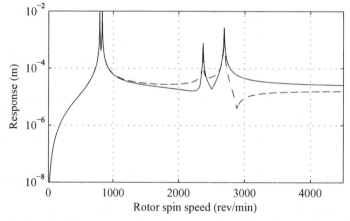

Figure 6.34. The amplitude of the response (the length of the semimajor axis of the orbit) of the rotor with anisotropic supports (Example 6.3.2(b)) at nodes 3 (solid) and 5 (dashed) due to out-of-balance forces at nodes 3 and 5.

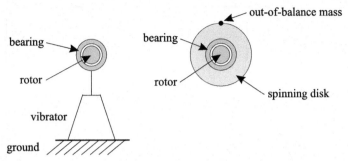

Figure 6.35. The excitation force applied through auxiliary bearings using a fixed shaker or a spinner.

6.3.4 Response to Forces Applied through Auxiliary Bearings

For test purposes, excitation is sometimes applied to the rotor through an auxiliary bearing. A split bearing can be attached to the shaft and a source of excitation applied to it. The excitation can be generated vertically or horizontally by a vibration exciter (or shaker) connected to the outer case of the bearing. The shaker may be mounted rigidly so that it can react forces to ground or else the forces may be reacted inertially. An alternative method of excitation is to mount a disk (called a *spinner*) carrying an out-of-balance mass on the auxiliary-bearing housing and then rotate the disk independently of the rotor. In either arrangement, the rotor can be spun at any speed and excited at any frequency through the auxiliary bearing so that it is possible to determine the frequency response map of the rotor over a range of speeds. These two forms of excitation are shown diagrammatically in Figure 6.35.

We now consider the response of a rotor to such external forces. The equations of motion for the system become

$$\mathbf{M}\ddot{\mathbf{q}} + \Omega\mathbf{G}\dot{\mathbf{q}} + \mathbf{C}\dot{\mathbf{q}} + \mathbf{K}\mathbf{q} = \mathbf{Q}(t) \tag{6.61}$$

where $\mathbf{Q}(t)$ is a vector of harmonic forces at frequency ω. Suppose that an out-of-balance force acts at an auxiliary bearing at node k on the rotor. Then, the forces at degrees of freedom $4k - 3, 4k - 2, 4k - 1$, and $4k$ are

$$\mathbf{Q}_k(t) = \Re\left(\omega^2 \begin{Bmatrix} m_0\varepsilon_0 e^{J\alpha} \\ -Jm_0\varepsilon_0 e^{J\alpha} \\ 0 \\ 0 \end{Bmatrix} e^{J\omega t}\right)$$

where m_0 is the out-of-balance mass at a radius ε_0 and is rotating at a speed of ω. The phase angle is α. The spinner can be spun in either direction – that is, in the same or in the opposite direction as that of the rotor.

If one or more shakers are attached to the auxiliary bearing, then for a harmonic excitation, the force vector at node k is given by

$$\mathbf{Q}_k(t) = \Re\left(\begin{Bmatrix} f_x \\ f_y \\ 0 \\ 0 \end{Bmatrix} e^{J\omega t}\right)$$

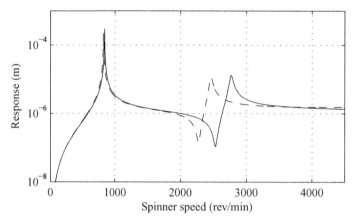

Figure 6.36. The machine response at node 3 to excitation by a spinner rotating either forward (solid) or backward (dashed) (Example 6.3.3 (a)).

where f_x and f_y are complex and represent the amplitudes and phases of the harmonic forces generated in the x and y directions, respectively, at frequency ω. More general forces may be analyzed using the methods discussed in Chapter 2.

Write $\mathbf{Q}(t) = \mathbf{Q}_0 e^{j\omega t}$ and let $\mathbf{q}(t) = \mathbf{q}_0 e^{j\omega t}$, where $\mathbf{Q}_0(\omega)$ and $\mathbf{q}_0(\omega)$ are the full-length force and displacement vectors, respectively, and both are complex in general. Then

$$\mathbf{q}_0 = \left[-\omega^2 \mathbf{M} + j\omega (\Omega \mathbf{G} + \mathbf{C}) + \mathbf{K} \right]^{-1} \mathbf{Q}_0 \qquad (6.62)$$

The system matrices also may depend on the shaft speed, Ω.

EXAMPLE 6.3.3. The rotor–bearing system described in Example 6.3.1 (and shown in Figure 6.18) rotates at 3,000 rev/min. Each bearing has a stiffness of 1 MN/m and a damping of 100 Ns/m in both the x and y directions. For test purposes, the rotor is excited via an auxiliary bearing that is attached to the rotor at midspan. Determine the response of the disk at node 3 to forces acting on an auxiliary bearing. The system is modeled using six elements, giving 28 degrees of freedom. The forces at the auxiliary bearing are (a) a rotating out-of-balance of 0.0001 kg m, and (b) a harmonic force of 10 N.

Solution

(a) *Out-of-balance.* When the out-of-balance at the auxiliary bearing rotates in the same direction as the rotor spin, only forward modes can be excited. This is shown in Figure 6.36 in which modes at approximately 836 and 2,767 rev/min are excited. However, if the out-of-balance rotates in the opposite direction to the rotor spin, then the backward modes at 819 and 2,461 rev/min are excited. This is also shown in Figure 6.36.

(b) *Shaker.* Figure 6.37 shows the response of the rotor due to a harmonic excitation at the auxiliary bearing in the vertical direction only. As expected, the shaker excites both forward and backward modes of vibration at 13.64, 13.93, 41.00, and 46.13 Hz, equivalent to 819.0, 835.8, 2,461, and 2,767 rev/min, respectively.

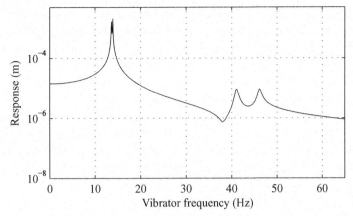

Figure 6.37. The machine response at node 3 to excitation by a vertical harmonic excitation at the auxiliary bearing (Example 6.3.3 (b)).

If the vibration test is undertaken at a range of rotor speeds rather than only one, it is possible to create a three-dimensional plot showing how the response varies with rotor speed and vibration frequency. Such a plot is shown in Figure 6.38 and the resonances are shown clearly. In particular, the resonance at approximately 40 Hz splits into two as the rotor speed increases, which illustrates the separation of two natural frequencies with increased rotor speed due to gyroscopic effects. Plots such as Figure 6.38 that illustrate the variation of vibration amplitude with frequency and either time or rotor spin speed are often referred to as *waterfall* or *cascade plots*. When this plot is presented in a two-dimensional form using color to indicate vibration amplitude, it is referred to as a *Z-mod plot*.

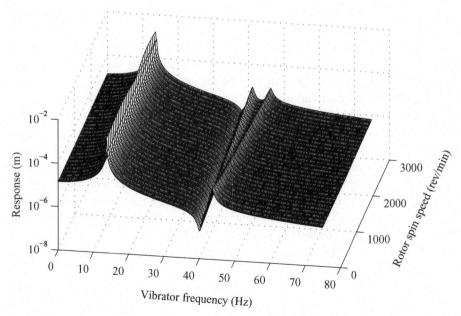

Figure 6.38. The machine response at node 3 to excitation by a shaker for a range of shaft speeds (Example 6.3.3 (b)).

Often, the response of a machine at the rotor spin speed and the harmonics is required. The spin speed may be estimated from reference marks on the shaft, and the measured sample time does not usually contain an integer number of revolutions of the rotor, leading to leakage (see Section 2.6.3). One option is to sample based on the shaft angle rather than the time increments by either hardware devices to trigger sampling based on the rotor position or resampling the measured time histories. The result is the response referred to shaft orders rather than frequency, and the process is called *order tracking*. Because the sampling is related to rotor position, the approach can cope with variations in the rotor speed or give valuable insight during the runup or rundown of the machine (Blough, 2003; Vold and Leuridan, 1995).

6.4 Forces on the Supports Due to Rotor Vibration

For any rotating machine, there are numerous interfaces among parts, and it is often of practical interest to determine the forces and moments transmitted across those interfaces. Among the interfaces, bearings are one obvious case in which the transmitted forces and moments are of interest. If excessive loads arise across a bearing, its integrity may be compromised. If the complete rotor is an assembly of parts, the forces and moments across the interfaces between rotor components also may be of interest. The forces and moments of interest here are those transmitted into the foundation; these are important in a number of practical circumstances. In the case of a power station, the steam turbine and generator are mounted on concrete or steel structures and the forces transmitted into these structures excite vibration that affects the operation. In a ship or submarine propulsion system, a prime-mover attached to the hull transmits oscillatory forces into the hull causing vibration that results in sound waves radiating into the water. In aircraft, the engines ordinarily are attached to the wings via pylons, and excitation results in vibration being transmitted into the fuselage and manifested as cabin noise.

Depending on whether we consider a system to consist of a rotor, bearings, and foundation or a rotor, bearings, stator, and foundation, the interface of interest may be the set of bearings or it may be the surface between the stator and the foundation. The same methods can be used to determine the transmitted forces, which are described herein. Section 2.5 explains a general technique for determining the response of a system that has constraints applied. Two subsystems can be coupled together by imposing constraints to force the degrees of freedom at either side of the interface to have identical values. The set of constraint equations in this case takes a form similar to Equations (2.122) and (2.123), in which each column of the constraint matrix, $\mathbf{E}$, contains the number $+1$ in one location, -1 in another, and zeros elsewhere. From this constraint matrix, a suitable transformation matrix, $\mathbf{T}$, is formed that satisfies the requirements that $\mathbf{T}^\top \mathbf{E} = \mathbf{0}$ and that $\mathbf{Y}\mathbf{Y}^\top$ is nonsingular (where $\mathbf{Y} = [\,\mathbf{E}\ \mathbf{T}\,]$). Having selected an appropriate transformation, $\mathbf{T}$, the reduced-dimension mass, damping, and stiffness matrices are calculated according to Equation (2.125); the response of the coupled system can be determined from Equation (2.128). This computed response then can be expanded again to determine the full displacement vectors for each of the two systems using Equation (2.127). Finally, the connection forces are determined using Equation (2.126).

6.5 Response to Ground Vibration

A large fixed installation may experience ground motion in the form of vibration caused by another plant in proximity or by earth tremors. Similarly, gas turbines in aircraft experience motion at bearing housings as the aircraft maneuvers. Assuming that the ground or supports are sufficiently large so that the dynamics of the machine do not influence this motion, the vibration may be modeled by imposing foundation displacements. Consequently, due to the flexibility of the bearing, this ground vibration acts as a force in a fixed direction on the rotor.

If a dynamic force is applied to a rotor in a fixed direction, it excites both forward and backward modes of vibration in the rotor, even when the system is isotropic. This can be explained as follows: Suppose that the harmonic force $f(t) = f_0 \cos(\omega t)$ acts on the rotor in a fixed direction. This can be expressed as $f(t) = f_0(e^{j\omega t} + e^{-j\omega t})/2$. Now, $f_0 e^{j\omega t}/2$ represents a counterclockwise rotating vector and $f_0 e^{-j\omega t}/2$ represents a clockwise rotating vector. Thus, the sum of the forward (i.e., counterclockwise) and the backward (i.e., clockwise) rotating forces, each equal to $f_0/2$, is equivalent to the original harmonic force. The counterclockwise rotating force excites the forward modes of vibration of the rotor; similarly, the clockwise rotating force excites the backward modes.

The forces are given in terms of the relative displacements and velocities within the bearings, assuming that the bearings have the property of damping and stiffness but not inertia. Equation (5.78) explicitly derives this force for a single, short bearing with two degrees of freedom; however, in general, each bearing has two displacements and two rotations, giving four degrees of freedom. The forces acting on the shaft due to the bearings are

$$\mathbf{Q}_s = -\mathbf{K}_b(\Omega)(\mathbf{q}_s - \mathbf{q}_f) - \mathbf{C}_b(\Omega)(\dot{\mathbf{q}}_s - \dot{\mathbf{q}}_f) \tag{6.63}$$

In this equation, the subscript s refers to the shaft, the subscript f refers to the foundation (or bearing housing), and the subscript b refers to the bearings. The displacement vector $\mathbf{q}_s$ denotes the displacement of the degrees of freedom on the shaft that determine the bearing force. $\mathbf{K}_b$ is an $m \times m$ stiffness matrix consisting of n_b blocks of size 4×4, where $m = 4n_b$ and n_b is the number of bearings excited through the foundation. The damping matrix $\mathbf{C}_b$ is of a similar form. Normally, the displacement at the foundation, $\mathbf{q}_f$, is set to zero, but this displacement is used to define the foundation excitation. When the bearings are connected to the rotor, using the force on the shaft, $\mathbf{f}_s$, the terms involving the shaft displacement, $\mathbf{q}_s$, are assembled into the global damping and stiffness matrices as before. The terms involving the foundation displacement, $\mathbf{q}_f$, remain as a force and the resulting equation is

$$\mathbf{M}\ddot{\mathbf{q}} + \Omega\mathbf{G}\dot{\mathbf{q}} + \mathbf{C}\dot{\mathbf{q}} + \mathbf{K}\mathbf{q} = \mathbf{T}_b\mathbf{C}_b\dot{\mathbf{q}}_f + \mathbf{T}_b\mathbf{K}_b\mathbf{q}_f \tag{6.64}$$

The matrix $\mathbf{T}_b$ selects those shaft degrees of freedom where the bearing is connected. If the rotor model has n degrees of freedom, then $\mathbf{T}_b$ is an $n \times m$ matrix. $\mathbf{C}$ and $\mathbf{K}$ contain contributions from the bearings.

The forces acting at the foundation produce a vector of translations (and possibly rotations), $\mathbf{q}_f$, that we now use as the input to our system. Thus, we can determine the response of the rotor to any prescribed displacement and velocities at the foundation or bearing housing. Vector $\mathbf{q}_f$ has m elements, many of which may be zero. A frequency-domain approach can be used to calculate the response of the rotor. However, the displacement at the foundation or bearing housing is unlikely to be harmonic and is more likely to arise from an earthquake or ground vibrations caused by the operation of another large machine. Given the foundation displacement as a function of time, often it is convenient to solve numerically the ordinary differential equation (i.e., Equation (6.64)) to determine the shaft displacement, $\mathbf{q}$, at various instances of time (see Section 2.6.1). The fourth-order Runge-Kutta method may be used, although this method requires the second-order differential equations to be written in state–space form. An alternative approach uses Newmark's method (Bathe and Wilson, 1976; Newmark, 1959; Petyt, 1990), which was developed specifically to solve second-order systems. A number of difficulties arise: If the displacement is given only at discrete intervals in time, then interpolation must be used for the intermediate times. The velocity would have to be estimated numerically using finite-difference methods. For systems with a large number of degrees of freedom, the time step must be determined by the highest natural frequency. Often, the response of the machine is required only for the lower frequencies; hence, the short step length requires significantly higher computational effort. One way around this is to reduce the model before performing the numerical integration. Model reduction has the double benefit of reducing the maximum natural frequency of the model, thus allowing a larger time step and also reducing the number of arithmetic operations per time step because of the reduced matrix size.

EXAMPLE 6.5.1. The rotor–bearing system described in Example 6.3.1 (and shown in Figure 6.18) rotates at 3,000 rev/min. Each bearing has a stiffness of 1 MN/m and a damping of 100 Ns/m in both the x and y directions. The system is modeled using six elements, giving 28 degrees of freedom. The bearings are excited by a ground vibration applied in the vertical direction to the foundation and in-phase at each bearing.

(a) Determine the response of the disks to a harmonic excitation at the foundation of 10 μm in the frequency range of 0 to 55 Hz.

(b) Determine the response of the disks to a half-sine pulse of 25 ms duration and 1 mm amplitude applied to the foundation in the vertical direction.

Solution. Unlike previous examples, the excitation force in this case is only in the y direction. Thus, if there were no dynamic coupling between the x and y directions, there would be no response in the x direction. In fact, there is dynamic coupling in the form of gyroscopic effects that link the x and y directions. Also, there can be cross-coupling in the bearing or internal damping in the rotor, but neither is present in this example.

(a) The response of the rotor at nodes 3 and 5 is shown in Figure 6.39. The figure includes an enlarged portion in the frequency range of 12 to 18 Hz. At 3,000 rev/min, the system natural frequencies are 13.64 Hz (backward), 13.93 Hz (forward), 41.00 Hz (backward), and 46.13 Hz (forward), and it is clear that these frequencies are being excited.

(b) The half-sine pulse displacement of duration 0.025 s (equivalent to 20 Hz) and maximum displacement of 1 mm is shown in Figure 6.40. The response of the rotor at nodes 3 and 5 (the disks) in the y direction is shown in

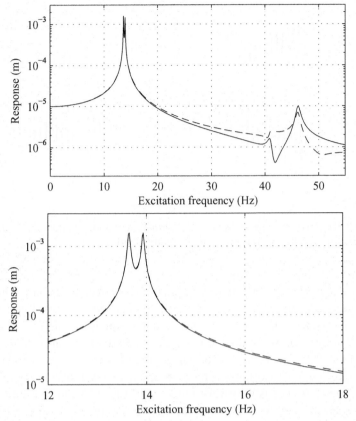

Figure 6.39. The response amplitude of the rotor at nodes 3 (solid) and 5 (dashed) to a vertical harmonic displacement of $10\,\mu$m at its foundation (Example 6.5.1 (a)).

Figure 6.41. The time series of the response is computed using a fourth-order Runge-Kutta method (see Section 2.6.1) using 65,536 samples with a time step of 0.00006 s, which gives a duration of 3.932 s. The highest natural frequency in the 28 degrees of freedom model – ignoring damping and gyroscopic effects – is 8,268 Hz; thus, the sample rate for the foundation

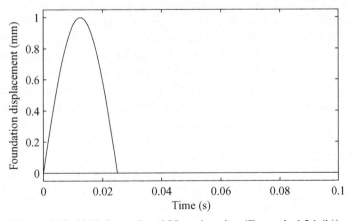

Figure 6.40. Half-sine pulse of 25 ms duration (Example 6.5.1 (b)).

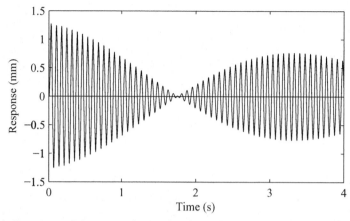

Figure 6.41. Response of the system in the Oy direction at nodes 3 and 5 to a half-sine pulse (Example 6.5.1 (b)). The two responses are virtually identical.

displacement is more than twice the highest resonance frequency. The time series shows that the response is primarily composed of a harmonic waveform with a mean frequency of 13.8 Hz and a beat frequency of 0.3 Hz. This suggests that the waveform is dominated by two close frequency components of similar amplitude at $13.8 \pm 0.3/2 = 13.65$ and 13.95 Hz. These frequencies correspond closely to the first two natural frequencies of the system. One is a backward mode and the other is a forward mode, and it is clear that both modes of vibration are excited. Note also that the response at the two nodes in Figure 6.41 is slightly different, although it is difficult to discern from the figure. This difference occurs because the machine is not exactly symmetric because the disks have different diameters.

To analyze these data further, we must transform the time data into the frequency domain using the DFT (see Section 2.6.3). The frequency resolution in the DFT is determined by the duration of the sampled data, which in this example is 3.932 s, giving a frequency resolution of 0.2543 Hz. This is barely adequate to resolve the two frequency components in the region of 13.8 Hz. Running the simulation for a longer time span requires excessive computation because the maximum time step is fixed by the highest natural frequency of the model. However, only the lower resonances of the machine are excited, so the model may be reduced by neglecting the higher frequency modes, thus allowing a larger time step. The model therefore is reduced to 10 degrees of freedom by retaining only the lower modes of the system, neglecting damping and gyroscopic effects. The highest frequency in the 10 degrees of freedom model is 268 Hz and the time step can be increased to 0.001 s. The 16,384 samples are taken with a total record time of 16.383 s; the frequency resolution in the DFT is then 0.061 Hz. Figure 6.42 shows the size of the semimajor axis of the orbital ellipse at each frequency component in the DFT of the data. The following resonance (i.e., maximum response) frequencies can be identified: 13.67, 13.92, 41, and 46 Hz. The resonance frequencies are not exactly the same for each response measurement because of the presence of nonproportional damping. Figure 6.43 gives the DFT of the half-sine pulse foundation displacement and

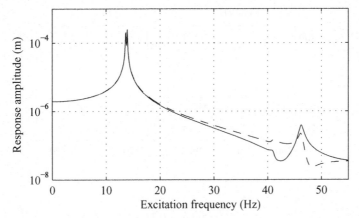

Figure 6.42. The magnitude of the frequency-domain response of the rotor at nodes 3 (solid) and 5 (dashed) to a half-sine pulse (Example 6.5.1 (b)).

shows the amplitude of the displacement at each frequency. Figure 6.42 cannot be directly compared to the frequency domain response given in Figure 6.39 because the foundation displacement at each frequency is not $10\,\mu$m. However, if the magnitude of the frequency response of the rotor shown in Figure 6.42 is divided by the magnitude of the DFT of the foundation displacement shown in Figure 6.43 at each frequency and multiplied by $10\,\mu$m, the results are identical to Figure 6.39.

Figure 6.44 shows the orbits at four frequencies, obtained from the DFT. At 13.67 Hz, the orbits are clockwise (i.e., backward whirl); at 14.28 Hz, the orbits are counterclockwise (i.e., forward whirl); at 40.96 Hz, one orbit is clockwise and the other is counterclockwise; and at 46.76 Hz, both orbits are counterclockwise (i.e., forward whirl).

6.6 Co-axial Rotors

Many machines consist of co-rotating (i.e., rotating in the same direction) or counter-rotating (i.e., rotating in the opposite direction) co-axial rotors. For

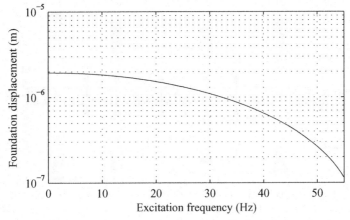

Figure 6.43. The magnitude of the DFT of the vertical displacement at the foundation (Example 6.5.1 (b)).

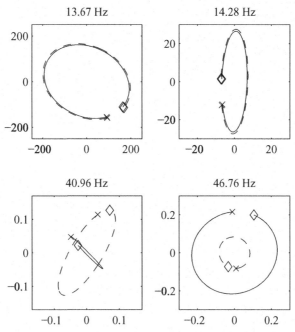

Figure 6.44. Whirl orbits for the foundation excitation at the left disk (node 3, solid) and right disk (node 5, dashed) obtained by the DFT (Example 6.5.1 (b)). The cross denotes the start of the orbit and the diamond denotes the end. The dimensions are μm.

example, in an aircraft gas turbine, the co-axial shafts co-rotate at different speeds or counter-rotate as in some propfans. A co-axial rotor is shown in Figure 6.45. Although the analysis of a co-axial rotor does not present any particular difficulties or require any new analysis tools, the results of such analysis reveal interesting forms of machine behavior. The following example illustrates some of the features of co-axial-rotor systems.

EXAMPLE 6.6.1. The co-axial rotor shown in Figure 6.45 is similar to one analyzed by Glasgow and Nelson (1980), but some of the dimensions have been changed.

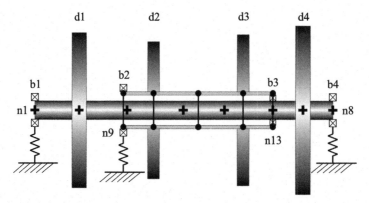

Figure 6.45. The example co-axial rotor. The nodes are indicated by n, the bearings by b, and the disks by d, followed by the number. The nodes on the inner rotor are indicated by a cross and those on the outer rotor by dots on the outer sections linked by lines.

Table 6.1. *Disk properties for the co-axial example in Figure 6.45 (Example 6.6.1)*

Disk	Mass (kg)	I_d (kg m^2)	I_p (kg m^2)	Node
d1	10.5	0.043	0.086	2
d2	7.0	0.021	0.042	10
d3	3.5	0.013	0.026	12
d4	7.0	0.034	0.068	7

The outer rotor spin speed is 1.5 times the spin speed of the inner rotor and it co-rotates. The system has the following dimensions and properties: The shaft of the inner rotor is solid with a diameter of 30 mm. The shaft of the outer rotor has an inside diameter of 50 mm and an outside diameter of 60 mm. Relative to the left bearing of the inner rotor, the positions of nodes 1 through 13 are 0, 0.076, 0.159, 0.254, 0.324, 0.406, 0.457, 0.508, 0.152, 0.203, 0.279, 0.356, and 0.406 m, respectively. Disks d1, d2, d3, and d4 are located at nodes 2, 10, 12, and 7, respectively. Table 6.1 lists the disk masses and inertias. Bearings b1, b2, and b4 are located at nodes 1, 9, and 8, respectively, and the inter-shaft bearing, b3, connects nodes 6 and 13. Table 6.2 lists the bearing properties. The rotors are made from steel with the properties $E = 207$ GPa and $\rho = 8{,}300$ kg/m^3. Plot the Campbell diagram for this system and calculate the response to unbalance forces of magnitude 0.0001 kg m on each of disks d1 and d2 in turn.

Solution. Using FEA, the inner rotor is modeled with seven elements (32 degrees of freedom) and the outer rotor with four elements (20 degrees of freedom); thus, the model has 52 degrees of freedom in total. The Campbell diagram for the system, plotted against the inner-rotor speed, is shown in Figure 6.46. Shown on the Campbell diagram are the lines giving correspondence between frequency and the inner-rotor speed and frequency and the outer-rotor speed. Thus, 10,000 rev/min (i.e., inner-rotor speed) is equivalent to a frequency of 166.67 Hz. If the inner-rotor speed is 10,000 rev/min, then the outer-rotor speed is 15,000 rev/min, or 250 Hz. In this Campbell diagram, there are four examples of mode veering, in which natural frequency lines approach but never touch or cross one another (see Section 3.7).

Figure 6.47 shows the response at each disk when an out-of-balance mass of 0.0001 kg m is attached to the rotor at node n2 (disk d1) on the inner

Table 6.2. *Bearing properties for the co-axial example in Figure 6.45 (Example 6.6.1)*

Bearing	k_{xx} (MN/m)	k_{yy} (MN/m)	Node
b1	26	52	1
b2	18	36	9
b3	9	9	6 and 13
b4	18	36	8

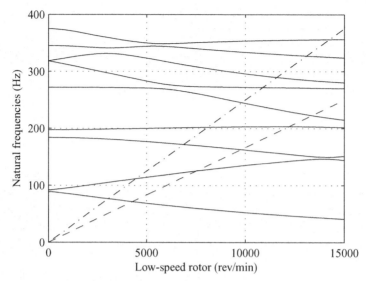

Figure 6.46. Campbell diagram for the co-axial rotor of Figure 6.45 (Example 6.6.1). The dashed line corresponds to the frequency of the low-speed rotor and the dash-dot line corresponds to the frequency of the high-speed rotor.

(i.e., lower-speed) rotor. Because the bearings are anisotropic, both forward and backward modes are excited; there are five peaks in the frequency response functions, where the rotor speed is equal to a natural frequency.

Figure 6.48 shows the response at each disk when an out-of-balance mass of 0.0001 kg m is attached to the rotor at node n10 (disk d2) on the outer (higher-speed) rotor. As the inner-rotor speed is increased to 15,000 rev/min, the outer-rotor speed increases to 22,500 rev/min and nine frequencies are excited (see Figure 6.48). Although the speed of the inner rotor is used to define the machine speed, the unbalance response is actually at the frequency of the outer (higher-speed) rotor.

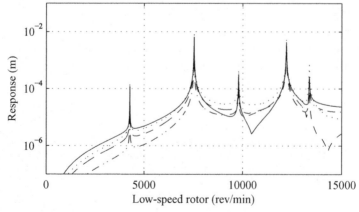

Figure 6.47. The response of the co-axial rotor (Example 6.6.1) at disks d1 (solid), d2 (dash-dot), d3 (dot), and d4 (dashed) to an unbalance response at disk d1.

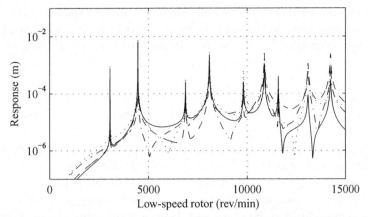

Figure 6.48. Response of the coaxial rotor machine, Example 6.6.1, at disks d1 (solid), d2 (dash-dot), d3 (dot), and d4 (dashed) to an unbalance response at disk d2.

6.7 Formal Definitions of Critical Speeds

In Section 6.1, we state that a critical speed is a rotational speed of a machine at which some combination of vibration displacements and/or forces reaches a (local) maximum. For most practical purposes, this is a perfectly workable definition. However, there are some problems with this definition, as follows:

- This definition depends on the distribution of forcing (usually unbalance) applied. In analyzing machine dynamics, a clear distinction is usually drawn between system properties such as stiffness, damping, and mass (which are features of a particular design) and the forcing applied (which is normally designed to be near zero). Most forcing comes about as a result of imperfections in the machine construction; the level of these imperfections may be reduced but not eliminated by the design.
- Depending on which vibration quantities are selected for consideration as well as weighting attached to each one, different critical speeds are detected. Also, in systems with general (i.e., nonproportional) damping, the peaks in the response may occur at slightly different rotational speeds at different degrees of freedom.
- Local maxima in some vibration measurements may exist at speeds that are by no means *critical* in that they are neither fundamental characteristics of the system nor speeds at which the vibration of the machine is likely to reach problematic levels.

For these reasons, a more useful definition of *critical speeds* is that it is a rotational speed of a machine at which one of the frequency components of forcing coincides with a natural frequency of the system. This definition is more satisfactory than the previous one in that it is a true property of the system and does not depend on either the spatial distribution of forcing on the machine or the measurement quantities considered. In this book, we choose to adopt this definition, recognizing at the same time that it identifies many speeds as being critical that might never be of any concern to an engineer who designs machines. The single most important source of forcing in most rotating machines is related to unbalance. For single-rotor machines with axisymmetric rotors, unbalance forces occur at one frequency: the rotor-spin speed. Furthermore, rotating unbalance always produces a force that rotates in the same sense as the shaft. In the rotordynamic analysis of machines, the direction of rotation of a force is often as important as the absolute value of its frequency. For example, if an axisymmetric rotor is supported on isotropic bearings, then the modes are circular and either purely forward-whirling or purely backward-whirling; in this case, the unbalance force does not excite the backward-whirling modes, even though the corresponding natural frequencies are calculated as critical speeds.

For damped systems, there remains the question of whether the required frequency for the system is the natural frequency, the damped natural frequency, or the resonance frequency (i.e., the frequency in which the peak response occurs). The response of heavily damped systems is likely to be insignificant, so lightly damped systems are more likely to be of interest. However, for lightly damped systems, all of the frequencies are very close; therefore, for practical purposes, any one of them may be used to determine critical speeds.

At this point, it is worth highlighting the difference between critical speeds and resonance frequencies in static structures. In static structures, a large response can occur at excitation frequencies close to the natural frequencies. The main difference with rotating machinery is that the natural frequencies change with the speed of rotation. Therefore, a single set of natural frequencies cannot be calculated and checked to see if any coincides to the forcing frequency. Both the forcing frequency and the natural frequency are dependent on the rotational speed yielding a coupled problem that must be solved as such.

6.8 Computing Critical Speeds

For simple systems, methods based on either computing the peak response or the Campbell diagram are straightforward to implement, and this approach is demonstrated in Section 6.2 for simple rotor–bearing systems. In the *local maximum in response* criterion, for each speed in the range of interest, the response of the machine to synchronous unbalance is computed. This involves computing the dynamic stiffness and finding the response to a nominal unbalance force for each speed, Ω. Peaks in the response are noted and recorded as critical speeds. For a simple system, it may be possible to obtain an analytical solution for the frequency of this peak response.

The Campbell diagram is the most general method to determine critical speeds. There is no circumstance in which the Campbell diagram does not yield all of the critical speeds of a machine over a given speed range. The completeness and the conceptual clarity of the Campbell diagram as a method for obtaining critical speeds initially appears to obviate the need for any other method. To compute the Campbell diagram for each speed, Ω, in the speed range of interest, mass, damping, and stiffness matrices $\mathbf{M}(\Omega)$, $\mathbf{C}(\Omega)$, and $\mathbf{K}(\Omega)$ are derived, and the characteristic roots of this combination of matrices are found by solving the associated eigenvalue problem at each rotational speed. A Campbell diagram can be drawn showing the variation of the damped natural frequencies with rotational speed. This can be computationally intensive and presents the possibility that the curves may be confused. The intersection of each natural frequency curve with the line through the origin on which the rotational speed equals the natural frequency defines one of the critical speeds.

Plotting the frequency components of forcing is simple because they are usually known as explicit functions of rotational speed.

Because of their importance, it is reasonable to question whether there is a more efficient way to compute critical speeds for machines with complex models as an alternative to computing the response of the system or its eigenvalues over its complete running range. For sensitivity studies or design optimization, the savings in computational time can be significant. Two methods to calculate critical speeds are described in the following subsections.

6.8.1 A Direct Approach

If the bearing coefficients do not depend on rotor speed, then the critical speeds may be calculated directly if the forcing frequency can be written in terms of the rotational speed. This approach is adopted for the simple rotor models in Section 6.2, where the forcing due to unbalance occurs at the rotational speed. The critical speeds then are obtained when a natural frequency is equal to the speed of rotation, and they are calculated as the solution to an eigenvalue problem.

Suppose that for any one frequency component of forcing, the frequency, ω_f, can be written in terms of rotational speed as

$$\omega_f = n\Omega \tag{6.65}$$

By far, the most common case of Equation (6.65) applies to unbalance forcing and, in this case, $n = 1$. If the force is $\mathbf{Q}(t) = \mathbf{Q}_0 e^{J\omega_f t}$, then looking for solutions of the form $\mathbf{q}(t) = \mathbf{q}_0 e^{J\omega_f t}$ in Equation (6.56) leads to

$$\left(-\Omega^2 \left[n^2\mathbf{M} + Jn\mathbf{G}\right] + J\Omega n\mathbf{C} + \mathbf{K}\right)\mathbf{q}_0 = \mathbf{Q}_0 \tag{6.66}$$

The critical speeds are obtained when the response is large, which occurs when the matrix is close to singular. The solution is obtained by setting $\mathbf{Q}_0 = \mathbf{0}$, which results in an eigenvalue problem for Ω, given by

$$\left(-\Omega^2 \left[n^2\mathbf{M} + Jn\mathbf{G}\right] + J\Omega n\mathbf{C} + \mathbf{K}\right)\mathbf{q}_0 = \mathbf{0} \tag{6.67}$$

Although the resulting equation looks more complex than those obtained in Chapter 5, it has precisely the same form. Because of their identical forms, the same techniques used to obtain the eigenvalues in Section 5.8 are equally applicable here as a method to find the critical speeds resulting from one particular frequency component of forcing. The direction of whirl of each mode may be obtained using the methods described in Sections 3.6.1 and 6.2.6.

If the solution for Ω is real, then the response at the critical speed is infinite. Often, the solution is complex with a small imaginary part; in this case, the response is large but not infinite. In this case, the real part of Ω is taken as the critical speed and Equation (6.65) is only an approximation.

$\mathbf{K}$ and $\mathbf{C}$ can be dependent on speed. It can be shown that if $\mathbf{K}$ can be adequately approximated as $\mathbf{K} \approx \mathbf{K}_0 + \Omega\mathbf{K}_1 + \Omega^2\mathbf{K}_2$ and $\mathbf{C} \approx \mathbf{C}_0 + \Omega\mathbf{C}_1$, then this dependence also may be incorporated into Equation (6.66) as

$$\left(-\Omega^2 \left[n^2\mathbf{M} + Jn\mathbf{G} - Jn\mathbf{C}_1 - \mathbf{K}_2\right] + J\Omega \left[n\mathbf{C}_0 - J\mathbf{K}_1\right] + \mathbf{K}_0\right)\mathbf{q}_0 = \mathbf{Q}_0 \tag{6.68}$$

with virtually no additional cost in computation terms. This facilitates the direct calculation of critical speeds for rotors supported on hydrodynamic bearings.

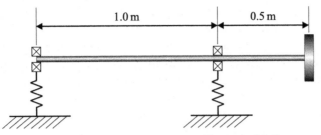

Figure 6.49. The overhung rotor (Example 6.8.1).

6.8.2 An Iterative Approach

When the system matrices are not independent of speed, the direct method cannot be used to determine the critical speeds and must be replaced by an iterative search. In this method, the first critical speed is estimated as $\Omega_1^{(0)}$. Then, $\mathbf{K}\left(\Omega_1^{(0)}\right)$, $\Omega_1^{(0)}\mathbf{G}$, $\mathbf{C}\left(\Omega_1^{(0)}\right)$, and $\mathbf{M}\left(\Omega_1^{(0)}\right)$ are determined and the eigenvalues s_i are computed from Equation (5.113). We now obtain a new estimate of the first critical speed as $\Omega_1^{(1)} = \Im(s_1)$. The process is repeated with this new estimate of the first critical speed, $\Omega_1^{(1)}$, until satisfactory convergence is obtained. The Campbell diagram may be used to choose initial estimates of the critical speeds and to ensure that no critical speeds were missed. The iterative procedure refines these estimates. Example 6.8.1 illustrates this process in action and convergence is generally very fast. Higher critical speeds are obtained by choosing the imaginary parts of the higher eigenvalues as a better approximation to the critical speed at each iteration.

6.8.3 Features of Critical Speeds

The critical speeds of rotating systems have a number of important features, many of which are similar to the features of natural frequencies described in Section 5.8.1; therefore, they are not repeated here. In a cyclically symmetrical rotor with isotropic support characteristics, the critical speeds occur in pairs, each corresponding to a set of four eigenvalues. These critical speeds separate due to bearing and support anisotropy and gyroscopic effects. Often, when engineers discuss the *first critical speed* of a machine, they are referring to a pair of critical speeds, corresponding to four roots of the characteristic equation.

In Example 6.3.1, the rotational speed caused changes in the calculated natural frequencies of the system. These changes arise because gyroscopic couples increase the resonance frequencies of the forward-whirling modes of the machine and the same gyroscopic couples reduce the frequencies of the backward-whirling modes. In general, rotating machines have several effects that can cause system matrices to depend on speed, as described in Chapter 5. These features are now demonstrated using an example.

> **EXAMPLE 6.8.1.** Consider the model of a simple overhung rotor, 1.5 m long with bearings at 0.0 m and 1.0 m, and shown in Figure 6.49. The bearings are short in that they present insignificant angular stiffness to the shaft, but they present finite translational stiffness. The shaft is 25 mm in diameter and the disk at the overhung end is 250 mm in diameter and 40 mm thick. The shaft and disk are

Table 6.3. *The bearing and support characteristics for the overhung rotor (Example 6.8.1) (All cross terms are zero)*

	Stiffness (MN/m)		Damping (kNs/m)	
Case	k_{xx}	k_{yy}	c_{xx}	c_{yy}
1	10	10	0	0
2	10	20	0	0
3	10	20	60	60
4	10	20	400	400
5	0.2	0.4	0	0

made of steel, and a mass density $\rho = 7,810 \, \text{kg/m}^3$, a modulus of elasticity $E = 211 \, \text{GPa}$, and a Poisson's ratio of 0.3 are assumed. Several different variants of this system are considered in which the differences lie in the bearing (and bearing-support) properties, shown in Table 6.3, and the steady rotational speed of the shaft. Calculate the eigenvalues for various rotor spin speeds and estimate the critical speeds for unbalance excitation.

Solution. Combining the geometric and material properties of the shaft gives $EI = 4,046 \, \text{Nm}^2$, $GA = 39.84 \, \text{MN}$, and $\rho A = 3.83 \, \text{kg/m}$. The disk has a mass of $m = 15.3 \, \text{kg}$, a polar inertia of $I_p = 0.120 \, \text{kg m}^2$, and a diametral inertia of $I_d = 0.062 \, \text{kg m}^2$. The shaft is split into six equal-length elements, giving a model with seven nodes and 28 degrees of freedom. Table 6.4 lists the characteristic roots for a selection of bearing characteristics and rotor spin speeds.

The damping in the bearings is sufficiently large in bearing cases 3 and 4 such that some purely real characteristic roots exist. Whenever the right-bearing characteristics are set to those of case 3, two real roots at approximately -169 and -340 are found. Whenever the right-bearing characteristics are set to those of case 4, the two real roots change to approximately -25 and -50. These real roots are not listed in Table 6.4.

Table 6.4 illustrates the important points about the eigenvalues of a rotating machine discussed previously and in Section 5.8.1. The effect of the support

Table 6.4. *Characteristic roots computed for a number of configurations (Example 6.8.1)*

Speed (rev/min)	Left	Right	First	Pair of Roots (rad/s) Second	Third	Fourth
0	1	1	$0 \pm 44.28\jmath$	$0 \pm 44.28\jmath$	$0 \pm 397.6\jmath$	$0 \pm 397.6\jmath$
0	2	2	$0 \pm 44.28\jmath$	$0 \pm 44.36\jmath$	$0 \pm 397.6\jmath$	$0 \pm 400.2\jmath$
0	2	3	$-0.038 \pm 44.29\jmath$	$-0.010 \pm 44.36\jmath$	$-1.220 \pm 400.6\jmath$	$-0.851 \pm 401.2\jmath$
0	2	4	$-0.066 \pm 44.39\jmath$	$-0.038 \pm 44.40\jmath$	$-0.211 \pm 401.1\jmath$	$-0.209 \pm 401.9\jmath$
0	5	5	$0 \pm 5.983\jmath$	$0 \pm 6.465\jmath$	$0 \pm 36.01\jmath$	$0 \pm 44.86\jmath$
1,000	1	1	$0 \pm 42.22\jmath$	$0 \pm 46.35\jmath$	$0 \pm 389.9\jmath$	$0 \pm 402.3\jmath$
1,000	2	3	$-0.020 \pm 42.26\jmath$	$-0.028 \pm 46.40\jmath$	$-0.619 \pm 392.0\jmath$	$-1.358 \pm 406.6\jmath$
2,000	2	3	$-0.017 \pm 40.22\jmath$	$-0.033 \pm 48.47\jmath$	$-0.190 \pm 377.9\jmath$	$-1.605 \pm 410.3\jmath$
3,000	2	3	$-0.139 \pm 38.23\jmath$	$-0.039 \pm 50.51\jmath$	$-0.0004 \pm 357.3\jmath$	$-1.786 \pm 413.0\jmath$

Note: The left/right columns refer to the bearing characteristics in Table 6.3; real eigenvalues are not listed.

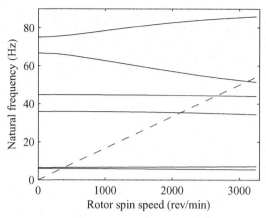

Figure 6.50. The Campbell diagram for the overhung rotor (Example 6.8.1). The dashed line represents $\omega = \Omega$.

anisotropy is insignificant because the bearings are quite stiff compared to the rotor.

Figure 6.50 shows the Campbell diagram when both the left and right bearings have characteristics of case 5 (see Table 6.3). Modes 1, 3, and 5 are predominantly backward-whirling, although some locations on the rotor are whirling in the forward direction. Similarly, modes 2, 4, and 6 are predominantly forward-whirling. There is only one forcing frequency line on this plot and it corresponds to unbalance forcing. At every intersection of a forcing-frequency line with a resonance-frequency line, there is – by definition – a critical speed. Clearly, there are four critical speeds below 3,000 rev/min and one critical speed just above 3,000 rev/min.

For this example, the critical speeds also may be calculated using the iterative and direct approaches, which yield the same result. Table 6.5 provides the results of applying the iterative approach and shows the convergence of the first five critical speeds. The initial estimates of the critical speeds were taken as the corresponding natural frequencies at a rotor spin speed of 1,000 rev/min. Clearly, the convergence is very fast and these critical speeds correspond to those depicted on the Campbell diagram. The direct approach computes the same five critical speeds without any iteration.

Table 6.5. *The first five critical speeds computed by the iterative method (Example 6.8.1)*

Iteration	Critical speeds (rev/min)				
	1	2	3	4	5
0	352.91	393.69	2,152.9	2,687.8	3,767.2
1	358.08	388.95	2,122.9	2,658.4	3,000.1
2	358.05	388.92	2,124.0	2,659.3	3,145.9
3	358.05	388.92	2,124.0	2,659.3	3,113.5
4					3,120.5
5					3,119.0
6					3,119.3
7					3,119.3

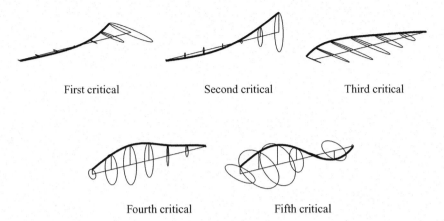

First critical   Second critical   Third critical

Fourth critical   Fifth critical

Figure 6.51. Mode shapes for the critical speeds for the overhung rotor (Example 6.9.1).

6.9 Mode Shapes Associated with Critical Speeds

Chapter 5 computes the mode shapes or eigenvectors for a constant rotor speed. It is obvious that the mode shape corresponding to a critical speed is important in determining how the rotor system might vibrate when the critical speed is excited. This mode shape is calculated easily by setting the rotor speed equal to the critical speed and computing the right eigenvector by the methods described in Chapter 5.

> **EXAMPLE 6.9.1.** Find the critical speeds and associated mode shapes for the system defined in Example 6.8.1 with bearing characteristic case 5, defined in Table 6.3.
>
> *Solution.* Example 6.8.1 shows that the overhung rotor has four critical speeds for rotor speeds below 3,000 rev/min and one speed just above, shown in Table 6.5 and Figure 6.50. The mode shapes corresponding to these critical speeds are shown in Figure 6.51. Because the orbits are not circular, even the backward-whirling modes in this example are excited by unbalance forcing.

6.10 Maps of Critical Speeds and Mode Shapes

Frequently, in the design of machines, some of the parameters that directly affect the rotordynamics of the machine are not accurately known. In particular, bearing-support stiffness is one such parameter. However, modeling is invariably an expensive exercise. If machine builders can determine that they do not need to model bearing-support stiffness, for example, in any better way than simply recalling an empirical value, this is a positive outcome. Bearing-support stiffness is a particularly relevant example of such a variable parameter because this quantity may depend not only on the design and manufacture of a particular machine, it also can depend strongly on the way in which that machine is mounted.

Maps of critical speed and mode shapes enable an engineer to obtain rapidly an impression of how the uncertainty in one of the parameters affects the behavior of the machine. The most relevant concern is that the parameter cannot assume a

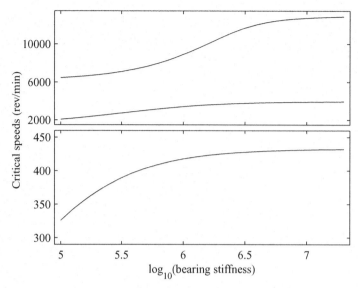

Figure 6.52. The variation of the lower positive critical speeds with bearing stiffness (Example 6.10.1).

value that causes one of the machine critical speeds to occur at an undesirable part of the running range of speeds.

EXAMPLE 6.10.1. Consider the overhung rotor of Example 6.8.1. Suppose that the bearings are isotropic and that we are interested only in the forward-whirling critical speeds. The left bearing has a stiffness of 10 MN/m in both directions (i.e., bearing characteristic 1 in Table 6.3), whereas the stiffness of the right bearing varies between 0.1 and 20 MN/m. Damping is neglected. Generate the critical speed and mode-shape maps.

Solution. Figure 6.52 shows the variation of the first three positive critical speeds as the bearing stiffness changes. Figures 6.53 through 6.55 show how the first three mode shapes vary with the bearing stiffness. At low bearing stiffnesses, the rotor is hardly constrained, whereas the bearing presents an almost pinned constraint at high stiffnesses.

6.11 Running through Critical Speeds

If a single degree of freedom system is subjected to an oscillating force whose frequency coincides exactly with the resonance frequency of the system, then the steady-state response of that system is governed only by the damping. Theoretically, if the damping is zero, the response can be infinite. However, this infinite response is not achieved instantaneously and the vibration may not reach unacceptable levels if the excitation is applied for a sufficiently short period. In the operation of rotating machines, it is not possible to alter the frequency or magnitude of applied forcing at will because both quantities are normally determined by the speed of rotation of the shaft through unbalance, and there is normally a practical limit to how quickly the shaft speed can be changed. However, in general, the faster a system can be run

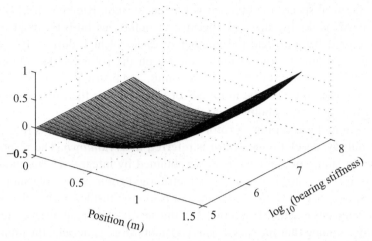

Figure 6.53. The variation of the first mode shape with bearing stiffness (Example 6.10.1).

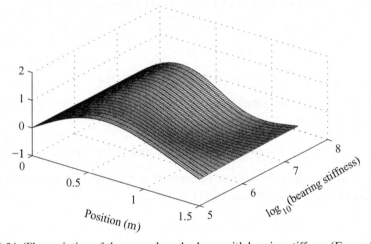

Figure 6.54. The variation of the second mode shape with bearing stiffness (Example 6.10.1).

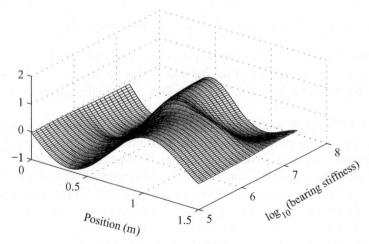

Figure 6.55. The variation of the third mode shape with bearing stiffness (Example 6.10.1).

through a critical speed, the better it is for that system. Running quickly through critical speeds is doubly desirable because it minimizes both the peak response (i.e., force or deflection) and the number of cycles of high duress. The static response of a machine to known forcing acting on the rotor is relatively simple to estimate from the stiffness of the shaft, bearings, and bearing supports. The steady-state dynamic response can be hundreds of times greater than the static response (even neglecting the possibility of instability).

The action of accelerating a rotor to its operating speed often is referred to as *runup*; similarly, decelerating a rotor is referred to as *rundown*. The response of a rotor during runup or rundown may be obtained by integrating the equations of motion with respect to time. For a symmetric rigid rotor on flexible supports, the lateral response decouples from the angular response; furthermore, only the lateral response becomes excited. In Section 6.2, the steady-state response of the rotor is computed assuming that the speed, $\dot{\phi}$, is constant. More generally, the equations of motion are

$$
\begin{aligned}
m\ddot{u} + c\dot{u} + ku &= m\varepsilon\dot{\phi}(t)^2\cos\phi(t) + m\varepsilon\ddot{\phi}(t)\sin\phi(t) \\
m\ddot{v} + c\dot{v} + kv &= m\varepsilon\dot{\phi}(t)^2\sin\phi(t) - m\varepsilon\ddot{\phi}(t)\cos\phi(t)
\end{aligned}
\tag{6.69}
$$

or, in complex notation, where $r = u + jv$

$$
m\ddot{r} + c\dot{r} + kr = m\varepsilon\left\{\dot{\phi}(t)^2 - j\ddot{\phi}(t)\right\}e^{j\phi(t)}
\tag{6.70}
$$

If we assume a linear runup from zero speed so that the rotor speed is a linear function of time, then

$$
\phi(t) = \frac{1}{2}\alpha t^2, \quad \dot{\phi}(t) = \alpha t \quad \text{and} \quad \ddot{\phi}(t) = \alpha
\tag{6.71}
$$

for some constant α. To demonstrate the important features of the runup, it is useful to transform Equations (6.70) and (6.71), to give

$$
\frac{d^2\bar{r}}{d\tau^2} + 2\zeta\frac{d\bar{r}}{d\tau} + \bar{r} = \left\{\bar{\alpha}^2\tau^2 - j\bar{\alpha}\right\}e^{j\bar{\alpha}\tau^2/2}
\tag{6.72}
$$

where

$$
\tau = \omega_n t, \quad \bar{\alpha} = \frac{\alpha}{\omega_n^2}, \quad \bar{r} = \frac{r}{\varepsilon} \quad \text{and} \quad \omega_n = \sqrt{\frac{k}{m}}
\tag{6.73}
$$

The damping ratio is ζ. Equation (6.72) clearly shows that the key factors in determining the transient response of the rotor as it runs through the critical speed are the damping ratio and ratio of the runup acceleration to the square of the natural frequency, $\bar{\alpha}$.

Lewis (1932) addressed the problem of predicting the envelope of shaft response given the rate of acceleration of the shaft through the critical speed. He obtained this envelope by analytically integrating the equations of motion. With modern computational power, it is much easier to numerically integrate Equation (6.72) for various values of runup acceleration and damping ratio. This integration may be performed using the fourth-order Runge-Kutta variable-step-length integration or any other suitable method (see Section 2.6.1), where the tolerance is set to ensure that there is no perceptible error in the plots. Figure 6.56 shows a typical response and excitation force; the initial displacement and velocity are assumed to be zero.

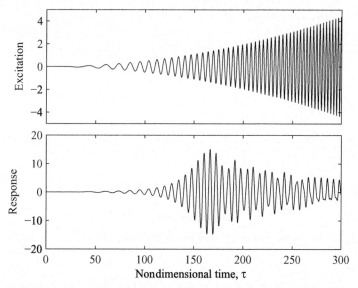

Figure 6.56. The runup through a resonance for $\bar{\alpha} = 0.007$ and $\zeta = 0.01$. The response is the real part of $\bar{r}$ and the excitation is the real part of the right side of Equation (6.72).

The effect of the runup acceleration may be demonstrated by performing this integration for various values of $\bar{\alpha}$. Figure 6.57 shows the variation of peak response attained as a function of runup acceleration, $\bar{\alpha}$, for zero damping and for a damping ratio of 0.01. Clearly, increasing either the damping or the runup acceleration reduces the maximum response.

This analysis is simple and does not account for the fact that the resonance can change with shaft speed and, in some cases, the *effective acceleration rate* for the purposes of knowing maximum response may be different from the actual rate. If the system has a resonance that varies with speed and therefore the resonance is changing in the same direction as the speed (i.e., excitation frequency), then the effective acceleration is lower. For example, suppose a system has a critical speed at 100 Hz and the dependence of system resonance on shaft speed (near the critical) is 0.2

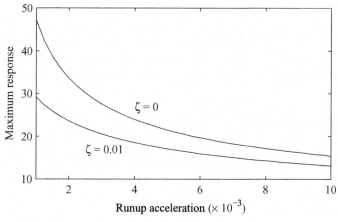

Figure 6.57. Maximum response, $|\bar{r}|$, as a function of runup acceleration, $\bar{\alpha}$.

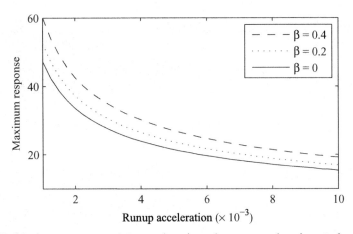

Figure 6.58. Maximum response, $|\bar{r}|$, as a function of runup acceleration, $\bar{\alpha}$, for $\beta = 0, 0.2,$ and 0.4, and with $\zeta = 0$.

(Hz/Hz). Then, if the shaft is accelerating at 10 Hz/s, the system resonance is moving upward at 2 Hz/s and the effective acceleration is approximately 8 Hz/s. Thus, the effective acceleration might be defined as the rate at which the gap between system resonance and shaft speed is changing.

Consider the symmetrical rigid rotor on flexible supports in Equation (6.69). Suppose that near the critical speed, the natural frequency may be approximated as a linear function of rotor speed, so

$$\omega_n = \left(\omega_c - \beta\omega_c + \beta\dot{\phi}(t)\right) \qquad (6.74)$$

where ω_c is the critical speed, ω_n is the speed-dependent natural frequency, and β is the slope of the Campbell diagram at the critical speed. Substituting this into Equation (6.70) and assuming a constant rotor acceleration given by Equation (6.71), the transformed equation of motion equivalent to Equation (6.72) is

$$\frac{d^2\bar{r}}{d\tau^2} + 2\zeta\frac{d\bar{r}}{d\tau} + (1 - \beta + \bar{\alpha}\beta\tau)^2\,\bar{r} = \left(\bar{\alpha}^2\tau^2 - j\bar{\alpha}\right)e^{j\bar{\alpha}\tau^2/2} \qquad (6.75)$$

where

$$\tau = \omega_c t \quad \text{and} \quad \bar{\alpha} = \frac{\alpha}{\omega_c^2} \qquad (6.76)$$

Figure 6.58 shows the variation of peak response attained as a function of runup acceleration, $\bar{\alpha}$, for $\beta = 0, 0.2,$ and 0.4, with zero damping. The previous argument suggests that the effective $\bar{\alpha}$ is

$$\bar{\alpha}_{\text{eff}} \approx \bar{\alpha}\,(1 - \beta) \qquad (6.77)$$

and Figure 6.58 shows that this is a reasonable approximation.

For complicated rotor systems, it is possible to integrate the equations of motion to obtain the time response in much the same way as for the simple system shown in Figure 6.56. Speed-dependent bearing properties are easily incorporated by changing the linear stiffness and damping matrices as the shaft speed varies. However, even for models of small size or for large models reduced in size, the numerical integration becomes computationally expensive.

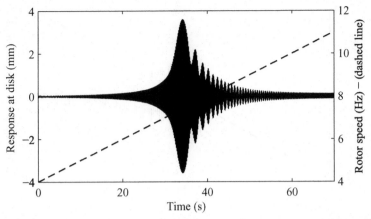

Figure 6.59. The runup through the first resonance for the overhung rotor (Example 6.11.1) with $\alpha = 0.10$ Hz/s.

For very large machines (e.g., power-station turbines), the runup or rundown is very slow, typically taking of the order of 20 minutes. With such a slow runup or rundown, the unbalance response may be taken as the steady-state response, defined as the *quasi steady state*, which may be computed efficiently in the frequency domain.

Gasch et al. (1979) show that the effect of the torsional flexibility of the shaft on the lateral vibration during runup is negligible. Genta and Delprete (1995) consider the effect of the variable rotor spin speed on the model of the machine. Goodwin et al. (1984) and Wauer and Suherman (1998) consider using a variable support stiffness to suppress the vibrations of flexible rotor systems passing through critical speeds.

EXAMPLE 6.11.1. Consider the overhung rotor of Example 6.8.1, with bearing characteristics given by case 4 in Table 6.3 and an unbalance of magnitude 10^{-3} kg m on the disk. Calculate the response as the machine is run up through the first critical speed with constant acceleration for the two cases $\alpha = 0.10$ Hz/s and $\alpha = 0.40$ Hz/s.

Solution. The machine has 28 degrees of freedom and only the lower-frequency dynamics become excited by the unbalance force. However, the time step must be determined by the highest natural frequency of the machine (see Section 2.6.1). Thus, a larger step size may be used if the model is reduced and only the lower frequency modes are retained. This also has the advantage that the size of the model matrices is reduced. Hence, the 28 degrees of freedom is reduced to four degrees of freedom using a transformation based on the lowest four modes obtained by neglecting damping and gyroscopic effects (see Section 2.5). Figures 6.59 and 6.60 show the response at the disk for $\alpha = 0.10$ and 0.40 Hz/s, respectively, starting at a rotor spin speed of 4 Hz, or 240 rev/min. Clearly, the response is reduced by running through the critical speed more quickly. The beating phenomenon arises because the gyroscopic effects cause the lowest pair of resonances to split. In steady state, the unbalance excites only the forward-whirling mode; however, during the runup, the steady state is never achieved, so both modes are excited.

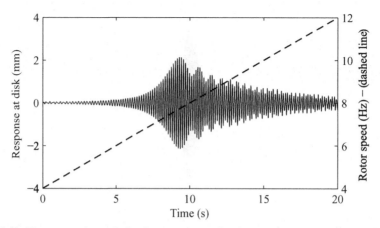

Figure 6.60. The runup through the first resonance for the overhung rotor (Example 6.11.1) with $\alpha = 0.40$ Hz/s.

6.12 Stresses in a Rotor

In addition to determining the deflection of a rotor due to out-of-balance and other forces, it is sometimes necessary to determine the resultant internal stresses in the rotor. Because the rotor is spinning, centrifugal forces act on the rotor and give rise to radial and hoop stresses. Deformation of the rotor in the lateral direction also causes bending and shear stresses. We now examine each of these in turn. In Chapter 10, we show that these stresses can affect the dynamics of a machine.

6.12.1 Radial and Hoop Stresses due to Spin

As a rotor spins, the radial stress, σ_r and hoop stress, σ_ϕ, in a uniform axially thin disk at a distance r from the axis of rotation (Fenner, 1989) are approximately

$$\sigma_r(r) = \rho\Omega^2 \left(\frac{3+\nu}{8}\right) \left(R_1^2 + R_2^2 - \frac{R_1^2 R_2^2}{r^2} - r^2\right) \tag{6.78}$$

$$\sigma_\phi(r) = \rho\Omega^2 \left(\frac{3+\nu}{8}\right) \left(R_1^2 + R_2^2 + \frac{R_1^2 R_2^2}{r^2} - \frac{1+3\nu}{3+\nu}r^2\right) \tag{6.79}$$

where ρ is the disk density, Ω is the rotational speed, R_1 and R_2 are the inside and outside radii of the disk, and ν is Poisson's ratio. The stresses are independent of disk thickness if the thickness is small compared to the diameter. The maximum radial stress is found by setting $d\sigma_r/dr$ to zero. This gives a maximum radial stress of

$$(\sigma_r)_{max} = \rho\Omega^2 \left(\frac{3+\nu}{8}\right)(R_2 - R_1)^2 \quad \text{when} \quad r = \sqrt{R_1 R_2} \tag{6.80}$$

It is shown that increasing the outside radius, R_2, increases the stress for a given rotational speed. Thus, these stresses are more likely to be significant for large-diameter disks than for shaft sections. The maximum hoop stress occurs at the inside radius, R_1. Then,

$$(\sigma_\phi)_{max} = 2\rho\Omega^2 \left(\frac{3+\nu}{8}\right) \left(R_2^2 + \frac{1-\nu}{3+\nu} R_1^2\right) \tag{6.81}$$

In the case of a solid rotor or disk, the inside radius, R_1, is zero. From Equations (6.78) and (6.79)

$$\sigma_r(r) = \rho\Omega^2 \left(\frac{3+\nu}{8}\right) \left(R_2^2 - r^2\right) \tag{6.82}$$

$$\sigma_\phi(r) = \rho\Omega^2 \left(\frac{3+\nu}{8}\right) \left(R_2^2 - \frac{1+3\nu}{3+\nu} r^2\right) \tag{6.83}$$

The maximum values occur when $r = 0$ and are

$$(\sigma_r)_{max} = (\sigma_\phi)_{max} = \rho\Omega^2 \left(\frac{3+\nu}{8}\right) R_2^2 \tag{6.84}$$

Similar equations can be used to determine the stresses in a rotor with a thickness that is substantially greater than the diameter. In this case, ν must be replaced by $\nu/(1-\nu)$.

In the design of many rotors, one of the ultimate limitations is the rupture strength of the rotor. Great care is taken to ensure that the radial and hoop stresses can be sustained by the material.

EXAMPLE 6.12.1. A disk with an outside diameter of 1.6 m rotates at 3,000 rev/min. Given that $\rho = 7,850 \text{ kg/m}^3$ and $\nu = 0.3$, determine the maximum hoop and radial stresses in the disk (a) if the disk has an inside diameter of 0.1 m, and (b) if the disk has no central hole.

Solution. $\Omega = 3,000 \text{ rev/min} = 3,000 \times 2\pi/60 \text{ rad/s} = 314 \text{ rad/s}$. Thus, $\rho\Omega^2 (3 + \nu)/8 = 7,850 \times 314^2 \times (3+0.3)/8 = 426.1 \text{ MN/m}^4$.

(a) $(\sigma_r)_{max} = 426.1 \times (0.8 - 0.05)^2 = 239.7 \text{ MPa}$, at $r = \sqrt{0.8 \times 0.05} = 0.2 \text{ m}$.
$(\sigma_\phi)_{max} = 2 \times 426.1 \times \left\{0.8^2 + \left(\frac{1-0.3}{3+0.3}\right) 0.05^2\right\} = 545.9 \text{ MPa}$.

(b) $(\sigma_r)_{max} = (\sigma_\phi)_{max} = 426.1 \times 0.8^2 = 272.7 \text{ MPa}$.

6.12.2 Axial Stresses due to Lateral Deformation of the Rotor

It has been shown that due to the effect of lateral and other forces, flexible shafts bend as they rotate; that is, they whirl. The axial stress due to bending at a radius r is given by $\sigma_z(r) = \dfrac{Mr}{I}$, where M is the bending moment acting on a rotor section and I is the second moment of area of the shaft. Ignoring shear, the bending moment acting on the rotor at z is $M(z) = -EIu''(z)$, where $u''(z)$ is the second derivative of the beam deflection with respect to position along the rotor; physically, it is the curvature of the rotor. Thus, $\sigma_z(r, z) = -Eru''(z)$ and

$$\sigma_z(R, z) = (\sigma_z)_{max} = -ERu''(z) \tag{6.85}$$

where R is the radius of the shaft.

Assuming that the rotor is modeled using finite elements, it is relatively simple to determine the curvature of the rotor. The deflected shape for a beam or shaft element in the x direction, Equation (5.10), is

$$u_e(\xi) = \left(1 - 3\frac{\xi^2}{\ell_e^2} + 2\frac{\xi^3}{\ell_e^3}\right) u_1 + \ell_e \left(\frac{\xi}{\ell_e} - 2\frac{\xi^2}{\ell_e^2} + \frac{\xi^3}{\ell_e^3}\right) \psi_1$$

$$+ \left(3\frac{\xi^2}{\ell_e^2} - 2\frac{\xi^3}{\ell_e^3}\right) u_2 + \ell_e \left(-\frac{\xi^2}{\ell_e^2} + \frac{\xi^3}{\ell_e^3}\right) \psi_2$$

Because the deflections have been determined, we can determine u_e'' for any element:

$$u_e''(\xi) = \left(-\frac{6}{\ell_e^2} + 12\frac{\xi}{\ell_e^3}\right) u_1 + \ell_e \left(-\frac{4}{\ell_e^2} + 6\frac{\xi}{\ell_e^3}\right) \psi_1$$

$$+ \left(\frac{6}{\ell_e^2} - 12\frac{\xi}{\ell_e^3}\right) u_2 + \ell_e \left(-\frac{2}{\ell_e^2} + 6\frac{\xi}{\ell_e^3}\right) \psi_2 \qquad (6.86)$$

Letting $\xi = 0$ in Equation (6.86) allows us to determine the bending moment at the left end of the beam element; letting $\xi = \ell_e$, we can determine the bending moment at the right end of the beam element. This gives

$$u_{z1}'' = (-6u_1 + 6u_2 - 4\ell_e\psi_1 - 2\ell_e\psi_2)/\ell_e^2$$

$$u_{z2}'' = (6u_1 - 6u_2 + 2\ell_e\psi_1 + 4\ell_e\psi_2)/\ell_e^2$$

$$v_{z1}'' = (-6v_1 + 6v_2 + 4\ell_e\theta_1 + 2\ell_e\theta_2)/\ell_e^2 \qquad (6.87)$$

$$v_{z2}'' = (6v_1 - 6v_2 - 2\ell_e\theta_1 - 4\ell_e\theta_2)/\ell_e^2$$

Because cubic-shape functions are used, the curvature varies linearly between the nodes; hence, the maximum curvature is at one of the two nodes.

This bending stress might be constant or it might oscillate at some frequency. If the rotor spins at Ω and vibrates in fixed coordinates at ω, then the frequency of oscillation of the stress is $\Omega - \omega$. If the rotor is in pure forward synchronous whirl, then the stresses in the rotor are constant. This is the case when a rotor on isotropic supports is excited by a rotating out-of-balance force and is in forward whirl. If, however, the rotor is in backward whirl, then the frequency of the stress variation is twice the rotor speed. When the rotor is excited by a nonsynchronous force – for example, by foundation excitation – then it is unlikely that $\Omega = \omega$ and the frequency of the stress oscillation can have components at $\Omega - \omega$ and $\Omega + \omega$. Where oscillatory stresses occur in a rotor, fatigue is likely to be a major concern. If the sum of the mean and oscillatory stresses does not exceed the endurance stress of the material, then the rotor has an infinite life.

EXAMPLE 6.12.2. For the rotor of Example 6.3.2, determine the maximum axial stress when it is rotating at 496 rev/min.

Solution. From the deflected shape of the rotor at 496 rev/min (determined from the FE model), we can determine the curvature using Equation (6.87). The maximum curvature is $1.958 \times 10^{-5}\,\text{m}^{-1}$ at node 3. Because the diameter of the shaft is 0.05 m, then using Equation (6.85), we have $\sigma_z = 2.11 \times 10^{11} \times (0.05/2) \times 1.958 \times 10^{-5} = 10.33\,\text{kPa}$.

6.13 Summary

In this chapter, we show how the response of a rotor to various excitations can be computed, for both simple systems and more complex rotors represented by FE models and supported by isotropic or anisotropic bearings. The nature of the response of rotors to out-of-balance forces and to forces acting through the bearings is discussed, as is the response of a rotor to a bent shaft. We also introduce the concept of critical speeds for rotor–bearing systems and demonstrate methods to calculate those speeds. Reducing the response of a system to out-of-balance forces leads naturally to Chapter 8, wherein rotor balancing is described.

The response of rotors during runup and rundown also is examined. We examine the stresses induced in rotors during operation.

The models in this chapter are limited to systems with axisymmetric rotors. In Chapter 7, the dynamics, the response, and the stability of systems with asymmetric rotors are considered.

6.14 Problems

6.1 The rigid rotor described in Chapter 3, Problem 3.1, is carried by bearings as shown in Figure 3.26. The bearing-support stiffness at the right end is 1 MN/m in both the x and y directions. Determine the critical speeds of the system. Which critical speed is excited by unbalance?

An out-of-balance mass of 0.1 kg at a radius of 100 mm is attached to the right end. Determine the radius of the whirl orbit at the right end when the rotor is spinning at 1,500 rev/min.

6.2 The rotor described in Chapter 3, Problem 3.1, is carried in bearings as shown in Figure 3.26. The bearing-support stiffness at the right end in both the x and y directions is $k(\Omega)$, where $k(\Omega) = k_0 + k_1\Omega + k_2\Omega^2$ and Ω, the speed of rotation of the rotor, is positive. In this equation, $k_0 = 10^6$ N/m, $k_1 = 1.5 \times 10^3$ Ns/m, and $k_2 = 3$ Ns^2/m. Determine the critical speeds for this system.

An out-of-balance moment rotates synchronously with the rotation of the rotor. The moment is $5 \times 10^{-3}\Omega^2 e^{j\Omega t}$, where Ω is the spin speed of the rotor. Determine the radius of the whirl orbit at the right end when the rotor is spinning at 1,500 rev/min.

6.3 The rigid rotor, described in Chapter 3, Problem 3.1, is carried in bearings as shown in Figure 3.26. The bearing-support stiffness at the number 2 end is 1 MN/m in the y direction and 1.3 MN/m in the x direction.

Determine the critical speeds for this machine. Using this information, together with the natural frequencies derived by solving Problem 3.2 in Chapter 3, sketch the Campbell diagram for the system. Hint: There are three points to plot on each frequency line.

6.4 An overhung rotor system is shown in the Figure 6.61. The rotor is effectively cantilevered from a pair of stiff bearings and extends 400 mm from the nearest bearing, as shown in the figure. The rotor shaft is 30 mm in diameter and is made from steel with a Young's modulus of 210 GPa. The mass of the shaft is negligible compared to the mass of the disk on the end. The disk has a diameter of 300 mm, is 25 mm thick, and is made from steel with a density of 7,800 kg/m^3. Determine the critical speeds for this rotor system.

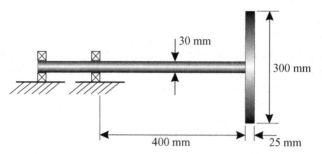

Figure 6.61. The overhung rotor on stiff supports (Problem 6.4).

Determine the maximum radius of the whirl orbit of the disk when the rotor spins at 2,000 rev/min, if the maximum allowable radial misalignment of the disk on the shaft is 0.1 mm.

If the radius of the whirl orbit is not allowed to exceed 3 mm, estimate the minimum acceleration required through the critical speed when the radial misalignment of the disk on the shaft is 0.1 mm. Use the graphical data provided in Figure 6.57 and assume that damping of the rotor is negligible. (Note: Figure 6.57 was generated based on the translational response of a symmetric rotor, and we assume that the relationship between the maximum response and the runup acceleration is similar for the system considered here.) What torque is required to produce this acceleration?

Determine the maximum allowable angular misalignment of the disk axis relative to the shaft axis if the radius of the resulting whirl orbit of the disk is equal to the whirl orbit caused by a radial misalignment of 0.1 mm.

6.5 Consider a Jeffcott rotor consisting of a flexible steel shaft supported in short, rigid bearings. The shaft is 800 mm long with a diametral second moment of area of $4 \times 10^{-8}\,\text{m}^4$. The mass of the shaft is negligible compared to the mass of a disk at the center. The disk has a polar moment of inertia of $0.155\,\text{kg}\,\text{m}^2$ and a diametral moment of inertia of $0.078\,\text{kg}\,\text{m}^2$. Young's modulus for steel is 210 GPa. Determine the maximum value of θ or ψ if the disk has an angular misalignment of $\beta = 1°$ and the rotor is spinning at 6,000 rev/min.

6.6 A 1 m-long steel rotor is supported in short bearings at each end and carries a central steel disk. The shaft is 30 mm in diameter and the central disk is 150 mm in diameter and 50 mm thick. It is discovered that during storage, the rotor has become distorted; at the midspan, it bows by 0.5 mm. Determine the amplitude of the whirl when the rotor spins at 1,000 rev/min. Neglect the mass of the shaft. For steel, $E = 200\,\text{GPa}$ and $\rho = 7,850\,\text{kg/m}^3$. By examining the mass of the shaft, determine why the results should be viewed with considerable caution.

To overcome the whirl due to bow, it is decided to add a small out-of-balance mass at a radius of 75 mm, diametrally opposite to the bow. Determine the mass to be added if the rotor is to be perfectly balanced when it spins at 1,000 rev/min. Neglect the mass of the shaft. What is the response when the machine runs at 1,500 rev/min?

6.7 A Jeffcott rotor is supported at each end by ball bearings with a high stiffness. The stiffness of the bearing supports was measured at 5MN/m in both the horizontal and vertical directions with no cross-coupling (i.e., a horizontal force

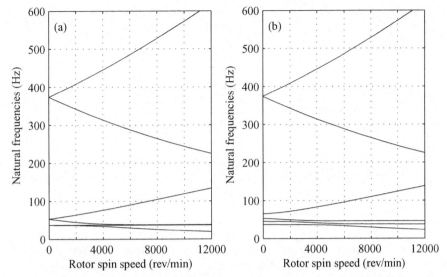

Figure 6.62. The natural frequency maps for two machines (Problem 6.9).

on the bearing support produces a zero vertical deflection and vice versa). The shaft has a total length of 1 m, an outside diameter of 40 mm, and an inside diameter of 36 mm. The disk is mounted at the center of the shaft and has a thickness of 50 mm and a diameter of 150 mm. The entire rotor is made from steel with a density of 7,800 kg/m^3 and a modulus of elasticity of 200 GPa. Estimate the first critical speed of this machine.

If the damping in the rotor is small and there is an unbalance on the periphery of the disk of 0.2 kg, what is the response of the disk when the shaft is spinning at 1,500 rev/min? Using this information and the shaft and bearing stiffnesses, estimate the response of the shaft at the bearings.

Suppose that the single ball bearings are replaced by long roller bearings or double-race ball bearings; the length of the free shaft between the bearings remains at 1 m. What is the maximum amount by which the first critical speed could be raised?

6.8 For the rotating machine described in Chapter 3, Problem 3.6, determine the forward-whirling critical speeds of the system. It might be reasonable to expect that two of the speeds are real and positive. Why, in this case, is only one of the speeds real and positive?

6.9 Figure 6.62 shows the natural frequency maps for two particular rotor–bearing systems. Identify which of the rotor–bearing systems shown in the figure has an isotropic and which has an anisotropic bearing and/or foundation.

Use these natural frequency maps to estimate the following for the systems in Figure 6.62:

(a) the critical speeds due to synchronous unbalance

(b) the critical speeds due to a force that rotates in the same direction as the rotor spins at three times the shaft speed

(c) the peaks in the response plot caused by a harmonic force acting horizontally at one of the bearings and sweeping through the frequency range of 0 to 600 Hz, when the machine operates with a shaft speed of 10,000 rev/min

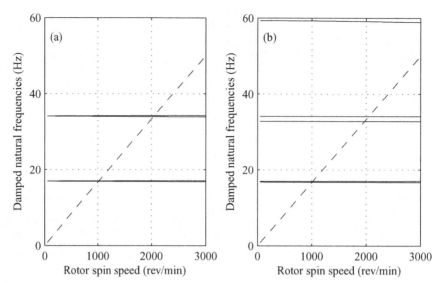

Figure 6.63. The Campbell diagrams for the machine with isotropic and anisotropic flexible bearings (Problem 6.11 (a) and (b)). The dashed line represents $\omega = \Omega$.

6.10 Consider the machine described in Problem 5.2, modeled with eight elements. Plot the Campbell diagram for this system up to the maximum rotor spin speed of 5,000 rev/min, and estimate the critical speeds that may be excited by unbalance. Calculate the response at the disks to the following out-of-balance mass distributions over the full range of spin speeds:
(a) 20 gm at 0° on the left disk, 20 gm at 0° on the center disk, and 20 gm at 0° on the right disk
(b) 20 gm at 0° on the left disk and 20 gm at 180° on the right disk
(c) 20 gm at 0° on the left disk only
By comparing the shapes of the first two modes, comment on the relative amplitudes of the response for the different out-of-balance mass distributions.

6.11 This problem is based on Problem 5.9, except that the outer bearings are located at the ends of the shaft. Figure 6.63 shows the Campbell diagrams for the machine mounted on flexible bearings with isotropic and anisotropic properties, corresponding to parts (a) and (b) of Problem 5.9. Estimate the critical speeds of this machine on these supports. Calculate the response to an out-of-balance mass placed at the disk located 2.0 m from the left end. Compare the frequencies where peaks occur in this response to the estimated critical speeds. Figure 6.64 shows the Campbell diagrams for the machine mounted on oil-film bearings, corresponding to (c) in Problem 5.9 with equal load on each bearing. Estimate the critical speeds and calculate the unbalance response. Compare the frequencies where the peaks occur with the critical speeds and comment.

6.12 Consider the machine described in Problem 5.1 with the solid shaft. Estimate the response at the disks up to a maximum rotor spin speed of 5,000 rev/min to an out-of-balance mass distribution given by 200 gm at 0° on the left disk and 200 gm at 0° on the right disk. Compare this response to that for a bent shaft, where the bend is given by $u(z) = 0.0025 \sin(\pi z / 1.6)$, where z and u are given in meters.

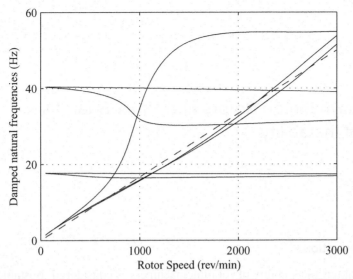

Figure 6.64. The Campbell diagram for the machine on oil-film bearings (Problem 6.11 (c)). The dashed line represents $\omega = \Omega$.

6.13 Consider the model of Problem 3.1 but with a ground displacement of $u_g(t)$ in the x direction at the flexible support. Show that the equation of motion is identical to that in Problem 3.1 with the addition of a forcing term on the right side given by $\begin{bmatrix} 0 & k_x L u_g(t) \end{bmatrix}^\top$, where L is the length of the rotor and is 0.5 m. Write the equation of motion in state–space form (see Section 5.8) and calculate the left and right normalized eigenvectors when the rotor is spinning at 3,000 rev/min. Diagonalize the state–space equation to give four decoupled equations of the form

$$\dot{p}_i - s_i p_i = f_i u_g(t)$$

where s_i is the ith eigenvalue and f_i is a complex constant. The solution to this equation, with zero initial condition, is given by the Duhamel integral as

$$p_i(t) = \int_0^t f_i u_g(\tau) e^{s_i(t-\tau)} d\tau$$

The ground displacement is a sudden change followed by an exponential decay given by $u_g(t) = u_0 e^{-\beta t}$ for constants u_0 and β. Show that if the rotor is initially at rest, the ith modal response is

$$p_i(t) = f_i u_0 \left[\frac{1}{s_i + \beta} \left(e^{s_i t} - e^{-\beta t} \right) \right]$$

Use these modal responses and the eigenvectors to calculate the response of the disk in the x and y directions when $\beta = 2\,\mathrm{s}^{-1}$ and $u_0 = 1\,\mathrm{mm}$. Plot the orbit and interpret the response.

What is the effect of adding proportional damping obtained from the stiffness matrix with a factor of $8 \times 10^{-5}\mathrm{s}^{-1}$?

7 Asymmetric Rotors and Other Sources of Instability

7.1 Introduction

In this chapter, the stability of rotating machinery is considered. In stable systems, an initial disturbance decays to zero in the absence of excitation forces. By contrast, in an unstable system, the response grows, producing a large and undesirable response that may damage a machine. A simple example of instability is the motion of a pendulum. One equilibrium position is when the pendulum hangs vertically downward. This position is stable because if the pendulum is slightly displaced, it returns to the equilibrium position. In contrast, there is an equilibrium position when the pendulum is balanced vertically upward. This position is unstable because any slight disturbance from the vertical causes the pendulum to move away from the vertical and, in fact, rotate to the lower equilibrium position. For a linear system with constant coefficients, instability may be determined by considering the eigenvalues, computed in the usual way. Thus, the same or similar calculations used to determine eigenvalues of a system also provide a user with information about the stability of the system. As demonstrated in Chapter 2, the imaginary part of the eigenvalue gives the frequency of free oscillations, whereas the real part determines how rapidly the oscillations decay. The oscillations decay only if the real part of the eigenvalue is negative. A zero real part of the eigenvalue gives an undamped response in which the magnitude of the free oscillation remains constant and a positive real part causes the oscillation to grow. Thus, a system is unstable if the real part of any of its eigenvalues is positive. In the vibration of static structures, with no active control or externally applied force, the structure is always stable because there is no source to provide the energy necessary to increase amplitudes of vibration. In contrast, rotating machines have such a source: the energy of rotation. Even the presence of gravity forces can cause instability in a rotating asymmetric system. In this chapter, three causes of instability are discussed: rotor asymmetry, internal damping, and bearings with nonsymmetric stiffness matrices.

Many rotors are *rotationally symmetric* about their centerline so that the rotor appears identical after rotation by any angle. However, asymmetric rotors are relatively common. For example, in certain types of electrical machines, axial slots are cut into the rotor to carry electrical conductors. This has the effect of making the transverse stiffness of the rotor in one plane differ from that in other planes. If the system is analyzed in stationary coordinates, as in Chapters 3, 5, and 6, then the resultant equations of motion have coefficients that vary sinusoidally, and the result is a system that may be parametrically excited. Such a problem cannot be solved using a standard eigenvalue problem, which is demonstrated for a simple system in Example 7.3.1. To avoid this problem for a machine with an asymmetric rotor, if the bearings and supports are isotropic, it is more convenient to work in a set of coordinates that are fixed to and rotate with the rotor. This results in equations of motion with constant coefficients and produces a standard eigenvalue problem. When the rotor is *asymmetric* and the supports are *anisotropic*, the equations of motion have coefficients that vary sinusoidally in both the stationary and rotating coordinate systems. In this case, stability must be determined using Floquet theory (see Section 7.4) because a standard eigenvalue problem does not exist in either coordinate system. This is demonstrated for a simple system in Section 7.4.

All rotors have internal damping and one intuitively expects that damping reduces the amplitude of any vibration. In fact, under certain circumstances, this is not the case; it is shown that damping in a flexible shaft can *cause* instability when the rotor spin speed exceeds at least one critical speed of a machine.

The nonsymmetry in the stiffness matrix of journal bearings also may cause the rotor system to become unstable, as Example 5.9.6 demonstrates. In this chapter, the destabilizing effects of this bearing-matrix nonsymmetry are discussed further using simple bearing models.

7.2 Rotating Coordinate Systems

In rotating systems, many instabilities arise because of the effects of the rotation. Therefore, we must be able to transform between the response in the rotating and stationary frames of reference. The transformations are developed by looking at displacements and forces in the x and y directions; however, rotations and moments about the Ox and Oy axes use the same transformations. Coordinates in the rotating frame are denoted by a tilde; therefore, u and v are the displacements in the x and y directions in the stationary frame, and $\tilde{u}$ and $\tilde{v}$ are the displacements in the $\tilde{x}$ and $\tilde{y}$ directions in the rotating frame. Once per revolution, these axis sets coincide. With reference to Figure 7.1, we can express the displacement in the rotating frame in terms of the displacement in the stationary frame as

$$u = \tilde{u}\cos\phi - \tilde{v}\sin\phi$$
$$v = \tilde{u}\sin\phi + \tilde{v}\cos\phi$$

(7.1)

where ϕ is the angle between the two frames. Only the derivation for v is illustrated in Figure 7.1; the derivation for u is similar. Equation (7.1) may be written in matrix terms as

$$\begin{Bmatrix} u \\ v \end{Bmatrix} = \begin{bmatrix} \cos\phi & -\sin\phi \\ \sin\phi & \cos\phi \end{bmatrix} \begin{Bmatrix} \tilde{u} \\ \tilde{v} \end{Bmatrix} = \mathbf{T} \begin{Bmatrix} \tilde{u} \\ \tilde{v} \end{Bmatrix}$$

(7.2)

The 2×2 matrix is called the *transformation matrix* and is denoted by $\mathbf{T}$. This matrix is orthogonal; hence,

$$\mathbf{T}\mathbf{T}^{\top} = \mathbf{I} \qquad \text{and, thus,} \qquad \mathbf{T}^{-1} = \mathbf{T}^{\top}$$

(7.3)

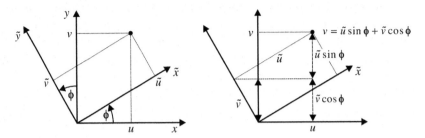

Figure 7.1. The rotating $(\tilde{x}, \tilde{y})$ and stationary (x, y) coordinate systems.

where $\mathbf{I}$ is the identity matrix. The displacements in the rotating frame may be obtained from Equation (7.2) as

$$\begin{Bmatrix} \tilde{u} \\ \tilde{v} \end{Bmatrix} = \mathbf{T}^{-1} \begin{Bmatrix} u \\ v \end{Bmatrix} = \mathbf{T}^{\top} \begin{Bmatrix} u \\ v \end{Bmatrix} = \begin{bmatrix} \cos\phi & \sin\phi \\ -\sin\phi & \cos\phi \end{bmatrix} \begin{Bmatrix} u \\ v \end{Bmatrix} \tag{7.4}$$

Thus, the transformation from the stationary to the rotating frame is given by $\mathbf{T}^{\top}$.

Suppose that the moving coordinate frame rotates relative to the stationary frame with a constant angular velocity Ω. Then, $\phi = \Omega t$ and

$$\mathbf{T} = \begin{bmatrix} \cos\Omega t & -\sin\Omega t \\ \sin\Omega t & \cos\Omega t \end{bmatrix} \tag{7.5}$$

Differentiating Equation (7.2) with respect to time, with $\mathbf{T}$ defined by Equation (7.5), gives

$$\begin{Bmatrix} \dot{u} \\ \dot{v} \end{Bmatrix} = \mathbf{T} \begin{Bmatrix} \dot{\tilde{u}} \\ \dot{\tilde{v}} \end{Bmatrix} + \dot{\mathbf{T}} \begin{Bmatrix} \tilde{u} \\ \tilde{v} \end{Bmatrix} \tag{7.6}$$

where

$$\dot{\mathbf{T}} = \Omega \begin{bmatrix} -\sin\Omega t & -\cos\Omega t \\ \cos\Omega t & -\sin\Omega t \end{bmatrix} \tag{7.7}$$

Note that

$$\mathbf{T}^{\top}\dot{\mathbf{T}} = \dot{\mathbf{T}}\mathbf{T}^{\top} = -\Omega\mathbf{J}, \qquad \text{where} \qquad \mathbf{J} = \begin{bmatrix} 0 & 1 \\ -1 & 0 \end{bmatrix} \tag{7.8}$$

Because

$$\ddot{\mathbf{T}} = \Omega^2 \begin{bmatrix} -\cos\Omega t & \sin\Omega t \\ -\sin\Omega t & -\cos\Omega t \end{bmatrix} = -\Omega^2\mathbf{T} \tag{7.9}$$

differentiating Equation (7.6) gives

$$\begin{Bmatrix} \ddot{u} \\ \ddot{v} \end{Bmatrix} = \mathbf{T} \left(\begin{Bmatrix} \ddot{\tilde{u}} \\ \ddot{\tilde{v}} \end{Bmatrix} - \Omega^2 \begin{Bmatrix} \tilde{u} \\ \tilde{v} \end{Bmatrix} \right) + 2\dot{\mathbf{T}} \begin{Bmatrix} \dot{\tilde{u}} \\ \dot{\tilde{v}} \end{Bmatrix} \tag{7.10}$$

Similarly, differentiating Equation (7.4), with $\mathbf{T}$ defined by Equation (7.5), gives

$$\begin{Bmatrix} \dot{\tilde{u}} \\ \dot{\tilde{v}} \end{Bmatrix} = \mathbf{T}^{\top} \begin{Bmatrix} \dot{u} \\ \dot{v} \end{Bmatrix} + \dot{\mathbf{T}}^{\top} \begin{Bmatrix} u \\ v \end{Bmatrix} \tag{7.11}$$

Differentiating again gives

$$\begin{Bmatrix} \ddot{\tilde{u}} \\ \ddot{\tilde{v}} \end{Bmatrix} = \mathbf{T}^\top \left(\begin{Bmatrix} \ddot{u} \\ \ddot{v} \end{Bmatrix} - \Omega^2 \begin{Bmatrix} u \\ v \end{Bmatrix} \right) + 2\dot{\mathbf{T}}^\top \begin{Bmatrix} \dot{u} \\ \dot{v} \end{Bmatrix} \qquad (7.12)$$

Having developed the transformation matrices, we can now convert between the stationary and rotating frames as required.

7.3 Rotor Asymmetry with Isotropic Supports: Simple Rotors

We now analyze the behavior of an asymmetric rotor and consider a simple model of a light, asymmetric shaft supported in isotropic bearings and carrying a central, axisymmetric disk. For simplicity, gyroscopic effects are ignored. Because the bearings and supports are isotropic, it is more convenient to work in a set of coordinates that are fixed to and rotate with the rotor.

The bearings and supports are modeled as springs, the stiffness of which is constant in all directions. The elastic forces in the rotating coordinates are

$$\begin{Bmatrix} f_{\tilde{x}} \\ f_{\tilde{y}} \end{Bmatrix}_{\text{elastic}} = - \begin{bmatrix} k_{\tilde{x}} & 0 \\ 0 & k_{\tilde{y}} \end{bmatrix} \begin{Bmatrix} \tilde{u} \\ \tilde{v} \end{Bmatrix} \qquad (7.13)$$

The forces are easy to derive in coordinates that rotate with the shaft. Because the bearing stiffness is constant is all directions, the total stiffness is a combination of the bearing and shaft stiffnesses, using the expression for springs in series.

Newton's second law must be expressed first in terms of the stationary coordinate system and then transformed to the moving frame. Because the central disk is assumed to be axisymmetric, we have

$$m \begin{Bmatrix} \ddot{u} \\ \ddot{v} \end{Bmatrix} = \begin{Bmatrix} f_x \\ f_y \end{Bmatrix} = \mathbf{f} \qquad (7.14)$$

where f_x and f_y are the forces acting on the disk.

The accelerations are now converted from the fixed coordinate system to one that rotates with the shaft. Using Equation (7.10)

$$m\mathbf{T} \left(\begin{Bmatrix} \ddot{\tilde{u}} \\ \ddot{\tilde{v}} \end{Bmatrix} - \Omega^2 \begin{Bmatrix} \tilde{u} \\ \tilde{v} \end{Bmatrix} \right) + 2m\dot{\mathbf{T}} \begin{Bmatrix} \dot{\tilde{u}} \\ \dot{\tilde{v}} \end{Bmatrix} = \begin{Bmatrix} f_x \\ f_y \end{Bmatrix} \qquad (7.15)$$

Transforming the force in the fixed coordinates to an equivalent force in the rotating coordinates requires the force equivalent of Equation (7.4)

$$\begin{Bmatrix} f_{\tilde{x}} \\ f_{\tilde{y}} \end{Bmatrix} = \mathbf{T}^\top \begin{Bmatrix} f_x \\ f_y \end{Bmatrix} \qquad (7.16)$$

Combining Equations (7.15) and (7.16) and using Equation (7.3) gives

$$m\mathbf{T}^\top\mathbf{T} \left(\begin{Bmatrix} \ddot{\tilde{u}} \\ \ddot{\tilde{v}} \end{Bmatrix} - \Omega^2 \begin{Bmatrix} \tilde{u} \\ \tilde{v} \end{Bmatrix} \right) + 2m\mathbf{T}^\top\dot{\mathbf{T}} \begin{Bmatrix} \dot{\tilde{u}} \\ \dot{\tilde{v}} \end{Bmatrix} = \begin{Bmatrix} f_{\tilde{x}} \\ f_{\tilde{y}} \end{Bmatrix}$$

and, thus,

$$m \begin{Bmatrix} \ddot{\tilde{u}} \\ \ddot{\tilde{v}} \end{Bmatrix} - 2m\Omega\mathbf{J} \begin{Bmatrix} \dot{\tilde{u}} \\ \dot{\tilde{v}} \end{Bmatrix} - m\Omega^2 \begin{Bmatrix} \tilde{u} \\ \tilde{v} \end{Bmatrix} = \begin{Bmatrix} f_{\tilde{x}} \\ f_{\tilde{y}} \end{Bmatrix} \qquad (7.17)$$

where $\mathbf{J}$ is defined in Equation (7.8).

Applying Newton's second law for free vibrations, where only elastic forces are acting on the disk, we have

$$m \left\{ \begin{matrix} \ddot{\tilde{u}} \\ \ddot{\tilde{v}} \end{matrix} \right\} - 2m\Omega \mathbf{J} \left\{ \begin{matrix} \dot{\tilde{u}} \\ \dot{\tilde{v}} \end{matrix} \right\} - m\Omega^2 \left\{ \begin{matrix} \tilde{u} \\ \tilde{v} \end{matrix} \right\} + \begin{bmatrix} k_{\tilde{x}} & 0 \\ 0 & k_{\tilde{y}} \end{bmatrix} \left\{ \begin{matrix} \tilde{u} \\ \tilde{v} \end{matrix} \right\} = \left\{ \begin{matrix} 0 \\ 0 \end{matrix} \right\} \tag{7.18}$$

This is the equation of motion for the system in terms of the rotating coordinates. The second term is the *Coriolis acceleration* and the third term is called *spin softening*, or *centripetal softening*. These effects arise because we are expressing the equations of motion in rotating coordinates.

Dividing Equation (7.18) by m, we have

$$\left\{ \begin{matrix} \ddot{\tilde{u}} \\ \ddot{\tilde{v}} \end{matrix} \right\} + 2\Omega \begin{bmatrix} 0 & -1 \\ 1 & 0 \end{bmatrix} \left\{ \begin{matrix} \dot{\tilde{u}} \\ \dot{\tilde{v}} \end{matrix} \right\} + \begin{bmatrix} \omega_{\tilde{x}}^2 - \Omega^2 & 0 \\ 0 & \omega_{\tilde{y}}^2 - \Omega^2 \end{bmatrix} \left\{ \begin{matrix} \tilde{u} \\ \tilde{v} \end{matrix} \right\} = \left\{ \begin{matrix} 0 \\ 0 \end{matrix} \right\} \tag{7.19}$$

where $\omega_{\tilde{x}}^2 = k_{\tilde{x}}/m$ and $\omega_{\tilde{y}}^2 = k_{\tilde{y}}/m$ are the natural frequencies of the stationary rotor in the two directions.

For the free response, we look for solutions of the form $\tilde{u}(t) = \tilde{u}_0 e^{st}$ and $\tilde{v}(t) = \tilde{v}_0 e^{st}$. Thus, Equation (7.19) becomes

$$\begin{bmatrix} s^2 + \omega_{\tilde{x}}^2 - \Omega^2 & -2\Omega s \\ 2\Omega s & s^2 + \omega_{\tilde{y}}^2 - \Omega^2 \end{bmatrix} \left\{ \begin{matrix} \tilde{u}_0 \\ \tilde{v}_0 \end{matrix} \right\} = \left\{ \begin{matrix} 0 \\ 0 \end{matrix} \right\} \tag{7.20}$$

The eigenvalues of the system in rotating coordinates are obtained from the solution of

$$\det \begin{bmatrix} s^2 + \omega_{\tilde{x}}^2 - \Omega^2 & -2\Omega s \\ 2\Omega s & s^2 + \omega_{\tilde{y}}^2 - \Omega^2 \end{bmatrix} = 0 \tag{7.21}$$

which produces the following quartic equation in s:

$$s^4 + \left(\omega_{\tilde{x}}^2 + \omega_{\tilde{y}}^2 + 2\Omega^2 \right) s^2 + \left(\omega_{\tilde{x}}^2 - \Omega^2 \right) \left(\omega_{\tilde{y}}^2 - \Omega^2 \right) = 0 \tag{7.22}$$

with roots

$$s_{1,2}^2 = -\frac{1}{2} \left(\omega_{\tilde{x}}^2 + \omega_{\tilde{y}}^2 + 2\Omega^2 \right)$$
$$\pm \frac{1}{2} \sqrt{ \left(\omega_{\tilde{x}}^2 + \omega_{\tilde{y}}^2 + 2\Omega^2 \right)^2 - 4 \left(\omega_{\tilde{x}}^2 - \Omega^2 \right) \left(\omega_{\tilde{y}}^2 - \Omega^2 \right) } \tag{7.23}$$

The root s_1^2 actually corresponds to two eigenvalues denoted by s_1 and s_3, where $s_1 = -s_3$ and, thus, $s_1^2 = s_3^2$. Hence, these two eigenvalues correspond to the same root in Equation (7.23). Similarly, s_2^2 corresponds to the two eigenvalues, s_2 and s_4. Simplifying Equation (7.23) produces

$$s_{1,2}^2 = \frac{1}{2} \left\{ - \left(\omega_{\tilde{x}}^2 + \omega_{\tilde{y}}^2 + 2\Omega^2 \right) \pm \sqrt{ \left(\omega_{\tilde{x}}^2 - \omega_{\tilde{y}}^2 \right)^2 + 8\Omega^2 \left(\omega_{\tilde{x}}^2 + \omega_{\tilde{y}}^2 \right) } \right\} \tag{7.24}$$

Both s_1^2 and s_2^2 are real because it is clear in Equation (7.24) that the quantity under the square root is always positive and at least one root is negative. Thus, the eigenvalues are either real or purely imaginary.

7.3.1 Relating Frequencies in the Stationary and Rotating Coordinate Systems

Solving the eigenvalue problem corresponding to Equation (7.18) gives the eigenvalues and, hence, natural frequencies in the frame of reference fixed to the rotor. A stationary observer – that is, one observing the motion relative to a stationary frame – sees a slightly different motion and, consequently, different resonance frequencies. Consider a purely imaginary complex conjugate pair of eigenvalues, $s = \pm j\sigma$, where σ is a natural frequency in the rotating frame of reference. Considering only this natural frequency, the system response is of the form

$$\begin{Bmatrix} \tilde{u} \\ \tilde{v} \end{Bmatrix} = \begin{Bmatrix} \tilde{u}_0 \cos(\sigma t + \phi_{\tilde{u}}) \\ \tilde{v}_0 \cos(\sigma t + \phi_{\tilde{v}}) \end{Bmatrix} \tag{7.25}$$

where the mode shape is rewritten in terms of magnitudes, $\tilde{u}_0$ and $\tilde{v}_0$, and phase angles, $\phi_{\tilde{u}}$ and $\phi_{\tilde{v}}$. The response in the fixed frame of reference is then

$$\begin{aligned} \begin{Bmatrix} u \\ v \end{Bmatrix} &= \begin{bmatrix} \cos\Omega t & -\sin\Omega t \\ \sin\Omega t & \cos\Omega t \end{bmatrix} \begin{Bmatrix} \tilde{u} \\ \tilde{v} \end{Bmatrix} \\ &= \begin{Bmatrix} \tilde{u}_0 \cos\Omega t \cos(\sigma t + \phi_{\tilde{u}}) - \tilde{v}_0 \sin\Omega t \cos(\sigma t + \phi_{\tilde{v}}) \\ \tilde{u}_0 \sin\Omega t \cos(\sigma t + \phi_{\tilde{u}}) + \tilde{v}_0 \cos\Omega t \cos(\sigma t + \phi_{\tilde{v}}) \end{Bmatrix} \end{aligned} \tag{7.26}$$

Using the sum and difference formulae for sine and cosine shows that the frequencies $(\Omega \pm \sigma)$ occur in the solution. For example, u becomes

$$u = \frac{1}{2}\tilde{u}_0 \left\{ \cos((\Omega + \sigma)t + \phi_{\tilde{u}}) + \cos((\Omega - \sigma)t - \phi_{\tilde{u}}) \right\}$$

$$- \frac{1}{2}\tilde{v}_0 \left\{ \sin((\Omega + \sigma)t + \phi_{\tilde{v}}) + \sin((\Omega - \sigma)t - \phi_{\tilde{v}}) \right\} \tag{7.27}$$

This shows that the natural frequencies σ in the rotating frame are transformed to *pseudonatural frequencies* $|\Omega \pm \sigma|$ in the stationary frame. The term *pseudonatural frequencies* is used because they are not natural frequencies in the usual sense. The eigenvalues of undamped systems with real coefficient matrices occur as complex conjugate pairs allowing for a free response at a single frequency. Here, the free response at a single frequency may occur only in rotating coordinates, leading to a response at a pair of pseudonatural frequencies in the stationary frame. If we solve the problem in the stationary frame, the equations of motion have time-varying coefficients; for an example, see Equation (7.55). Such equations cannot be solved as a standard eigenvalue problem and therefore do not have natural frequencies in the normal way. Thus, any eigenvalue, λ, in the rotating frame – whether complex or real – is transformed to a complex root in the stationary frame, $\lambda + j\Omega$. The response to forcing in the stationary frame is considered further in Section 7.3.6.

An interesting question arises in the analysis of a symmetric rotor on isotropic supports – that is, conveniently analyzed in either the stationary or rotating coordinate systems using standard eigenvalue problems. The previous analysis highlights that each natural frequency in the rotating frame produces two pseudonatural frequencies in the stationary frame. In the case of a symmetric rotor on isotropic supports, natural frequencies exist in both frames of reference and their number should

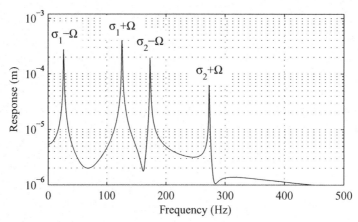

Figure 7.2. The spectrum of the response at the bearings in the stationary coordinates for the asymmetric rotor (Example 7.3.1).

be equal. The explanation to this apparent paradox is related to the direction of whirl in such cases. Section 6.2.2 clearly shows that a symmetric rotor on isotropic supports has circular orbits and that unbalance excites only the forward-whirling modes. A similar situation occurs here, in which the transient response in the stationary frame occurs only at a selection of the pseudonatural frequencies that correspond to actual natural frequencies in the stationary frame. Which natural frequencies appear is determined by the direction of the modes in the rotating frame. For forward-whirling modes in the rotating coordinate system, the rotor spin speed is added to the natural frequency; whereas, for backward-whirling modes, the rotor spin speed is subtracted from the natural frequency.

EXAMPLE 7.3.1. Consider the simple machine modeled with two degrees of freedom by Equation (7.18). The disk mass is $m = 1\,\text{kg}$ and the stiffnesses of the shaft in two orthogonal directions are 0.4 and 1.6 MN/m. The rotor spin speed is 3,000 rev/min. Calculate the natural frequencies of the system in the rotating frame. Assuming that the disk initially is at rest with a displacement of 1 mm in the most flexible direction, determine the transient response in the rotating frame. Transform this response to the stationary frame and calculate the spectrum. Repeat this for a symmetric rotor with a stiffness of 1.0 MN/m.

Solution. The eigenvalue problem corresponding to Equation (7.18) for the asymmetric rotor in the rotating coordinate system gives natural frequencies of $\sigma_1 = 76.31\,\text{Hz}$ and $\sigma_2 = 223.24\,\text{Hz}$. The pseudonatural frequencies in the stationary frame are 26.31, 126.31, 173.24, and 273.24 Hz. Figure 7.2 shows the spectrum of 1 s of the transient response in the stationary frame of reference and highlights that the response occurs at the pseudonatural frequencies. For the symmetric rotor, the natural frequencies in the rotating coordinate system are $\sigma_1 = 109.15\,\text{Hz}$ and $\sigma_2 = 209.15\,\text{Hz}$, corresponding to forward- and backward-whirling modes, respectively. The pseudonatural frequencies are 59.15, 159.15 (twice), and 259.15 Hz; however, because the rotor is symmetric, the only frequency occurring in the transient response is 159.15 Hz, which corresponds to the repeated natural frequency in the stationary frame. This is demonstrated

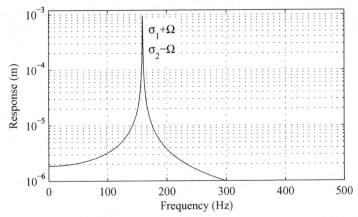

Figure 7.3. The spectrum of the response at the bearings in the stationary coordinates for the symmetric rotor (Example 7.3.1).

by the spectrum of the transient response in the stationary frame of reference shown in Figure 7.3.

EXAMPLE 7.3.2. The transition between the symmetric and asymmetric rotors provides valuable insight to the dynamics of a machine. The shaft asymmetry is defined as the ratio of the shaft stiffnesses, $k_{\bar{y}}/k_{\bar{x}}$, and this ratio is 1 for a symmetric rotor and 4 for the asymmetric rotor for the system described in Example 7.3.1. For the machine in Example 7.3.1, determine the pseudonatural frequencies and the peak responses as the shaft-stiffness asymmetry varies between 1 and 4. To obtain finite response peaks, add damping in the stationary frame, which has a damping ratio of 1 percent based on the symmetric shaft stiffness.

Solution. Figure 7.4 shows the variation of the pseudonatural frequencies with the shaft-stiffness asymmetry, clearly indicating four pseudonatural frequencies even for the symmetric shaft. Figure 7.5 clearly shows the variation in the peak responses with the shaft-stiffness asymmetry. There is no peak at the smallest and largest pseudonatural frequencies for the symmetric shaft because the response peak has reduced to zero. The simulation time is increased to 8 s to reduce the frequency spacing. Hence, the peak responses cannot be compared directly to those in Figures 7.2 and 7.3. However, the responses at the smallest and largest pseudonatural frequencies clearly reduce as the shaft asymmetry reduces.

7.3.2 Stability of Asymmetric Rotors

Stability is a major consideration in the operation of machines with asymmetric rotors. Equation (7.24) gives the eigenvalues for a simple asymmetric rotor. Because the constant term in Equation (7.22) is $s_1^2 s_2^2$, it is readily apparent that if

$$\left(\omega_{\bar{x}}^2 - \Omega^2\right)\left(\omega_{\bar{y}}^2 - \Omega^2\right) < 0 \tag{7.28}$$

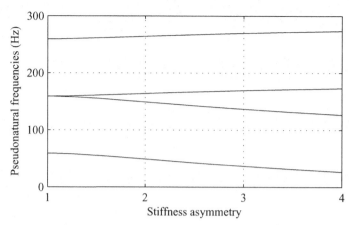

Figure 7.4. The variation of the pseudonatural frequencies with the shaft-stiffness asymmetry (Example 7.3.1).

then one root (e.g., s_2^2) is negative and the other root (e.g., s_1^2) is positive. If Equation (7.28) holds, then s_1^2 can be positive, which makes the eigenvalues, s_1 and s_3, real. A positive real root implies an unstable system. However, if Equation (7.28) does not hold, then both roots are negative and the eigenvalues are purely imaginary, which indicates a stable solution.

Assume that $\omega_{\bar{x}}^2 < \omega_{\bar{y}}^2$. If $\Omega^2 < \omega_{\bar{x}}^2 < \omega_{\bar{y}}^2$, then both bracketed terms of Equation (7.28) are positive and their product is positive. Thus, the system is stable. If $\omega_{\bar{x}}^2 < \Omega^2 < \omega_{\bar{y}}^2$, then one bracketed term in Equation (7.28) is positive, the other is negative, and their product is negative. Thus, the system is unstable. Finally, if $\omega_{\bar{x}}^2 < \omega_{\bar{y}}^2 < \Omega^2$, then both bracketed terms in Equation (7.28) are negative and, again, the system is stable.

Summarizing, we expect the system to be unstable when the running speed is between $\omega_{\bar{x}}$ and $\omega_{\bar{y}}$. In this case, two eigenvalues are purely imaginary and two eigenvalues are purely real, with one positive and one negative. When the running speed is less than $\omega_{\bar{x}}$ or greater than $\omega_{\bar{y}}$, the system is stable and all four eigenvalues are purely imaginary. In practice, damping may help to stabilize the system, which is

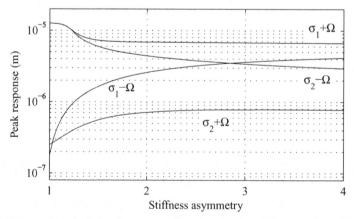

Figure 7.5. The variation of the response at the pseudonatural frequencies with the shaft-stiffness asymmetry (Example 7.3.1).

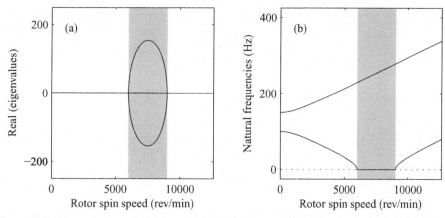

Figure 7.6. The real parts of the eigenvalues and the natural frequencies in the rotating frame for an asymmetric rotor without damping (Example 7.3.3). The shading denotes unstable speeds.

considered in Section 7.3.3. Also, if the unstable region is passed through sufficiently quickly, large amplitudes of vibration may not have time to occur.

EXAMPLE 7.3.3. Investigate the stability of a machine consisting of an asymmetric rotor, on rigid bearings, and carrying a central disk, where $\omega_{\bar{x}} = 100$ Hz and $\omega_{\bar{y}} = 150$ Hz (equivalent to 6,000 and 9,000 rev/min, respectively). The expressions for the eigenvalues, Equation (7.24), involve only the natural frequencies in the two directions when the rotor is stationary; hence, this is all the information required.

Solution. Figure 7.6 shows the real parts of the eigenvalues and the natural frequencies in the rotating frame, obtained from Equation (7.24), as the rotational speed varies. The positive real parts of the eigenvalues indicate unstable solutions, which are shown by the shaded region on the plots. Figure 7.7 shows the real parts of the eigenvalues and the pseudonatural frequencies in the stationary frame. The dashed and dash–dot lines represent the frequencies at once and twice the rotor spin speed, respectively. There is a rotor spin speed at approximately 3,500 rev/min, at which the resonance frequency is zero. This is called the *gravity critical* and is discussed further in Section 7.3.5.

7.3.3 The Effect of External Damping on the Asymmetric Rotor

Damping is now added to the stationary part of the model discussed in Section 7.3 and as shown in Figure 7.8. To add external damping to this simple model requires a somewhat contrived system consisting of a bearing of negligible mass and clearance around the disk and connected to ground through damping elements in the Ox and Oy directions. Assuming that the damping is isotropic, the damping forces acting on the disk may be obtained in the fixed frame as

$$\begin{Bmatrix} f_x \\ f_y \end{Bmatrix}_{\text{damping}} = -c \begin{Bmatrix} \dot{u} \\ \dot{v} \end{Bmatrix} \tag{7.29}$$

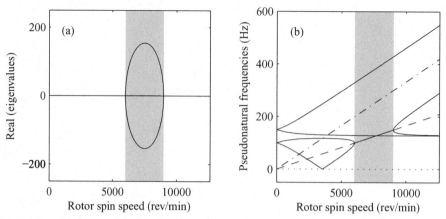

Figure 7.7. The real parts of the eigenvalues and the pseudonatural frequencies in the stationary frame for an asymmetric rotor without damping (Example 7.3.3). The shading denotes unstable speeds. On the right plot, the dashed line represents 1X and the dash–dot line represents 2X.

Transforming to the rotating frame, using Equations (7.6) and (7.16), gives

$$
\begin{Bmatrix} f_{\tilde{x}} \\ f_{\tilde{y}} \end{Bmatrix}_{\text{damping}} = -c\mathbf{T}^{\top} \begin{Bmatrix} \dot{u} \\ \dot{v} \end{Bmatrix} = -c\mathbf{T}^{\top}\left(\mathbf{T} \begin{Bmatrix} \dot{\tilde{u}} \\ \dot{\tilde{v}} \end{Bmatrix} + \dot{\mathbf{T}} \begin{Bmatrix} \tilde{u} \\ \tilde{v} \end{Bmatrix} \right)
$$
$$
= -c \begin{Bmatrix} \dot{\tilde{u}} \\ \dot{\tilde{v}} \end{Bmatrix} + c\Omega\mathbf{J} \begin{Bmatrix} \tilde{u} \\ \tilde{v} \end{Bmatrix}
\tag{7.30}
$$

This expression for the damping force may be added to the elastic force in Equation (7.18) to give

$$
m \begin{Bmatrix} \ddot{\tilde{u}} \\ \ddot{\tilde{v}} \end{Bmatrix} + \begin{bmatrix} c & -2\Omega m \\ 2\Omega m & c \end{bmatrix} \begin{Bmatrix} \dot{\tilde{u}} \\ \dot{\tilde{v}} \end{Bmatrix}
$$
$$
+ \begin{bmatrix} k_{\tilde{x}} - m\Omega^2 & -c\Omega \\ c\Omega & k_{\tilde{y}} - m\Omega^2 \end{bmatrix} \begin{Bmatrix} \tilde{u} \\ \tilde{v} \end{Bmatrix} = \begin{Bmatrix} 0 \\ 0 \end{Bmatrix} \tag{7.31}
$$

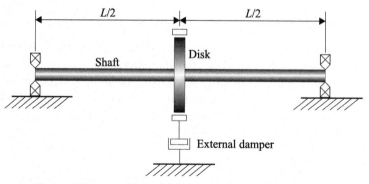

Figure 7.8. A machine with internal and external damping.

Equation (7.31) may be written in terms of the natural frequencies by dividing by m, and the eigenvalues may be obtained from the solution (compare with Equation (7.21)) of

$$\det \begin{bmatrix} s^2 + c_0 s + \omega_{\bar{x}}^2 - \Omega^2 & -2\Omega s - c_0 \Omega \\ 2\Omega s + c_0 \Omega & s^2 + c_0 s + \omega_{\bar{y}}^2 - \Omega^2 \end{bmatrix} = 0 \qquad (7.32)$$

where $c_0 = c/m$. Calculating the determinant in Equation (7.32) gives the following quartic equation in s:

$$s^4 + 2c_0 s^3 + \left(\omega_{\bar{x}}^2 + \omega_{\bar{y}}^2 + 2\Omega^2 + c_0^2\right) s^2$$
$$+ c_0 \left(\omega_{\bar{x}}^2 + \omega_{\bar{y}}^2 + 2\Omega^2\right) s + \left(\omega_{\bar{x}}^2 - \Omega^2\right)\left(\omega_{\bar{y}}^2 - \Omega^2\right) + c_0^2 \Omega^2 = 0 \qquad (7.33)$$

The eigenvalues are obtained as solutions of Equation (7.33). Alternatively, by applying the Routh-Hurwitz stability criterion (Dorf and Bishop, 2008; Tondl, 1965), it is possible to show that the rotor system is stable if the constant term in Equation (7.33) is positive or, equivalently, the machine is unstable if

$$\left(\omega_{\bar{x}}^2 - \Omega^2\right)\left(\omega_{\bar{y}}^2 - \Omega^2\right) + c_0^2 \Omega^2 < 0 \qquad (7.34)$$

Clearly, if the damping is zero, this condition reduces to Equation (7.28). By writing Equation (7.34) as a quadratic in Ω^2, we can deduce that there are no real solutions to Equation (7.34), and hence the machine is stable at all speeds, if

$$c_0 > |\omega_{\bar{x}} - \omega_{\bar{y}}| \qquad (7.35)$$

In a manner similar to the undamped case, the eigenvalues in the rotating frame, $\xi \pm j\sigma$, are transformed to complex roots in the stationary frame, $\xi + j(\Omega \pm \sigma)$, and the pseudonatural frequencies in the stationary frame are abs$(\Omega \pm \sigma)$.

EXAMPLE 7.3.4. Add damping to Example 7.3.3, when $c_0 = 40\,\mathrm{Hz}$ and $c_0 = 50\,\mathrm{Hz}$, and investigate the effect on the stability.

Solution. Figure 7.9 shows the real parts of the resulting eigenvalues and the pseudonatural frequencies in the stationary frame when $c_0 = 40\,\mathrm{Hz}$. The pseudonatural frequencies are similar to those in Figure 7.7; however, the real parts of the eigenvalues in Figure 7.9 were shifted by the damping. This has the effect of reducing the speed range over which the machine is unstable. By adding more external damping, it is possible to make the unstable region disappear completely. Figure 7.10 shows the case when $c_0 = 50\,\mathrm{Hz}$, which is the stability boundary of the machine given by Equation (7.35).

7.3.4 Unbalance Response

An *unbalance force* is a force that rotates with the rotor at the spin speed. Therefore, in the rotating frame, the unbalance is represented by a static force. If the stator is isotropic, then the steady-state response in the rotating frame is a static deflection.

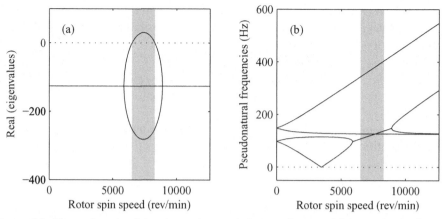

Figure 7.9. The real parts of the eigenvalues and the pseudonatural frequencies in the stationary frame for an asymmetric rotor with damping (Example 7.3.4), $c_0 = 40$ Hz. The shading denotes unstable speeds.

Adding the unbalance force to Equation (7.31) and setting $\dot{\tilde{u}} = \dot{\tilde{v}} = 0$ and $\ddot{\tilde{u}} = \ddot{\tilde{v}} = 0$, the static deflection may be obtained as

$$
\begin{aligned}
\begin{Bmatrix} \tilde{u} \\ \tilde{v} \end{Bmatrix} &= \begin{bmatrix} k_{\tilde{x}} - m\Omega^2 & -c\Omega \\ c\Omega & k_{\tilde{y}} - m\Omega^2 \end{bmatrix}^{-1} \begin{Bmatrix} f_{\tilde{u}} \\ f_{\tilde{v}} \end{Bmatrix}_{\text{unbal}} \\
&= \frac{1}{m} \begin{bmatrix} \omega_{\tilde{x}}^2 - \Omega^2 & -c_0\Omega \\ c_0\Omega & \omega_{\tilde{y}}^2 - \Omega^2 \end{bmatrix}^{-1} \begin{Bmatrix} f_{\tilde{u}} \\ f_{\tilde{v}} \end{Bmatrix}_{\text{unbal}}
\end{aligned}
\tag{7.36}
$$

where $f_{\tilde{u}}$ and $f_{\tilde{v}}$ are the constant unbalance forces for a given speed in the $\tilde{x}$ and $\tilde{y}$ directions. If the machine is stable, then the condition given in Equation (7.34) shows that the matrix in Equation (7.36) is nonsingular and the response is bounded. If the rotor is unstable for some rotor spin speeds, then the response is unbounded at those speeds. For stable systems, the determinant of the matrix in Equation (7.36) is positive and is a quadratic in Ω^2 with a single minimum. Because the supports are isotropic, the orbit is circular.

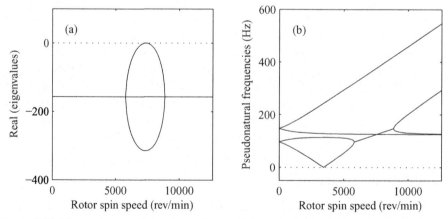

Figure 7.10. The real parts of the eigenvalues and the pseudonatural frequencies in the stationary frame for an asymmetric rotor with damping (Example 7.3.4), $c_0 = 50$ Hz.

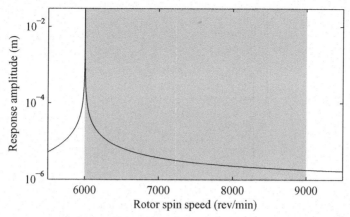

Figure 7.11. The unbalance response amplitude for the undamped asymmetric rotor (Example 7.3.5), with $\beta = 0°$. The shading denotes the unstable speed range.

EXAMPLE 7.3.5. Find the unbalance response for Examples 7.3.3 and 7.3.4 when $c_0 = 0$, 40, 50, and 60 Hz. The unbalance is generated by an offset disk with eccentricity $\varepsilon = 1\,\mu$m, with various angular positions.

Solution. Suppose that the unbalance force is

$$\left\{ \begin{matrix} f_{\tilde{u}} \\ f_{\tilde{v}} \end{matrix} \right\}_{\text{unbal}} = m\varepsilon\Omega^2 \left\{ \begin{matrix} \cos\beta \\ \sin\beta \end{matrix} \right\}$$

This implies that the unbalance mass is located at an angle β relative to the direction of the highest flexibility of the shaft. The mass is not specified, but it cancels out because the natural frequencies of the stationary rotor are given in Example 7.3.3. Figures 7.11 and 7.12 show the unbalance response for $\beta = 0°$ and $\beta = 90°$, respectively. The shaded regions denote the unstable rotor speeds. The responses given are steady-state solutions. Although steady-state solutions are obtained in the unstable speed ranges, the instability means that these solutions are not obtained in practice. When the out-of-balance is placed in the direction of minimum shaft stiffness,

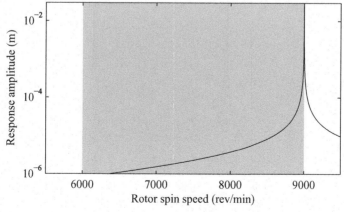

Figure 7.12. The unbalance response amplitude for the undamped asymmetric rotor (Example 7.3.5), with $\beta = 90°$. The shading denotes the unstable speed range.

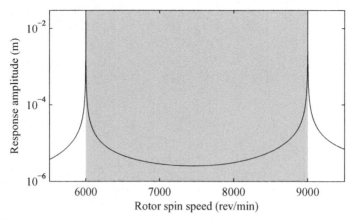

Figure 7.13. The unbalance response amplitude for the undamped asymmetric rotor (Example 7.3.5), with $\beta = 45°$. The shading denotes the unstable speed range.

the peak in the response occurs at the left side of the unstable speed range; when placed in the direction of maximum shaft stiffness, the peak occurs at the right side of the range. These are special cases and, for any other angular position of the out-of-balance mass, peaks occur at both ends of the unstable speed range, as shown in Figure 7.13 for $\beta = 45°$. In this general case, it is clear that the unbalance response increases as the rotor speed approaches the instability region.

Figures 7.14 and 7.15 show the unbalance response magnitude over the rotor speed range that includes the two natural frequencies of the rotor: when $c_0 = 40$ Hz and $c_0 = 50$ Hz, respectively. In Figure 7.14, the shaded region denotes the unstable rotor speeds, which means that the steady-state solutions are not obtained in practice. Figure 7.14 is similar to Figure 7.13 except that the unstable speed range is reduced. In Figure 7.15, two peaks have coalesced into a single peak and the rotor is stable at all speeds. However, the peak amplitude is unbounded because the real part of the eigenvalues is zero at the corresponding rotational speed. This may be compared to the forced response of an undamped oscillator. For higher levels

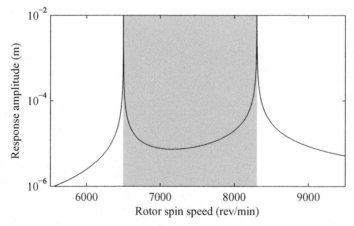

Figure 7.14. The unbalance response amplitude for the asymmetric rotor with damping (Example 7.3.5), $c_0 = 40$ Hz and $\beta = 90°$. The shading denotes the unstable speed range.

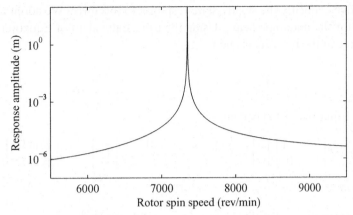

Figure 7.15. The unbalance response amplitude for the asymmetric rotor with damping (Example 7.3.5), $c_0 = 50\,\text{Hz}$ and $\beta = 90°$.

of damping, the response becomes finite at all speeds, as shown in Figure 7.16 for $c_0 = 60\,\text{Hz}$.

7.3.5 The Gravity Critical Speed

It is clear from Figures 7.7 through 7.10 for Examples 7.3.3 and 7.3.4 that there is a rotational speed (at 3,530 rev/min) at which the imaginary part of one of the eigenvalues is zero. This means that any static lateral force excites the structure and, for a horizontal rotor, the force due to gravity is such a force. This is called the *gravity critical speed* and is clearly excited in many, if not most, applications. Another interpretation is that the frequency of the backward-traveling mode of a machine coincides with spin speed. Of course, the level of excitation is related to the significance of the asymmetry in the rotating parts of a machine. The frequency of this gravity critical speed may be calculated by looking for the solutions of Equation (7.22) or (7.33) with zero frequency in the stationary frame (i.e., by setting $s = j\Omega$ in the rotating frame). The alternative is to transform the gravity force to the rotating frame

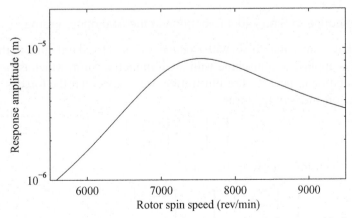

Figure 7.16. The unbalance response amplitude for the asymmetric rotor with damping (Example 7.3.5), $c_0 = 60\,\text{Hz}$ and $\beta = 90°$.

and to derive explicitly the steady-state response to this force. We adopt the second approach for the undamped case. Using the force equivalent of Equation (7.4), the gravity force in the rotating frame is

$$\left\{ \begin{matrix} f_{\tilde{x}} \\ f_{\tilde{y}} \end{matrix} \right\}_{gravity} = \mathbf{T}^\top \left\{ \begin{matrix} 0 \\ -mg \end{matrix} \right\} = -mg \left\{ \begin{matrix} \sin \Omega t \\ \cos \Omega t \end{matrix} \right\} \tag{7.37}$$

and, thus, Equation (7.19) becomes

$$\left\{ \begin{matrix} \ddot{\tilde{u}} \\ \ddot{\tilde{v}} \end{matrix} \right\} + 2\Omega \begin{bmatrix} 0 & -1 \\ 1 & 0 \end{bmatrix} \left\{ \begin{matrix} \dot{\tilde{u}} \\ \dot{\tilde{v}} \end{matrix} \right\} + \begin{bmatrix} \omega_{\tilde{x}}^2 - \Omega^2 & 0 \\ 0 & \omega_{\tilde{y}}^2 - \Omega^2 \end{bmatrix} \left\{ \begin{matrix} \tilde{u} \\ \tilde{v} \end{matrix} \right\} = -g \left\{ \begin{matrix} \sin \Omega t \\ \cos \Omega t \end{matrix} \right\} \tag{7.38}$$

The steady-state solution is of the form

$$\tilde{u} = \tilde{u}_0 \sin \Omega t, \qquad\qquad \tilde{v} = \tilde{v}_0 \cos \Omega t \tag{7.39}$$

for some real constants $\tilde{u}_0$ and $\tilde{v}_0$. Substituting Equation (7.39) into Equation (7.38) produces the equations

$$\begin{bmatrix} \omega_{\tilde{x}}^2 - 2\Omega^2 & 2\Omega^2 \\ 2\Omega^2 & \omega_{\tilde{y}}^2 - 2\Omega^2 \end{bmatrix} \left\{ \begin{matrix} \tilde{u}_0 \\ \tilde{v}_0 \end{matrix} \right\} = -g \left\{ \begin{matrix} 1 \\ 1 \end{matrix} \right\} \tag{7.40}$$

with solution

$$\left\{ \begin{matrix} \tilde{u}_0 \\ \tilde{v}_0 \end{matrix} \right\} = \frac{-g}{\omega_{\tilde{x}}^2 \omega_{\tilde{y}}^2 - 2\Omega^2 \left(\omega_{\tilde{x}}^2 + \omega_{\tilde{y}}^2 \right)} \left\{ \begin{matrix} \omega_{\tilde{y}}^2 - 4\Omega^2 \\ \omega_{\tilde{x}}^2 - 4\Omega^2 \end{matrix} \right\} \tag{7.41}$$

The rotor speeds at which the response becomes infinite are the critical speeds and are obtained by setting the denominator in Equation (7.41) to be zero. Thus,

$$\Omega^2 = \frac{\omega_{\tilde{x}}^2 \omega_{\tilde{y}}^2}{2 \left(\omega_{\tilde{x}}^2 + \omega_{\tilde{y}}^2 \right)} \tag{7.42}$$

If the asymmetry is relatively small so that $\omega_{\tilde{x}}^2$ and $\omega_{\tilde{y}}^2$ are very close, then the critical speed is approximately half of the mean natural frequency of the machine. The emergence of this critical speed may be used to monitor asymmetries in the rotor – for example, the detection of cracks (see Section 10.14).

7.3.6 Response to Sinusoidal Excitation in the Stationary Frame

Suppose that a sinusoidal force, with frequency ω, in a fixed direction in the stationary frame is applied to a machine with an asymmetric rotor, which is running at a constant rotational speed Ω. For illustration, we assume that the force is applied in the x direction and is of the form

$$\left\{ \begin{matrix} f_x \\ f_y \end{matrix} \right\} = f_0 \left\{ \begin{matrix} \cos \omega t \\ 0 \end{matrix} \right\} \tag{7.43}$$

In the rotating frame, this force becomes

$$\begin{aligned} \left\{ \begin{matrix} f_{\tilde{x}} \\ f_{\tilde{y}} \end{matrix} \right\} &= \mathbf{T}^\top \left\{ \begin{matrix} f_x \\ f_y \end{matrix} \right\} = f_0 \left\{ \begin{matrix} \cos \Omega t \cos \omega t \\ -\sin \Omega t \cos \omega t \end{matrix} \right\} \\ &= \frac{1}{2} f_0 \left\{ \begin{matrix} \cos (\omega + \Omega) t + \cos (\omega - \Omega) t \\ -\sin (\omega + \Omega) t + \sin (\omega - \Omega) t \end{matrix} \right\} \end{aligned} \tag{7.44}$$

This force may be applied to the equations of motion in the rotating frame, either Equation (7.18) or Equation (7.31) for the undamped or damped case, respectively. Without damping, the response is of the form

$$\begin{Bmatrix} \tilde{u} \\ \tilde{v} \end{Bmatrix} = \begin{Bmatrix} \tilde{u}_{0p} \cos{(\omega + \Omega)}t + \tilde{u}_{0m} \cos{(\omega - \Omega)}t \\ \tilde{v}_{0p} \sin{(\omega + \Omega)}t + \tilde{v}_{0m} \sin{(\omega - \Omega)}t \end{Bmatrix} \tag{7.45}$$

To obtain $\tilde{u}_{0p}$ and $\tilde{v}_{0p}$ in Equation (7.45), the coefficients of $\cos(\omega + \Omega)t$ and $\sin(\omega + \Omega)t$ are equated, using the equation motion derived from Equations (7.18) and (7.44). This gives

$$\begin{bmatrix} \omega_{\tilde{x}}^2 - (\omega + \Omega)^2 - \Omega^2 & -2\Omega(\omega + \Omega) \\ -2\Omega(\omega + \Omega) & \omega_{\tilde{y}}^2 - (\omega + \Omega)^2 - \Omega^2 \end{bmatrix} \begin{Bmatrix} \tilde{u}_{0p} \\ \tilde{v}_{0p} \end{Bmatrix} = \frac{1}{2} \frac{f_0}{m} \begin{Bmatrix} 1 \\ -1 \end{Bmatrix} \tag{7.46}$$

Similarly, $\tilde{u}_{0m}$ and $\tilde{v}_{0m}$ are obtained by equating coefficients of $\cos(\omega - \Omega)t$ and $\sin(\omega - \Omega)t$ and solving

$$\begin{bmatrix} \omega_{\tilde{x}}^2 - (\omega - \Omega)^2 - \Omega^2 & -2\Omega(\omega - \Omega) \\ -2\Omega(\omega - \Omega) & \omega_{\tilde{y}}^2 - (\omega - \Omega)^2 - \Omega^2 \end{bmatrix} \begin{Bmatrix} \tilde{u}_{0m} \\ \tilde{v}_{0m} \end{Bmatrix} = \frac{1}{2} \frac{f_0}{m} \begin{Bmatrix} 1 \\ 1 \end{Bmatrix} \tag{7.47}$$

Transforming the response in the rotating frame, Equation (7.45), to the stationary frame gives

$$\begin{Bmatrix} u \\ v \end{Bmatrix} = \mathbf{T} \begin{Bmatrix} \tilde{u} \\ \tilde{v} \end{Bmatrix} \tag{7.48}$$

Using trigonometric formulae, this yields responses in the stationary frame at frequencies ω and $|\omega \pm 2\Omega|$. If the rotor is symmetric, then the responses in the stationary frame at $|\omega \pm 2\Omega|$ sum to zero. Adding damping causes a change in the amplitude and phase of the response; however, the response frequencies are the same.

EXAMPLE 7.3.6. Find the response of the machine in Example 7.3.4 to a sinusoidal excitation of frequency $\omega = 125\,\text{Hz}$ in the x direction of magnitude $100m\,\text{N}$, where m is the rotor mass. The damping coefficient is $c_0 = 40\,\text{Hz}$ and the rotational speed is $\Omega = 5,400\,\text{rev/min}$ (equivalent to 90 Hz).

Solution. Figure 7.17 shows the amplitude of the response in the x direction in the stationary frame. The response can be determined from Equations (7.45) through (7.48), although the resulting expressions are tedious to derive and provide little insight. Instead, the force and response is simulated in the time domain and transformed to the frequency domain using the DFT (see Section 2.6.3). The peaks occur at frequencies $\omega = 125\,\text{Hz}$, $|\omega + 2\Omega| = 305\,\text{Hz}$, and $|\omega - 2\Omega| = 55\,\text{Hz}$. In theory, the response should be non-zero only at those frequencies; however, leakage has caused some energy to spread into adjacent frequencies (see Section 2.6.3).

7.3.7 Response to General Excitation in the Stationary Frame

Because the system is linear, the approach adopted in Chapter 2 may be used. Initially, the Fourier transform of the excitation force is obtained, and then the response to the force at each frequency component is derived using the method described in Section 7.3.6. Each excitation frequency component produces a response

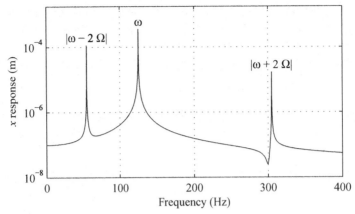

Figure 7.17. The magnitude of the response in the x direction in the stationary frame to a sinusoidal force of frequency 125 Hz; also in the x direction, for the asymmetric rotor with damping (Example 7.3.6), $c_0 = 40$ Hz.

at multiple frequencies (as described in Section 7.3.6); hence, these responses must be summed to give the total response to the general force. This makes it a difficult approach to implement; simulating the response in the time domain and transforming to the frequency domain using the DFT (see Section 2.6.3) usually is much easier.

EXAMPLE 7.3.7. Find the response of the machine in Example 7.3.3 to an impulse in the x direction when the rotor mass is $m = 1$ kg and the rotational speed is $\Omega = 5,400$ rev/min (equivalent to 90 Hz). Neglect damping and model the impulse as an initial non-zero velocity in the x direction of magnitude 1 mm/s.

Solution. An impulse has a component at every frequency and therefore excites the natural frequencies of the system. The response is simulated in the time domain and transformed to the frequency domain. Figure 7.18 shows the response in the stationary frame in the x direction. The peaks in the response occur at frequencies of 66.16, 113.8, 129.4, and 309.4 Hz. The solution of Equation (7.32) gives the natural frequencies in the rotating frame as $\omega_1 = 23.84$ Hz and $\omega_2 = 219.4$ Hz; hence, the peaks in the response in the stationary frame occur at $|\omega_i \pm \Omega|$.

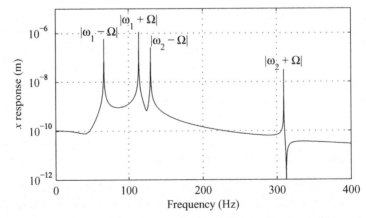

Figure 7.18. The magnitude of the response in the x direction in the stationary frame to an impulse in the same direction for the asymmetric rotor without damping (Example 7.3.7).

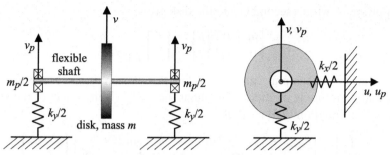

Figure 7.19. A simple machine with four degrees of freedom.

7.4 Asymmetric Rotors Supported by Anisotropic Bearing: Simple Rotors

When both the rotor and the bearings and support structure contain asymmetries, then the resulting equations contain time-dependent parameters, regardless of which frame of reference is employed. The parameters vary sinusoidally and produce an effect referred to as *parametric excitation* (Cartmell, 1990). This occurs in both the stationary and the rotating frames, and transforming from one to the other does not help. A full analysis of such equations is beyond the scope of this book; however, the equations of motion are generated for the simple example shown in Figure 7.19, and some stability results are given. The rotor has a central disk and the shaft has asymmetric stiffness properties but negligible mass. The foundation is represented by springs with unequal stiffnesses, k_x and k_y, in the horizontal and vertical directions. The bearings are assumed to be rigid with mass, m_p. Assuming that the only external excitation is an unbalance force acting on the disk, then the response at both bearings is identical. Thus, four degrees of freedom are required to describe the system: the lateral response of the disk in two directions, denoted u and v; the lateral response at the bearings, again in two directions, denoted u_p and v_p.

The equations of motion are written in the stationary frame and therefore the elastic force due to the asymmetric shaft stiffness must be transformed to the stationary frame. Hence,

$$\begin{Bmatrix} f_x \\ f_y \end{Bmatrix}_{\text{shaft}} = \mathbf{T} \begin{Bmatrix} f_{\tilde{x}} \\ f_{\tilde{y}} \end{Bmatrix}_{\text{shaft}} = -\mathbf{T} \begin{bmatrix} k_{\tilde{x}} & 0 \\ 0 & k_{\tilde{y}} \end{bmatrix} \begin{Bmatrix} \tilde{u} - \tilde{u}_p \\ \tilde{v} - \tilde{v}_p \end{Bmatrix}$$
$$= -\mathbf{T} \begin{bmatrix} k_{\tilde{x}} & 0 \\ 0 & k_{\tilde{y}} \end{bmatrix} \mathbf{T}^\top \begin{Bmatrix} u - u_p \\ v - v_p \end{Bmatrix} \tag{7.49}$$

where $k_{\tilde{x}}$ and $k_{\tilde{y}}$ represent the shaft stiffness in the principal axes. The force is based on the relative displacement between the disk and the bearings. Equation (7.49) may be simplified to give

$$\begin{Bmatrix} f_x \\ f_y \end{Bmatrix}_{\text{shaft}} = -\begin{bmatrix} \tilde{k}_{\text{mean}} + \tilde{k}_{\text{dev}} \cos 2\Omega t & \tilde{k}_{\text{dev}} \sin 2\Omega t \\ \tilde{k}_{\text{dev}} \sin 2\Omega t & \tilde{k}_{\text{mean}} - \tilde{k}_{\text{dev}} \cos 2\Omega t \end{bmatrix} \begin{Bmatrix} u - u_p \\ v - v_p \end{Bmatrix} \tag{7.50}$$

where the mean stiffness is $\tilde{k}_{\text{mean}} = (k_{\tilde{x}} + k_{\tilde{y}})/2$ and the deviatoric stiffness is $\tilde{k}_{\text{dev}} = (k_{\tilde{x}} - k_{\tilde{y}})/2$. If the shaft is symmetric, then $\tilde{k}_{\text{dev}} = 0$ and the parametric terms in Equation (7.50) are zero.

Applying Newton's second law to the disk gives

$$\begin{bmatrix} m & 0 \\ 0 & m \end{bmatrix} \begin{Bmatrix} \ddot{u} \\ \ddot{v} \end{Bmatrix} = \begin{Bmatrix} f_x \\ f_y \end{Bmatrix}_{\text{shaft}} \tag{7.51}$$

Although shaft-damping can be included, this often reduces the stability of the system and therefore is omitted to highlight the destabilizing effect of the stiffness asymmetries. Equation (7.51) is thus

$$\begin{bmatrix} m & 0 \\ 0 & m \end{bmatrix} \begin{Bmatrix} \ddot{u} \\ \ddot{v} \end{Bmatrix} + \begin{bmatrix} \tilde{k}_{\text{mean}} + \tilde{k}_{\text{dev}} \cos 2\Omega t & \tilde{k}_{\text{dev}} \sin 2\Omega t \\ \tilde{k}_{\text{dev}} \sin 2\Omega t & \tilde{k}_{\text{mean}} - \tilde{k}_{\text{dev}} \cos 2\Omega t \end{bmatrix} \begin{Bmatrix} u - u_p \\ v - v_p \end{Bmatrix} = \begin{Bmatrix} 0 \\ 0 \end{Bmatrix} \tag{7.52}$$

Applying Newton's second law to the pedestal mass, m_p, and including pedestal-damping with coefficient c_p gives

$$\begin{bmatrix} m_p & 0 \\ 0 & m_p \end{bmatrix} \begin{Bmatrix} \ddot{u}_p \\ \ddot{v}_p \end{Bmatrix} = -\begin{bmatrix} c_p & 0 \\ 0 & c_p \end{bmatrix} \begin{Bmatrix} \dot{u}_p \\ \dot{v}_p \end{Bmatrix} - \begin{bmatrix} k_x & 0 \\ 0 & k_y \end{bmatrix} \begin{Bmatrix} u_p \\ v_p \end{Bmatrix} - \begin{Bmatrix} f_x \\ f_y \end{Bmatrix}_{\text{shaft}} \tag{7.53}$$

or

$$\begin{bmatrix} m_p & 0 \\ 0 & m_p \end{bmatrix} \begin{Bmatrix} \ddot{u}_p \\ \ddot{v}_p \end{Bmatrix} + \begin{bmatrix} c_p & 0 \\ 0 & c_p \end{bmatrix} \begin{Bmatrix} \dot{u}_p \\ \dot{v}_p \end{Bmatrix} + \begin{bmatrix} k_x & 0 \\ 0 & k_y \end{bmatrix} \begin{Bmatrix} u_p \\ v_p \end{Bmatrix}$$
$$- \begin{bmatrix} \tilde{k}_{\text{mean}} + \tilde{k}_{\text{dev}} \cos 2\Omega t & \tilde{k}_{\text{dev}} \sin 2\Omega t \\ \tilde{k}_{\text{dev}} \sin 2\Omega t & \tilde{k}_{\text{mean}} - \tilde{k}_{\text{dev}} \cos 2\Omega t \end{bmatrix} \begin{Bmatrix} u - u_p \\ v - v_p \end{Bmatrix} = \begin{Bmatrix} 0 \\ 0 \end{Bmatrix} \tag{7.54}$$

It is clear that Equations (7.52) and (7.54) represent the equations of motion of a four degrees of freedom system in which the stiffness coefficient varies sinusoidally. Assembling these equations gives

$$\mathbf{M}\ddot{\mathbf{q}} + \mathbf{C}\dot{\mathbf{q}} + [\mathbf{K}_0 + \mathbf{K}_c \cos 2\Omega t + \mathbf{K}_s \sin 2\Omega t]\,\mathbf{q} = \mathbf{0} \tag{7.55}$$

where $\mathbf{q} = \begin{bmatrix} u & v & u_p & v_p \end{bmatrix}^\top$, $\mathbf{M} = \text{diag}\,[m\ m\ m_p\ m_p]$, $\mathbf{C} = \text{diag}\,[0\ 0\ c_p\ c_p]$,

$$\mathbf{K}_0 = \tilde{k}_{\text{mean}} \begin{bmatrix} 1 & 0 & -1 & 0 \\ 0 & 1 & 0 & -1 \\ -1 & 0 & 1 & 0 \\ 0 & -1 & 0 & 1 \end{bmatrix} + \begin{bmatrix} 0 & 0 & 0 & 0 \\ 0 & 0 & 0 & 0 \\ 0 & 0 & k_x & 0 \\ 0 & 0 & 0 & k_y \end{bmatrix}$$

$$\mathbf{K}_c = \tilde{k}_{\text{dev}} \begin{bmatrix} 1 & 0 & -1 & 0 \\ 0 & -1 & 0 & 1 \\ -1 & 0 & 1 & 0 \\ 0 & 1 & 0 & -1 \end{bmatrix}, \qquad \mathbf{K}_s = \tilde{k}_{\text{dev}} \begin{bmatrix} 0 & 1 & 0 & -1 \\ 1 & 0 & -1 & 0 \\ 0 & -1 & 0 & 1 \\ -1 & 0 & 1 & 0 \end{bmatrix}$$

For machines excited by unbalance, the steady-state response is periodic and the stability of this response then must be investigated. Floquet theory or various approximate techniques (Jordan and Smith, 1977; Tondl, 1965; Tondl et al., 2000) enable the stability to be assessed by looking at perturbations from the steady-state solution. If any of these perturbations can grow during one time period, then the solution is unstable. Because Equation (7.55) is a linear differential equation with time-dependent coefficients, the stability may be assessed without computing the steady-state solution. This equation is transformed to state–space with state vector

$\mathbf{x} = \begin{Bmatrix} \mathbf{q} \\ \dot{\mathbf{q}} \end{Bmatrix}$, of length $2n$, where in this case $n = 4$, to give a differential equation of the form

$$\dot{\mathbf{x}} = \mathbf{P}(t)\mathbf{x} \qquad (7.56)$$

where $\mathbf{P}(t + T) = \mathbf{P}(t)$ and T is the time period of the parametric excitation. The *fundamental matrix* is denoted by $\boldsymbol{\Phi}(t)$ and consists of columns that are $2n$ independent solutions of the differential equation. Jordan and Smith (1977) showed that

$$\boldsymbol{\Phi}(t + T) = \boldsymbol{\Phi}(t)\mathbf{E} \qquad (7.57)$$

for some $2n \times 2n$ constant matrix $\mathbf{E}$, called the *Floquet transition matrix* or *monodromy matrix*. If any of the eigenvalues of $\mathbf{E}$ have magnitude greater than 1, then at least one solution to the differential equation grows and the system is unstable. The most convenient method for constructing $\mathbf{E}$ is to integrate the differential equations between time $t = t_0$ and time $t = t_0 + T$, for $2n$ linearly independent initial conditions. If these initial conditions are chosen such that $\boldsymbol{\Phi}(t_0) = \mathbf{I}$, then $\boldsymbol{\Phi}(t_0 + T) = \mathbf{E}$. Thus, the initial condition $\mathbf{x}(t_0) = \mathbf{e}_i, i = 1, \ldots, 2n$, where $\mathbf{e}_i$ is the i th column of the identity matrix, immediately gives the i th column of $\mathbf{E}$ as $\mathbf{x}(t_0 + T)$. The integration may be performed numerically (see Section 2.6.1 for a discussion of suitable methods in more detail). Floquet theory provides the stability of the system for a given set of parameters.

A simple example is now given to show the results that may be obtained.

EXAMPLE 7.4.1. Investigate the stability of a machine with the following parameters:

$$m = 1\,\text{kg}, \qquad m_p = 10\,\text{kg}, \qquad c_p = 300\,\text{kg/s},$$

$$\tilde{k}_{\text{mean}} = 10\,\text{kN/m}, \quad k_x = 16\,\text{kN/m}, \quad k_y = 400\,\text{kN/m}$$

as the rotor asymmetry, $\tilde{k}_{\text{dev}}$, varies between 0 and 2 kN/m, where $\tilde{k}_{\text{mean}}$ and $\tilde{k}_{\text{dev}}$ are defined in Equation (7.50). The difference in pedestal stiffnesses in the x and y directions is very large in order to demonstrate clearly the machine stability. This example is motivated by the frequent occurrence of horizontal rotors mounted on flexible bearings or supports that have high vertical stiffness but relatively low horizontal stiffness.

Solution. The stability of this machine is assessed for various shaft speeds and increasing shaft asymmetry. For example, when $\tilde{k}_{\text{dev}} = 1\,\text{kN/m}$, the eigenvalues of $\mathbf{E}$ when the rotor spins at 900 rev/min are $0.9777 \pm 0.1651\jmath$, $0.6938 \pm 0.5637\jmath$, $0.2174 \pm 0.3013\jmath$, and $-0.2957 \pm 0.2854\jmath$, with magnitudes of $0.9915, 0.8940, 0.3715$, and 0.4109. Because the magnitudes of all eigenvalues are less than 1, the machine is stable. However, when the rotor spins at 950 rev/min, the eigenvalues of $\mathbf{E}$ are $-0.2702 \pm 0.3353\jmath$, $0.8525 \pm 0.2129\jmath$, $0.3807 \pm 0.0910\jmath$, 0.8810, and 1.1699, with magnitudes $0.4306, 0.8787, 0.3914$, 0.8810, and 1.1699. The last eigenvalue is real with a magnitude greater than 1, showing that the machine is unstable.

The results for a range of rotor spin speeds and rotor asymmetry are summarized in Figure 7.20. The rotor asymmetry is normalized and defined as $\dfrac{\tilde{k}_{\text{dev}}}{\tilde{k}_{\text{mean}}}$. When

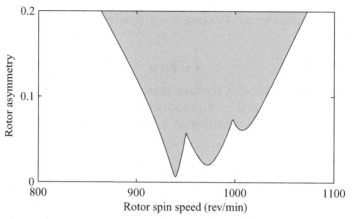

Figure 7.20. The stability of the simple example with combined rotor and stator asymmetry. The shaded area represents unstable motion.

the shaft is symmetric, the system is stable at all speeds because of the damping. As the rotor asymmetry increases, the stability reduces until instability occurs at approximately 940 rev/min for a rotor asymmetry of approximately 0.006. As the rotor asymmetry increases, there is a range of speeds at which the machine is unstable; indeed, other unstable regions also occur that merge for high rotor asymmetry.

7.5 Internal Rotor Damping: Simple Rotors

We now illustrate the effect of internal rotor damping for the simple system described in Section 3.9. Consider a disk mounted on an axisymmetric flexible shaft with internal damping. Unlike any damping that may exist in the rotor supports, the internal damping produces a damping force that rotates with the rotor. This damping arises from many sources – such as material damping, friction in bolted joints, and laminated rotors – and modeling these effects is difficult. To illustrate the effect of internal damping, viscous damping is assumed.

The viscous damping force in the rotating coordinate frame, $\tilde{\mathbf{f}}_{damping}$, is modeled as

$$\tilde{\mathbf{f}}_{damping} = -c_i \begin{Bmatrix} \dot{\tilde{u}} \\ \dot{\tilde{v}} \end{Bmatrix} \tag{7.58}$$

where c_i is the damping coefficient internal to the rotor and $\begin{Bmatrix} \dot{\tilde{u}} \\ \dot{\tilde{v}} \end{Bmatrix}$ are the velocities in the rotating frame. The forces in the stationary frame may be obtained using the transformation in Equation (7.2)

$$\mathbf{f}_{damping} = \mathbf{T}\tilde{\mathbf{f}}_{damping} = -c_i \mathbf{T} \begin{Bmatrix} \dot{\tilde{u}} \\ \dot{\tilde{v}} \end{Bmatrix} \tag{7.59}$$

In terms of the displacements and velocities in the stationary frame, using Equation (7.11), this force is

$$\mathbf{f}_{damping} = -c_i \mathbf{T} \left(\mathbf{T}^\top \begin{Bmatrix} \dot{u} \\ \dot{v} \end{Bmatrix} + \dot{\mathbf{T}}^\top \begin{Bmatrix} u \\ v \end{Bmatrix} \right) = -c_i \begin{Bmatrix} \dot{u} \\ \dot{v} \end{Bmatrix} - c_i \Omega \mathbf{J} \begin{Bmatrix} u \\ v \end{Bmatrix} \tag{7.60}$$

The equation of motion for free vibrations in the fixed coordinate frame is then

$$m \begin{Bmatrix} \ddot{u} \\ \ddot{v} \end{Bmatrix} + (c_e + c_i) \begin{Bmatrix} \dot{u} \\ \dot{v} \end{Bmatrix} + \begin{bmatrix} k & c_i \Omega \\ -c_i \Omega & k \end{bmatrix} \begin{Bmatrix} u \\ v \end{Bmatrix} = \mathbf{0} \tag{7.61}$$

where c_e is the damping external to the rotor – that is, in the bearing and foundation. The source of this damping is not immediately apparent if the flexible shaft is supported on rigid bearings, but the external damping is included based on the discussion in Section 7.3.3 and the system shown in Figure 7.8. Such damping may arise from air resistance or small displacements at the bearing, but this source is not of concern here.

Dividing this equation by m, we have

$$\begin{Bmatrix} \ddot{u} \\ \ddot{v} \end{Bmatrix} + 2\omega_n (\zeta_e + \zeta_i) \begin{Bmatrix} \dot{u} \\ \dot{v} \end{Bmatrix} + \begin{bmatrix} \omega_n^2 & 2\omega_n \Omega \zeta_i \\ -2\omega_n \Omega \zeta_i & \omega_n^2 \end{bmatrix} \begin{Bmatrix} u \\ v \end{Bmatrix} = \mathbf{0} \tag{7.62}$$

where $\omega_n = \sqrt{k/m}$, $\zeta_e = \dfrac{c_e}{2m\omega_n}$, and $\zeta_i = \dfrac{c_i}{2m\omega_n}$. We can determine a solution for this equation by letting $\begin{Bmatrix} u(t) \\ v(t) \end{Bmatrix} = \begin{Bmatrix} u_0 \\ v_0 \end{Bmatrix} e^{st}$. Thus, with $R = \Omega/\omega_n$, we have

$$\begin{bmatrix} s^2 + 2\omega_n (\zeta_e + \zeta_i) s + \omega_n^2 & 2R\zeta_i \omega_n^2 \\ -2R\zeta_i \omega_n^2 & s^2 + 2\omega_n (\zeta_e + \zeta_i) s + \omega_n^2 \end{bmatrix} \begin{Bmatrix} u_0 \\ v_0 \end{Bmatrix} = \mathbf{0} \tag{7.63}$$

For a nontrivial solution, the determinant of the matrix must be zero; thus,

$$\left(s^2 + 2(\zeta_e + \zeta_i)\omega_n s + \omega_n^2 \right)^2 + \left(2\zeta_i R\omega_n^2 \right)^2 = 0 \tag{7.64}$$

Moving the last term to the right side and taking the square root gives

$$s^2 + 2(\zeta_e + \zeta_i)\omega_n s + \omega_n^2 = \pm_J \left(2\zeta_i R\omega_n^2 \right) \tag{7.65}$$

where the $\pm_J$ term arises because of the square root of a negative term. Thus, roots of Equation (7.64) are given by

$$s^2 + 2(\zeta_e + \zeta_i)\omega_n s + \omega_n^2 (1 \pm 2_J \zeta_i R) = 0 \tag{7.66}$$

By solving Equation (7.66), the eigenvalues are

$$s_1, s_2, s_3, s_4 = -(\zeta_e + \zeta_i)\omega_n \pm \omega_n \sqrt{(\zeta_e + \zeta_i)^2 - (1 \pm 2_J \zeta_i R)} \tag{7.67}$$

and the response is

$$\begin{Bmatrix} u(t) \\ v(t) \end{Bmatrix} = \begin{Bmatrix} u_{01} \\ v_{01} \end{Bmatrix} e^{s_1 t} + \begin{Bmatrix} u_{02} \\ v_{02} \end{Bmatrix} e^{s_2 t} + \begin{Bmatrix} u_{03} \\ v_{03} \end{Bmatrix} e^{s_3 t} + \begin{Bmatrix} u_{04} \\ v_{04} \end{Bmatrix} e^{s_4 t} \tag{7.68}$$

where coefficients u_{0i} and v_{0i} depend on the initial conditions. Because the original eigenvalue problem, Equation (7.63), has real coefficients, any complex eigenvalues and corresponding eigenvectors occur in complex conjugate pairs; thus, the response in Equation (7.68) is real.

If the eigenvalues have negative real parts, the solution is stable; that is, the vibrations die away after any initial disturbance. However, the presence of an imaginary term in the square root can lead to a positive real part in these eigenvalues, thereby producing an unstable system. The conditions for stability can be obtained directly from Equation (7.64) using the Routh-Hurwitz criterion (Dorf and Bishop,

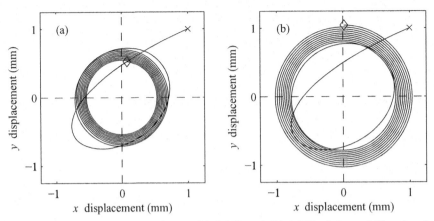

Figure 7.21. Motion of rotor running at (a) 0.975 and (b) 1.025 of the stability threshold speed (Example 7.5.1).

2008). Alternatively, the stability boundary may be obtained from Equation (7.66). At the stability boundary, the real part of the eigenvalues is zero; therefore, $s = j\omega$ for some ω. Substituting this expression into Equation (7.66) gives

$$-\omega^2 + 2\left(\zeta_e + \zeta_i\right)\omega_n j\omega + \omega_n^2\left(1 \pm 2j\zeta_i R\right) = 0 \qquad (7.69)$$

Equating the real parts of this equation gives

$$\omega = \pm\omega_n \qquad (7.70)$$

and the imaginary parts produce

$$2\left(\zeta_e + \zeta_i\right)\omega_n\omega \pm 2\zeta_i R\omega_n^2 = 0 \qquad (7.71)$$

Substituting Equation (7.70) into Equation (7.71) and realizing that all of the quantities in Equation (7.71) are positive (thus, only the equation with the negative sign has a nontrivial solution), the stability boundary is

$$\zeta_i R = \left(\zeta_e + \zeta_i\right) \qquad (7.72)$$

Because the rotor is obviously stable at zero rotational speed, the rotor is stable if

$$R < 1 + \frac{\zeta_e}{\zeta_i} \quad \text{or, equivalently, if} \quad \Omega < \omega_n\left(1 + \frac{\zeta_e}{\zeta_i}\right) \qquad (7.73)$$

Clearly, by increasing the damping external to the rotor or by decreasing the damping internal to the rotor, the stability boundary is at a higher rotational speed. The worst situation arises when $\zeta_e = 0$, in which case the machine is stable only below its critical speed.

EXAMPLE 7.5.1. Consider a system with $\omega_n = 100\,\text{rad/s}$ (15.92 Hz) and $\zeta_e = \zeta_i = 0.1$ so that the threshold of stability is given when the nondimensional spin speed parameter $R = 2$. The rotor is disturbed with initial displacements in both the x and y directions of 1 mm. Plot the orbits for $R = 1.95$ and $R = 2.05$. Calculate the eigenvalues and determine whether the forward- or backward-whirling mode is destabilized.

Solution. Figure 7.21(a) shows the orbit of a rotor running with $R = 1.95$ obtained from Equation (7.68). The cross and diamond denote the initial and final

Table 7.1. *Eigenvalues for the simple rotor with internal damping (Example 7.5.1)*

	Eigenvalues (rad/s)	
Rotor speed (R)	First pair (forward whirl)	Second pair (backward whirl)
1.95	$-0.48 \pm 99.90\jmath$	$-39.52 \pm 99.90\jmath$
2.05	$0.48 \pm 100.10\jmath$	$-40.48 \pm 100.10\jmath$

positions of the rotor, respectively. The figure shows that the motion of the rotor decays back to the equilibrium position. Figure 7.21(b) shows the behavior of the same rotor when disturbed by the same initial displacement but running at $R = 2.05$. Although the amplitude of the orbit initially reduces, it is shown in the figure that it is inexorably increasing and the system is unstable. Table 7.1 shows the eigenvalues for both cases; also given are the directions of whirl obtained from the eigenvectors using the method described in Section 3.6.1. These results show that the system is unstable when the real parts of the eigenvalues corresponding to the forward-whirling mode become positive.

7.6 Rotor Asymmetry with Isotropic Supports: Complex Rotors

The analysis of general asymmetric rotors with symmetric bearings is similar to that for simple asymmetric rotors in that the equations of motion must be transformed into the rotating coordinate frame (Kang et al., 1992). This may be done separately for the shaft elements, disks, and bearings, and the complete system model assembled in the usual way. The element matrices are derived assuming that the rotating coordinate frame has axes that coincide with the principal axes of inertia or stiffness. If this is not the case, then it is easy to transform the element matrices using a rotation based on the angle between the actual axis directions and the principal axes.

Each node now has four degrees of freedom because the rotations, as well as the displacements, must be included. The transformation used in Section 7.2 must be extended so that

$$\begin{Bmatrix} u \\ v \\ \theta \\ \psi \end{Bmatrix} = \mathbf{T}_2 \begin{Bmatrix} \tilde{u} \\ \tilde{v} \\ \tilde{\theta} \\ \tilde{\psi} \end{Bmatrix} \tag{7.74}$$

where

$$\mathbf{T}_2 = \begin{bmatrix} \mathbf{T} & \mathbf{0} \\ \mathbf{0} & \mathbf{T} \end{bmatrix} = \begin{bmatrix} \cos \Omega t & -\sin \Omega t & 0 & 0 \\ \sin \Omega t & \cos \Omega t & 0 & 0 \\ 0 & 0 & \cos \Omega t & -\sin \Omega t \\ 0 & 0 & \sin \Omega t & \cos \Omega t \end{bmatrix} \tag{7.75}$$

For a single node, the degrees of freedom are denoted by the vector $\mathbf{q}_n = \begin{bmatrix} u & v & \theta & \psi \end{bmatrix}^\top$ and, similarly, by $\tilde{\mathbf{q}}_n$ in rotating coordinates; these coordinates are transformed using $\mathbf{T}_2$. Shaft elements have eight degrees of freedom denoted by the vector $\mathbf{q}_e = \begin{Bmatrix} \mathbf{q}_{n_1} \\ \mathbf{q}_{n_2} \end{Bmatrix}$, where n_1 and n_2 are the two nodes at each end of the shaft element. The degrees of freedom for the shaft element must be transformed by

$$\mathbf{T}_4 = \begin{bmatrix} \mathbf{T} & \mathbf{0} & \mathbf{0} & \mathbf{0} \\ \mathbf{0} & \mathbf{T} & \mathbf{0} & \mathbf{0} \\ \mathbf{0} & \mathbf{0} & \mathbf{T} & \mathbf{0} \\ \mathbf{0} & \mathbf{0} & \mathbf{0} & \mathbf{T} \end{bmatrix} \tag{7.76}$$

With the coordinates and transformations defined, we may now consider the elements of the model.

7.6.1 Disks

The diametral inertia of the disk is assumed to be different in the $\tilde{x}$ and $\tilde{y}$ directions – denoted by $I_{d\tilde{x}}$ and $I_{d\tilde{y}}$ – and the axes fixed in the disk are assumed to coincide with the principal axes of inertia. The kinetic energy of the disk is

$$T_{\text{disk}} = \frac{1}{2}m\left(\dot{u}^2 + \dot{v}^2\right) + \frac{1}{2}I_{d\tilde{x}}\omega_{\tilde{x}}^2 + \frac{1}{2}I_{d\tilde{y}}\omega_{\tilde{y}}^2 + \frac{1}{2}I_p\omega_{\tilde{z}}^2 \tag{7.77}$$

where $\omega_{\tilde{x}}$, $\omega_{\tilde{y}}$, and $\omega_{\tilde{z}}$ are the instantaneous angular velocities about the $O\tilde{x}$, $O\tilde{y}$, and $O\tilde{z}$ axes fixed in the disk. The term involving the polar moment of inertia I_p produces the gyroscopic terms in the equations of motion in exactly the same way as the symmetric disk introduced in Equation (5.2).

The equations of motion must be developed in the stationary frame and transformed to the rotating frame. From Equation (5.4), for small rotations in the stationary frame θ and ψ

$$\begin{Bmatrix} \omega_{\tilde{x}} \\ \omega_{\tilde{y}} \end{Bmatrix} = \mathbf{T}^\top \begin{Bmatrix} \dot{\theta} \\ \dot{\psi} \end{Bmatrix} \tag{7.78}$$

and $\omega_{\tilde{z}} = \Omega - \dot{\psi}\theta$. Writing the kinetic energy in terms of velocities in the stationary frame gives

$$T_{\text{disk}} = \frac{1}{2}\dot{\mathbf{q}}_n^\top \begin{bmatrix} \mathbf{I} & \mathbf{0} \\ \mathbf{0} & \mathbf{T} \end{bmatrix} \tilde{\mathbf{M}}_d \begin{bmatrix} \mathbf{I} & \mathbf{0} \\ \mathbf{0} & \mathbf{T} \end{bmatrix}^\top \dot{\mathbf{q}}_n + \frac{1}{2}I_p\left(\Omega - \dot{\psi}\theta\right)^2 \tag{7.79}$$

where

$$\tilde{\mathbf{M}}_d = \begin{bmatrix} m & 0 & 0 & 0 \\ 0 & m & 0 & 0 \\ 0 & 0 & I_{d\tilde{x}} & 0 \\ 0 & 0 & 0 & I_{d\tilde{y}} \end{bmatrix} \tag{7.80}$$

The equations of motion for the disks in the stationary frame of reference are then obtained using Lagrange's equations

$$\frac{\mathrm{d}}{\mathrm{d}t}\left\{\frac{\partial T_{\text{disk}}}{\partial \dot{q}_i}\right\} - \frac{\partial T_{\text{disk}}}{\partial q_i} = Q_i \tag{7.81}$$

where Q_i is the generalized force at the ith degree of freedom. The term involving the transformation $\mathbf{T}$ must be differentiated with respect to time by the product rule, resulting in derivatives of the transformation matrix. The equations of motion then are transformed to the rotating frame of reference by premultiplying by $\mathbf{T}_2^{\mathsf{T}}$ and transforming the disk degrees of freedom from $\mathbf{q}_n$ to $\tilde{\mathbf{q}}_n$ using Equation (7.74). Identities for the transformation matrix, such as Equations (7.3) and (7.8), are used to simplify the result. The details are tedious and not provided here. The resulting equation of motion in the rotating frame is

$$\tilde{\mathbf{M}}_d \ddot{\tilde{\mathbf{q}}}_n + \Omega \left(\tilde{\mathbf{M}}_{1d} + \tilde{\mathbf{G}}_d \right) \dot{\tilde{\mathbf{q}}}_n + \Omega^2 \left(\tilde{\mathbf{M}}_{2d} + \tilde{\mathbf{G}}_{1d} \right) \tilde{\mathbf{q}}_n = \tilde{\mathbf{Q}} \tag{7.82}$$

where

$$\tilde{\mathbf{M}}_{1d} = \begin{bmatrix} 0 & -2m & 0 & 0 \\ 2m & 0 & 0 & 0 \\ 0 & 0 & 0 & -(I_{d\tilde{x}} + I_{d\tilde{y}}) \\ 0 & 0 & I_{d\tilde{x}} + I_{d\tilde{y}} & 0 \end{bmatrix}, \tag{7.83}$$

$$\tilde{\mathbf{M}}_{2d} = - \begin{bmatrix} m & 0 & 0 & 0 \\ 0 & m & 0 & 0 \\ 0 & 0 & I_{d\tilde{y}} & 0 \\ 0 & 0 & 0 & I_{d\tilde{x}} \end{bmatrix}, \tag{7.84}$$

$$\mathbf{G}_d = \tilde{\mathbf{G}}_d = \begin{bmatrix} 0 & 0 & 0 & 0 \\ 0 & 0 & 0 & 0 \\ 0 & 0 & 0 & I_p \\ 0 & 0 & -I_p & 0 \end{bmatrix}, \qquad \tilde{\mathbf{G}}_{1d} = \begin{bmatrix} 0 & 0 & 0 & 0 \\ 0 & 0 & 0 & 0 \\ 0 & 0 & I_p & 0 \\ 0 & 0 & 0 & I_p \end{bmatrix} \tag{7.85}$$

The parameters $I_{d\tilde{x}}$ and $I_{d\tilde{y}}$ interchange their positions between $\tilde{\mathbf{M}}_{2d}$ and $\tilde{\mathbf{M}}_d$.

7.6.2 Shaft Elements

The stiffness matrix for the shaft elements does not need to be transformed because the asymmetry is fixed in the rotating frame. However, the stiffness matrices of a circular shaft must be extended to include the asymmetry. Assume that the principal axes of stiffness symmetry are in the $\tilde{x}$ and $\tilde{y}$ directions. This means that a force in the $\tilde{x}$ direction does not produce a displacement in the $\tilde{y}$ direction and vice versa. The stiffness matrix for the beam element then is defined by Equation (5.41) with parameters $EI_{e\tilde{x}}$ and $\Phi_{e\tilde{x}}$ for the $O\tilde{x}\tilde{z}$ plane and $EI_{e\tilde{y}}$ and $\Phi_{e\tilde{y}}$ for the $O\tilde{y}\tilde{z}$ plane. Thus, the stiffness matrix for a Euler-Bernoulli beam element is

$$\tilde{\mathbf{K}}_e = \tilde{\mathbf{K}}_{e\tilde{x}} + \tilde{\mathbf{K}}_{e\tilde{y}} \tag{7.86}$$

where

$$\tilde{\mathbf{K}}_{e\tilde{x}} = \frac{EI_{e\tilde{x}}}{\ell_e^3} \begin{bmatrix} 12 & 0 & 0 & 6\ell_e & -12 & 0 & 0 & 6\ell_e \\ 0 & 0 & 0 & 0 & 0 & 0 & 0 & 0 \\ 0 & 0 & 0 & 0 & 0 & 0 & 0 & 0 \\ 6\ell_e & 0 & 0 & 4\ell_e^2 & -6\ell_e & 0 & 0 & 2\ell_e^2 \\ -12 & 0 & 0 & -6\ell_e & 12 & 0 & 0 & -6\ell_e \\ 0 & 0 & 0 & 0 & 0 & 0 & 0 & 0 \\ 0 & 0 & 0 & 0 & 0 & 0 & 0 & 0 \\ 6\ell_e & 0 & 0 & 2\ell_e^2 & -6\ell_e & 0 & 0 & 4\ell_e^2 \end{bmatrix}$$

and

$$\tilde{\mathbf{K}}_{e\bar{y}} = \frac{EI_{e\bar{y}}}{\ell_e^3}
\begin{bmatrix}
0 & 0 & 0 & 0 & 0 & 0 & 0 & 0 \\
0 & 12 & -6\ell_e & 0 & 0 & -12 & -6\ell_e & 0 \\
0 & -6\ell_e & 4\ell_e^2 & 0 & 0 & 6\ell_e & 2\ell_e^2 & 0 \\
0 & 0 & 0 & 0 & 0 & 0 & 0 & 0 \\
0 & 0 & 0 & 0 & 0 & 0 & 0 & 0 \\
0 & -12 & 6\ell_e & 0 & 0 & 12 & 6\ell_e & 0 \\
0 & -6\ell_e & 2\ell_e^2 & 0 & 0 & 6\ell_e & 4\ell_e^2 & 0 \\
0 & 0 & 0 & 0 & 0 & 0 & 0 & 0
\end{bmatrix}$$

Shear effects are included easily by using the stiffness matrix given in Equation (5.41) for each plane of motion.

To obtain the mass matrices for the shaft elements, an approach similar to that used to obtain the inertia terms due to a disk must be used. However, for most shafts, the asymmetry in the diametral inertia is small and the shaft mass is assumed to be symmetric. Consider first a Euler-Bernoulli beam element, the mass matrix of which is given in Equation (5.50). In this case, the mass matrix depends only on the product of density and cross-sectional area of the rotor. The transformation between the stationary and rotating frames is given by Equation (7.74) for each node. Thus, the equation of motion for the element is

$$\tilde{\mathbf{M}}_e \ddot{\tilde{\mathbf{q}}}_e + \Omega \left(\tilde{\mathbf{M}}_{1e} + \tilde{\mathbf{G}}_e \right) \dot{\tilde{\mathbf{q}}}_e + \Omega^2 \left(-\tilde{\mathbf{M}}_e + \tilde{\mathbf{G}}_{1e} \right) \tilde{\mathbf{q}}_e = \tilde{\mathbf{Q}}_e \qquad (7.87)$$

where $\tilde{\mathbf{q}}_e$ is the usual vector of linear and rotational displacements at the two nodes (i.e., used to compute the element matrices for the symmetric shaft; see Section 5.4.4) and $\tilde{\mathbf{Q}}_e$ is the applied force. It is straightforward to demonstrate that $\tilde{\mathbf{M}}_e = \mathbf{M}_e$. The matrix $\tilde{\mathbf{M}}_{1e}$ is skew-symmetric and easily computed from the transformation and the element-mass matrix as

$$\tilde{\mathbf{M}}_{1e} = \frac{\rho_e A_e \ell_e}{210}
\begin{bmatrix}
0 & -156 & 22\ell_e & 0 & 0 & -54 & -13\ell_e & 0 \\
156 & 0 & 0 & 22\ell_e & 54 & 0 & 0 & -13\ell_e \\
-22\ell_e & 0 & 0 & -4\ell_e^2 & -13\ell_e & 0 & 0 & 3\ell_e^2 \\
0 & -22\ell_e & 4\ell_e^2 & 0 & 0 & -13\ell_e & -3\ell_e^2 & 0 \\
0 & -54 & 13\ell_e & 0 & 0 & -156 & -22\ell_e & 0 \\
54 & 0 & 0 & 13\ell_e & 156 & 0 & 0 & -22\ell_e \\
13\ell_e & 0 & 0 & 3\ell_e^2 & 22\ell_e & 0 & 0 & -4\ell_e^2 \\
0 & 13\ell_e & -3\ell_e^2 & 0 & 0 & 22\ell_e & 4\ell_e^2 & 0
\end{bmatrix}$$

$$(7.88)$$

$\mathbf{G}_e$ is given by Equation (5.83), $\tilde{\mathbf{G}}_e = \mathbf{G}_e$, and $\tilde{\mathbf{G}}_{1e}$ is

$$\tilde{\mathbf{G}}_{1e} = \frac{2\rho_e I_e}{15\ell_e}
\begin{bmatrix}
36 & 0 & 0 & 3\ell_e & -36 & 0 & 0 & 3\ell_e \\
0 & 36 & -3\ell_e & 0 & 0 & -36 & -3\ell_e & 0 \\
0 & -3\ell_e & 4\ell_e^2 & 0 & 0 & 3\ell_e & -\ell_e^2 & 0 \\
3\ell_e & 0 & 0 & 4\ell_e^2 & -3\ell_e & 0 & 0 & -\ell_e^2 \\
-36 & 0 & 0 & -3\ell_e & 36 & 0 & 0 & -3\ell_e \\
0 & -36 & 3\ell_e & 0 & 0 & 36 & 3\ell_e & 0 \\
0 & -3\ell_e & -\ell_e^2 & 0 & 0 & 3\ell_e & 4\ell_e^2 & 0 \\
3\ell_e & 0 & 0 & -\ell_e^2 & -3\ell_e & 0 & 0 & 4\ell_e^2
\end{bmatrix} \qquad (7.89)$$

The shear and rotary inertia effects may be added similarly using the element matrices obtained in Chapter 5 and the transformation in Equation (7.87). The rotary inertia effects are incorporated based on the average inertia properties of the shaft section in the two directions. Although not strictly correct, the errors introduced are minor because most of the mass is usually in the disks. If we neglect shear deformations, the effect of rotary inertia is to add the following matrix to $\tilde{\mathbf{M}}_e$:

$$
\tilde{\mathbf{M}}_{re} = \frac{\rho_e I_e}{30\ell_e}
\begin{bmatrix}
36 & 0 & 0 & 3\ell_e & -36 & 0 & 0 & 3\ell_e \\
0 & 36 & -3\ell_e & 0 & 0 & -36 & -3\ell_e & 0 \\
0 & -3\ell_e & 4\ell_e^2 & 0 & 0 & 3\ell_e & -\ell_e^2 & 0 \\
3\ell_e & 0 & 0 & 4\ell_e^2 & -3\ell_e & 0 & 0 & -\ell_e^2 \\
-36 & 0 & 0 & -3\ell_e & 36 & 0 & 0 & -3\ell_e \\
0 & -36 & 3\ell_e & 0 & 0 & 36 & 3\ell_e & 0 \\
0 & -3\ell_e & -\ell_e^2 & 0 & 0 & 3\ell_e & 4\ell_e^2 & 0 \\
3\ell_e & 0 & 0 & -\ell_e^2 & -3\ell_e & 0 & 0 & 4\ell_e^2
\end{bmatrix}
\tag{7.90}
$$

and the following matrix to $\tilde{\mathbf{M}}_{1e}$:

$$
\tilde{\mathbf{M}}_{r1e} = \frac{\rho_e I_e}{15\ell_e}
\begin{bmatrix}
0 & -36 & 3\ell_e & 0 & 0 & 36 & 3\ell_e & 0 \\
36 & 0 & 0 & 3\ell_e & -36 & 0 & 0 & 3\ell_e \\
-3\ell_e & 0 & 0 & -4\ell_e^2 & 3\ell_e & 0 & 0 & \ell_e^2 \\
0 & -3\ell_e & 4\ell_e^2 & 0 & 0 & 3\ell_e & -\ell_e^2 & 0 \\
0 & 36 & -3\ell_e & 0 & 0 & -36 & -3\ell_e & 0 \\
-36 & 0 & 0 & -3\ell_e & 36 & 0 & 0 & -3\ell_e \\
-3\ell_e & 0 & 0 & \ell_e^2 & 3\ell_e & 0 & 0 & -4\ell_e^2 \\
0 & -3\ell_e & -\ell_e^2 & 0 & 0 & 3\ell_e & 4\ell_e^2 & 0
\end{bmatrix}
\tag{7.91}
$$

Hence, we have the model for the rotating parts of the machine; it remains to derive models for the bearings and stator.

7.6.3 Bearings and Foundations

The bearing and foundation models are usually given in the stationary frame and must be transformed to the rotating frame. The transformation given by Equation (7.76) is used to transform both the bearing force and the displacement and/or velocity to the rotating frame. The analysis is simplified if the bearings are such that the transformed matrices do not depend on time. Diagonal stiffness and damping matrices for the bearing, with identical stiffness and damping in the x and y directions, satisfy this condition and are used to demonstrate the process. The stiffness matrix is then the same in stationary and rotating coordinates. The damping force in the stationary frame is

$$
\mathbf{Q}_{\text{damping}} = -\mathbf{C}_0 \dot{\mathbf{q}}_n = -
\begin{bmatrix}
c_1 & 0 & 0 & 0 \\
0 & c_1 & 0 & 0 \\
0 & 0 & c_2 & 0 \\
0 & 0 & 0 & c_2
\end{bmatrix}
\begin{Bmatrix}
\dot{u} \\
\dot{v} \\
\dot{\theta} \\
\dot{\psi}
\end{Bmatrix}
\tag{7.92}
$$

The velocity is transformed to give $\dot{\mathbf{q}}_n = \mathbf{T}_2\dot{\tilde{\mathbf{q}}}_n + \dot{\mathbf{T}}_2\tilde{\mathbf{q}}_n$; thus, the damping force in the rotating frame is

$$\begin{aligned}\tilde{\mathbf{Q}}_{\text{damping}} &= -\mathbf{T}_2^\top\mathbf{C}_0\left(\mathbf{T}_2\dot{\tilde{\mathbf{q}}}_n + \dot{\mathbf{T}}_2\tilde{\mathbf{q}}_n\right)\\ &= -\tilde{\mathbf{C}}_0\dot{\tilde{\mathbf{q}}}_n - \Omega\tilde{\mathbf{C}}_1\tilde{\mathbf{q}}_n\end{aligned} \tag{7.93}$$

where

$$\tilde{\mathbf{C}}_0 = \mathbf{C}_0 = \begin{bmatrix} c_1 & 0 & 0 & 0\\ 0 & c_1 & 0 & 0\\ 0 & 0 & c_2 & 0\\ 0 & 0 & 0 & c_2 \end{bmatrix}, \quad \tilde{\mathbf{C}}_1 = \begin{bmatrix} 0 & -c_1 & 0 & 0\\ c_1 & 0 & 0 & 0\\ 0 & 0 & 0 & -c_2\\ 0 & 0 & c_2 & 0 \end{bmatrix} \tag{7.94}$$

We are now in a position to assemble the full model in the rotating frame of reference.

7.6.4 The Equations of Motion

The previous sections derive the element matrices for disks, shafts, and bearings. In each case, the transformed force is a linear combination of the displacement, velocity, and acceleration in the rotating frame. The process of assembling the matrices is exactly the same as that used for models in the stationary frame of reference (see Section 5.7). One difference is that now each element of the model produces more than one matrix and contributes to the coefficient of generalized displacement, velocity, and acceleration. Also, these matrices often depend on rotational speed. However, the assembly procedure is exactly as before, although there are more global matrices to assemble. Combining all of the element matrices gives equations of motion of the form

$$\tilde{\mathbf{M}}_0\ddot{\tilde{\mathbf{q}}} + \left[\tilde{\mathbf{C}}_0 + \Omega\left(\tilde{\mathbf{G}}_0 + \tilde{\mathbf{M}}_1\right)\right]\dot{\tilde{\mathbf{q}}} + \left[\tilde{\mathbf{K}}_0 + \Omega\tilde{\mathbf{C}}_1 + \Omega^2\left(\tilde{\mathbf{G}}_1 + \tilde{\mathbf{M}}_2\right)\right]\tilde{\mathbf{q}} = \tilde{\mathbf{Q}} \tag{7.95}$$

where $\tilde{\mathbf{q}}$ represents the assembled generalized coordinate vector in the rotating frame and $\tilde{\mathbf{Q}}$ is the generalized force, also in the rotating frame. The matrices $\tilde{\mathbf{M}}_0$, $\tilde{\mathbf{C}}_0$, $\tilde{\mathbf{G}}_1$, $\tilde{\mathbf{K}}_0$, and $\tilde{\mathbf{M}}_2$ are symmetric and the matrices $\tilde{\mathbf{M}}_1$, $\tilde{\mathbf{C}}_1$, and $\tilde{\mathbf{G}}_0$ are skew-symmetric. $\tilde{\mathbf{K}}_0$ represents the combined stiffness of the shaft and the foundation, expressed in rotating coordinates.

> **EXAMPLE 7.6.1.** A 1.5 m-long shaft, shown in Figure 7.22, has a cross-sectional area of 0.0020 m². The shaft is asymmetric and its second moment of area has a maximum of 4.2043×10^{-7} m⁴ and a minimum of 2.4112×10^{-7} m⁴ in the perpendicular plane. The shaft carries two circular disks at 0.5 and 1 m from one end of the shaft. The left disk is 0.07 m thick with a diameter of 0.28 m; the right disk is 0.07 m thick with a diameter of 0.35 m. For the shaft, $E = 211$ GN/m². There is no internal shaft-damping. For both the shaft and the disks, $\rho = 7{,}810$ kg/m³. The shaft is supported at the ends by identical isotropic bearings with a stiffness of 1 MN/m in all directions.
>
> Create an FE model of the shaft using six Euler-Bernoulli beam elements and investigate the stability of the rotor–bearing system up to a speed of

Figure 7.22. The layout of the two-disk, two-bearing rotor (Example 7.6.1).

4,500 rev/min. Also derive a map of the system natural frequencies against running speed in rotating coordinates and a map of the pseudonatural frequencies against running speed in fixed coordinates, in each case to a maximum frequency of 150 Hz.

To improve the stability of the rotor, damping is introduced into the bearings. If the damping is 5,000 Ns/m in all directions at both bearings, is it sufficient to make the rotor stable at all speeds up to 4,500 rev/min? What is the value of any gravity critical speeds?

An out-of-balance of 0.001 kg m is attached to the left-hand disk. Determine the response of the system at the disks for various angular locations of the unbalance, both with and without the damping in the bearings.

Solution. Figure 7.23 shows the eigenvalues in the rotating frame for speeds up to 4,500 rev/min. The imaginary parts of the eigenvalues correspond to the damped natural frequencies. Figure 7.24 shows the real parts of the eigenvalues (identical to the plot in Figure 7.23) and the pseudonatural frequencies. There are four ranges of speeds (one is very narrow and barely visible at approximately 1,774 rev/min) at which the machine is unstable because one or more of the eigenvalues has a positive real part (in the rotating frame, the eigenvalues are real for two of the speed ranges). There are three speeds (i.e., 414, 1,342, and 3,917 rev/min) at which one of the pseudonatural frequencies is zero, producing three gravity critical speeds. Figures 7.25 and 7.26 show the eigenvalues

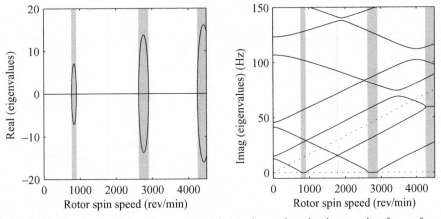

Figure 7.23. The real and imaginary parts of the eigenvalues in the rotating frame for an undamped asymmetric rotor (Example 7.6.1). The shading denotes the unstable speed ranges.

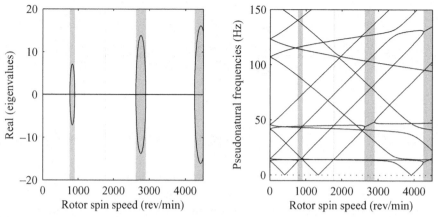

Figure 7.24. The real part of the eigenvalues and the pseudonatural frequencies in the stationary frame for an undamped asymmetric rotor (Example 7.6.1). The shading denotes the unstable speed ranges.

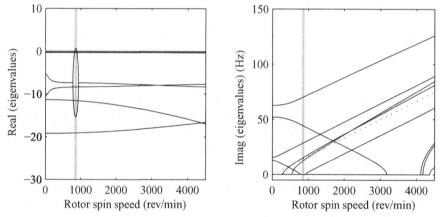

Figure 7.25. The real and imaginary parts of the eigenvalues in the rotating frame for an asymmetric rotor with damping in the bearings (Example 7.6.1). The shading denotes the unstable speed range.

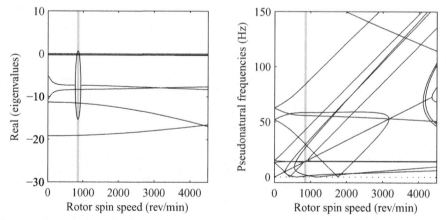

Figure 7.26. The real part of the eigenvalues and the pseudonatural frequencies in the stationary frame for an asymmetric rotor with damping in the bearings (Example 7.6.1). The shading denotes the unstable speed range.

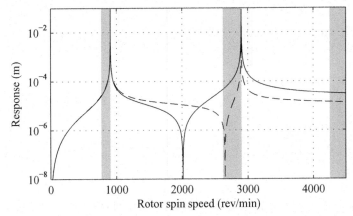

Figure 7.27. The response of the machine to an out-of-balance mass located in the maximum stiffness direction on the undamped asymmetric rotor (Example 7.6.1). The shading denotes the unstable speed ranges. The solid line gives the response at the left disk and the dashed line at the right disk.

and pseudonatural frequencies when the damping is included in the bearings. Clearly, the real parts of the eigenvalues have reduced and there is now only one unstable speed range. There are now four speeds (i.e., 422, 625, 1,192, and 1,765 rev/min) at which one of the pseudonatural frequencies is zero, producing gravity critical speeds.

Now consider the response to an out-of-balance. For the undamped case, Figures 7.27 and 7.28 show the response to an out-of-balance mass placed at angular positions corresponding to the maximum stiffness in the rotor and at 45° to this direction, respectively. The responses given are steady-state solutions; although solutions are obtained in the unstable speed ranges, the instability means that these solutions are not obtained in practice. When the out-of-balance is placed in the direction of maximum stiffness, the peaks in response occur at the upper extremes of two of the unstable speed ranges, which is similar to the

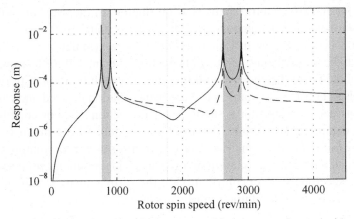

Figure 7.28. The response of the machine to an out-of-balance mass located midway between the minimum and maximum stiffness directions on the undamped asymmetric rotor (Example 7.6.1). The shading denotes the unstable speed ranges. The solid line gives the response at the left disk and the dashed line at the right disk.

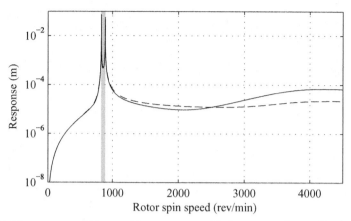

Figure 7.29. The response of the machine to an out-of-balance mass located in the maximum stiffness direction on the asymmetric rotor with damping in the bearings (Example 7.6.1). The shading denotes the unstable speed range. The solid line gives the response at the left disk and the dashed line at the right disk.

response of the two degrees of freedom system described in Example 7.3.5. If the out-of-balance mass were placed in the direction of the minimum stiffness, then the peaks occur at the lower extremes of the unstable speed ranges (the response is not shown here). When the out-of-balance mass is placed at other positions, peaks occur at both extremes of the two unstable speed ranges, as shown in Figure 7.28. Unbalance does not excite the rotor in the unstable speed ranges at 1,774 and above 4,252 rev/min. Figure 7.29 shows the response when damping is added to the bearings. The higher resonance is overdamped and peaks occur at both ends of the unstable speed range because of the coupling in the rotating frame caused by damping in the stationary frame.

7.7 Internal Rotor Damping: Complex Rotors

Internal rotor damping appears in two places in the equations of motion in fixed co-ordinates for an axisymmetric shaft: as a contribution to the usual damping matrix and as a skew-symmetric component of the stiffness matrix that increases linearly with rotor speed, as shown in Equation (7.61). Skew-symmetry in a stiffness matrix tends to be destabilizing, as shown in Sections 7.8 and 7.9 for bearing stiffness. Because of the linear dependence of this asymmetry with rotational speed, the machine tends to become less stable with increasing rotational speed and can become unstable, as demonstrated for a simple rotor in Section 7.5. For complex rotors, the mass, damping, and stiffness matrices are produced in terms of coordinates fixed in the rotor, and these matrices must be transformed from rotating to stationary coordinates. The transformation of a viscous-damping matrix produces a skew-symmetric, speed-dependent contribution to the stiffness matrix. Suppose that the damping force is given in rotating coordinates as

$$\tilde{\mathbf{Q}}_{\text{damping}} = -\tilde{\mathbf{C}}_i \dot{\tilde{\mathbf{q}}} \tag{7.96}$$

where $\tilde{\mathbf{q}}$ is the displacement in rotating coordinates. Following the node and element examples given in Section 7.6, the rotating coordinates may be written in

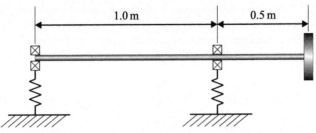

Figure 7.30. The overhung rotor (Example 7.7.1).

terms of the stationary coordinates, $\mathbf{q}$, as

$$\dot{\tilde{\mathbf{q}}} = \mathbf{T}_n^\top \dot{\mathbf{q}} + \dot{\mathbf{T}}_n^\top \mathbf{q} \tag{7.97}$$

for the time-dependent transformation matrix $\mathbf{T}_n(\Omega t)$. This transformation matrix has a dimension of $2n \times 2n$ and is block diagonal with the matrix $\mathbf{T}$ on the diagonal. Equation (7.76) gives $\mathbf{T}_4$ explicitly.

Transforming the damping force to the stationary frame of reference produces

$$\mathbf{Q}_{\text{damping}} = \mathbf{T}_n \tilde{\mathbf{Q}}_{\text{damping}} = -\mathbf{T}_n \tilde{\mathbf{C}}_i \mathbf{T}_n^\top \dot{\mathbf{q}} - \mathbf{T}_n \tilde{\mathbf{C}}_i \dot{\mathbf{T}}_n^\top \mathbf{q} \tag{7.98}$$

Generally, for a symmetric rotor, the first term in Equation (7.98) is invariant; that is, $\mathbf{C}_i = \mathbf{T}_n \tilde{\mathbf{C}}_i \mathbf{T}_n^\top = \tilde{\mathbf{C}}_i$. Modeling damping is difficult and a common approximation is to assume that the damping is distributed in a similar way to the stiffness, the so-called proportional damping approximation. Thus,

$$\tilde{\mathbf{C}}_i = \beta \tilde{\mathbf{K}} \tag{7.99}$$

for some scalar constant β. This approximation is used to demonstrate the effects of internal damping for the FE formulations described in Chapter 5. Equation (5.51) gives the element stiffness for a Euler-Bernoulli beam. Substituting the transformation Equation (7.76) into Equation (7.98) using the stiffness matrix for a symmetric shaft element, Equation (5.51), gives

$$\mathbf{Q}_{\text{damping}} = -\beta \mathbf{K}_e \dot{\mathbf{q}} - \beta \Omega \mathbf{K}_{1e} \mathbf{q} \tag{7.100}$$

where $\mathbf{K}_e$ is given by Equation (5.51) and

$$\mathbf{K}_{1e} = \frac{E_e I_e}{\ell_e^3} \begin{bmatrix} 0 & 12 & -6\ell_e & 0 & 0 & -12 & -6\ell_e & 0 \\ -12 & 0 & 0 & -6\ell_e & 12 & 0 & 0 & -6\ell_e \\ 6\ell_e & 0 & 0 & 4\ell_e^2 & -6\ell_e & 0 & 0 & 2\ell_e^2 \\ 0 & 6\ell_e & -4\ell_e^2 & 0 & 0 & -6\ell_e & -2\ell_e^2 & 0 \\ 0 & -12 & 6\ell_e & 0 & 0 & 12 & 6\ell_e & 0 \\ 12 & 0 & 0 & 6\ell_e & -12 & 0 & 0 & 6\ell_e \\ 6\ell_e & 0 & 0 & 2\ell_e^2 & -6\ell_e & 0 & 0 & 4\ell_e^2 \\ 0 & 6\ell_e & -2\ell_e^2 & 0 & 0 & -6\ell_e & -4\ell_e^2 & 0 \end{bmatrix} \tag{7.101}$$

$\mathbf{K}_{1e}$ provides a skew-symmetric contribution to the stiffness matrix that depends linearly on rotational speed.

EXAMPLE 7.7.1. Consider a simple overhung rotor, 1.5 m long with bearings at $z = 0.0$ m and $z = 1.0$ m, and shown in Figure 7.30. The bearings are short in that they present insignificant angular stiffness to the shaft but they present finite

Table 7.2. *The bearing and support characteristics for the*
overhung rotor (Example 7.7.1). All cross terms are zero

Case	Stiffness (MN/m)		Damping (kNs/m)	
	k_{xx}	k_{yy}	c_{xx}	c_{yy}
1	10	10	60	60
2	10	20	0	0
3	10	20	60	60

translational stiffness. The shaft is 25 mm in diameter, and the disk at the over-hung end of the shaft is 250 mm in diameter and 40 mm thick. The shaft and disk are made of steel, and a mass density $\rho = 7,810 \, \text{kg/m}^3$, a modulus of elasticity $E = 211 \, \text{GPa}$, and a Poisson's ratio of 0.3 are assumed. Table 7.2 gives the range of bearing characteristics to be considered. This system is used in Example 6.8.1 and has six elements and 28 degrees of freedom. The level of internal damp-ing is given by Equation (7.99) with $\beta = 10^{-5}$ s. Calculate the eigenvalues for a range of rotational speeds up to 3,000 rev/min and for various combinations of bearing characteristics to demonstrate the destabilizing effect of internal damping.

Solution. Table 7.3 shows the eigenvalues calculated at various rotor speeds for the overhung rotor, the bearing properties of which are given in Table 7.2. The first two cases in Table 7.3 have bearing characteristic 1, which is an isotropic bearing with some damping. In these two cases, the gyroscopic effects are ne-glected so that the effect of internal damping with rotor speed may be demon-strated more clearly. The results show that internal damping affects the real part of the eigenvalue and breaks the symmetry of the system, causing each group of four eigenvalues to split into two pairs. This effect also may be seen in the results for the systems with bearing characteristics 2 and 3 for the left and right bearings, respectively, shown in Table 7.3. Comparing the results with those without internal damping, also shown in Table 7.3, shows that internal damping affects the real part of the eigenvalues. Furthermore, the second mode of the system becomes unstable between 1,000 and 2,000 rev/min, when the real part of the eigenvalue becomes positive.

A convenient method to determine stability is to plot the eigenvalues of the system on the complex plane as the rotor speed varies. This plot is called a *root locus* and is commonly used in control engineering when the parameter being changed is usually a controller gain. A system becomes unstable if any of the eigenvalues cross the imaginary axis and the real parts become positive. Figure 7.31 shows the root locus as the rotor speed changes for the overhung ro-tor example, with bearing characteristics 2 and 3. The instability arises because the real part of the second pair of eigenvalues becomes positive between 1,000 and 2,000 rev/min. We may conclude that viscous damping in the rotor tends to separate the real parts associated with each resonance frequency, making one more negative and the other more positive (i.e., less stable). At a sufficiently high speed, viscous damping in the rotor causes instability by driving at least one of the roots into the right half-plane (i.e., positive real part).

Table 7.3. *Characteristic roots computed for a number of configurations, including internal rotor damping (Example 7.7.1)*

Rotor speed (rev/min)	Bearing		Eigenvalues (rad/s)			
	Left	Right	First pair	Second pair	Third pair	Fourth pair
0	1	1	$-0.052 \pm 44.29j$	$-0.052 \pm 44.29j$	$-2.615 \pm 402.0j$	$-2.615 \pm 402.0j$
1,000*	1	1	$-0.075 \pm 44.29j$	$-0.029 \pm 44.29j$	$-2.825 \pm 402.0j$	$-2.405 \pm 402.0j$
0	2	3	$-0.048 \pm 44.29j$	$-0.020 \pm 44.36j$	$-2.015 \pm 400.6j$	$-1.651 \pm 401.2j$
1,000	2	3	$-0.053 \pm 42.26j$	$-0.015 \pm 46.40j$	$-1.614 \pm 392.0j$	$-1.960 \pm 406.6j$
2,000	2	3	$-0.072 \pm 40.22j$	$0.002 \pm 48.47j$	$-1.397 \pm 377.9j$	$-2.008 \pm 410.3j$
3,000	2	3	$-0.091 \pm 38.23j$	$0.019 \pm 50.51j$	$-1.433 \pm 357.3j$	$-1.985 \pm 413.0j$
0†	2	3	$-0.038 \pm 44.29j$	$-0.010 \pm 44.36j$	$-1.220 \pm 400.6j$	$-0.851 \pm 401.2j$
1,000†	2	3	$-0.020 \pm 42.26j$	$-0.028 \pm 46.40j$	$-0.619 \pm 392.0j$	$-1.358 \pm 406.6j$
2,000†	2	3	$-0.017 \pm 40.22j$	$-0.033 \pm 48.47j$	$-0.190 \pm 377.9j$	$-1.605 \pm 410.3j$
3,000†	2	3	$-0.139 \pm 38.23j$	$-0.039 \pm 50.51j$	$-0.0004 \pm 357.3j$	$-1.786 \pm 413.0j$

Note: The columns labeled Left and Right are the bearing characteristics given in Table 7.2. The * indicates that the gyroscopic effects are neglected, and the † indicates that the internal damping is neglected.

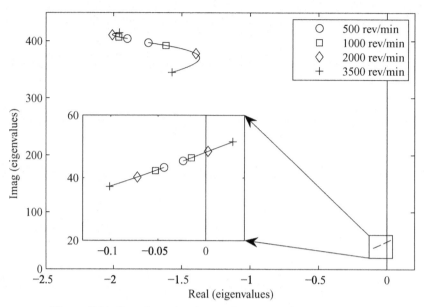

Figure 7.31. Root locus for the overhung rotor (Example 7.7.1).

7.8 Internal Cross-Coupling in the Bearing: Simple Rotors

Oil-film bearings have stiffness coupling between the vertical and horizontal directions as shown in Equation (5.87). Consider a machine with a single fluid bearing modeled with two translational degrees of freedom. Assuming isotropic damping, the equations of motion for such a system are

$$m\ddot{u} + c\dot{u} + k_{uu}u + k_{uv}v = 0$$
$$m\ddot{v} + c\dot{v} + k_{vu}u + k_{vv}v = 0$$

(7.102)

If the system is disturbed, then the orbit of the shaft center usually reduces in amplitude due to the presence of damping and the system is stable.

For a particular system, the stability of the system described by Equation (7.102) may be checked by calculating the eigenvalues. An unstable system has at least one eigenvalue with a positive real part. Alternatively, the Routh-Hurwitz criterion may be used to determine simple stability for the system (Dorf and Bishop, 2008). This is left as an exercise for readers but the result is given here. The direct stiffnesses, k_{uu} and k_{vv}, are usually positive. Then, the system is stable if

$$k_{uu}k_{vv} - k_{uv}k_{vu} > 0$$

(7.103)

and

$$m(k_{uu} - k_{vv})^2 + 4mk_{uv}k_{vu} + 2c^2(k_{uu} + k_{vv}) > 0$$

(7.104)

Equation (7.103) is required for a stable static system.

EXAMPLE 7.8.1. Consider the system of Equation (7.102) with $m = 1\,\text{kg}$, $c = 15\,\text{Ns/m}$, $k_{uu} = 50\,\text{kN/m}$, and $k_{vv} = 50\,\text{kN/m}$. Investigate the stability when (a) $k_{uv} = k_{vu} = -5\,\text{kN/m}$, and (b) $k_{uv} = 5\,\text{kN/m}$, $k_{vu} = -5\,\text{kN/m}$. The initial displacement is 1 mm in the x direction and zero in the y direction.

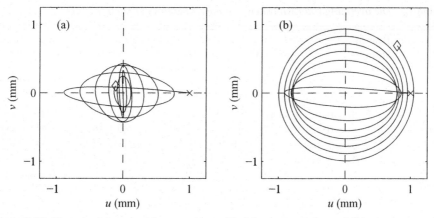

Figure 7.32. Transient response for a system with (a) a symmetric, and (b) an asymmetric-stiffness matrix (Example 7.8.1).

Solution

(a) Figure 7.32(a) shows the transient response of a machine, calculated using the method described in Section 2.6.1. The cross and diamond denote the initial and final positions of the rotor, respectively. The transient response decays, indicating that this machine is stable.

(b) In this case, one cross-coupling term is positive and the other is negative; the response is shown in Figure 7.32(b). The figure shows that the effect causes the system to be unstable, despite the presence of damping. Table 7.4 lists the eigenvalues in both cases, clearly showing that the example with skew-symmetric bearings is unstable. In this case, the criterion given by Equation (7.104) is not satisfied.

7.9 Internal Cross-Coupling in the Bearing: Complex Rotors

An anisotropic bearing with a nonsymmetric-stiffness matrix in a complex model of a machine can cause instability in a way similar to the case for the simple rotor.

EXAMPLE 7.9.1. Consider the overhung rotor in Example 7.7.1. The direct stiffness magnitudes of both bearings are 10 and 20 MN/m in the vertical and horizontal directions, respectively, and the direct damping is 10 kNs/m in both directions. Calculate the eigenvalues of the machine for different bearing parameters and rotor spin speeds of 0, 1,000, and 3,000 rev/min by varying the off-diagonal bearing-stiffness term, which is ±5 MN/m. Investigate the effect on stability.

Table 7.4. *Eigenvalues for the simple rotor with skew-symmetric bearing stiffness (Example 7.8.1)*

k_{12} (kN/m)	k_{21} (kN/m)	Eigenvalues (rad/s)	
		First pair	Second pair
−5	−5	$-7.50 \pm 212.00j$	$-7.50 \pm 234.40j$
5	−5	$-18.67 \pm 223.76j$	$3.67 \pm 223.76j$

Table 7.5. *The overhung rotor with skew-symmetric bearings (Example 7.9.1)*

Rotor speed (rev/min)	Bearing Stiffness (kN/m)		Eigenvalues (rad/s)			
	k_{uv}	k_{vu}	First pair	Second pair	Third pair	Fourth pair
0	−5	−5	$-0.0121 \pm 44.44_J$	$-0.0016 \pm 44.58_J$	$-2.676 \pm 397.5_J$	$-0.410 \pm 400.5_J$
1,000	−5	−5	$-0.0059 \pm 42.43_J$	$-0.0079 \pm 46.60_J$	$-1.259 \pm 390.9_J$	$-1.722 \pm 404.1_J$
3,000	−5	−5	$-0.0041 \pm 38.37_J$	$-0.0109 \pm 50.74_J$	$-0.211 \pm 357.3_J$	$-2.540 \pm 409.6_J$
0	5	−5	$-0.0034 \pm 44.54_J$	$-0.0034 \pm 44.54_J$	$-0.862 \pm 399.5_J$	$-0.862 \pm 399.5_J$
1,000	5	−5	$-0.0367 \pm 42.46_J$	$0.0391 \pm 46.64_J$	$-1.287 \pm 391.6_J$	$0.126 \pm 404.1_J$
3,000	5	−5	$-0.0282 \pm 38.39_J$	$0.0483 \pm 50.78_J$	$-0.2578 \pm 357.5_J$	$0.0091 \pm 409.6_J$

Solution. Table 7.5 lists the eigenvalues for the system with symmetric and asymmetric bearing-stiffness matrices for the different rotor speeds. The table clearly demonstrates that a large-magnitude skew-symmetric part of the stiffness matrix can make the machine unstable, as shown by the real part of the eigenvalue becoming positive.

7.10 Summary

In this chapter, we demonstrate that instabilities can arise in rotor–bearing systems due to rotor asymmetry, internal rotor damping, and any components in the stator with nonsymmetric stiffness properties. An oil-film bearing is an example of the latter case. For a linear system with constant coefficients, stability can be assessed by examining real parts of the system eigenvalues. For linear systems with periodic coefficients, Floquet theory may be used to assess stability.

It is shown that if the bearings and foundation are isotropic, the analysis of systems with an asymmetric rotor is best performed using a set of coordinates that are fixed to and rotate with the rotor. The resulting second-order differential equations have constant coefficients. In contrast, for systems with a symmetric rotor, the analysis is best performed in coordinates that are fixed in space (see Chapters 3, 5, and 6). The resulting second-order differential equations again have constant coefficients.

If the rotor is asymmetric and the bearing and foundation are anisotropic, then the resulting second-order differential equations have periodic coefficients with a period equivalent to twice the rotor speed, irrespective of the coordinate system used. These equations can be solved by using techniques such as numerical integration or the harmonic-balance method.

7.11 Problems

7.1 A slender shaft 0.7 m long is supported at each end in stiff, plain, self-aligning bearings and at the center carries a disk of mass 30 kg, diametral moment of inertia of 2 kg m², and polar moment of inertia of 3.75 kg m². The shaft is made of steel with a modulus of 210 GPa and the shaft diameter is 30 mm. Obtain equations of motion for the system in both coordinates fixed in space and rotating with the shaft, including gyroscopic effects but neglecting the mass of the shaft. Determine the system natural frequencies when the shaft rotates at 3,000 rev/min in both fixed and rotating coordinates. Compare the frequencies in the two reference frames, noting that the rotor spin speed is equivalent to 50 Hz.

7.2 A slender shaft 0.7 m long is supported at each end in stiff, plain, self-aligning bearings and carries a noncircular disk 0.2 m from the left end. The inertia properties of the disk are $I_{d\bar{x}} = 0.75\,\mathrm{kg\,m^2}$, $I_{d\bar{y}} = 1.5\,\mathrm{kg\,m^2}$, and $I_p = 2.1\,\mathrm{kg\,m^2}$, and the mass is 30 kg. The shaft is made of steel with a modulus of 210 GPa and the shaft diameter is 30 mm. Obtain equations of motion for the system in rotating coordinates, including gyroscopic effects but neglecting the mass of the shaft, and investigate the stability of the system when the shaft rotates at 2,400, 2,600, and 2,800 rev/min. This example demonstrates that mass asymmetry can cause instability.

7.3 A slender shaft 0.7 m long is supported at each end in stiff, plain, self-aligning bearings and at the center carries a disk of mass 30 kg, diametral moment of inertia of 2 kg m², and polar moment of inertia of 3.75 kg m². The shaft is made of steel with a modulus of 210 GPa. It is elliptical in cross section with a major axis of 32 mm and a minor axis of 30 mm. Obtain equations of motion for the system in rotating coordinates, including gyroscopic effects but neglecting the mass of the shaft, and investigate the stability of the system when the shaft rotates at 1,900, 2,000, and 2,100 rev/min.

The second moments of area for an ellipse are $I_{xx} = \dfrac{\pi d_y d_x^3}{64}$, $I_{yy} = \dfrac{\pi d_y^3 d_x}{64}$, where d_x and d_y are the major and minor axes of the ellipse, respectively.

7.4 A slender shaft 0.7 m long is supported at each end in plain, self-aligning bearings and at the center carries a disk of mass 30 kg, diametral moment of inertia of 2 kg m², and polar moment of inertia of 3.75 kg m². The shaft is made of steel with a modulus of 210 GPa and the diameter is 30 mm. The effective viscous-damping coefficient for the shaft due to internal damping is 50 Ns/m when the disk displaces along the x- or y-axis and 8 Nms/rad when the disk rotates about these axes. In addition, the motion of the central disk is constrained by an external damper with a viscous-damping coefficient of 10 Ns/m, as shown in Figure 7.8. This damper acts when the disk displaces along the x- and y-axes; it has no influence on the rotations of the disk about these axes. What are the damping factors due to both internal and external damping when the rotor is at rest? Obtain equations of motion for the system, including gyroscopic effects but neglecting the mass of the shaft, and investigate the stability of the system when the shaft rotates at 2,200 and 2,300 rev/min.

7.5 The disk in the rotor system described in Problem 7.3 has an out-of-balance of 5×10^{-4} kg m. Determine the response of the rotor at the disk when the rotor spins at 1,900 rev/min if the out-of-balance is (a) in the axis of the major diameter of the elliptical shaft cross section, (b) in the axis of the minor diameter of the elliptical shaft cross section, and (c) at 45° to the major diameter axis.

7.6 Determine the gravity critical speed for the system described in Problem 7.3 by using Equation (7.42). Verify this result by letting $s = j\Omega$ in Equation (7.22).

7.7 A Jeffcott rotor consists of a shaft carrying a disk at the midspan. The disk is of mass m, diametral moment of inertia I_d, and polar moment of inertia I_p. The stiffnesses of the shaft at the midspan relating forces and displacements along the Ox and Oy axes are k_T and the stiffnesses relating moments and rotations about the Ox and Oy axes are k_R. Neglecting the mass of the shaft shows that $s_r = s_f \pm j\Omega$, where s_r is a root of the equations of motion in rotating coordinates, s_f is a root of the equations of motion in coordinates fixed in space, and Ω is the speed of rotation of the shaft. Which conditions determine whether the rotor spin speed is added or subtracted from roots in fixed coordinates?

7.8 The rigid rotor shown in Figure 3.26 is 0.5 m long. Bearing 1 is a rigidly supported, short, self-aligning bearing and bearing 2 is a flexibly supported, short, self-aligning bearing. The rotor has a polar moment of inertia of 0.8 kg m^2 and has maximum and minimum diametral moments of inertia about bearing 1 of 10.6 and 10.2 kg m^2, respectively. The bearing-support stiffnesses are 1 MN/m in both the Ox and Oy directions. Determine the equations of motion for this system in rotating coordinates, in terms of the angles θ and ψ. Include gyroscopic effects in the analysis. Then:

(a) Determine the stability of the system when it spins at 1,500 and 1,540 rev/min.

(b) Determine the critical speed due to gravity.

(c) The center of mass of the rotor is halfway along the rotor and slightly offset from the centerline of the bearings; therefore, the product of the rotor mass and the offset is 0.004 kg m. The offset is in a direction 45° from the direction of the maximum diametral moment of inertia. Determine the response when the rotor spins at 1,500 rev/min.

7.9 A rigid rotor, shown in Figure 3.26, is 0.5 m long. Bearing 1 is a short, rigid bearing and bearing 2 is a short, flexibly supported bearing. The rotor has a polar moment of inertia of 0.6 kg m^2 and a diametral moment of inertia about bearing 1 of 10 kg m^2. The bearing-support stiffness, k, is 1 MN/m in both the x and y directions. There is also a cross-coupling stiffness, k_c, between the x and y directions of 0.2 MN/m. The forces acting on the bearing in the x and y directions are $f_x = ku + k_c v$ and $f_y = -k_c u + kv$, respectively, where u and v are the displacements at bearing 2 in the x and y directions. The difference between this problem and Problem 3.4 is the negative term in the expression for f_y.

Calculate the eigenvalues and natural frequencies when the rotor is stationary and when it rotates at 3,000 rev/min. Comment on the stability of the system and compare your solutions to those of Problem 3.4.

7.10 Consider the machine in Problem 5.2, where the short, rigid bearings are replaced with short, isotropic, flexible bearings of stiffness 5 MN/m and damping of 100 Ns/m. The rotor has internal damping proportional to the rotor stiffness with a factor of 10^{-5} s^{-1}. Using eight Euler-Bernoulli elements, calculate the rotor spin speed when the rotor first becomes unstable.

7.11 Consider the machine in Problem 5.2. The rotor is now assumed to be asymmetric and modeled by increasing the shaft stiffness in one direction by 10 percent. Using eight Euler-Bernoulli elements, calculate the ranges of rotor spin speed below 5,000 rev/min, at which the machine is unstable. At 3,000 rev/min, what are the first three natural frequencies of the system in rotating coordinates and the equivalent pseudonatural frequencies in fixed coordinates?

8 Balancing

8.1 Introduction

In all rotating machinery, some degree of mass unbalance is always present. Chapter 6 shows that small deviations in mass symmetry about the axis of rotation can lead to significant unbalance forces being exerted on the bearings. It is imperative to minimize these forces because they can lead to damaging vibration levels in the rotor, bearings, supporting structure, and ancillary equipment. Mass unbalance also can cause large shaft responses, leading to rotor–stator rubs or high stresses in the rotor. This unbalance is controlled primarily by attention to tolerances in rotor manufacture, but this alone is rarely sufficient and some means of further reducing vibration levels on the complete machine is necessary.

In this chapter, we discuss methods to achieve a satisfactory state of balance for a machine by adjusting the distribution of mass on the rotor, by either adding or removing correction masses at specific locations. Here, we concentrate on how to determine which unbalance corrections are appropriate; we do not focus on how, practically, these corrections are achieved. There are several approaches to the balancing of a rotor and all are based on the assumption of system linearity. Occasionally, iteration is required in the case of machines the bearings of which exhibit some degree of nonlinearity. The fundamental concepts of balancing are to monitor the effect on the synchronous (i.e., 1X) response of the machine due to one or more small masses attached to the rotor and then to scale those influences to determine the extent of the unbalance present in the rotor. Usually, these masses are added at predetermined axial locations, called *balance planes*. This is simple enough for the case of a rigid rotor, where no bending deflection of the rotor occurs; this is the first case discussed. A similar approach may be adopted for a flexible rotor; however in this case, several alternative strategies are available. We see that what may be a benign state of unbalance for a rotor operating in one situation can produce rather severe vibrations in a different situation.

In most cases in which balancing is attempted on a rotor, some type of *shaft marker* is used to generate a signal to act as a phase reference for the synchronous components of various vibration, stress, or force signals that may be measured. A feature of rotor balancing is that signals of many different origins may be used to both determine which unbalance correction masses should be applied and to assess how successful the unbalance correction was. This chapter includes a brief discussion of the shaft marker and an outline of methods that may be used to balance rotors without using phase information.

Ordinarily, balancing takes place in several stages. At the design stage, the distribution of mass in the rotor is known and designers must try to ensure that the center of mass of the rotor is on the axis of rotation (see Section 8.2). When a rotor is still in the manufacture stage, some balance adjustments are usually made. Under these circumstances, most of the rotor is readily accessible and the objective is to balance it sufficiently well for the machine to run up to the rated rotational speed without endangering integrity. Finally, *trim-balancing* is performed on the machine in the running state. Typically, a finer degree of balance is now required, but access to different parts of the rotor is more restricted. Frequently, only one or two balance planes are accessible, often situated close to the bearings. The techniques of factory and trim balancing are identical and are termed *field balancing*.

It is important to highlight some terminology before proceeding. A *flexible machine* is one that operates close to or above the first natural frequency. However, it is important to note that such a machine may have a rigid rotor, such that the rotor moves as a rigid body. This situation is common in smaller machines. In recent decades, the trend has been toward *flexible rotors*, which deform under the action of unbalance forces and moments; these are often found in larger units such as turbogenerators.

This chapter provides only a brief introduction to the theory of balancing. Ehrich (1999), Muszyńska (2005), Parkinson et al. (1980), and Rieger (1986) should be consulted for further details of the theory and practice of balancing. Foiles et al. (1998) provide an overview of the state of the art in balancing and list many papers in the literature. Many of the techniques were introduced in the 1950s and 1960s and the papers of Bishop and Gladwell (1959) and Parkinson (1965, 1966, 1967) provide an historical perspective, particularly with respect to modal balancing. Parkinson (1967, 1991) summarized much of this work. The International Organization for Standardization (ISO) (1997, 1998, 2003, 2007) also provides an overview of balancing approaches.

8.2 Balancing Rigid Rotors at the Design Stage

A rotating machine typically is designed in such a way as to make the net out-of-balance forces and moments zero. In practice, this can never be achieved because even if a rotor is designed to have perfect balance, rotor dimensions and build tolerances cause a small amount of unbalance to exist when the rotor is assembled. If this unbalance is too large for the safe operation of a machine, it can be reduced to an acceptable level by field balancing. This process is described in Sections 8.4 and 8.5.

Assuming that the magnitudes and distribution of masses in the rotor are known, then balancing a rigid rotor at the design stage is relatively straightforward. Suppose that the rotor consists of P masses, each with known radial positions and distributed along the rotor at specific planes. Then, the out-of-balance force, f_{ub}, and the out-of-balance moment, M_{ub}, generated by the rotor when it spins with angular velocity Ω are

$$f_{ub} = \left(\sum_{i=1}^{P} m_i \varepsilon_i \right) \Omega^2 \quad \text{and} \quad M_{ub} = \left(\sum_{i=1}^{P} m_i \varepsilon_i z_i \right) \Omega^2 \qquad (8.1)$$

where the center of mass of m_i is at a radius ε_i and lies in a plane at a distance z_i from some arbitrary chosen *reference plane*. The radius ε_i is complex so that both the angular position of the mass and its distance from the centerline can be specified by this single complex quantity. Correspondingly, the force f_{ub} and the moment M_{ub} are also complex.

We now introduce two balance planes that may or may not coincide with any of the planes in which the original rotor masses are situated. To make the out-of-balance force zero requires

$$\left(m_{b1} \varepsilon_{b1} + m_{b2} \varepsilon_{b2} + \sum_{i=1}^{P} m_i \varepsilon_i \right) \Omega^2 = 0 \qquad (8.2)$$

and to make the out-of-balance moments zero requires

$$\left(m_{b1} \varepsilon_{b1} z_{b1} + m_{b2} \varepsilon_{b2} z_{b2} + \sum_{i=1}^{P} m_i \varepsilon_i z_i \right) \Omega^2 = 0 \qquad (8.3)$$

where m_{b1} and m_{b2} represent the masses added at the two balance planes, ε_{b1} and ε_{b2} denote the positions of the two correction masses within the planes, and z_{b1} and z_{b2} are the axial locations of the two balancing planes, respectively, relative to the reference plane. Rearranging Equations (8.2) and (8.3), canceling the Ω^2 multiplier, and writing the resulting equations in matrix form give

$$\begin{bmatrix} 1 & 1 \\ z_{b1} & z_{b2} \end{bmatrix} \begin{Bmatrix} b_{b1} \\ b_{b2} \end{Bmatrix} = - \sum_{i=1}^{P} m_i \varepsilon_i \begin{Bmatrix} 1 \\ z_i \end{Bmatrix} \qquad (8.4)$$

where $b_{b1} = m_{b1} \varepsilon_{b1}$ and $b_{b2} = m_{b2} \varepsilon_{b2}$. Inverting the 2×2 matrix gives

$$\begin{Bmatrix} b_{b1} \\ b_{b2} \end{Bmatrix} = \frac{1}{z_{b1} - z_{b2}} \begin{bmatrix} z_{b2} & -1 \\ -z_{b1} & 1 \end{bmatrix} \left(\sum_{i=1}^{P} m_i \varepsilon_i \begin{Bmatrix} 1 \\ z_i \end{Bmatrix} \right) \qquad (8.5)$$

Thus, the required unbalance corrections, b_{b1} and b_{b2}, may be determined. In many practical situations, the radii at which unbalance corrections can be applied are predetermined; hence, $|\varepsilon_{b1}|$ and $|\varepsilon_{b2}|$ are known. The correction masses then are determined uniquely. In the remainder of this chapter, determining appropriate unbalance corrections, such as b_{b1} and b_{b2}, to minimize some measure of response is emphasized.

EXAMPLE 8.2.1. When all the mass of a rigid rotor is accounted for, it is found that most of it is distributed symmetrically about the axis of rotation; however, three discrete masses are not symmetrically distributed, as listed in Table 8.1. Balance plane 1 is provided at the reference plane $z = 0$, and balance plane 2 is 1.5 m from it, as shown in Figure 8.1. Determine the sizes and angular positions of the balance masses that must be added to the rotor at a radius of 0.2 m to balance this rotor.

Table 8.1. *The mass distribution of the assembled rotor*

Mass (kg)	Radius (m)	Angle ($°$)	Distance from reference plane (m)
1.0	0.2	0	0.45
1.1	0.2	126	0.90
1.0	0.2	258	1.35

Solution. In this case, $z_{b1} = 0$ and $z_{b2} = 1.5$ m. To obtain the complex unbalance, recall that $e^{j\beta} = \cos\beta + j\sin\beta$ for phase angle β given in radians. From Table 8.1, the complex radial positions of the masses are $\varepsilon_1 = 0.2e^{j0} = 0.2$ m, $\varepsilon_2 = 0.2e^{j126\pi/180} = (-0.1176 + 0.1618j)$ m, and $\varepsilon_3 = 0.2e^{j258\pi/180} = (-0.04158 - 0.1956j)$ m. The axial positions of the masses are $z_1 = 0.45$ m, $z_2 = 0.90$ m, and $z_3 = 1.35$ m. Solving Equation (8.5) gives

$$b_{b1} = (-0.08412 - 0.05163j)\,\text{kg m} \quad \text{and} \quad b_{b2} = (0.05501 + 0.06928j)\,\text{kg m}$$

Thus, the required unbalance correction at balance plane 1 has magnitude $|b_{b1}| = 0.09870$ kg m and phase $-148.46°$. The unbalance correction needed at balance plane 2 has magnitude $|b_{b2}| = 0.08846$ kg m and phase $51.55°$. Because $b_{b1} = m_{b1}\varepsilon_{b1}$ and $b_{b2} = m_{b2}\varepsilon_{b2}$, then with $|\varepsilon_{b1}| = |\varepsilon_{b2}| = 0.2$ m, m_{b1} is 0.4935 kg and m_{b2} is 0.4423 kg.

8.3 The Shaft Marker and the Phase of Response Signals

Most balancing operations require that there is a once-per-revolution signal available to provide a phase reference for other signals that are used for balancing, which is variously referred to as either a *key phasor* or a *shaft marker*. This reference signal is generated by a feature fixed on the rotor detected by a sensor on the stator – for example, a notch (or protrusion) detected by a proximity sensor, a strip of reflective tape detected by an optical sensor, or an isolated conducting patch detected by pair of contacting brushes. In each case, a signal is generated that is perfectly periodic provided that the spin speed of the shaft remains constant. The signal generally has a sharp upward or downward leading edge that can be used to define $\phi = 0$. Without this signal, only the magnitudes of vibration response can be measured directly.

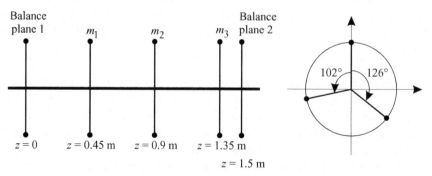

Figure 8.1. The layout of the assembled rotor (Example 8.2.1).

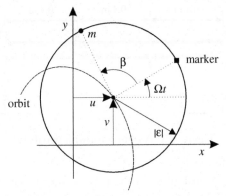

Figure 8.2. The definition of the marker position and unbalance mass location on a disk.

With the shaft-marker signal, a unique value also can be associated with the phase of any synchronous signal.

The phase angle of any response signal can be obtained directly by evaluating the fundamental (i.e., once-per-revolution) components of the Fourier series for both the response and the shaft-marker signals. If a response measurement, $r(t)$, is purely sinusoidal with frequency equal to the rotational speed, Ω, then for a magnitude c_r and a phase angle γ_r

$$r(t) = c_r \cos(\Omega(t - t_0) + \gamma_r) \tag{8.6}$$

The time t_0 represents the instant at which the recording of the signals began and it may be regarded as an unknown.

The shaft-marker signal, $s(t)$, is not generally a pure sinusoid but it can be described by a Fourier series because it is periodic. This consists of a constant component, c_{s0}; a fundamental component with magnitude c_{s1} and phase angle γ_{s1}; and higher-order components. Thus,

$$s(t) = c_{s0} + \sum_{i=1}^{\infty} c_{si} \, \cos(\Omega(t - t_0) + \gamma_{si}) \tag{8.7}$$

The phase of the response signal relative to the shaft marker, denoted by γ_{rs}, is calculated as

$$\gamma_{rs} = \gamma_r - \gamma_{s1} \tag{8.8}$$

The angular positions of the unbalance masses also may be defined relative to the shaft marker. Figure 8.2 shows the case of a single unbalance mass, m, where the unbalance is given as

$$b = m|\varepsilon|e^{J\beta} \tag{8.9}$$

8.4 Field Balancing of Rigid Rotors

Rigid rotors present the simplest balancing problems. The instantaneous set of transverse displacements of any rigid rotor can be specified fully by four displacement coordinates, and these may be defined in many different ways. One logical

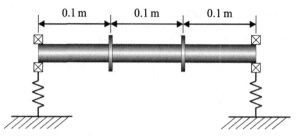

Figure 8.3. A rigid rotor with two disks.

definition includes the translations of the rotor center of gravity in the x and y directions and the rotations of the rotor about the axes through the center of gravity parallel to the x and y directions, respectively. The terms *bounce mode* and *tilt mode* often are applied to the pure translation of the rotor in a transverse direction and pure rotation of the rotor about a transverse axis through the center of gravity, respectively. These terms are slightly misleading in the sense that neither is necessarily a natural mode of the system. Strictly speaking, they refer to patterns of displacement of the rotor.

Any pattern of unbalance on a rotor that can directly excite (i.e., do work on) a bounce mode is referred to as *static unbalance*. Static unbalance is present on a rotor when any single unbalance mass is present, whether or not the unbalance is located in the same plane as the center of gravity of the rotor. Any pattern of unbalance on a rotor that can directly excite a tilt mode is referred to as *dynamic unbalance*. Dynamic unbalance may be present when two unbalance masses are present on a rotor located in different planes. The terms *static* and *dynamic unbalance* may seem to imply in one case that the level of excitation is independent of the rotational speed, whereas in the other case, the level of excitation depends on the rotational speed. In fact, the excitation levels of both are proportional to the square of rotational speed. The term *static unbalance* originates from the fact that if a horizontal-axis rotor is supported on perfect frictionless bearings, gravity causes it to rotate such that the center of gravity of the rotor ultimately moves to the lowest possible point. The example with which most people are familiar is a bicycle wheel lifted off the ground; it rotates, until the center of gravity (usually on the same radial line as the tire valve) finds the lowest possible position. Even very large rotors supported on hydrostatic bearings behave similarly, although the time required to discover the static unbalance to worthwhile accuracy can be many tens of hours. By contrast, gravity does not cause any movement at all on a rotor that has only dynamic unbalance.

Figure 8.3 illustrates a rigid rotor with a length of 0.3 m and a diameter of 50 mm, with two thin disks with a diameter of 100 mm and a thickness of 10 mm. The disks are mounted at 100 mm from each end and the rotor is supported on identical flexible bearings at either end of the shaft. The bearings are short with a translational stiffness of 10 kN/m in each transverse direction and a damping coefficient of 20 Ns/m. The rotational stiffnesses of the bearings are zero and no damping elements are shown in Figure 8.3. The rotor has a mass of 5.51 kg and a moment of inertia about the center of gravity of 0.0405 kg m^2. One natural mode shape of this rotor is the bounce mode, in which all points on the rotor move with the same magnitude and phase; another is the tilt mode, in which opposite ends of the rotor move

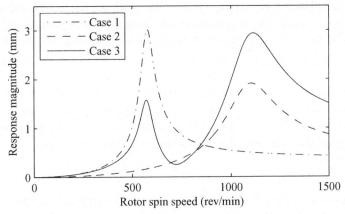

Figure 8.4. Response of the rigid rotor at $z = 0$ to the different unbalance conditions.

in diametrically opposite directions with the same magnitude. In this example, the bounce mode has a lower natural frequency than the tilt mode.

Figure 8.4 shows the effects of the three different unbalance configurations given in Table 8.2. In the first case, the unbalance forces at the two disks are in-phase; therefore, only the bounce mode is excited. In contrast, in the second case, the two unbalance forces are 180° out-of-phase; hence, there is no net force to excite the bounce mode – there is, however, a couple that excites the tilt mode. The third case has an unbalance distribution that is a combination of the first two cases; therefore, it is not surprising that the response curve shows peaks at both critical speeds. With a rigid rotor on nonisotropic supports, there can be four peaks in the overall unbalance response (see Section 6.2.5). A rigid rotor can be balanced perfectly by adding appropriate unbalance correction masses at two distinct planes, irrespective of the dynamic properties of the support structure. For this reason, most current balancing operations treat the rotor as a rigid object.

For many machines, such as short axial fans or *pancake motors* (in which the rotor has a large diameter compared to its length), an acceptable behavior can be achieved by applying unbalance corrections in only a single plane.

8.4.1 Single-Plane Balancing

To illustrate the principles of single-plane balancing, the motion of a symmetric rigid rotor is considered. The rotor is mounted on general anisotropic supports; that is, the support stiffnesses may be different in the x and y directions but, for illustration

Table 8.2. *Unbalance cases for the rigid rotor*

Case	Unbalance at disk 1		Unbalance at disk 2	
	Magnitude (kg m)	Phase (°)	Magnitude (kg m)	Phase (°)
1	0.001	0	0.001	0
2	0.001	0	0.001	180
3	0.002	0	0.001	180

purposes, the angular degrees of freedom of the rotor are neglected so that the rotor may be modeled using only two translational degrees of freedom. In effect, we are assuming that any unbalance present lies on a plane of symmetry normal to the axis of the rotor and that the stiffnesses of the supports on either side of this plane are identical. Unbalance corrections also are applied at this plane.

Suppose that the initial response of the rotor, at spin speed Ω, is r_0. If an additional unbalance b_c $(=m_c\varepsilon_c)$ is fixed to the balancing plane, a new response, r_c, is measured after the rotor is accelerated to the same speed, Ω. Assuming that the rotor system behaves linearly, this new response can be written as

$$r_c = r_0 + R(\Omega)b_c \tag{8.10}$$

where $R(\Omega)$ is a response function independent of the trial balance added. The objective is to determine the appropriate unbalance correction, b_c, such that $r_c = 0$. Evidently, this requires knowledge of $R(\Omega)$. If a reliable numerical model is available, it may be used; however, this model must represent not only the mechanical dynamics of the rotor system but also any dynamics and gains of the measurement system used. In practice, it is preferable to establish this response function by testing.

A *trial balance* b_1 $(= m_1\varepsilon_1)$ is fixed to the balancing plane and a new response, r_1, is measured obeying

$$r_1 = r_0 + R(\Omega)b_1 \tag{8.11}$$

Define

$$r_d = (r_1 - r_0) \tag{8.12}$$

Then, from Equation (8.11)

$$R(\Omega) = \frac{r_d}{b_1} \tag{8.13}$$

Having determined the response function, $R(\Omega)$, from Equation (8.10) with $r_c = 0$

$$b_c = \frac{-r_0}{R(\Omega)} = \frac{-r_0 \times b_1}{r_d} \tag{8.14}$$

A practical issue is evident in Equation (8.14). Calculating the corrective unbalance involves a division by the quantity r_d. If this quantity is small in magnitude compared to the magnitude of r_0, then the degree of uncertainty on this number due to noise, sensor errors, quantization errors, or numerical rounding may be significant. It is standard practice to select b_1 such that $|r_d|$ is similar to $|r_0|$.

In this simple illustration of balancing, the same unbalance correction, b_c, is obtained irrespective of the properties of the supports; however, they do affect the responses produced. To see this, the initial response can be written as

$$r_0 = R(\Omega)b_0 \tag{8.15}$$

for an initial unbalance, b_0. Then, the unbalance correction that we have computed here is exact and $b_c = -b_0$. In the more general situation of multiplane balancing, the optimal set of unbalance corrections can depend on the dynamic stiffness properties of the stator. This approach to single-plane balancing also may be implemented graphically by plotting the responses on the complex plane. This is demonstrated in Example 8.4.1.

EXAMPLE 8.4.1. The initial *as-found response* of an axial fan was measured. The response was 15 μm with phase 30° ahead of (leading) the shaft-marker signal. A trial mass of 0.1 kg was added to the fan at a radius of 200 mm and an angular position 165° ahead of the reference mark. The response was then measured as 20 μm at 105°. What mass should be added to the fan at radius 100 mm to balance it and at what angular position?

Solution. Because $30° = \pi/6$ rad and $105° = 7\pi/12$ rad, the responses may be written in complex form as

$$r_0 = 15 \times e^{J\pi/6} = (12.99 + 7.50J) \ \mu\text{m}$$

and

$$r_1 = 20 \times e^{7J\pi/12} = (-5.18 + 19.32J) \ \mu\text{m}$$

Hence,

$$r_d = (-18.17 + 11.82J) \ \mu\text{m}$$

The trial balance is

$$b_1 = m_1\varepsilon_1 = 0.02 \times (-0.9659 + 0.2588J) = (-0.019318 + 0.005176J) \ \text{kg m}$$

and, thus, the response function, $R(\Omega)$, is found as

$$R(\Omega) = \frac{r_d}{b_1} = (1030.36 - 335.717J) \ \mu\text{m/kg m}$$

From Equation (8.14), the unbalance correction is calculated to be

$$b_c = (-0.00925 - 0.01029J) \ \text{kg m}$$

Thus, the magnitude of the unbalance correction is estimated to be 0.0138 kg m and the phase angle is 228° ahead of the shaft marker. Because this correction is fixed at a radius $|\varepsilon_c| = 0.1$ m, the magnitude of the correction mass is 0.138 kg.

Figure 8.5 shows the graphical solution to this problem. Usually, the response and balance-mass diagrams are combined on the same axes; however, they have been separated here for clarity. The response plot shows the effect of the trial-balance mass, given by r_d and the desired effect to balance the machine, $-r_0$. Thus, the machine is balanced when the trial balance is rotated by 63° and the magnitude is multiplied by a factor of $|r_0| / |r_d| = 0.692$. This is shown in the balance-mass plot, Figure 8.5(b).

8.4.2 Two-Plane Balancing

Two-plane balancing, in which unbalance-correction masses are added to two balance planes, is sufficient to balance a machine with a rigid rotor. Although mass is distributed along the shaft, only the position of the center of mass determines the net unbalance force. The net unbalance moment is determined by the misalignment

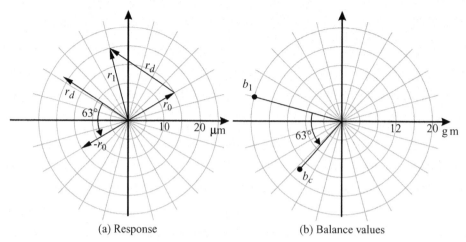

(a) Response   (b) Balance values

Figure 8.5. The graphical solution for single-plane balancing of an axial fan. Note that each sector is 15°.

between the rotor axis and the principal axes of inertia (see Section 6.2.1). Both the net unbalance force and the net unbalance moment can be adjusted by changing the distribution of mass at two axially separated planes.

For a linear system, it is always possible to write a linear relationship between the displacements and the applied forces. When the applied forces originate from unbalance only, we can write

$$\mathbf{r} = \mathbf{R}(\Omega)\,\mathbf{b} \tag{8.16}$$

where $\mathbf{r}$ represents a vector of displacements, $\mathbf{b}$ represents a vector of unbalance quantities, and $\mathbf{R}(\Omega)$ is a response-function matrix. All of these are complex in general. Where the measured responses are displacements, the entries of $\mathbf{R}(\Omega)$ can be found from the *receptance matrix* for the complete system (multiplied by Ω^2) (see Section 6.3.2). However, as for single-plane balancing, numerical models are generally not sufficiently accurate and $\mathbf{R}(\Omega)$ should be estimated from extra trial runs. For two-plane balancing, we use at least two independent response quantities. Initially, the method is described using two responses, and the general method is described in Section 8.5.1. In the case of the system in Figure 8.6, Equation (8.16) takes the form

$$\begin{Bmatrix} r_1 \\ r_2 \end{Bmatrix} = \mathbf{R}(\Omega) \begin{Bmatrix} b_1 \\ b_2 \end{Bmatrix} \tag{8.17}$$

where b_1 and b_2 represent the initial unbalances on the two disks according to

$$\begin{aligned} b_1 &= m_1\varepsilon_1 \quad \text{for the first disk} \\ b_2 &= m_2\varepsilon_2 \quad \text{for the second disk} \end{aligned} \tag{8.18}$$

Here, m_1 and m_2 are the unbalance masses at the two planes and ε_1 and ε_2 are complex numbers representing their locations within those planes. In Equation (8.18), r_1 and r_2 represent independent response quantities that should be measured at two different planes.

Letting $\mathbf{r}_0$ represent the vector of initial responses, Equation (8.16) can be used to express the response vector $\mathbf{r}_c$ in the presence of any set of correction unbalances

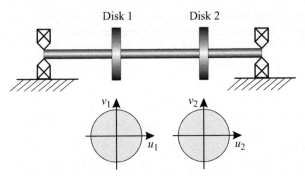

Figure 8.6. A rigid rotor with two disks (balance planes).

$\mathbf{b}_c$ according to

$$\mathbf{r}_c = \mathbf{r}_0 + \mathbf{R}\left(\Omega\right)\mathbf{b}_c \tag{8.19}$$

If the response matrix, $\mathbf{R}\left(\Omega\right)$, is known, then an appropriate unbalance-correction vector, $\mathbf{b}_c$, can be determined directly. Unfortunately, $\mathbf{R}\left(\Omega\right)$ is not yet known and trial balances must be added to estimate this matrix. For single-plane balancing, $\mathbf{R}\left(\Omega\right)$ is a 1×1 matrix (i.e., the scalar, $R(\Omega)$) and to determine this requires the addition of one trial-balance mass. For two-plane balancing, $\mathbf{R}\left(\Omega\right)$ is a 2×2 matrix and determining this requires the addition of two sets of trial-balance masses. With two balance planes, a single trial-balance mass may be added to each plane in turn. Alternatively, trial-balance masses may be added to both planes for both trial runs. If the trial masses are added one plane at a time, it is guaranteed that all of the trial-balance vectors, $\mathbf{b}$, are independent. Where the trial-balance masses are added in combinations, care must be taken to ensure that the various vectors are independent, for reasons that will become apparent.

Consider that a combination of trial balances, $\mathbf{b}_1$, is added to the rotor. Responses $\mathbf{r}_1$ are measured and, because the same response matrix applies

$$\mathbf{r}_1 = \mathbf{r}_0 + \mathbf{R}\left(\Omega\right)\mathbf{b}_1 \tag{8.20}$$

Now, the first set of trial balances is removed and a different set, $\mathbf{b}_2$, is added. The vector of responses, $\mathbf{r}_2$, is measured and we have

$$\mathbf{r}_2 = \mathbf{r}_0 + \mathbf{R}\left(\Omega\right)\mathbf{b}_2 \tag{8.21}$$

Defining

$$\mathbf{r}_{d1} = \mathbf{r}_1 - \mathbf{r}_0, \qquad \mathbf{r}_{d2} = \mathbf{r}_2 - \mathbf{r}_0 \tag{8.22}$$

we obtain from Equations (8.20) and (8.21)

$$\mathbf{r}_{d1} = \mathbf{R}\left(\Omega\right)\mathbf{b}_1 \quad \text{and} \quad \mathbf{r}_{d2} = \mathbf{R}\left(\Omega\right)\mathbf{b}_2 \tag{8.23}$$

Equation (8.23) represents four scalar equations for the four unknowns in $\mathbf{R}\left(\Omega\right)$. Combining these equations gives

$$\left[\mathbf{r}_{d1}\ \mathbf{r}_{d2}\right] = \mathbf{R}\left(\Omega\right)\left[\mathbf{b}_1\ \mathbf{b}_2\right] \tag{8.24}$$

and, from this, it is evident that

$$\mathbf{R}\left(\Omega\right) = \left[\mathbf{r}_{d1}\ \mathbf{r}_{d2}\right]\left[\mathbf{b}_1\ \mathbf{b}_2\right]^{-1} \tag{8.25}$$

Table 8.3. *Rotor responses for the two-plane balancing example*

Trial balance	Disk 1		Disk 2	
	Amplitude (mm)	Phase (°)	Amplitude (mm)	Phase (°)
As found	0.4	120	0.3	240
100 g at 150 mm, 0°, Disk 1	0.5	140	0.35	150
100 g at 150 mm, 0°, Disk 2	0.45	50	0.4	300

Using this to substitute for $\mathbf{R}(\Omega)$ in Equation (8.19) with $\mathbf{r}_c = \mathbf{0}$ yields

$$\mathbf{b}_c = -\mathbf{R}(\Omega)^{-1}\mathbf{r}_0 = -\begin{bmatrix} \mathbf{b}_1 & \mathbf{b}_2 \end{bmatrix}\begin{bmatrix} \mathbf{r}_{d1} & \mathbf{r}_{d2} \end{bmatrix}^{-1}\mathbf{r}_0 \qquad (8.26)$$

Equation (8.26) is analogous to the single-plane case given in Equation (8.14). This approach works for anisotropic and coupled supports in which the response is measured only in a single direction at each of the balance planes.

For this method to be successful, it is necessary that the estimated unbalance correction, $\mathbf{b}_c$, is not sensitive to small quantities of measurement noise on the response vectors $\mathbf{r}_0$, $\mathbf{r}_1$, and $\mathbf{r}_2$. A major concern is that the changes in response $\mathbf{r}_{d1}$ and $\mathbf{r}_{d2}$ are significant compared with $\mathbf{r}_0$ (to reduce the effect of measurement noise) and different from each other to ensure that the inversion of $\begin{bmatrix} \mathbf{r}_{d1} & \mathbf{r}_{d2} \end{bmatrix}$ in Equation (8.26) is well conditioned. This requires a suitable choice of the trial-balance masses $\mathbf{b}_1$ and $\mathbf{b}_2$. Indeed, if the matrix $\begin{bmatrix} \mathbf{b}_1 & \mathbf{b}_2 \end{bmatrix}$ is poorly conditioned, then the matrix $\begin{bmatrix} \mathbf{r}_{d1} & \mathbf{r}_{d2} \end{bmatrix}$ also will be poorly conditioned from Equation (8.24).

EXAMPLE 8.4.2. Table 8.3 shows the measurements taken at two balance planes or disks. The data consist of the amplitude and phase of the response at each disk from the machine that requires balancing and from the machine after two distinct balance masses were added. Determine appropriate unbalance corrections.

Solution. From these data, the following values may be derived:

$$\mathbf{r}_0 = \begin{Bmatrix} 0.4 \times e^{j \times 120 \times \pi/180} \\ 0.3 \times e^{j \times 240 \times \pi/180} \end{Bmatrix} = \begin{Bmatrix} -0.200 + 0.346j \\ -0.150 - 0.2560j \end{Bmatrix} \text{ mm}$$

$$\mathbf{r}_1 = \begin{Bmatrix} 0.5 \times e^{j \times 140 \times \pi/180} \\ 0.35 \times e^{j \times 150 \times \pi/180} \end{Bmatrix} = \begin{Bmatrix} -0.383 + 0.321j \\ -0.303 + 0.175j \end{Bmatrix} \text{ mm}$$

$$\mathbf{r}_2 = \begin{Bmatrix} 0.45 \times e^{j \times 50 \times \pi/180} \\ 0.4 \times e^{j \times 300 \times \pi/180} \end{Bmatrix} = \begin{Bmatrix} 0.289 + 0.345j \\ 0.200 - 0.346j \end{Bmatrix} \text{ mm}$$

$$\mathbf{r}_{d1} = \begin{Bmatrix} -0.183 - 0.025j \\ -0.153 + 0.435j \end{Bmatrix} \text{ mm}, \qquad \mathbf{r}_{d2} = \begin{Bmatrix} 0.489 - 0.002j \\ 0.350 - 0.087j \end{Bmatrix} \text{ mm}$$

and, thus,

$$[\mathbf{r}_{d1} \ \mathbf{r}_{d2}]^{-1} = \begin{bmatrix} 0.486 + 1.681j & -0.100 - 2.372j \\ 2.137 + 0.661j & 0.087 - 0.892j \end{bmatrix} \text{mm}^{-1}$$

Because

$$\mathbf{b}_1 = \begin{Bmatrix} 0.1 \times 0.15 \\ 0 \end{Bmatrix} \text{kg m}, \quad \mathbf{b}_2 = \begin{Bmatrix} 0 \\ 0.1 \times 0.15 \end{Bmatrix} \text{kg m}$$

the 2×2 response matrix, $\mathbf{R}$, is calculated using Equation (8.25) as

$$\mathbf{R} = \begin{bmatrix} 0.0073 + 0.0252j & -0.0015 - 0.0356j \\ 0.0321 + 0.0099j & 0.0013 - 0.0134j \end{bmatrix} \text{mm/kg m}$$

Then, the unbalance correction is obtained from Equation (8.26) as

$$\mathbf{b}_c = \begin{Bmatrix} 0.0192 - 0.0032j \\ 0.0135 - 0.0108j \end{Bmatrix} \text{kg m} = \begin{Bmatrix} 0.130 \times 0.150 \times e^{j \times 351 \times \pi/180} \\ 0.115 \times 0.150 \times e^{j \times 321 \times \pi/180} \end{Bmatrix} \text{kg m}$$

Thus, to balance the machine, a mass of 130 g at a radius of 150 mm should be added to disk 1 at an angle of 351° ($= -9°$), and a mass of 115 g at a radius of 150 mm should be added to disk 2 at 321° ($= -39°$).

8.5 Field Balancing of Flexible Rotors

Many large systems have flexible rotors that undergo significant flexural deformation during operation. Balancing such systems requires more care than balancing rigid rotors. The most obvious change is that a machine with a flexible rotor can have several critical speeds within the range of operating speeds. For the rigid rotors that have been discussed thus far, if the rotor is balanced using two planes at any given speed, then it is balanced at all speeds. This is not the case with a flexible rotor. The method discussed in Section 8.4.2 is extended readily to become the *influence-coefficient method* for flexible rotors, and this approach is discussed herein. In the standard version of influence-coefficient balancing, the balancing is undertaken at a single speed. However, good balance at one single speed does not guarantee that a rotor is well behaved as it runs up or down through the critical speeds. The alternative is to extend the influence-coefficient method to multiple speeds or to apply the modal approach to balance machines with flexible rotors. These approaches are described in the next two subsections.

8.5.1 The Influence-Coefficient Method

With a flexible rotor, it may be necessary to balance in several planes; the actual number required is dependent on the number of critical speeds that significantly influence the response in the operating range. With N independent balancing planes, it is possible (in theory) to zero any N discrete outputs from that machine. These outputs can be the responses at N different locations on the machine at a single rotational speed or the responses at a small number of points for several different spin speeds. A more general approach is to minimize some quantity that involves M

machine outputs, at each of m spin speeds, where $Mm \geq N$ and where N is equal to the number of balance planes.

In all cases, the influence-coefficient method relies on estimating a response matrix, $\mathbf{R}(\Omega)$, relating changes in the vector of responses to changes in the applied unbalance. Difficulties can arise if the response matrix is ill-conditioned. This occurs if adding mass to each balance plane does not produce a significantly different response (e.g., if the balance planes are too close together or if the shaft speed is too close to a critical speed) or if the balance planes do not have any effect on one or more responses.

We begin by considering a widely used approach, in which the balancing is performed at a single speed Ω and also $M = N$ (i.e., the number of independent measurable quantities is identical to the number of balancing planes). Let the measured machine-response vector be denoted by $\mathbf{r}$. At spin speed Ω, some response, $\mathbf{r}_0$, is measured arising from initial unbalance on the rotor. This initial unbalance generally does not lie in the planes in which unbalance corrections can be applied. We will see that this fact can be important for flexible rotors, especially when the rotor is run at speeds other than the balancing speed or when the dynamic properties of the supporting structure may change. When an additional (i.e., correction) unbalance, $\mathbf{b}_c$ is applied, the measured response changes to $\mathbf{r}_c$ given by

$$\mathbf{r}_c = \mathbf{r}_0 + \mathbf{R}(\Omega)\mathbf{b}_c \qquad (8.27)$$

The response vector has length M and the unbalance vector has length N; thus, $\mathbf{R}(\Omega)$ is an $M \times N$ matrix. In the present case, $M = N$ and $\mathbf{R}(\Omega)$ is square. Equation (8.19) of the two-plane balancing section is a special case of Equation (8.27) with $M = N = 2$.

Suppose that p trial runs are performed. For each trial run, masses are added or removed to obtain a different vector of additional unbalance $\mathbf{b}_\ell$ for $\ell = 1, \ldots, p$. In each case, the unbalance response is $\mathbf{r}_\ell$, where

$$\mathbf{r}_\ell = \mathbf{r}_0 + \mathbf{R}(\Omega)\mathbf{b}_\ell \qquad (8.28)$$

Defining $\mathbf{r}_{d\ell}$ as

$$\mathbf{r}_{d\ell} = \mathbf{r}_\ell - \mathbf{r}_0 \qquad (8.29)$$

we find

$$\mathbf{r}_{d\ell} = \mathbf{R}(\Omega)\mathbf{b}_\ell \qquad (8.30)$$

If the number of trial runs equals the number of balance planes (i.e., $p = N$), then the response matrix, $\mathbf{R}(\Omega)$, may be estimated as

$$\mathbf{R}(\Omega) = \begin{bmatrix} \mathbf{r}_{d1} & \mathbf{r}_{d2} & \ldots & \mathbf{r}_{dN} \end{bmatrix}\begin{bmatrix} \mathbf{b}_1 & \mathbf{b}_2 & \ldots & \mathbf{b}_N \end{bmatrix}^{-1} \qquad (8.31)$$

From Equation (8.27), to obtain $\mathbf{r}_c = \mathbf{0}$, we require

$$\mathbf{b}_c = -(\mathbf{R}(\Omega))^{-1}\mathbf{r}_0 = -\begin{bmatrix} \mathbf{b}_1 & \mathbf{b}_2 & \ldots & \mathbf{b}_N \end{bmatrix}\begin{bmatrix} \mathbf{r}_{d1} & \mathbf{r}_{d2} & \ldots & \mathbf{r}_{dN} \end{bmatrix}^{-1}\mathbf{r}_0 \qquad (8.32)$$

Because only one spin speed was used to estimate the initial unbalance, the unbalance-correction vector obtained is strictly accurate only at the speed examined, and the response is zero only at the measured degrees of freedom. If a sufficient number of balance planes is used, the unbalance-correction procedure is likely to reduce the response of the rotor even at unmeasured degrees of freedom and at speeds other than those from which balancing information was extracted.

It is assumed here that the number of balance planes is equal to the number of unbalance responses to minimize. For a given shaft speed, this leads to a set of equations in which the number of equations equals the number of unknowns and the balancing objectives may be met exactly; that is, the balanced response is zero. Often, however, there are more independent measurements available than balance planes ($M > N$) or else the measurements may be taken at more than one speed, with the result that $Mm > N$. In this case, it is not possible to reduce all of the responses to zero, but a weighted average of the squared response amplitudes may be minimized. With multiple spin speeds, Ω_i for $i = 1, \ldots, m$, multiple response matrices, $\mathbf{R}(\Omega_i)$, are involved, but the same unbalance-correction vector, $\mathbf{b}_c$, applies at all spin speeds. For practical reasons, it is common practice to record the same set of responses for each different spin speed.

The general influence-coefficient method uses the assumption of linearity to write the vector of responses, $\mathbf{r}_c(\Omega_i)$, as a linear function of the unbalance corrections, $\mathbf{b}_c$, applied at the balance planes. The same correction is applied at each of the m different balancing speeds, Ω_i, so that

$$\mathbf{r}_c(\Omega_i) = \mathbf{r}_0(\Omega_i) + \mathbf{R}(\Omega_i)\,\mathbf{b}_c, \quad \text{for all } i = 1, \ldots, m \tag{8.33}$$

Defining

$$\mathbf{s}_c = \begin{Bmatrix} \mathbf{r}_c(\Omega_1) \\ \mathbf{r}_c(\Omega_2) \\ \vdots \\ \vdots \\ \mathbf{r}_c(\Omega_m) \end{Bmatrix}, \quad \mathbf{s}_0 = \begin{Bmatrix} \mathbf{r}_0(\Omega_1) \\ \mathbf{r}_0(\Omega_2) \\ \vdots \\ \vdots \\ \mathbf{r}_0(\Omega_m) \end{Bmatrix}, \quad \text{and} \quad \mathbf{S} = \begin{bmatrix} \mathbf{R}(\Omega_1) \\ \mathbf{R}(\Omega_2) \\ \vdots \\ \vdots \\ \mathbf{R}(\Omega_m) \end{bmatrix} \tag{8.34}$$

then Equation (8.33) can be replaced by

$$\mathbf{s}_c = \mathbf{s}_0 + \mathbf{S}\mathbf{b}_c \tag{8.35}$$

where $\mathbf{s}_0$ and $\mathbf{s}_c$ are vectors of length Mm, the matrix $\mathbf{S}$ has dimensions $Mm \times N$, and the unbalance correction $\mathbf{b}_c$ is a vector of length N.

If the number of trial runs equals the number of balance planes ($p = N$), then $\mathbf{S}$ may be obtained using the same procedure for obtaining $\mathbf{R}$ given in Equation (8.31). Here, we consider the general case when potentially there are more trial-balance vectors than balance planes ($p \geq N$). Suppose that $\mathbf{s}_\ell$ represents the response in the presence of the ℓth unbalance vector, $\mathbf{b}_\ell$, for $\ell = 1, \ldots, p$ and defining $\mathbf{s}_{d\ell} = \mathbf{s}_\ell - \mathbf{s}_0$; then

$$\begin{bmatrix} \mathbf{s}_{d1} & \mathbf{s}_{d2} & \ldots & \mathbf{s}_{dp} \end{bmatrix} = \mathbf{S}\begin{bmatrix} \mathbf{b}_1 & \mathbf{b}_2 & \ldots & \mathbf{b}_p \end{bmatrix} \tag{8.36}$$

The matrix containing the trial-balance vectors is rectangular and cannot be simply inverted to obtain $\mathbf{S}$ using an expression equivalent to Equation (8.31). However, the solution for $\mathbf{S}$ that minimizes the least squares error in Equation (8.36) is obtained as

$$\mathbf{S} = \begin{bmatrix} \mathbf{s}_{d1} & \mathbf{s}_{d2} & \ldots & \mathbf{s}_{dp} \end{bmatrix}\begin{bmatrix} \mathbf{b}_1 & \mathbf{b}_2 & \ldots & \mathbf{b}_p \end{bmatrix}^{\dagger} \tag{8.37}$$

where † denotes the right pseudoinverse of a matrix and is defined by $\mathbf{A}^{\dagger} = \mathbf{A}^{\top}\left[\mathbf{A}\mathbf{A}^{\top}\right]^{-1}$.

Given $\mathbf{S}$, the optimal unbalance-correction vector, $\mathbf{b}_c$, can be defined as the vector of correction unbalances that minimizes the following scalar *cost function*:

$$J\left(\mathbf{b}_c\right) = \mathbf{s}_c^{\mathrm{H}}\mathbf{W}^2\mathbf{s}_c = \left(\mathbf{s}_0 + \mathbf{S}\mathbf{b}_c\right)^{\mathrm{H}}\mathbf{W}^2\left(\mathbf{s}_0 + \mathbf{S}\mathbf{b}_c\right) \tag{8.38}$$

where $\mathbf{W}$ is a real diagonal matrix of weights and the superscript H denotes the complex conjugate transpose. This is a weighted least squares linear estimation problem (Golub and van Loan, 1996) in the unknown optimal unbalance correction $\mathbf{b}_c$. The solution may be obtained as

$$\mathbf{b}_c = -\left(\mathbf{S}^{\mathrm{H}}\mathbf{W}^2\mathbf{S}\right)^{-1}\left(\mathbf{S}^{\mathrm{H}}\mathbf{W}^2\mathbf{s}_0\right) \tag{8.39}$$

or by other direct optimization methods (Bigret, 2004; Untaroiu et al., 2008). Because the weighting factors in $\mathbf{W}$ may have different units, it is possible to combine multiple different response signals into the balancing objective. These may include absolute displacements, stresses, displacements of the rotor relative to the stator, loads transmitted at the bearings, and so forth.

In general, utilizing more balance planes improves the quality of the balancing. Using more than two balance planes may be possible during manufacture, but it is often impractical for an operational plant. Often, only two balance planes are used, which is frequently satisfactory for machines operating at speeds well below the third critical speed. The machine may be balanced using more balancing planes than the number of significant modes, but this can lead to nonunique solutions for the unbalance correction. Using more balancing speeds also tends to improve the quality of balancing, up to a point. However, the improvements achieved by including increasingly more speeds in a given range diminish rapidly because the information acquired at the many different balancing speeds ceases to be independent.

The quality of the estimated unbalance corrections from the influence-coefficient method depends critically on trial masses causing a significant change in the response of a machine. This change in the response depends on the dynamics of a machine and also significantly on the position of the balance planes where the trial balance and, ultimately, the unbalance correction are added. Usually, problems occur if the balance planes are too close, causing a similar change in response.

Three examples are now presented.

EXAMPLE 8.5.1. To illustrate the influence-coefficient method, consider the rotor shown in Figure 8.7. In practice, the measurements are taken directly from the machine and a model is neither available nor required. However, in this example, the measured vibration responses are simulated using an FE model with 12 Euler-Bernoulli shaft elements. For simplicity, the combined support and bearing stiffnesses are equal in the vertical and horizontal directions and are 1 MN/m. The bearings have damping coefficients of 1 kNs/m in the vertical and horizontal directions. The shaft is 1.5 m long and 30 mm in diameter. Each of the two disks has a diameter of 0.25 m and a thickness of 20 mm. The shaft and disks are made of steel with $E = 211\,\mathrm{GPa}$ and $\rho = 7{,}810\,\mathrm{kg/m}^3$. Figure 8.8 shows the response during a rundown for this machine in the as-found condition, in which the measurements are taken at the disks, which are also the balance planes. The

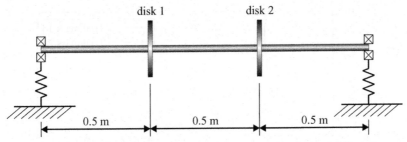

Figure 8.7. A machine with a flexible rotor (Example 8.5.1).

initial unbalance is modeled using three masses placed along the rotor at axial positions 0.375, 0.75, and 1.375 m from the left bearing. The initial unbalances have magnitudes of 0.6, 0.4, and 0.2 g m with phases of 90°, 120°, and 0°, respectively.

Trial-balance masses of 0.2 and 0.3 g m are added, in turn, to the two balance planes shown by the disks in Figure 8.7. For a simple illustration, both are added at the zero phase; however, in general, this is not a requirement. Figure 8.9 shows the new responses during rundowns with these balance weights fitted. From these three curves, the amplitude of vibration at any rotational speed may be determined. Using the exact simulated response during a rundown from the as-found condition and with the additional trial-balance masses, perform two-plane balancing at 3,000 and 1,000 rev/min.

In practice, the response cannot be measured exactly. Tables 8.4 and 8.5 list the amplitude (to three significant figures) and phase (to the nearest degree) response for the different responses at the two speeds. Balance the machine using these approximate data and demonstrate the effect of measurement errors on the unbalance correction and the balanced response.

Solution. The machine has critical speeds at 812 and 3,074 rev/min, obtained by equating the rotor spin speed to a forward-whirling, damped natural frequency. When this machine operates at 3,000 rev/min, it is running near the

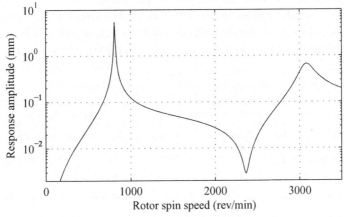

Figure 8.8. The response at disk 1 of the machine with a flexible rotor in the as-found condition (Example 8.5.1).

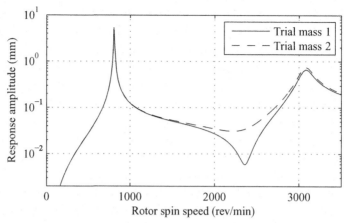

Figure 8.9. The responses at disk 1 during two rundowns with the trial masses fitted for the machine with a flexible rotor (Example 8.5.1)

second critical speed and the dynamics are influenced mainly by the first two forward-whirling flexural modes. This is a common situation in large machinery. In many instances, the running speed is above the second critical speed, in which case the third forward-whirling mode also may have significant influence.

The approach used to calculate the balance masses is exactly the same as that used for two-plane balancing, given by Equation (8.26). For the exact simulated data, the resulting unbalance corrections are 0.8248 g m on disk 1 at $-84.7°$ and 0.1552 g m on disk 2 at $-116.2°$. When the calculated values are added to the rotor at the required phases, the new unbalance distribution yields the response during a rundown shown in Figure 8.10. The scales are different between Figure 8.10 and Figures 8.8 and 8.9 and the vibration responses were reduced substantially over the speed range. Also, the vibration response at 3,000 rev/min was reduced to zero, as required. The new curve does illustrates the important distinction between rigid and flexible rotors. Although the response levels at the balancing speed of 3,000 rev/min are close to zero, there is substantial vibration as the machine goes through the first critical speed.

Now, consider balancing the rotor at 1,000 rev/min. Often, it is necessary to balance a rotor at a speed other than the operational speed, as is the case here. Using the exact simulated responses at 1,000 rev/min and the two-plane

Table 8.4. *Vibration response at 3,000 rev/min (Example 8.5.1)*

Trial balance	Disk 1 response		Disk 2 response	
	Amplitude (mm)	Phase (°)	Amplitude (mm)	Phase (°)
As found	0.442	52	0.510	−120
0.2 g m at 0°, disk 1	0.436	35	0.500	−136
0.3 g m at 0°, disk 2	0.502	76	0.597	−103

Table 8.5. *Vibration response at 1,000 rev/min (Example 8.5.1)*

Trial balance	Disk 1 response		Disk 2 response	
	Amplitude (mm)	Phase (°)	Amplitude (mm)	Phase (°)
As found	0.125	−79	0.133	−79
0.2 g m at 0°, disk 1	0.123	−92	0.131	−92
0.3 g m at 0°, disk 2	0.125	−100	0.132	−97

balancing procedure, the required correction unbalances are 0.8871 g m on disk 1 at −81.1° and 0.0243 g m on disk 2 at −7.2°. The addition of these correction masses leads to the response during a rundown shown in Figure 8.11. Comparison of the responses of these two balanced machines (i.e., Figures 8.10 and 8.11) shows significant differences, but this should not be surprising. The response simply was minimized at different speeds and, because we are using two-plane balancing, the speed dependence is correctly represented only if the machine has two forward-whirling modes contributing significantly to the vibration responses over the range of running speeds. However, the responses of the balanced machine in both cases are significantly reduced from the initial response, by a factor of approximately 100. It is clear that in the speed range of interest, two forward-whirling modes dominate, which is why the calculated balance masses are similar for balancing at 1,000 and 3,000 rev/min. Also, when the machine is balanced at 3,000 rev/min, the reduction in the response near the second resonance is significantly greater than the reduction in the first resonance because the second forward-whirling mode dominates the response at this speed. The opposite happens at 1,000 rev/min, at which the first forward-whirling mode dominates.

Suppose that the machine is balanced at 3,000 rev/min and then operated at this speed. The responses are zero at the disks, as shown in Figure 8.10, because

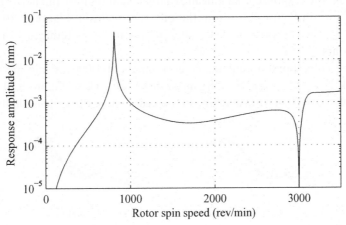

Figure 8.10. The response at disk 1 of the machine with a flexible rotor, balanced at 3,000 rev/min, Example 8.5.1.

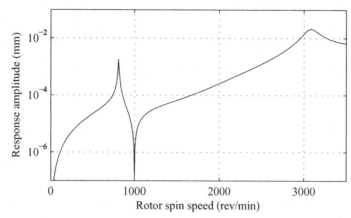

Figure 8.11. The response at disk 1 of the machine with a flexible rotor, balanced at 1,000 rev/min (Example) 8.5.1.

they were used to balance the machine. Figure 8.12 shows the ODS of the rotor when it is operating at 3,000 rev/min, after being balanced at 3,000 rev/min. The response at the disks is zero; however, the response is clearly non-zero at other locations on the rotor. Also shown is the result when the machine is both balanced and operated at 1,000 rev/min.

The use of exact response data in the balancing procedure enables a machine to be balanced perfectly. However, in practice, the response cannot be measured exactly; and therefore, a zero response is never fully achieved. Tables 8.4 and 8.5 list the approximate initial response of the machine and the approximate responses with the two trial-balance masses added. If the balancing exercise is repeated with these data, then the unbalance corrections for 3,000 rev/min are calculated as 0.8239 g m on disk 1 at −85.2° and 0.1560 g m on disk 2 at −117.4°. The accuracy of the unbalance correction is limited by the constraints of the practical implementation. Figure 8.13 shows the response of the balanced machine in this case, where the scale is identical to that for the initial response given in Figure 8.8. Clearly, the response at 3,000 rev/min is non-zero, and the response is significantly larger than that for the perfect balancing case shown in Figure 8.10 (note the different scales in Figures 8.10 and 8.13). However, the response at disk 1 at 3,000 rev/min was reduced from 0.442 to 0.0226 mm.

The corresponding unbalance corrections at 1,000 rev/min are 0.8871 g m on disk 1 at −81.2° and 0.0238 g m on disk 2 at −11.2°. Figure 8.14 shows the response of the balanced machine in this case, where the scale is identical to that

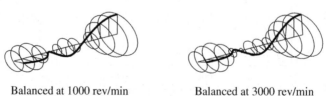

Balanced at 1000 rev/min   Balanced at 3000 rev/min

Figure 8.12. The operating deflection shapes of the machine with a flexible rotor at 1,000 and 3,000 rev/min (Example 8.5.1). The machine is balanced at the running speed in each case.

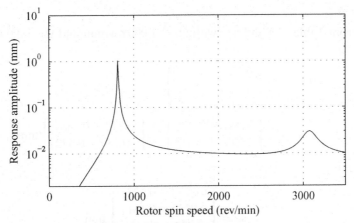

Figure 8.13. The response at disk 1 of the machine with a flexible rotor, balanced at 3,000 rev/min using the approximate data in Table 8.4 (Example 8.5.1).

for Figure 8.8. Similarly, by comparing Figures 8.11 and 8.14 (note the different scales), the performance of the balancing exercise has been compromised by the errors in the measurements. However, the response at disk 1 at 1,000 rev/min was reduced from 0.125 to 0.0040 mm.

EXAMPLE 8.5.2. Repeat Example 8.5.1 but use the response at different rotational speeds and for the same two sets of trial-balance masses. Consider the following cases:

(1) at 1,000 and 3,000 rev/min for both disks, using the approximate data in Tables 8.4 and 8.5
(2) at 1,000 and 3,000 rev/min for both disks, using the exact simulated responses
(3) for speeds 0 to 3,500 rev/min in increments of 10 rev/min for disk 1 only, using the exact simulated responses

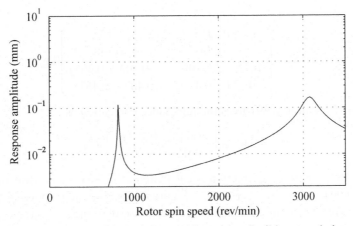

Figure 8.14. The response at disk 1 of the machine with a flexible rotor, balanced at 1,000 rev/min using the approximate data in Table 8.5 (Example 8.5.1).

Solution

(1). From Tables 8.4 and 8.5, with $\Omega_1 = 1,000\,\text{rev/min}$ and $\Omega_2 = 3,000\,\text{rev/min}$, we have

$$\mathbf{r}_0(\Omega_1) = \begin{Bmatrix} 0.125e^{-J79\pi/180} \\ 0.133e^{-J79\pi/180} \end{Bmatrix} = \begin{Bmatrix} 0.0239 - 0.123J \\ 0.0254 - 0.131J \end{Bmatrix} \text{mm},$$

$$\mathbf{r}_0(\Omega_2) = \begin{Bmatrix} 0.442e^{J52\pi/180} \\ 0.510e^{-J120\pi/180} \end{Bmatrix} = \begin{Bmatrix} 0.272 + 0.348J \\ -0.255 - 0.442J \end{Bmatrix} \text{mm},$$

and, thus,

$$\mathbf{s}_0 = \begin{Bmatrix} 0.0239 - 0.123J \\ 0.0254 - 0.131J \\ 0.272 + 0.348J \\ -0.255 - 0.442J \end{Bmatrix} \text{mm}$$

Similarly,

$$\mathbf{s}_1 = \begin{Bmatrix} -0.00429 - 0.123J \\ -0.00457 - 0.131J \\ 0.357 + 0.250J \\ -0.360 - 0.347J \end{Bmatrix} \text{mm} \quad \text{and} \quad \mathbf{s}_2 = \begin{Bmatrix} -0.0217 - 0.123J \\ -0.0161 - 0.131J \\ 0.121 + 0.487J \\ -0.134 - 0.582J \end{Bmatrix} \text{mm}$$

Thus,

$$\begin{bmatrix} \mathbf{s}_{d1} & \mathbf{s}_{d2} \end{bmatrix} = \begin{bmatrix} -0.0281 & -0.0456 \\ -0.0299 & -0.0415 \\ 0.0850 - 0.09822J & -0.151 + 0.139J \\ -0.105 + 0.0943J & 0.121 - 0.140J \end{bmatrix} \text{mm}$$

The trial-balance vectors are combined to give

$$\begin{bmatrix} \mathbf{b}_1 & \mathbf{b}_2 \end{bmatrix} = \begin{bmatrix} 0.2 & 0 \\ 0 & 0.3 \end{bmatrix} \text{g m}$$

The matrix $\mathbf{S}$ is then estimated from Equation (8.37), noting that $\begin{bmatrix} \mathbf{b}_1 & \mathbf{b}_2 \end{bmatrix}$ is square, as

$$\mathbf{S} = \begin{bmatrix} -0.141 & -0.152 \\ -0.150 & -0.138 \\ 0.425 - 0.491J & -0.502 + 0.463J \\ -0.523 + 0.472J & 0.402 - 0.467J \end{bmatrix} \text{kg}^{-1}$$

Solving Equation (8.39) gives the unbalance correction as

$$\mathbf{b}_c = \begin{bmatrix} 0.143 - 0.783J \\ 0.00125 - 0.0994J \end{bmatrix} \text{g m}$$

Thus, the unbalance correction is calculated as 0.7964 g m on disk 1 at $-79.6°$ and 0.0994 g m on disk 2 at $-89.3°$. Figure 8.15 shows the response at disk 1 for the balanced machine. The maximum response over the speed range is smaller when the machine is balanced using two speeds simultaneously, as opposed to a single speed.

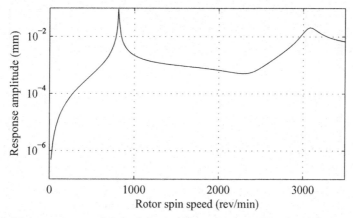

Figure 8.15. The response at disk 1 of the machine with a flexible rotor, balanced using responses at 1,000 and 3,000 rev/min and using the approximate data in Tables 8.4 and 8.5 (Example 8.5.2).

(2). When the exact simulated responses are used, then the unbalance correction is calculated as 0.8124 g m on disk 1 at $-79.3°$ and 0.0946 g m on disk 2 at $-86.0°$. Figure 8.16 shows the response at disk 1 for the balanced machine. The response is smaller than for the machine balanced using approximate data, particularly at the two balance speeds.

(3). Minimizing the response at disk 1 for all rotational speeds shows the required unbalances to be 0.8127 g m on disk 1 at $-79.3°$ and 0.0935 g m on disk 2 at $-85.8°$. These balance-mass values are similar to those obtained in Example 8.5.1 because two forward-whirling modes dominate in the frequency range of interest. Figure 8.17 shows the balanced response of the machine and should be compared to Figures 8.10 and 8.11. It is clear that using the entire frequency range reduced the maximum response of the rotor within that frequency range.

EXAMPLE 8.5.3. The effect of the position of the balancing planes is now demonstrated. Figure 8.18 shows the machine of Example 8.5.1, but with two massless,

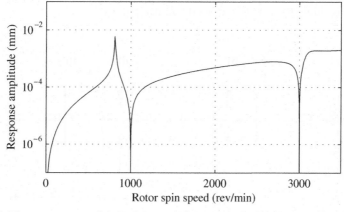

Figure 8.16. The response at disk 1 of the machine with a flexible rotor, balanced using responses at 1,000 and 3,000 rev/min and using the exact simulated responses (Example 8.5.2).

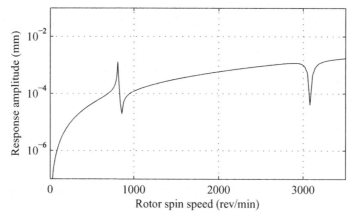

Figure 8.17. The response at disk 1 of the machine with a flexible rotor, balanced using the response over the whole frequency range (Example 8.5.2).

rigid disks added; 0.375 and 1.250 m from the left end. The support, shaft, and disk properties are unchanged; therefore, Tables 8.4 and 8.5 give the response of this system to the trial masses. Table 8.6 shows these data at disk 1 as well as the response at disk 1 when the trial mass is added to the new balance planes. The amplitude data are given to three significant figures and the phase to the nearest degree and, hence, contain measurement errors.

Balance the machine using the approximate response at disk 1 given in Table 8.6 at both rotor spin speeds for pairs of balance planes 2 and 3, 1 and 4, and 1 and 2. Compare the simulated balanced responses in these cases.

Solution. The machine is balanced using three pairs of balance planes. The first set consists of balance planes 2 and 3, which represents the planes used in Example 8.5.2. The second set consists of balance planes 1 and 4, which have maximum spacing and are likely to lead to good unbalance-correction estimates. In contrast, the final set consists of balance planes 1 and 2, which are very close and likely to lead to poor estimates. Figure 8.19 shows the unbalance response at disk 1 for the as-found condition and the machine balanced using these three sets of balance planes. The figure also shows that the balanced response is smaller for widely spaced balance planes.

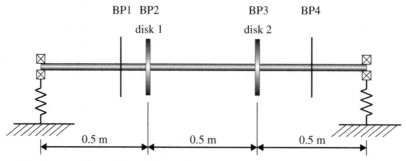

Figure 8.18. The machine with a flexible rotor and four balance planes (Example 8.5.3). BP denotes the balance plane.

Table 8.6. *Vibration response of disk 1 at 1,000 and 3,000 rev/min, with trial mass added to all balance planes (Example 8.5.3)*

Trial balance	1,000 rev/min		3,000 rev/min	
	Amplitude (mm)	Phase (°)	Amplitude (mm)	Phase (°)
As found	0.125	−79	0.442	52
0.2 g m at 0°, BP 1	0.123	−89	0.439	33
0.2 g m at 0°, BP 2	0.123	−92	0.436	35
0.3 g m at 0°, BP 3	0.125	−100	0.502	76
0.2 g m at 0°, BP 4	0.123	−92	0.517	76

BP = blance plane.

8.5.2 Modal Balancing

The influence-coefficient approach to balancing remains the most common approach to field balancing of large machines. There is an alternative, the *modal method*, which offers a number of advantages but requires knowledge of the rotor's dynamic properties a priori. In particular, the deflections of the rotor must be known for each of the low-frequency mode shapes of the machine. In the influence-coefficient approach, several trial runs are made at a single rotational speed, with correction masses applied independently. By contrast, in the modal approach, the rotor is balanced at or close to each critical speed in turn, and the objective is to eliminate the excitation of the modes in turn rather than to force the rotor response to zero at a single speed. The additional work involved in these runs is reduced by

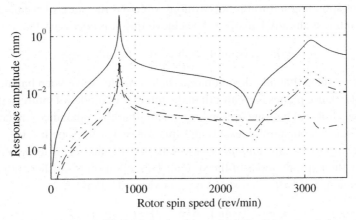

Figure 8.19. The response at disk 1 of the machine with a flexible rotor, balanced using the data in Table 8.6 (Example 8.5.3). The solid line is the as-found response, and the other responses represent balancing using planes 2 and 3 (dashed), 1 and 4 (dash-dot), and 1 and 2 (dotted).

choosing combinations of balance masses to excite only the particular mode under consideration. In this way, the balancing at each critical speed is equivalent to a single-plane balancing exercise. Only those modes that are excited are of interest; thus, the backward-whirling modes for a machine on isotropic supports are neglected.

If **b** represents the total unbalance distribution on the rotor due to the initial unbalance and any unbalance corrections subsequently applied, then to balance the ℓth mode of interest, we require

$$\mathbf{u}_\ell^\top \mathbf{b} = 0 \tag{8.40}$$

where $\mathbf{u}_\ell$ is the ℓth mode shape. In cases in which there are some components of skew-symmetry in the system matrices (e.g., machines with significant gyroscopic effects or machines on relatively flexible fluid bearings) there is a distinction between the left and the right eigenvectors corresponding to any given eigenvalue. In the case of Equation (8.40), $\mathbf{u}_\ell$ represents the left eigenvector associated with the ℓth eigenvalue.

Modal balancing generally begins by balancing the lower-frequency modes first and then attending to higher-frequency modes without disturbing the state of balance of the lower modes. There is a sound reason for this. To drive the net excitation of any one mode to zero, it is necessary to determine that excitation. This usually involves running the rotor at a rotational speed close to the resonance frequency of the mode of interest so that this one mode tends to dominate the response. However, if there is strong excitation of the lower modes already present, then it is often impossible to accelerate the rotor through these modes without causing damage.

The total unbalance correction, $\mathbf{b}_c$, consists of the sum of unbalance corrections of each mode of interest, $\{\mathbf{b}_{c1}, \mathbf{b}_{c2}, \ldots\}$; thus,

$$\mathbf{b}_c = \mathbf{b}_{c1} + \mathbf{b}_{c2} + \ldots \tag{8.41}$$

and each modal unbalance correction is chosen so that it adjusts the net excitation of the corresponding natural mode without affecting the state of balance of those modes of interest already balanced. To achieve this, it is necessary that

$$\mathbf{u}_i^\top \mathbf{b}_{ci} \neq 0 \quad \text{and} \quad \mathbf{u}_j^\top \mathbf{b}_{ci} = 0 \quad \text{for all } j < i \tag{8.42}$$

Here, it is assumed that all relevant modes below mode i have been balanced. If the number of balance planes is greater than the number of modes of interest, then there is still some choice in $\mathbf{b}_{ci}$; one option is to ensure that the unbalance correction does not excite selected higher modes. Using Equation (8.42), vectors $\{\mathbf{b}_{c1}, \mathbf{b}_{c2}, \ldots\}$ can be determined *up to a scalar multiplier* before the balancing begins using information about only the natural modes of the system. This equation is satisfied automatically if some sequence of *basis vectors* $\{\mathbf{e}_1, \mathbf{e}_2, \ldots\}$ exists such that

$$\mathbf{u}_i^\top \mathbf{e}_i = 1 \quad \text{and} \quad \mathbf{u}_j^\top \mathbf{e}_i = 0 \quad \text{for all } j < i \tag{8.43}$$

and if, for some arbitrary (i.e., complex) scalars, α_i

$$\mathbf{b}_{ci} = \alpha_i \mathbf{e}_i \quad \text{for all } i \tag{8.44}$$

Because it is assumed here that the levels of unbalance on a rotor are relatively small, the natural modes of the system are not affected by this lack of balance. Balancing the general ith mode of interest of the machine involves determining the scaling factor α_i, which can be found using what is, in effect, an influence-coefficient approach. Initially, the unbalance correction on the rotor is characterized by $\alpha_1 = 0$, $\alpha_2 = 0$, and so forth. Measuring any one response of the rotor at a rotational speed close to the first critical speed, in theory, is sufficient to determine α_1. In practice, a vector of responses, $\mathbf{r}_{01}$, is measured for the uncorrected rotor and then a trial balance of $\mathbf{b}_1 = \delta_1 \mathbf{e}_1$ is added to the rotor, where δ_1 is an arbitrary positive real scalar. The new rotor response, $\mathbf{r}_1$, is measured and the difference $\mathbf{r}_{d1} = \mathbf{r}_1 - \mathbf{r}_{01}$ is determined. An appropriate choice of δ_1 must be made so that $|\mathbf{r}_{d1}|$ is the same order of magnitude as $|\mathbf{r}_{01}|$. Then, α_1 is determined from

$$\alpha_1 = -\delta_1 \frac{\mathbf{r}_{d1}^H \, \mathbf{r}_{01}}{\mathbf{r}_{d1}^H \, \mathbf{r}_{d1}} \tag{8.45}$$

where the superscript H denotes the Hermitian transpose. This equation is an extension of single-plane balancing, Equation (8.14), and a particular case of the general influence-coefficient method, Equation (8.39).

This method for calculating α_1 ensures that the response contributions present in $\mathbf{r}_{01}$ due to higher modes tend to have minimal effect on the calculated α_1. The unbalance correction, $\mathbf{b}_{c1} = \alpha_1 \mathbf{e}_1$, is applied to the rotor, which is left in place. Having balanced the first mode of interest, the second mode of interest is then addressed. The rotor is run up to a speed at which the contributions from this second mode are dominant. A new vector, $\mathbf{r}_{02}$, is measured at this speed. Then, a trial balance of $\mathbf{b}_2 = \delta_2 \mathbf{e}_2$ is added to the rotor, where δ_2 is another arbitrary real positive number. The new rotor response, $\mathbf{r}_2$, is measured and used to determine $\mathbf{r}_{d2} = \mathbf{r}_2 - \mathbf{r}_{02}$. Then, α_2 is determined from

$$\alpha_2 = -\delta_2 \frac{\mathbf{r}_{d2}^H \, \mathbf{r}_{02}}{\mathbf{r}_{d2}^H \, \mathbf{r}_{d2}} \tag{8.46}$$

and a further unbalance correction, $\mathbf{b}_{c2} = \alpha_2 \mathbf{e}_2$, is applied to the rotor and left in place. The process continues until all modes of interest are balanced.

Although modal balancing involves prior knowledge of the system, it is easily implemented. If masses are added to each plane in proportion to the rotor mode shape (or, strictly speaking, orthogonal to the other modes), then each mode can be balanced effectively using the single-plane approach. There are practical issues in modal balancing that are now discussed. Knowledge of the rotor mode shapes is vital, which normally are obtained from an FE model or from measurements. Often, the mode shapes are estimated by ignoring gyroscopic effects and assuming that the axisymmetric rotor is carried on isotropic supports. The corresponding mode shapes are then planar, which is strictly invalid for machines with nonclassical damping or significant gyroscopic effects. However, for many machines, the assumption of planar mode shapes is sufficient for successful balancing. The number of balance planes to use is also important. In the influence-coefficient method, the number of planes required is related to the number of measurement points. However, in the modal method, the number of balance planes required is equal to the number of modes to be balanced. Although multiple-balance planes are required to balance

First mode of interest   Second mode of interest

Figure 8.20. The first two mode shapes in the vertical plane of a machine with a flexible rotor shown in Figure 8.7 and modeled using 12 elements (Example 8.5.4).

multiple modes, because each mode is effectively balanced as a single plane, exercise measurements are required only at a single location. In practice, measurements at multiple locations are used to reduce the effect of noise. The following example illustrates the approach.

EXAMPLE 8.5.4. Consider the rotor described in Example 8.5.1. The first two mode shapes are obtained from the FE analysis by ignoring gyroscopic effects and are shown in Figure 8.20. Balance these first two modes, using the two balance planes at the two disks and the responses at 820 and 3,075 rev/min. These speeds are close to the first two critical speeds of the machine at 812 and 3,074 rev/min, as illustrated by the rundown response shown in Figure 8.8. The response measurements are taken at disk 1, and Table 8.7 gives the initial responses, r_{01} and r_{02}, at 820 and 3,075 rev/min.

Solution. The mode shapes shown in Figure 8.20 at the disks are given by $\mathbf{u}_1 = [1\ 1]^T$ and $\mathbf{u}_2 = [1\ -1]^T$. Thus, to balance the first mode of interest, $\mathbf{e}_1 = [0.5\ 0.5]^T$ is a suitable choice; in this case, the trial masses at the two chosen planes are equal and with the same phase. As originally proposed, modal balancing of the first mode requires mass to be added only to a single balancing plane so that one possible choice is $\mathbf{e}_1 = [1\ 0]^T$. The reason for not choosing this $\mathbf{e}_1$ is practical: it is likely to increase the unbalance response near the second critical speed. For the second mode of interest, $\mathbf{e}_2 = [0.5\ -0.5]^T$ is a suitable choice; hence, the trial masses at the two chosen planes are equal but have opposite phases. Choosing the trial balances in this way dictates that each trial influences the response in only one mode. Balancing a third mode requires a third balance plane.

We now consider the two-plane example whose mode shapes of interest are given in Figure 8.20. The vibration is measured at 820 rev/min, which is close to the first critical speed, on the basis that this response contains predominantly the response due to the first forward-whirling mode. For the trial run, because of the

Table 8.7. *Responses at disk 1 and calculated unbalance corrections for the modal method (Example 8.5.4.)*

	Mode 1 ($i = 1$)	Mode 2 ($i = 2$)
Rotor spin speed (rev/min)	820	3,075
r_{0i} (mm) – measured	$1.0261 - 1.8580j$	$0.6545 + 0.1526j$
δ_i (g m)	0.4	0.4
r_i (mm) – measured	$0.1380 - 2.1620j$	$0.6652 - 0.2208j$
r_{di} (mm)	$-0.8881 - 0.3040j$	$0.0107 - 0.3734j$
α_i (g m)	$0.1573 - 0.8907j$	$0.1433 - 0.7052j$

Table 8.8. *Unbalance corrections for the modal method (Example 8.5.4)*

	Disk 1		Disk 2	
	Magnitude (g m)	Phase (°)	Magnitude (g m)	Phase (°)
Mode 1	0.452	280	0.452	280
Mode 2	0.360	281	0.360	101
Total	0.812	281	0.093	274

symmetry in the mode shape, the two added unbalances must be equal in magnitude and phase. In this case, we choose both added unbalance masses to be 0.2 g m at zero phase, making $\mathbf{b}_1 = [\,0.2\ \ 0.2\,]^\top$ g m and, hence, $\delta_1 = 0.4$ g m. The trial masses are added and the response at 820 rev/min is measured. Because the machine is operating near the first critical speed, the spatial response approximates the first forward-whirling mode. Thus, the responses at the balance planes should be equal in magnitude and phase. Because the balancing exercise is essentially a single-plane balance, we need to measure only one response; hence, Table 8.7 gives the response at disk 1, r_1. Using this response and the initial response, r_{01}, we calculate $r_{d1} = r_1 - r_{01} = -0.8881 - 0.3040\jmath$. Because the responses at only one disk are measured, Equation (8.45) simplifies to give $\alpha_1 = -\delta_1 r_{01}/r_{d1} = 0.1573 - 0.8907\jmath$. The vector of required unbalance corrections in complex form is $\mathbf{b}_{c1} = \alpha_i \mathbf{e}_1$. Hence, the actual balance mass must be added to both balance planes and is shown in Table 8.8. Figure 8.21 shows the response of the rotor after the correction required to balance the first mode of interest is added, clearly showing that the response in the first forward-whirling mode is substantially reduced; the response is zero at the rotor spin speed used for the balancing – that is, 820 rev/min.

The rotor spin speed is now increased to 3,075 rev/min, which is close to the second critical speed. After noting the vibration level, two test unbalance masses (again, 0.2 g m) were added to each plane. However, this time they are out-of-phase because the second mode of interest is being considered and

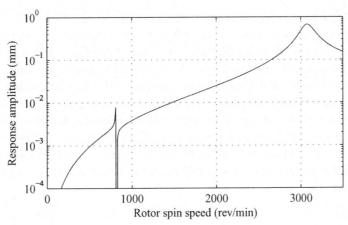

Figure 8.21. The response at disk 1 of the flexible machine after balancing the first mode (Example 8.5.4).

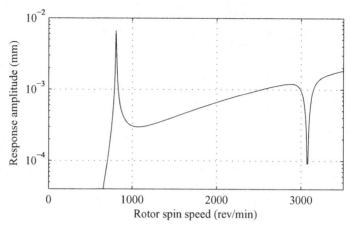

Figure 8.22. The response at disk 1 of the flexible machine after balancing two modes (Example 8.5.4).

$\mathbf{e}_2 = [0.5 \ -0.5]^\top$. The responses at disk 1 are given in Table 8.7, and a single-plane balance gives the required unbalance correction as 0.360 g m at 281° for disk 1 and at 101° for disk 2. These results for the unbalance corrections are summarized in Table 8.8, in which the total unbalance obtained by vector addition is close to that obtained by the influence-coefficient method because, in this example, the first two critical speeds dominate. Figure 8.22 shows the response after the required balance masses are added. There is no longer any speed close to the first critical speed where the response is zero, mainly because the mode shapes used in the balancing are not exact.

With the rotor under study, balancing is performed near the two critical speeds. Figure 8.22 shows the vibration levels of the balanced rotor; using this method, the vibration levels are reduced throughout the speed range. However, it is important to appreciate that if a wider range of speed is considered, a resonance at the third critical speed may become significant because the higher modes are not balanced and high vibration levels may result.

EXAMPLE 8.5.5. Consider the machine shown in Figure 8.23, which is 1.5 m long with short bearings located at the left end of the shaft and at 1 m from the left end. The steel shaft has a diameter of 30 mm, a Young's modulus of 211 GPa, and a mass density of 7,810 kg/m^3. A steel disk 250 mm in diameter and 20 mm in thickness is located 0.5 m from the left end of the shaft and a disk 200 mm in diameter and 10 mm in thickness is located at the right end. The shaft is modeled using 12 elements. The inherent unbalance is modeled as unbalance magnitudes of 0.6, 0.4, and 0.2 g m, with phases 180°, 120°, and 0°, at locations 0.375, 0.75, and 1.375 m from the left end of the shaft. Apply modal balancing to the first two forward-whirling modes of this machine. Assume that the balance planes are located at the disk positions.

Solution. Figure 8.24 shows the first two mode shapes of interest of the stationary machine. For the first mode, the ratio of the displacement at the right disk to the displacement at the left disk is −2.943; for the second mode, this ratio is

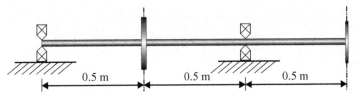

Figure 8.23. An overhung flexible machine that requires modal balancing (Example 8.5.5).

1.098. Figure 8.25 shows the response at the left disk to the inherent unbalance and shows that the first two critical speeds occur at 1,223 and 2,751 rev/min. To balance the first forward-whirling mode, a trial mass with balance magnitude 0.2 g m and 0° phase is placed on the right disk and a trial mass with balance magnitude 1.098×0.2 g m and 180° phase is placed on the left disk. Based on the ratio of modal displacements at the disks, this set of trial masses does not excite the second forward-whirling mode. The balance mass required to zero the response of the left disk at 1,200 rev/min is computed using the single-plane balance method. The required balance corrections are listed in Table 8.9 and Figure 8.25 shows the resulting response. As expected, the response is zero at 1,200 rev/min and the response due to the second mode is unaltered. The second forward-whirling mode is now balanced using a trial mass with balance magnitude 0.2 g m and 0° phase at the right disk, and a trial mass with balance magnitude 2.943×0.2 g m and also 0° phase at the left disk. Based on the ratio of modal displacements at the disks, this set of trial masses does not excite the first forward-whirling mode. The balance mass required to zero the response of the left disk at 2,700 rev/min is computed using the single-plane balance method. The required balance corrections are given in Table 8.9 and Figure 8.25 shows the resulting response. At 2,700 rev/min, the response is zero; hence, the response due to the second forward-whirling mode is significantly reduced. The response due to the first forward-whirling mode has changed little; however, as expected, the response is no longer zero at 1,200 rev/min because the assumed modes do not account for gyroscopic effects.

8.6 Balancing Machines without a Phase Reference

Often, a phase reference is used to control the spin speed of the machine and therefore is available. However, for situations in which a phase reference is not available, Somervaille (1954) suggested a graphical method for single-plane balancing based on measuring only the amplitudes of responses. When using only amplitudes, the response must be predominantly synchronous. Subsequent papers (e.g., Foiles and

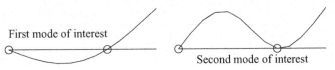

Figure 8.24. The first two mode shapes in the vertical plane of an overhung machine (Example 8.5.5) modeled using 12 elements. The circles denote the bearing locations.

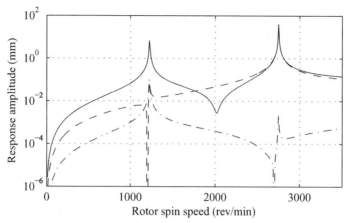

Figure 8.25. The responses at the left disk of an overhung machine (Example 8.5.5) due to the inherent unbalance (solid), after balancing the first mode (dashed), and after balancing both the first and second modes (dot-dashed).

Allaire, 2006; Foiles and Bently, 1998) outline how the method can be extended to the case of multiplane balancing. At the root of all of these methods is the fact that more trial runs must take place than are needed if phase information were also available. For single-plane balancing, using the notation in Section 8.4.1, suppose that the amplitude of the initial response at rotor spin speed Ω is measured and given by $|r_0|$. From Equation (8.10), the rotor is balanced by a correction b_c if

$$r_0 + R(\Omega)b_c = 0 \tag{8.47}$$

where $R(\Omega)$ is the response function. Because the rotor is balanced, $r_c = 0$ in Equation (8.10). Rearranging and taking the amplitude gives

$$|r_0| = |R(\Omega)||b_c| \tag{8.48}$$

Assuming that the machine behaves linearly, then the response when a trial balance b_i is added is

$$r_i = r_0 + R(\Omega)b_i = R(\Omega)(b_i - b_c) \tag{8.49}$$

Because we are able to measure only the response amplitude $|r_i|$, this gives

$$|r_i| = |R(\Omega)||b_i - b_c| \tag{8.50}$$

There are three unknown real quantities in Equations (8.48) and (8.50): the amplitude of the response function, $|R(\Omega)|$; and the real and imaginary parts of the

Table 8.9. *Unbalance corrections for the modal method applied to the overhung machine (Example 8.5.5)*

	Left disk		Right disk	
	Magnitude (g m)	Phase (°)	Magnitude (g m)	Phase (°)
Mode 1	0.317	346	0.289	166
Mode 2	0.421	343	0.143	343
Total	0.738	344	0.146	169

unbalance correction, b_c. It may be thought that two trial runs are sufficient because this would give three real equations; one from Equation (8.48) and two from Equation (8.50) for $i = 1$ and $i = 2$. However, the equations are essentially quadratic in nature (Foiles and Allaire, 2006) and, if only two trial runs are used, there are usually two distinct possible solutions. Thus, for single-plane balancing without phase, a minimum of three different balance trials is required in addition to the original trial, in which no balance masses were added to the rotor. For multiplane balancing without phase information, three different balance trials are required for each plane in which unbalance correction is to be determined, in addition to the original as-found response.

There are techniques to produce closed-form solutions of the resulting equations (Foiles and Allaire, 2006). The alternative is to apply numerical methods to minimize the sum of squares of errors in responses given by Equations (8.48) and (8.50). For a general optimization approach, an initial estimate of the solution is required, which may be obtained by establishing which of the three trial-balance configurations accounts for the largest (or smallest) response and then estimating that this response is perfectly in-phase (or out-of-phase) with the original response. Then, by making minor changes to the estimate for the unbalance correction phase and magnitude (or, equivalently, real and imaginary parts), a best fit solution is quickly achieved by minimizing

$$J(\sigma_1, \sigma_2, \sigma_3) = \sum_{i=0}^{p} \left(|r_i| - \sigma_1 |b_i - (\sigma_2 + j\sigma_3)| \right)^2 \qquad (8.51)$$

where p is the number of trial runs, $|R(\Omega)| = \sigma_1$, $b_c = \sigma_2 + j\sigma_3$, and we define $b_0 = 0$. The phase information for the unbalance correction is derived from the phase of the trial-balance masses. We illustrate the process with a single-plane balancing example.

EXAMPLE 8.6.1. A given machine is to be trim-balanced by adding unbalance correction masses at a single disk. The only response measurement available is a horizontal velocity at one bearing housing. Although velocity is used here, any response measurement (e.g., displacement) can be used. Before balancing, the magnitude of the velocity response is 2.0 mm/s. A trial balance of 20 g m is fitted to the disk at an angular orientation of 0°; when the machine is brought back up to speed, the magnitude of response is 1.5573 mm/s. When the same trial balance is advanced by 120°, the response measured is 1.9709 mm/s; when it is advanced by 240°, the response measured is 2.6759 mm/s. Determine the magnitude and phase of the unbalance correction required to reduce the bearing-velocity measurement to zero.

Solution. The first task is to determine an initial estimate of the solution. The largest difference between the original response magnitude and the modified response occurred when the trial balance was fitted at 240°, which corresponds to an increase of 0.6759 mm/s. To a first approximation, therefore, the effect of adding 20 g m of unbalance adjustment to the balancing plane is to modify the response by approxmately 0.6759 mm/s. Thus, the first approximation of the magnitude of the correction to balance the response of 2.0 mm/s is

$20 \times 2/0.6759 = 59.18\,\text{g m}$. Because this trial balance increases the response, the initial estimate of the phase is $240 - 180 = 60°$. An initial value for $|R(\Omega)|$ is approximated by the ratio of response to unbalance amplitude and, in this case, is $2.0/59.18 = 0.0338\,\text{kg}^{-1}\,\text{s}^{-1}$.

Using these initial values in an optimization scheme (Lindfield and Penny, 2000), we can iteratively determine improved estimates by seeking to reduce the differences among the four measured response magnitudes (i.e., one initial response and three with trial balances attached) and the corresponding calculated response magnitudes. The calculated unbalance correction is $b_c = (42.14 + 38.60\jmath)\,\text{g m}$. This correction has a magnitude of $57.14\,\text{g m}$ and a phase of $42.49°$. The magnitude of the complex response function is also determined as $|R(\Omega)| = 0.0350\,\text{kg}^{-1}\text{s}^{-1}$.

This example shows that the influence coefficient relating any change in response to some discrete unbalance correction at a given plane can be found easily if three distinct trial runs are made. It is clear from the example that this is sufficient to achieve balancing at a single plane. In fact, it is sufficient to achieve multiplane balancing using either direct influence-coefficient methods or even modal-balancing methods.

8.7 Automatic Balancing Methods

Thus far, the focus is exclusively on one-off adjustments to the distribution of mass on a rotor. This chapter is incomplete without some discussion of automatic balancer devices and the use of active provisions such as magnetic bearings to adjust the synchronous response of a machine. Automatic balancer devices have been available for several decades and the most basic consists of a circular track within which two or more balls may be located. It can be shown that if the location of the automatic balancer and the spin speed of the rotor are such that a synchronous frequency external force results in rotor deflections in-phase with the applied force, the balls tend to locate such that their net unbalance forcing drives the rotor-displacement response toward zero at this axial location. At other speeds, these automatic balancers can be unstable, however (Chung and Ro, 1999; Green et al., 2008); as yet, their takeup is not widespread.

A different approach is to use an active balancing head consisting of various mechanisms to vary the unbalance distribution without interrupting the operation of the machine (Zhou and Shi, 2001). Van de Vegte and Lake (1978) and Van de Vegte (1981) used unbalanced disks driven by worm gears to vary the amplitude and phase of the unbalance. Alauze et al. (2001) referenced other approaches to realize the controlled unbalance. The response of the rotor must be measured and a suitable control algorithm employed to balance the machine. Lee et al. (1990), for example, used the influence-coefficient method to perform modal balancing as the machine passed though critical speeds.

Magnetic bearings are capable of exerting controlled forces on a rotor (see Section 5.5.4). The force required to keep a rotor suspended is a low-frequency force, and all magnetic bearings are equipped with controllers that achieve very high dynamic stiffness for low frequencies. At synchronous frequency, at least two distinct techniques are common in conjunction with magnetic bearings (Zhou and

Shi, 2001). One technique uses, so-called *feedforward* methods to add a periodic component of force onto the force automatically developed by the *feedback controller* of the magnetic bearing (Knospe et al., 1995, 1996). The parameters of the additional periodic force are allowed to vary only very slowly. In this way, there is a firm guarantee that no instability arises. The other method often mentioned deploys a tracking *notch-filter* into the feedback loop to prevent the magnetic bearings from exerting any significant forces at synchronous speed (Herzog et al., 1996; Lum et al., 1996). As a result, although the rotor is no more balanced than it was before, the bearing reaction forces are much reduced and the rotor shaft is allowed to execute small orbits within the clearance of the bearing.

8.8 Issues in Balancing Real Machines

Having briefly described two main approaches to rotor balancing, it is appropriate to discuss some of the issues and difficulties that can and do arise in real situations. These are addressed in the following discrete paragraphs:

- *Choosing the response quantity at each plane.* A key aspect of balancing is to use response measurements that are sensitive to changes in the unbalance and have a high signal-to-noise ratio. The response in a particular direction at each disk can be chosen. Alternatively, the response used might be a linear combination of the rotor displacements u and v in the x and y directions given by

$$r = pu + qv \tag{8.52}$$

for two complex constants, p and q. These constants should be chosen so that the unbalance response is always large. For a rotor on isotropic bearings and supports, Chapter 6 demonstrates that only the forward modes are excited by unbalance and that the orbits are circular. Thus, u and v have equal magnitude but u leads v by 90°. Hence, in the frequency domain, $r = u - jv$ is always zero and this is a poor choice of p and q. Conversely, $r = u + jv$ has maximum amplitude; hence, it is the best choice. For general bearing and supports, the situation is more complex, although $r = u + jv$ is still a good choice.
- *Choosing appropriate balancing speeds.* Whatever the approach, the essence of balancing procedures is the observation of the influence of a trial mass that is then scaled to determine the rotor unbalance. For this to be valid, however, the speed of the trials must be consistent, which is often difficult to achieve in practice. The choice of speed is important here in an attempt to satisfy two separate and often conflicting requirements. First, the speed must be chosen to give adequate sensitivity to balance changes; and second, the sensitivity to rotor speed should be small in order to minimize any errors due to lack of speed repeatability.
- *Nonlinearity.* A frequently encountered difficulty is a lack of linearity. This is obviously problematic because the basis of all balancing procedures relies on linearity of the system. However, the fact that many machines – particularly those using oil-journal bearings – do not behave in a linear manner is less troublesome than may be feared. The effect of the nonlinearities is to make the influence coefficients depend on the unbalance. The process must be performed

iteratively because the application of balance masses calculated in a single iteration cannot completely correct the response. To perform this efficiently requires some knowledge of the particular machine, but successful balancing often may be achieved with only two iterations.

- *The presence of a rotor bend and runout.* The rotor under study may have a bend. Chapter 6 shows that although a bend gives rise to a forward-synchronous force, it is not the same as unbalance. A bend may be compensated with a balance mass at a particular speed, but this may exacerbate the situation at other speeds. A good check is to monitor the shaft position at very low speeds and then subtract this eccentricity from all subsequent measurements. Balancing in this way reduces the vibration levels to be no more than the amplitude of the rotor bend at any speed. Proximity sensors commonly are used to detect rotor position in rotating machines. Errors in the signals from these sensors arise from variation in material properties, which are referred to as *runout*. Like rotor bends, these errors are functions of rotor angular position and can be corrected by subtracting the low-speed response.

- *Lack of rotor symmetry.* This chapter considers the balancing of symmetric rotors; Han (2007), Matsukura et al. (1979), and Parkinson (1965, 1966) considered the balancing of asymmetric machines and rotors. A standard balancing approach, such as the influence-coefficient method or modal balancing, may be applied if the anisotropy present is confined to the stator and bearings. For machines with asymmetric rotors but isotropic bearings and supports, Chapter 7 demonstrates that a steady-state response exists in the rotating frame and that the machine may be unstable over certain frequency ranges. If the machine is stable, the standard approaches are readily extended to this case. When both the stator and bearings are anisotropic and the rotor is asymmetric, then Section 7.4 explains that the equations of motion are parametrically excited and the response is often complicated. Here, balancing is possible using the measured synchronous response.

- *At any given speed, modes may be coupled by the damping matrix.* Because the overwhelming proportion of machine damping usually arises in bearings, whereas the strain energy is predominantly in the rotor or bearing supports, linear models of machines are not *classically damped*. The presence of gyroscopic couples further exacerbates this because the gyroscopic component of the *damping matrix* is skew-symmetric, whereas the mass matrix is invariably symmetric and the stiffness tends to be predominantly symmetric. As a result, the mode shapes of rotating machines are complex and they do not obey the simple orthogonality relationships that apply to the modes of undamped structures. Strictly speaking, modal balancing is rendered invalid. The modal balancing method can be extended to the so called *bimodal* approach (Kellenberger, and Rihak, 1988; Saito and Azuma, 1983), in which an additional set of orthogonalizing functions is required, but this is not discussed here. In practice, sufficiently accurate results often can be obtained using the standard modal method.

- *Modes of a machine are speed-dependent.* Because models of machines are dependent on rotational speed, the mode shapes also vary with speed. A vector that is orthogonal to mode $\mathbf{u}_1$ at one rotational speed may not be orthogonal at another. In practice, the modes do not change dramatically with respect to speed unless two resonance frequencies happen to cross over or become very close. With close resonance frequencies, the mode shapes often couple and may become sensitive to small changes in the machine condition or environment (Jei and Lee, 1992; Perkins and Mote, 1986). In this case, a relatively minor extension to the modal-balancing approach can be adopted, in which – instead of balancing single modes at a time before progressing to the next – pairs or even triples of modes may be balanced simultaneously. The requirement for this is so rare that we do not develop it further herein. Friswell et al. (1998b) calculated the critical speeds by introducing additional displacement variables into a model such that the speed-dependent properties of bearings became incorporated into a model with constant parameters; such an approach provides an alternative solution.

- *Optimal rotor balance depends on the stator.* Many rotors are balanced in a stator, the properties of which may not be the same as those of the final stator. In other cases, the properties of a given stator can change with time and temperature – for example, when flexible elastomeric bearing supports are deployed. In the case of modal balancing, this obviously affects the modes that are balanced. In the case of influence-coefficient balancing, it also may be significant when the actual initial unbalance on the rotor does not lie predominantly at the planes where unbalance corrections can be applied. Robust-balancing approaches, whereby the rotor balancing accounts for the spread of stator dynamic conditions have been presented to cater for this difficulty (Garvey et al., 2002; Li et al., (2008).

- *Individual components and modules of rotor assemblies are often balanced individually.* This raises two distinct issues: (1) geometric errors in the individual modules; and (2) the dynamic environment of the rotor module being balanced. It is now standard practice to use *dummy modules* to compose a complete rotor when any one significant component is being balanced so that the correct dynamic context is achieved (Schneider, 2000). Addressing geometric errors is more complex and different companies employ different practices. A common strategy is to attempt to index modules relative to one another such that the final build is as straight as possible.

- *Noise on response signals causes bias.* All real measurements contain some noise. As requirements for ever-better balance quality evolve, the relative magnitude of noise tends to grow. Through averaging, the effects of noise can be reduced but never eliminated. Because the relationship between the response measurements made and the correction unbalance computed is not linear, there can be bias in the unbalance correction vectors computed. Effects of noise are relatively easy to simulate, enabling the assessment of the extent of this bias. Provided that a sufficient number of averages is taken, this can always be made satisfactory (Garvey et al., 2005).

- *Extra rundowns are required.* Conventional balancing requires trial masses to be added and extra rundowns to be performed to identify the machine model, $\mathbf{R}(\Omega)$. Apart from the extra time required to perform the measurements, the trial mass also may increase the response to a dangerous level. If a reliable numerical model is available, it may be used to determine $\mathbf{R}(\Omega)$; however, this model must represent not only the mechanical dynamics of the rotor system but also any dynamics and gains of the measurement system used. When the measured responses are displacements, the entries of $\mathbf{R}(\Omega)$ can be computed from the receptance matrix for the complete system model (multiplied by Ω^2) (see Section 6.3.2). Determining an accurate model for many systems is difficult but methods have been proposed (e.g., Sinha et al., 2002) in which a single rundown of a machine can be used to both identify and update a machine model and determine the state of unbalance of a rotor.

8.9 Summary

This chapter outlines the most commonly used and most effective methods to balance a machine. In most cases, balancing involves understanding the relationship between the magnitudes and angular positions of correction unbalances fixed at certain available balancing planes and the magnitudes and phases of the changes in responses (of various types) that can be measured on a machine. Usually, this relationship is measured. If the relationship is understood and if the intention of the balancing also is understood, then balancing is at least conceptually simple. Methods for rigid and flexible rotors are described, including single-plane, two-plane, general influence-coefficient, and modal balancing. A method for balancing without phase information is also described. Some of the key practical difficulties encountered during rotor balancing are outlined in this chapter.

8.10 Problems

8.1 A rigid rotor carries four rigid disks 0.5 m apart, and the disks are labeled A, B, C, and D along the rotor from one end. It is discovered that in the plane of disk A, there is an unbalance of 0.4 kg at a radius of 200 mm; in the plane of disk B, there is an unbalance of 0.2 kg at a radius of 200 mm and a phase of 76°; in the plane of disk C, there is an unbalance of 0.7 kg at a radius of 100 mm and a phase of 132°; and in the plane of disk D, there is an unbalance of 0.2 kg at a radius of 300 mm and a phase of 212°. All of the angles are relative to the position of the out-of-balance in plane A.

Determine the magnitudes and angular positions of the unbalance-correction masses that must be added at a radius of 100 mm to the disks in planes B and D to balance the rotor.

8.2 Figure 8.26 shows a rotor lying on the z axis whose total length is 1.6 m. The rotor is supported on short, stiff bearings at $z = 0$ and at $z = 1.2$ m. Three rigid

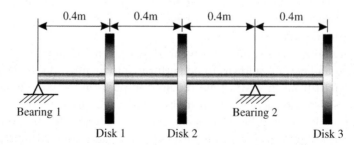

Figure 8.26. A machine with a light shaft and three disks (Problem 8.2).

Table 8.10. *The unbalance distribution for the three disks (Problem 8.2)*

Disk	Position (m)	Disk Mass (kg)	Eccentricity (mm)	Angular Orientation (degrees)
1	0.4	10	0.15	120
2	0.8	50	0.10	15
3	1.6	20	0.20	−45

disks, numbered 1 through 3, are fixed to the rotor and the mass center of each one is slightly eccentric. Table 8.10 lists the position of each disk and the details of the unbalance carried initially on each disk.

(a) Compute the magnitudes and phases of the bearing reaction forces (i.e., the forces that the bearings apply to the rotor) when the rotor is spinning at 800 rev/min, assuming that the rotor is rigid.

(b) Determine appropriate unbalance correction masses to be applied at disks 1 and 3.

8.3 A machine is to be balanced using the single-plane technique. Initially, the response was 0.02 mm (at 3,000 rev/min) with the maximum occurring 30° ahead of the shaft mark. The addition of 10 g m at a set radius and 180° ahead of the reference gave a response of amplitude 0.03 mm at 60° behind the mark. Calculate the magnitude and phase of the unbalance correction needed at the same radius to balance the machine.

8.4 A rigid rotor spins at 1,500 rev/min. Its responses at planes 1 and 2 due to some unknown out-of-balance are shown in the first row in Table 8.11. All phase angles are relative to a zero reference on the shaft. A trial balance of 0.5 g m is attached to plane 1 of the rotor at a phase of 0°. The responses at planes 1 and 2 are listed in the second row of Table 8.11. When the first trial balance is removed and a second trial balance of 0.5 g m is added to plane 2 of the rotor at a phase of 90°, the resulting responses at planes 1 and 2 are shown in the last row of Table 8.11. Determine the angular positions and magnitudes of the unbalance corrections that must be added at planes 1 and 2 to balance the rotor.

8.5 A machine with a flexible rotor is to be balanced at its running speed of 3,000 rev/min, which lies above its second critical speed. Table 8.12 lists the vibration measured in its original state and with two trial masses added at the same running speed. In this case, the second trial mass was added without removing the first. Determine the magnitudes and phases of the masses required

Table 8.11. *Measured responses at 1,500 rev/min of a rigid rotor with trial masses (Problem 8.4)*

	Balance Plane 1		Balance Plane 2	
	Magnitude (mm)	Phase (°)	Magnitude (mm)	Phase (°)
As found	0.35	−64	0.36	−89
0.5 g m at 0°, plane 1	0.35	−95	0.45	−122
0.5 g m at 90°, plane 2	0.59	−73	0.55	−86

Table 8.12. *Measured responses at 3,000 rev/min of a machine with trial masses (Problem 8.5)*

	Bearing 1		Bearing 2	
	Magnitude (mm)	Phase (°)	Magnitude (mm)	Phase (°)
As found	0.18	49	0.62	52
0.5 g m at 0°, plane 1	0.17	54	0.76	40
0.5 g m at 90°, plane 2 plus trial mass 1	0.10	−22	0.63	24

to balance the machine at this speed. Is the machine balanced at running speeds above 3,000 rev/min?

8.6 When the machine in Problem 8.5 is run down, it is found that at very low speed (i.e., 10 rev/min) there is still significant vibration. Explain the significance of this and suggest a possible cause.

8.7 At a speed of 1,500 rev/min, a rotor shows the vibration response given in Table 8.13. Calculate the magnitudes and angular locations of the masses, placed at 150 mm radius, required to balance the machine using a two-plane balancing procedure.

It is then decided to operate the machine at 3,000 rev/min; on increasing the spin speed, it is found that the machine is no longer balanced correctly. Explain why this may be the case. The response at the new speed is shown in Table 8.14. Calculate the magnitudes and angular locations of the masses, placed at 150 mm radius, required to balance the machine at this speed using two-plane balancing. Explain how to balance the machine to operate satisfactorily at both speeds and outline the information required.

8.8 A machine is mounted on two self-aligning bearings and carries one large disk at the midspan between bearings and another disk overhung beyond the second bearing. Balancing is to be performed at the machine's operating speed of 3,000 rev/min to reduce the bearing vibration levels by adding masses to the two disks. Both the trial and the correction masses are added at a radius of 250 mm. Table 8.15 shows the measured data. From these data, sketch the orbit of the shaft at bearing 1 for the three conditions. Calculate the sizes and positions of the correction masses to be added at the two disks, using all of the data. Compare this with an estimate using data only in the x direction.

8.9 In balancing the machine in Problem 8.8, at 3,000 rev/min, more accurate phase values are available, listed in Table 8.16. Calculate the sizes and positions of correction masses to be added at the two disks, using all of the data. Both the trial and the correction masses are added at a radius of 250 mm. Compare this with an estimate using data only in the x direction and the results of Problem 8.8.

8.10 An identical machine to the one considered in Problem 8.8 is to be balanced at 2,250 rev/min, using the data listed in Table 8.17. Both the trial and the correction masses are added at a radius of 250 mm. Sketch the orbit of the shaft at bearing 1 for the three cases. Calculate the sizes and positions of the correction masses to be added at the two disks, using all of the data. Compare this with an estimate using data only in the x direction.

Table 8.13. *The measured responses at 1,500 rev/min (Problem 8.7)*

	Bearing 1		Bearing 2	
	Magnitude (mm)	Phase (°)	Magnitude (mm)	Phase (°)
As found	0.2	150	0.3	120
0.1 kg at 150 mm, 0°, disk 1	0.1	140	0.2	150
0.1 kg at 150 mm, 90°, disk 2 plus trial mass 1	0.45	90	0.4	300

Table 8.14. *The measured responses at 3,000 rev/min (Problem 8.7)*

	Bearing 1		Bearing 2	
	Magnitude (mm)	Phase (°)	Magnitude (mm)	Phase (°)
As found	0.1	70	0.2	110
0.1 kg at 150 mm, 0°, disk 1	0.25	90	0.3	150
0.1 kg at 150 mm, 90°, disk 2 plus trial mass 1	0.3	130	0.4	300

Table 8.15. *The measured responses at 3,000 rev/min (Problem 8.8)*

		Bearing 1		Bearing 2	
		Magnitude (mm)	Phase (°)	Magnitude (mm)	Phase (°)
Original	x	0.042	0	0.045	10
	y	0.065	−40	0.067	−30
0.03 kg, 0°, disk 1	x	0.076	−20	0.080	−10
	y	0.120	−60	0.120	−60
0.03 kg, 0°, disk 2	x	0.048	−10	0.050	0
	y	0.074	−40	0.075	−40

Table 8.16. *The more accurate measured responses at 3,000 rev/min (Problem 8.9)*

		Bearing 1		Bearing 2	
		Magnitude (mm)	Phase (°)	Magnitude (mm)	Phase (°)
Original	x	0.042	0	0.045	9
	y	0.065	−39	0.067	−34
0.03 kg, 0°, disk 1	x	0.076	−22	0.080	−13
	y	0.120	−61	0.120	−56
0.03 kg, 0°, disk 2	x	0.048	−7	0.050	3
	y	0.074	−45	0.075	−40

Table 8.17. *The measured responses at 2,250 rev/min (Problem 8.10)*

		Bearing 1		Bearing 2	
		Magnitude (mm)	Phase (°)	Magnitude (mm)	Phase (°)
Original	x	0.031	0	0.051	20
	y	0.050	−30	0.072	−30
0.03 kg, 0°, disk 1	x	0.034	0	0.053	20
	y	0.055	−40	0.076	−30
0.03 kg, 0°, disk 2	x	0.032	0	0.051	20
	y	0.052	−30	0.072	−30

8.11 Repeat Example 8.5.5 in which the diameter of the left disk is increased from 0.25 to 0.35 m. The inherent unbalance on the machine is unchanged from the example. Calculate the mode shapes at the disks, and balance the machine at 1,110 and 2,315 rev/min for the first and second modes of interest, respectively.

8.12 This problem considers the effect of errors in the mode shapes used for modal balancing and is based on Example 8.5.4. The details of the model; the trial balances, δ_i; and the balancing speeds are exactly the same as in Example 8.5.4. Now suppose that the mode shapes have inaccuracies, arising either from measurement or modeling errors, and the mode shapes of interest at the disks are assumed to be $\mathbf{u}_1 = [1 \; 0.9]^T$ and $\mathbf{u}_2 = [0.95 \; -1]^T$. Determine $\mathbf{e}_i$; $\mathbf{e}_1$ is not unique – in this case, choose $\mathbf{e}_1$ to be orthogonal to $\mathbf{u}_2$. Perform modal balancing at 820 and 3,075 rev/min for the first and second modes of interest, respectively. What are the total estimated unbalance corrections at the two disks? If these unbalance corrections are added to the machine, what are the amplitudes of the responses at disk 1 at 820 and 3,075 rev/min? Compare these to the responses when the correct mode shapes are assumed and comment.

8.13 Re-examine Example 8.6.1 using only the initial response and the responses for the first two trial masses. Using the starting values given in the example, show that the same unbalance correction is obtained. Using a starting unbalance correction of 25 g m at 40°, show that a different but apparently valid solution is obtained. This illustrates that three trial masses are required to obtain a unique solution. This problem can be solved using appropriate optimization software.

8.14 Example 8.5.4 performed a modal balance for a first two modes of a machine using two balance planes. To balance the first mode, masses were added to both planes to avoid exciting the second mode. Repeat the balancing exercise, but this time add mass to only one of the disks to balance the first mode. Show

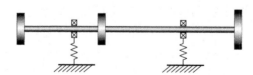

Figure 8.27. A schematic of the machine in Example 8.15.

Table 8.18. *The measured responses at various stages of the modal balancing of the machine in Figure 8.27 (Problem 8.15)*

Masses Added	Rotor Spin Speed rev/min	Disk 1		Disk 2		Disk 3	
		Magnitude (mm)	Phase (°)	Magnitude (mm)	Phase (°)	Magnitude (mm)	Phase (°)
None	830	1.616	81	0.633	−85	4.211	98
$\mathbf{b}_1$	830	1.726	72	0.651	−94	4.292	87
$\mathbf{b}_{c1}$	830	0.460	18	0.080	−103	0.200	124
$\mathbf{b}_{c1}$	1,080	9.142	−156	0.447	15	2.299	25
$\mathbf{b}_{c1} + \mathbf{b}_2$	1,080	13.664	−160	0.682	15	3.435	21
$\mathbf{b}_{c1} + \mathbf{b}_{c2}$	1,080	0.241	123	0.094	−83	0.059	−54
$\mathbf{b}_{c1} + \mathbf{b}_{c2}$	2,900	0.791	156	10.290	152	0.891	152
$\mathbf{b}_{c1} + \mathbf{b}_{c2} + \mathbf{b}_3$	2,900	0.823	169	10.731	164	0.929	165

that the total mass correction required to balance both modes is similar to that required in Example 8.5.4.

8.15 Figure 8.27 shows a machine with three disks supported on two flexible bearings. The first three modes of the machine are to be balanced using the three disks as balance planes. The first three pairs of critical speeds are 776/823, 1,023/1,063, and 2,822/2,862 rev/min, and the machine is to be balanced near the forward-whirling critical speeds at 830, 1,080, and 2,900 rev/min. The mode shapes at the balancing disks corresponding to these critical speeds are

$$\mathbf{u}_1 = \begin{Bmatrix} 0.3780 \\ -0.1526 \\ 1 \end{Bmatrix}, \quad \mathbf{u}_2 = \begin{Bmatrix} 1 \\ -0.0514 \\ -0.2513 \end{Bmatrix}, \quad \text{and} \quad \mathbf{u}_3 = \begin{Bmatrix} 0.0763 \\ 1 \\ 0.0866 \end{Bmatrix},$$

where the disks are ordered from left to right in Figure 8.27.

To balance the first mode, a trial-balance mass is added to the middle disk so that $\mathbf{b}_1 = \begin{Bmatrix} 0 \\ 5 \\ 0 \end{Bmatrix}$ g m at zero phase. Which values of δ_1 and $\mathbf{e}_1$ does this imply (see Section 8.5.2)? Table 8.18 lists the response to the initial unbalance and with the trial balance at 830 rev/min, which is close to the first forward-whirling critical speed. From the data in Table 8.18, calculate α_1 and, hence, $\mathbf{b}_{c1}$ from Equation (8.44). The third row in Table 8.18 (denoted $\mathbf{b}_{c1}$ at 830 rev/min) gives the result of this balancing; these data are not used in the balancing exercise and are for interest only.

To balance the second mode, a trial balance of $\mathbf{b}_2 = \begin{Bmatrix} 5 \\ 0 \\ -1.89 \end{Bmatrix}$ g m is chosen, in which the minus sign indicates the 180° phase. Calculate the values of δ_2 and $\mathbf{e}_2$ corresponding to the trial unbalance and verify that $\mathbf{e}_1$ and $\mathbf{e}_2$ satisfy Equation (8.43). Calculate α_2 and, hence, the unbalance correction $\mathbf{b}_{c2}$. The sixth row of Table 8.18 (denoted as $\mathbf{b}_{c1} + \mathbf{b}_{c2}$ at 1,080 rev/min) gives the result of this balancing; these data are not used in the balancing exercise and are for interest only.

For the third mode, the trial unbalance is $\mathbf{b}_3 = \begin{Bmatrix} 0.410 \\ 5 \\ -0.608 \end{Bmatrix}$ g m. Calculate $\mathbf{e}_3$ and verify that this trial unbalance does not excite the first two modes. From the data in Table 8.18, calculate α_3 and, hence, $\mathbf{b}_{c3}$.

9 Axial and Torsional Vibration

9.1 Introduction

In Chapter 1, it is noted that the dynamic behavior of many rotors can be divided into three different classes: lateral, axial, and torsional. This chapter addresses both the axial and torsional behaviors of rotors. Generally, these two categories of behavior do not interact with one another, except in worm drives and bevel gears. Treating axial and torsional behavior together is justifiable because, mathematically, they are close analogies of one another. For cyclically symmetrical rotors, the analysis of both axial and torsional behavior is relatively simple. Thus, it is possible to provide an overview treatment of both in the space of a single chapter.

The degree to which a single rotor can be analyzed in isolation is different for the three classes of vibration. Lateral vibrations of a rotor are usually strongly coupled to the vibrations of the supporting stator and structure. The only significant exception to this is where very flexible bearings are in use. Passive magnetic bearings often provide this condition. By contrast, there is usually little coupling between torsional or axial vibrations of a rotor and any motion of the stator, except in the case of a geared system. Although this fact is helpful in analyzing the rotor, it often has the undesirable effect that even very severe torsional or axial vibrations in a rotor easily may go undetected by vibration probes on the stator. The analysis of torsional behavior is often carried out for a complete shaft train, whereas it is sometimes possible to analyze separately the axial and lateral behavior of the individual rotors in a shaft train. The cause of this apparent anomaly is that it is often necessary for the rotors of individual machines to be connected through a coupling that is torsionally stiff but laterally flexible. As it happens, many such couplings are also axially flexible. This axial and lateral flexibility is deliberately designed so that moderate misalignment of the coupled shafts can be tolerated.

In the case of lateral vibrations, the resonance frequencies depend on the speed of rotation through gyroscopic effects and speed-dependent bearing properties. For torsional and axial vibrations, the resonance frequencies are generally independent of speed (to the first order, at least). Torsionally, systems almost always have one natural mode shape associated with the net rotation of the rotor (or set of rotors). This is a zero-frequency, rigid-body mode. Lateral and axial modes with zero frequency occur rarely, if ever.

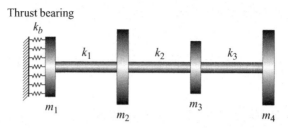

Figure 9.1. A rotor whose axial vibrations are to be modeled.

In analyzing lateral vibration, it is invariably important to consider synchronous forcing (i.e., a forcing frequency identical to rotational speed) because small amounts of unbalance can result in significant excitation (see Chapter 6). Generally, this is not the case for either torsional or axial vibration, although sometimes at least one frequency component of excitation is related to a multiple of rotational speed. Torsional systems are especially rich in sources of displacement- or velocity-driven excitation (see Section 9.7.3 for examples).

9.2 Simple System Models for Axial Vibrations

We begin by considering the axial vibrations of a typical rotor, shown in Figure 9.1. Initially, we assume that the rotor consists of a shaft, the mass of which is small compared to the mass of any disks connected to it. Then, each disk in Figure 9.1 may be represented as a rigid mass and each shaft section as a massless axial spring. A thrust bearing also can be modeled by a spring, sometimes with a parallel dashpot to provide representative damping.

The axial stiffness of a length of shaft with constant cross-sectional area, A, is determined from

$$k = \frac{F}{w} = \frac{EA}{L} \tag{9.1}$$

where F is the applied axial force, E is Young's modulus for the shaft material, L is the length, and w is the axial extension of the shaft section. This discrete component model is shown in Figure 9.2.

Discrete spring-mass systems are analyzed in Chapter 2 and the equations of motion for the previous system may be obtained by considering the free-body diagrams of each mass or via the assembly methods described in Chapter 4. Using matrix notation, the equations of motion (for free vibration) are

$$\mathbf{M\ddot{q}} + \mathbf{Kq} = \mathbf{0} \tag{9.2}$$

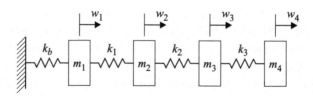

Figure 9.2. Idealization for axial behavior of the rotor in Figure 9.1.

where

$$\mathbf{M} = \begin{bmatrix} m_1 & 0 & 0 & 0 \\ 0 & m_2 & 0 & 0 \\ 0 & 0 & m_3 & 0 \\ 0 & 0 & 0 & m_4 \end{bmatrix}, \quad \mathbf{K} = \begin{bmatrix} k_b + k_1 & -k_1 & 0 & 0 \\ -k_1 & k_1 + k_2 & -k_2 & 0 \\ 0 & -k_2 & k_2 + k_3 & -k_3 \\ 0 & 0 & -k_3 & k_3 \end{bmatrix},$$

and $\mathbf{q} = \begin{bmatrix} w_1 & w_2 & w_3 & w_4 \end{bmatrix}^\top$.

We have not considered any damping in the rotor. Equation (9.2) represents a standard problem with constant mass and stiffness matrices that may be analyzed using the methods described in Chapter 2. In particular, the natural frequencies, ω_n, and mode shapes, $\mathbf{u}$, may be computed by solving the eigenvalue problem

$$\lambda \mathbf{M} \mathbf{u} = \mathbf{K} \mathbf{u}, \quad \text{where} \quad \lambda = \omega_n^2 \tag{9.3}$$

Equation (9.3) gives four eigenvalues (or four natural frequencies) and the associated four eigenvectors or mode shapes. To calculate the response of this system to known forcing is relatively simple. A forcing term is added to the right side of Equation (9.2) in place of the vector $\mathbf{0}$. The resulting dynamic equation may be solved by either integration in the time domain or regular matrix algebraic manipulations in the frequency domain, following methods outlined in Chapter 2.

EXAMPLE 9.2.1. A 0.9 m-long rotor consists of a hollow shaft onto which four identical disks are fixed at equal spacings along the shaft. The shaft has an outside diameter of 80 mm and a wall thickness of 20 mm. Each disk, is 50 mm thick and has an outside diameter of 400 mm. For the shaft and disks, the mass density and Young's modulus are $\rho = 7,850 \, \text{kg/m}^3$ and $E = 200 \, \text{GPa}$, respectively. The rotor is restrained at one end by a thrust bearing with an axial stiffness of 250 MN/m. The nonrotating side of the thrust bearing is rigidly located. Estimate the first four axial natural frequencies of this system.

Solution. The modeling of the mass and stiffness properties of the rotor is not as straightforward as may first appear. If the mass of the shaft is assumed to be negligible, then the mass of each disk is $m = \rho \pi h \left(D_o^2 - D_i^2 \right) / 4 = 47.350 \, \text{kg}$, where h is the disk thickness, D_o is the disk outside diameter, and D_i is the diameter of the hole through the center, which here is taken as the outside diameter of the shaft. However, if the mass at any location along the shaft is added to the mass of the nearest disk, then this extra mass is $\rho \pi L \left(d_o^2 - d_i^2 \right) / 4 = 8.878 \, \text{kg}$ for the inner disks and half this mass for the outer disks. Here, the length of each shaft section is $L = 0.9/3 = 0.3 \, \text{m}$, and d_i and d_o are the inside and outside shaft diameters. The effective discrete mass of the system is somewhere between these extremes.

The area for each shaft section is $A = \pi \left(d_o^2 - d_i^2 \right) / 4 = 3.7699 \times 10^{-3} \, \text{m}^2$. The length of each shaft section can be taken as the distance between the disk centers, or $L = 0.3 \, \text{m}$. Alternatively, the shaft under the disks can be assumed to be rigid and the length of the shaft elements taken as the distance between the disk faces, or $L = (0.9 - 0.15)/3 = 0.25 \, \text{m}$. In reality, the stiffness is somewhere between these extremes and depends on how the disk is attached to the shaft. Thus, $k_i = (EA/L) = 2,513.3 \, \text{MN/m}$ or $3,015.9 \, \text{MN/m}$, respectively, for

Table 9.1. *Natural frequencies (Hz) for the axial vibration of the rotor modeled with four degrees of freedom (Example 9.2.1)*

k_i (MN/m)		2,513.3	2,513.3	3,015.9	3,015.9
m_1, m_4 (kg)		47.350	51.789	47.350	51.789
m_2, m_3 (kg)		47.350	56.228	47.350	56.228
	1	175.13	163.91	176.38	165.09
Mode	2	919.65	873.54	1,001.6	951.40
	3	1,650.4	1,547.4	1,805.9	1,693.1
	4	2,144.9	1,981.2	2,349.2	2,169.8

the two shaft lengths for $i = 1, \ldots, 4$. Because $k_b = 250\,\text{MN/m}$, $k_b / k_i = 0.099472$ or 0.082893, respectively.

The eigenvalue problem, Equation (9.4), now can be formulated and solved. Table 9.1 lists the resulting natural frequencies for the different combinations of discrete mass and spring properties.

9.3 Shaft-Line Finite Element Models for Axial Vibrations

In Section 9.2, we model the axial vibrations of a rotor assuming that the shaft has no mass or that the shaft mass is included in the disk masses. If we model the rotor using the FEM, we can readily account for the mass of the shaft. We refer to such models as *shaft-line models* if the only degrees of freedom are axial displacements of nodes on the rotor centerline. The shaft behaves as a bar; therefore, the methods described in Section 4.4 may be applied directly.

EXAMPLE 9.3.1. Determine the natural frequencies of the rotor described in Example 9.2.1. Model the rotor using six elements.

Solution. Six elements are used, each of length 0.15 m, to model the shaft, as shown in Figure 9.3. Thus, the system then has seven degrees of freedom. Using appropriate software, we obtain the following system natural frequencies: 163.91, 880.62, 1,591.0, 2,077.6, 9,370.6, 9,589.4, and 9,796.7 Hz.

Comparing these natural frequencies with those derived by neglecting the mass of the shaft, given in the first column of Table 9.1, we see that differences in the first four natural frequencies are relatively small – the largest one being 6.8 percent for the first natural frequency. These differences arise mainly because the shaft mass is ignored completely. When the shaft mass is concentrated appropriately onto the disks, the natural frequencies are given in the second

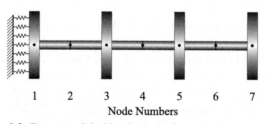

Figure 9.3. Rotor modeled by six shaft elements (Example 9.3.1).

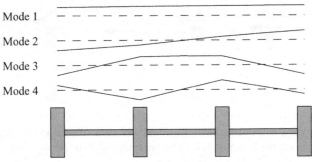

Figure 9.4. Axial vibration mode shapes for the system shown in Figure 9.3 (Example 9.3.1).

column of Table 9.1; the discrepancy in the first natural frequency is effectively zero (to five significant figures) and less than 1 percent for the second natural frequency. The net mass of the shaft in this case is just over 10 percent of the total mass of the system; because the first natural mode is similar to a rigid-body mode (in that the ratio of the difference in the displacement between any two points on the shaft to their mean displacement is very small), the frequency error made by ignoring the shaft mass is easily explained.

The fifth, sixth, and seventh frequencies are much higher than the first four resonances. These resonance frequencies are close to one another because they correspond to axial movements of the shaft between the disks whereas the disks remain relatively stationary. If we use 12 elements to model the system, then the first seven (of 12) natural frequencies are 163.91, 880.57, 1,590.8, 2,077.2, 8,895.8, 9,118.1 and 9,329.8 Hz. Whereas the lowest four frequencies have changed very little, the fifth, sixth, and seventh frequencies have changed substantially.

Figure 9.4 shows the mode shapes for the first four elastic modes, based on the 12-element model. The fact that the only observable bends in the mode-shape lines occur at the disks highlights that the kinetic energy in the lower frequency modes is dominated by the disks.

9.4 Simple System Models for Torsional Vibrations

There must be few rotor or machine applications that do not experience significant amounts of torsional excitation. Because the purpose of virtually all rotating machines involves consuming, producing, or transferring mechanical power, there is normally a substantial mean torque in the shaft; in most cases, this torque has some ripple on it. Even a small proportion of the rated torque of a machine can constitute a substantial level of torsional forcing. We begin by considering the rotor shown in Figure 9.1, which is redrawn in Figure 9.5 with the inertia and stiffness parameters for the torsional analysis. The thrust bearing is ignored in this case.

Assuming that the inertia of the shaft is small compared to that of the disks, then each disk in Figure 9.5 is modeled by an inertia and each shaft section can be modeled by a torsional spring. The relevant inertia in this case is the *polar moment of inertia*. By contrast, the *diametral moment of inertia* is vital for the lateral vibration discussed in Chapter 5, although the polar moment of inertia is also significant for the gyroscopic effects.

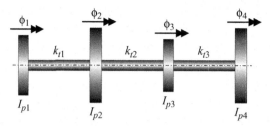

Figure 9.5. A typical rotor undergoing torsional vibration. The inertia of the ith disk is I_{pi}, the stiffness of the ith shaft section is k_{ti}, and the rotation of the ith disk is ϕ_i.

The torsional stiffness, k_t, of a shaft of length L and polar second moment of area J is given by

$$k_t = \frac{T}{\phi} = \frac{GJ}{L} \tag{9.4}$$

where T is the applied torque, G is the shear modulus (i.e., modulus of rigidity), and ϕ is the relative rotation of the ends of the shaft section. This expression is accurate only for circular (or annular) cross sections. For the torsional stiffness of noncircular sections, it is common to replace J in Equation 9.4 by aJ, where a is a correction factor that is related to the cross-sectional shape (Blevins, 1979). This is accurate only when the axial length over which the torque is sensibly constant is large compared with all of the cross-sectional dimensions. To be certain of avoiding significant errors, a two- or three-dimensional FE model of the rotor is recommended (see Chapter 10). For the remainder of this treatment on torsional oscillations, it is assumed that the torsional stiffness is given by Equation (9.4).

The discrete component model of the rotor is shown in Figure 9.5. Again, the equations of motion for free torsional vibration are independent of shaft speed. These equations are determined using the methods described in Chapter 2 and are essentially the same as Equation (9.2), with

$$\mathbf{M} = \begin{bmatrix} I_{p1} & 0 & 0 & 0 \\ 0 & I_{p2} & 0 & 0 \\ 0 & 0 & I_{p3} & 0 \\ 0 & 0 & 0 & I_{p4} \end{bmatrix}, \quad \mathbf{K} = \begin{bmatrix} k_{t1} & -k_{t1} & 0 & 0 \\ -k_{t1} & k_{t1} + k_{t2} & -k_{t2} & 0 \\ 0 & -k_{t2} & k_{t2} + k_{t3} & -k_{t3} \\ 0 & 0 & -k_{t3} & k_{t3} \end{bmatrix} \tag{9.5}$$

and $\mathbf{q} = \begin{bmatrix} \phi_1 & \phi_2 & \phi_3 & \phi_4 \end{bmatrix}^{\top}$. The eigenvalue problem, Equation (9.3), can be readily solved using the mass and stiffness matrices of Equation (9.5) to obtain the natural frequencies and mode shapes. Except for the effect of the thrust bearing (i.e, k_b in Equation (9.2)), the equations are identical in form but their parameters are different. We have not considered any damping in the rotor. The similarity with the axial system matrices is obviously strong; the principal difference is that the axial system almost never has any rigid-body modes, whereas the torsional system invariably has one.

EXAMPLE 9.4.1. Determine the first four torsional natural frequencies of the rotor described in Example 9.2.1. For convenience, the details are summarized here. The rotor is a single shaft that is modeled as a tube with an of outside diameter of 80 mm and an inside diameter of 40 mm, onto which four disks are attached, each with an outside diameter of 400 mm and an axial thickness of 50 mm. The central planes of the four disks are equally spaced at 0.3 m apart and the shaft does not extend past the outermost disks. Assume that the shear modulus is $G = 80$ GPa and $\rho = 7,850$ kg/m^3.

Solution. For each shaft section, $J = \pi \left(d_o^4 - d_i^4\right)/32 = 3.7699 \times 10^{-6}$ m^2 and $L = 0.3$ m. Thus, $k_t = GJ/L = 1.0053$ MNm. The disk polar moment of inertia is $I_p = \pi \rho h \left(D_o^4 - D_i^4\right)/32 = 0.98488$ kg m^2. The eigenvalue problem now can be formulated and solved to give the natural frequencies as 0, 123.07, 227.40, and 297.11 Hz. One frequency is zero and is called a rigid-body mode of vibration because in this mode, the rotor does not twist. This is a consequence of the lack of constraint in rotation that the rotor must have if it is to function as a machine. When vibrating at the other natural frequencies, the rotor twists. If the shaft inertia is included in the disk inertias, then the natural frequencies are 0, 122.75, 226.64, and 295.88 Hz. Including the shaft inertia has less effect on the natural frequencies than the shaft mass has on the axial natural frequencies. The shaft length between the disks may be defined as the distance between the disk centers or the disk faces in exactly the same way as for axial motion. The results given here assume the length to be defined using disk centers only.

This rotor has much higher natural frequencies for axial vibration (see Table 9.1) than torsional vibration. This is usually the case due to the values of the physical parameters involved. For a perfectly uniform circular shaft, the ratio of axial-resonance frequencies to torsional-resonance frequencies is $\sqrt{E/G}$, which is approximately 1.6 for most materials. For rotors with disks, such as the one studied in this example, the bulk of the kinetic energy in the low-frequency torsional and axial modes tends to be contained in the higher-diameter parts of the rotor. In contrast, the bulk of the strain energy is contained in the lower-diameter sections of the rotor. For axial vibrations, masses and stiffnesses are proportional to the square of diameter; whereas, for torsional vibrations, the inertias and torsional stiffnesses are proportional to the fourth power of diameter. In both cases, if the shaft length is constant, then the natural frequencies do not change with diameter for these simple models.

9.5 Shaft-Line Finite Element Analysis of Torsional Motion

The procedure for the FE modeling of the torsional motion of a rotor is similar to the analysis of the axial motion. For torsional problems, the quantity of interest is the twisting of the shaft about the axis, and the element matrices are given in Section 4.6.2. Assembly of the model follows exactly the same procedure as that for axial motion, described in Chapter 4.

EXAMPLE 9.5.1. Analyze the system of Example 9.4.1 using FEA with six elements to obtain the torsional natural frequencies.

Solution. Each of the three shaft sections is modeled by two elements, resulting a total of six elements. Running appropriate software gives the natural frequencies as 0, 122.81, 226.98, 296.64, 5,869.7, 5,876.7, and 5,883.6 Hz. Comparing these results with those from the simple model, Example 9.4.1, shows minor discrepancies. Proportionately, the errors made in natural frequency in torsion by ignoring shaft inertia are much smaller than the errors made on the natural frequencies of axial modes by ignoring shaft mass. In this example, the shaft accounts for only 1.11 percent of the total polar inertia of the rotor, but this is not always the case. When axial vibrations of this same system are considered, a shaft-line FE model with 12 elements shows a significant improvement over a six-element model. For torsional resonances of this model, the differences between the first four resonance frequencies computed from a six-element and a twelve-element model are smaller than 0.001 percent in all cases.

The mode shapes for this example are virtually identical to those for the axial motion shown in Figure 9.4, except that the motion shown is torsional rather than axial displacements. This similarity arises because the machine consists of identical disks that are evenly distributed on a uniform shaft. Furthermore, the thrust bearing is very flexible, giving a first axial mode shape that resembles a rigid-body mode.

9.6 Geared and Branched Systems

When a torsional model of a complete shaft train is being prepared, it is often necessary to deal with the presence of a gearbox. The effect of nonunity gear ratios on torsional vibration is examined. Initially, we focus on relatively intuitive means of handling the analysis of these systems; subsequently, we progress to a more formal approach that is also useful for understanding how to model the effects of important sources of excitations prevalent in gear systems. Gear connections can cause an interaction among torsional, axial, and lateral vibrations (see Section 10.9).

9.6.1 Applying Constraints for Geared Systems

Consider the torsional behavior in a system involving a gear drive, depicted schematically in Figure 9.6. The system is modeled using four rotary inertias and two torsional springs. The angular displacements, ϕ_2 and ϕ_3, of the inertias representing the gears are coupled through the gear mesh and therefore are not independent. Let R_2 and N_2 be the working radius and the number of teeth, respectively, for gear wheel 2. Similarly, let R_3 and N_3 be the working radius and the number of teeth, respectively, for gear wheel 3. If the gear pair is perfect, then the tangential velocities of the two gears are identical. For the present purposes, it is necessary only to recognize that $\dfrac{R_2}{N_2} = \dfrac{R_3}{N_3}$.

Equating the tangential velocities yields $R_3\dot{\phi}_3 = -R_2\dot{\phi}_2$, where $\dot{\phi}_2$ and $\dot{\phi}_3$ represent the instantaneous angular velocities of gear wheels 2 and 3, respectively. Defining γ to be the gear ratio $\gamma = \dfrac{N_2}{N_3} = \dfrac{R_2}{R_3}$, then $\dot{\phi}_3 = -\gamma\dot{\phi}_2$. The negative sign

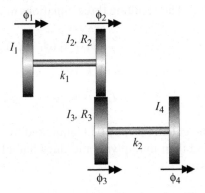

Figure 9.6. Idealization of a drive system with four inertias. The ith gear inertia is I_i, the working radius of gear i is R_i, the ith shaft stiffness is k_i, and the rotation of the ith gear is ϕ_i.

indicates that a positive $\dot\phi_2$ causes negative $\dot\phi_3$. With appropriate choices of initial conditions, we can deduce that $\phi_3 = -\gamma\phi_2$.

The four equations of motion of this simple system are

$$I_1\ddot\phi_1 + k_1(\phi_1 - \phi_2) = T_1$$
$$I_2\ddot\phi_2 + k_1(\phi_2 - \phi_1) = T_2 + R_2 F_{23}$$
$$I_3\ddot\phi_3 + k_2(\phi_3 - \phi_4) = T_3 + R_3 F_{23}$$
$$I_4\ddot\phi_4 + k_2(\phi_4 - \phi_3) = T_4$$

$$(9.6)$$

where F_{23} represents the instantaneous force acting between gears 2 and 3 along the common tangent and where $\{T_1, T_2, T_3, T_4\}$ are the instantaneous torques being exerted on inertias $\{1, 2, 3, 4\}$, respectively, from external sources. It appears anomalous at first that F_{23} contributes positively to both gear wheels, but inspection of the free-body diagrams for disks 2 and 3, shown in Figure 9.7, reveals that this is correct.

There are four rotation coordinates in Equation (9.6) but only three are independent because of the constraint between ϕ_2 and ϕ_3. We choose $\{\phi_1, \phi_2, \phi_4\}$ as independent coordinates, and ϕ_3 is eliminated using $\phi_3 = -\gamma\phi_2$. The forcing term F_{23} in Equation (9.6) is unknown and is removed by subtracting γ times the third

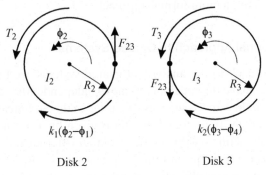

Disk 2   Disk 3

Figure 9.7. The free-body diagrams for disks 2 and 3 of the idealizations of a drive system with four inertias.

equation from the second. The resulting three equations in $\{\phi_1, \phi_2, \phi_4\}$ are

$$I_1\ddot{\phi}_1 + k_1 (\phi_1 - \phi_2) = T_1$$

$$\left(I_2 + \gamma^2 I_3\right)\ddot{\phi}_2 + k_1 (\phi_2 - \phi_1) + k_2 \left(\gamma^2\phi_2 + \gamma\phi_4\right) = T_2 - \gamma T_3 \qquad (9.7)$$

$$I_4\ddot{\phi}_4 + k_2 (\phi_4 + \gamma\phi_2) = T_4$$

The symmetry in the equations may be highlighted by making the further substitution, $\phi_4 = -\gamma\phi'_3$, and then multiplying the third line of Equation (9.7) by $-\gamma$. The resulting equations are

$$I_1\ddot{\phi}_1 + k_1 (\phi_1 - \phi_2) = T_1$$

$$\left(I_2 + \gamma^2 I_3\right)\ddot{\phi}_2 + k_1 (\phi_2 - \phi_1) + \gamma^2 k_2 (\phi_2 - \phi'_3) = T_2 - \gamma T_3 \qquad (9.8)$$

$$\gamma^2 I_4\ddot{\phi}'_3 + \gamma^2 k_2 (\phi'_3 - \phi_2) = -\gamma T_4$$

In matrix form, these equations are

$$
\begin{bmatrix} I_1 & 0 & 0 \\ 0 & I_2 + \gamma^2 I_3 & 0 \\ 0 & 0 & \gamma^2 I_4 \end{bmatrix} \begin{Bmatrix} \ddot{\phi}_1 \\ \ddot{\phi}_2 \\ \ddot{\phi}'_3 \end{Bmatrix}
$$
$$
+ \begin{bmatrix} k_1 & -k_1 & 0 \\ -k_1 & k_1 + \gamma^2 k_2 & -\gamma^2 k_2 \\ 0 & -\gamma^2 k_2 & \gamma^2 k_2 \end{bmatrix} \begin{Bmatrix} \phi_1 \\ \phi_2 \\ \phi'_3 \end{Bmatrix} = \begin{Bmatrix} T_1 \\ T_2 - \gamma T_3 \\ -\gamma T_4 \end{Bmatrix} \qquad (9.9)
$$

These equations are equivalent to those obtained from a three-inertia model of a simple rotor similar to that illustrated in Figure 9.5 (but consisting of only three inertias). We describe the torsional model of Equations (9.8) and (9.9) as the *referred form of the model*. In this case, the inertias and torsional stiffnesses on the lower shaft are all scaled by γ^2 and the torques and angles are been scaled by $-\gamma$. Suitable mass and stiffness matrices can be extracted from Equations (9.7) and (9.8) and used to find natural frequencies and mode shapes. A similar approach may be applied to any gear arrangement consisting of multiple gear ratios. However, for reasons specific to ease of understanding gear-meshing errors as a source of excitation, it is appropriate to consider a more formal approach.

9.6.2 A More Formal Approach to Geared Systems

In this formal approach to the analysis of geared systems, we begin with a set of inertias, all of which have the capability to move independently; that is, there are no constraints between the various inertias. We refer to this system as the *unconnected system*. The angular position of each inertia is given by a distinct variable, ϕ_{Ui}; a torque, T_{Ui}, may be acting on this inertia from an external source. The vector of all angular displacements is $\mathbf{q}_U$ and the vector of all externally applied torques is $\mathbf{Q}_U$. Here, the formulation is illustrated by reference to the four degrees of freedom example of Section 9.6.1. In this case, $\mathbf{q}_U = \begin{bmatrix} \phi_1 & \phi_2 & \phi_3 & \phi_4 \end{bmatrix}^\top$.

The equation of motion for the geared system in terms of the unconnected degree of freedom is of the form

$$\mathbf{M}_U \ddot{\mathbf{q}}_U + \mathbf{K}_U \mathbf{q}_U = \mathbf{Q}_U + \mathbf{Q}_C \qquad (9.10)$$

where $\mathbf{Q}_C$ are the internal forces that enforce the constraints between the unconnected degrees of freedom. For the four degrees of freedom example, from Equation (9.6), the constraint force is

$$\mathbf{Q}_C = \left\{ \begin{array}{c} 0 \\ R_2 \\ R_3 \\ 0 \end{array} \right\} F_{23}$$

The constraints on the system with perfect gear meshes ensure that the tangential velocities at the tooth contact for each gear are equal. These constraints are embodied in a matrix $\mathbf{E}$ through the equation

$$\mathbf{E}^\top \dot{\mathbf{q}}_U = \mathbf{0} \qquad (9.11)$$

In Section 9.6.1, the constraint is $R_3 \dot{\phi}_3 = -R_2 \dot{\phi}_2$; therefore

$$\mathbf{E} = \left[\begin{array}{c} 0 \\ R_2 \\ R_3 \\ 0 \end{array} \right]$$

Note that $\mathbf{Q}_C = \mathbf{E} F_{23}$; the constraint force is always the product of $\mathbf{E}$ and a vector of independent forces due to the action and reaction forces at the gear meshes. Integrating Equation (9.11) and setting the constant of integration to zero (because the reference angle for each inertia is arbitrary) gives

$$\mathbf{E}^\top \mathbf{q}_U = \mathbf{0} \qquad (9.12)$$

Section 2.5 describes one way by which these constraints may be enforced. It involves developing a transformation matrix, $\mathbf{T}$, relating $\mathbf{q}_U$ to a reduced dimension vector of coordinates, $\mathbf{q}_R$, through

$$\mathbf{q}_U = \mathbf{T}\mathbf{q}_R \qquad (9.13)$$

For the transformation to enforce the constraints

$$\mathbf{E}^T \mathbf{q}_U = \mathbf{E}^T \mathbf{T}\mathbf{q}_R = \mathbf{0} \qquad (9.14)$$

for all $\mathbf{q}_R$; hence,

$$\mathbf{E}^T \mathbf{T} = \mathbf{0} \qquad (9.15)$$

This also ensures that $\mathbf{T}^\top \mathbf{Q}_C = \mathbf{0}$. Furthermore, the reduced vector of coordinates, $\mathbf{q}_R$, should contain the same number of degrees of freedom as the constrained system. This may be expressed mathematically by the condition that $\mathbf{Y} = [\mathbf{E} \ \mathbf{T}]$ must be full rank, which is equivalent to the condition that $\mathbf{Y}\mathbf{Y}^\top$ is nonsingular. This latter condition is automatically satisfied if $\mathbf{Y}$ is nonsingular, but we encounter cases in which a square nonsingular $\mathbf{Y}$ cannot be achieved because the columns of $\mathbf{E}$ are not all independent. Any matrix $\mathbf{T}$ satisfying these two criteria is acceptable. For systems with few degrees of freedom and a small number of constraints, it is

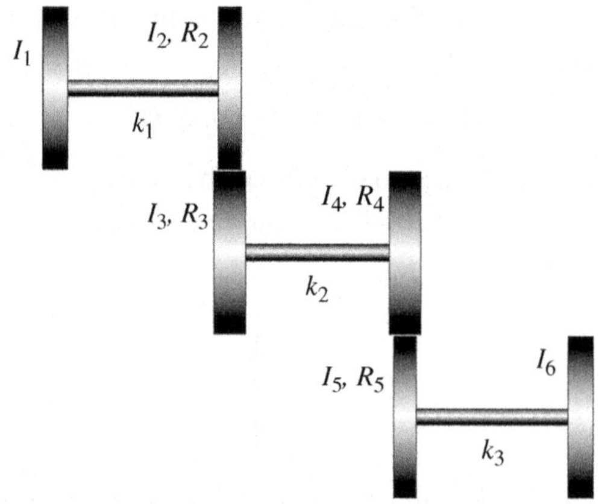

Figure 9.8. Idealized model with two gearboxes.

convenient to prepare such a matrix **T** by inspection. For the four degrees of freedom example of Section 9.6.1, a suitable transformation matrix is

$$\mathbf{T} = \begin{bmatrix} 1 & 0 & 0 \\ 0 & R_3 & 0 \\ 0 & -R_2 & 0 \\ 0 & 0 & 1 \end{bmatrix}$$

and it is straightforward to verify that these two criteria are satisfied for this choice. Formal methods of preparing a suitable matrix **T** are discussed in detail in Section 9.6.3.

Having found a suitable matrix **T**, the system is then transformed to

$$\mathbf{M}_R = \mathbf{T}^\top \mathbf{M}_U \mathbf{T}, \qquad \mathbf{K}_R = \mathbf{T}^\top \mathbf{K}_U \mathbf{T} \tag{9.16}$$

and a new equation of motion, $\mathbf{M}_R\ddot{\mathbf{q}}_R + \mathbf{K}_R\mathbf{q}_R = \mathbf{Q}_R$, is formed by transforming the original forcing vector using $\mathbf{Q}_R = \mathbf{T}^\top \mathbf{Q}_U$. The internal constraint forces, $\mathbf{Q}_C$, are eliminated because the reduced coordinates automatically satisfy the constraints; hence, $\mathbf{T}^\top \mathbf{Q}_C = \mathbf{0}$.

A commonly favored alternative version of **T** in this case is

$$\mathbf{T} = \begin{bmatrix} 1 & 0 & 0 \\ 0 & 1 & 0 \\ 0 & -\gamma & 0 \\ 0 & 0 & -\gamma \end{bmatrix} \tag{9.17}$$

where $\gamma = \dfrac{R_2}{R_3}$ as before. It is easily verified in this case that the transformation places the system in the referred form described previously with equations identical to Equation (9.9).

EXAMPLE 9.6.1. Determine the reduction-transformation matrix for the system with two gear meshes shown in Figure 9.8, which illustrates a six-inertia model.

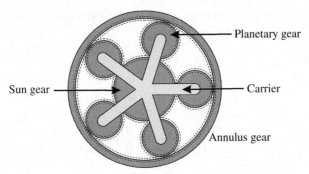

Figure 9.9. An epicyclic gearbox system.

Solution. In this case, one acceptable choice of matrices **E** and **T** is

$$
\mathbf{E} = \begin{bmatrix} 0 & 0 \\ R_2 & 0 \\ R_3 & 0 \\ 0 & R_4 \\ 0 & R_5 \\ 0 & 0 \end{bmatrix}, \qquad
\mathbf{T} = \begin{bmatrix} 1 & 0 & 0 & 0 \\ 0 & R_3 & 0 & 0 \\ 0 & -R_2 & 0 & 0 \\ 0 & 0 & R_5 & 0 \\ 0 & 0 & -R_4 & 0 \\ 0 & 0 & 0 & 1 \end{bmatrix}
\tag{9.18}
$$

It is simple to verify that the choice of **E** and **T** satisfies $\mathbf{E}^\top \mathbf{T} = 0$ and $[\mathbf{E} \; \mathbf{T}]$ nonsingular. The transformation matrix **T** does not produce the referred system. For that purpose, we choose

$$
\mathbf{T} = \begin{bmatrix} 1 & 0 & 0 & 0 \\ 0 & 1 & 0 & 0 \\ 0 & -\gamma_{23} & 0 & 0 \\ 0 & 0 & -\gamma_{23} & 0 \\ 0 & 0 & \gamma_{23}\gamma_{45} & 0 \\ 0 & 0 & 0 & \gamma_{23}\gamma_{45} \end{bmatrix}
\tag{9.19}
$$

where the two gear ratios are defined as $\gamma_{23} = \dfrac{R_2}{R_3}$ and $\gamma_{45} = \dfrac{R_4}{R_5}$.

EXAMPLE 9.6.2. We consider one more case for illustration purposes: an epicyclic gearbox shown schematically in Figure 9.9. A sun gear having radius R_S is engaged with five planetary gears, each of radius R_P. The planetary gears are all mounted on a carrier so that the centers are equally spaced on a pitch circle of radius $R_C = (R_S + R_P)$. A single annulus gear, radius $R_A = (R_S + 2R_P)$, engages with each of the planetary gears. Determine the reduction-transformation matrix for this case.

Solution. There are eight independent angles in the unconnected system,

$$\{\phi_S, \phi_C, \phi_{P1}, \phi_{P2}, \phi_{P3}, \phi_{P4}, \phi_{P5}, \phi_A\}$$

and each represents an absolute angle of rotation. The subscripts S and A refer to the sun and annulus gears, C refers to the carrier, and Pi refers to the ith planetary gear. Because there are 10 separate gear meshes in this gearbox, there are 10 constraint equations (on angular velocities) and, hence, 10 columns in the constraint matrix, $\mathbf{E}$. Thus,

$$\mathbf{E} = \begin{bmatrix} R_S & R_S & R_S & R_S & R_S & 0 & 0 & 0 & 0 & 0 \\ -R_C & -R_C & -R_C & -R_C & -R_C & R_C & R_C & R_C & R_C & R_C \\ R_P & 0 & 0 & 0 & 0 & R_P & 0 & 0 & 0 & 0 \\ 0 & R_P & 0 & 0 & 0 & 0 & R_P & 0 & 0 & 0 \\ 0 & 0 & R_P & 0 & 0 & 0 & 0 & R_P & 0 & 0 \\ 0 & 0 & 0 & R_P & 0 & 0 & 0 & 0 & R_P & 0 \\ 0 & 0 & 0 & 0 & R_P & 0 & 0 & 0 & 0 & R_P \\ 0 & 0 & 0 & 0 & 0 & -R_A & -R_A & -R_A & -R_A & -R_A \end{bmatrix} \quad (9.20)$$

Not all of these are independent; in fact, the last four columns of $\mathbf{E}$ can each be derived as a different linear combination of the first six columns. One transformation matrix, $\mathbf{T}$, with the requisite properties is

$$\mathbf{T} = \begin{bmatrix} 1 & (1/R_S) \\ 1 & 0 \\ 1 & -(1/R_P) \\ 1 & -(1/R_P) \\ 1 & -(1/R_P) \\ 1 & -(1/R_P) \\ 1 & -(1/R_P) \\ 1 & -(1/R_A) \end{bmatrix} \quad (9.21)$$

Each of the two columns of $\mathbf{T}$ represents a valid combination of the eight different angular velocities, and every possible combination of angular velocities achievable within the epicyclic gearbox can be constructed as a linear combination of these two columns. In this illustration, there are no torsional springs, so the stiffness matrix associated with the unconnected system is an 8×8 matrix of zeros. The mass matrix for the unconnected system is an 8×8 diagonal matrix

$$\text{diag}\,([I_S,\ I_C,\ I_{p1},\ I_{p2},\ I_{p3},\ I_{p4},\ I_{p5},\ I_A])$$

For this system, there is no single referred system model because two rotational speeds may be varied independently – for example, the mean rotational speeds of the sun gear and of the carrier.

EXAMPLE 9.6.3. The propulsion system for a small ship is illustrated in Figure 9.10, which shows that there are seven significant discrete inertias and three torsionally flexible shafts. The inertia values are summarized as (motor inertia) $I_{P0} = 3{,}200\,\text{kg m}^2$, (gear inertia) $I_{P1} = 200\,\text{kg m}^2$, (gear inertia) $I_{P2} = 800\,\text{kg m}^2$, and (propeller inertia) $I_{P3} = 450\,\text{kg m}^2$. The shaft dimensions are $L_1 = 2.5\,\text{m}$, $L_2 = 3.5\,\text{m}$, $D_1 = 0.2\,\text{m}$, and $D_2 = 0.25\,\text{m}$, and all shafts are made from an alloy of which the shear modulus, G, is 68 GPa. Each gear with inertia

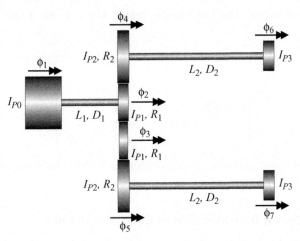

Figure 9.10. Idealization of a ship propulsion system (Example 9.6.3).

I_{P1} has 50 teeth ($N_1 = 50$) and a working radius of 0.5 m; whereas each gear with inertia I_{P2} has 80 teeth ($N_2 = 80$) and a working radius of 0.8 m. Evaluate the natural frequencies of this system.

Solution. The torsional stiffness of the motor shaft is 4.2726×10^6 Nm/rad and the stiffness of each propeller shaft is 7.4508×10^6 Nm/rad. Given the definition of the torsional degrees of freedom $\phi_1, \ldots, \phi_7$ shown in Figure 9.10, the mass and stiffness matrices of the unconnected system are

$$\mathbf{M}_U = \mathrm{diag}\left(\begin{bmatrix} 3{,}200 & 200 & 200 & 800 & 800 & 450 & 450 \end{bmatrix}\right) \mathrm{kg\,m^2}$$

$$\mathbf{K}_U = \begin{bmatrix} 4.27 & -4.27 & 0 & 0 & 0 & 0 & 0 \\ -4.27 & 4.27 & 0 & 0 & 0 & 0 & 0 \\ 0 & 0 & 0 & 0 & 0 & 0 & 0 \\ 0 & 0 & 0 & 7.45 & 0 & -7.45 & 0 \\ 0 & 0 & 0 & 0 & 7.45 & 0 & -7.45 \\ 0 & 0 & 0 & -7.45 & 0 & 7.45 & 0 \\ 0 & 0 & 0 & 0 & -7.45 & 0 & 7.45 \end{bmatrix} \mathrm{MN\,m/rad}$$

The matrix, $\mathbf{E}$, expressing the constraints and a suitable transformation matrix, $\mathbf{T}$, are

$$\mathbf{E} = \begin{bmatrix} 0 & 0 & 0 \\ 0.5 & 0.5 & 0 \\ 0.5 & 0 & 0.5 \\ 0 & 0.8 & 0 \\ 0 & 0 & 0.8 \\ 0 & 0 & 0 \\ 0 & 0 & 0 \end{bmatrix}, \quad \mathbf{T} = \begin{bmatrix} 1 & 0 & 0 & 0 \\ 0 & 1.6 & 0 & 0 \\ 0 & -1.6 & 0 & 0 \\ 0 & -1.0 & 0 & 0 \\ 0 & 1.0 & 0 & 0 \\ 0 & 0 & 1 & 0 \\ 0 & 0 & 0 & 1 \end{bmatrix}$$

The transformed mass and stiffness matrices, $\mathbf{M}_R = \mathbf{T}^\top \mathbf{M}_U \mathbf{T}$ and $\mathbf{K}_R = \mathbf{T}^\top \mathbf{K}_U \mathbf{T}$, are then

$$\mathbf{M}_R = \mathrm{diag}\left(\begin{bmatrix} 3{,}200, & 4{,}966, & 450, & 450 \end{bmatrix}\right) \mathrm{kg\,m}^2$$

$$\mathbf{K}_R = \begin{bmatrix} 4.273 & -6.836 & 0 & 0 \\ -6.836 & 25.839 & 7.451 & -7.451 \\ 0 & 7.451 & 7.451 & 0 \\ 0 & -7.451 & 0 & 7.451 \end{bmatrix} \mathrm{MN\,m/rad}$$

The natural frequencies are computed easily from these matrices as $0, 9.118$, 20.479, and $22.708\,\mathrm{Hz}$.

9.6.3 Developing a Transformation to Effect Constraints

In the previous subsection, constraints arising from geared connections are expressed as $\mathbf{E}^\top \mathbf{q}_U = \mathbf{0}$ and are automatically imposed by performing the coordinate transformation $\mathbf{q}_U = \mathbf{T}\mathbf{q}_R$, where $\mathbf{T}$ is any matrix selected so that $\mathbf{E}^T \mathbf{T} = \mathbf{0}$ and $\mathbf{Y}\mathbf{Y}^T$ is nonsingular where $\mathbf{Y} = [\mathbf{E}\ \mathbf{T}]$.

There are several methods by which such transformation matrices may be found and each results in a different matrix, $\mathbf{T}$. If any matrix $\mathbf{T}$ is found satisfying these criteria, a number of alternative transformations, $\hat{\mathbf{T}}$, may be obtained as $\hat{\mathbf{T}} = \mathbf{T}\mathbf{X}$, where $\mathbf{X}$ is any square invertible matrix of the appropriate dimension. We now describe one method commonly used in FE analysis for developing the transformation. This method is highly efficient and has good numerical stability. The algorithm applies one constraint at a time and the final transformation, $\mathbf{T}$, is developed as the product of p intermediate transformations, $\mathbf{T}_i$, where p is less than or equal to the number of constraints (columns in $\mathbf{E}$). Thus,

$$\mathbf{T} = \mathbf{T}_1 \times \mathbf{T}_2 \times \ldots \times \mathbf{T}_p \tag{9.22}$$

Each $\mathbf{T}_i$ reduces the dimension of the system by one and produces a new constraint matrix as

$$\mathbf{E}_i = \mathbf{T}_i^\top \mathbf{E}_{(i-1)} \tag{9.23}$$

Columns 1 to i of $\mathbf{E}_i$ contain only zeros and $\mathbf{E}_0$ implicitly represents the original constraint matrix, $\mathbf{E}$. The number of rows in $\mathbf{E}_i$ is denoted by n_i. There are n_0 degrees of freedom in the original system and, because each constraint applied reduces the number of degrees of freedom by one

$$n_i \equiv n_0 - i \tag{9.24}$$

In general, matrix $\mathbf{T}_i$ has dimensions $(n_{(i-1)} \times n_i)$ and it is determined by considering only the ith column of $\mathbf{E}_{i-1}$. $\mathbf{T}_i$ can be formed by the following steps:

(1) Set $\mathbf{T}_i = \mathbf{I}_{(n_i+1)}$ (the identity matrix of dimension $(n_i + 1)$).
(2) Define k as the position of the entry having the largest absolute value in the ith column of $\mathbf{E}_{i-1}$.
(3) Replace row k of $\mathbf{T}_i$ with the transpose of column i from $\mathbf{E}_{(i-1)}$.
(4) Divide this row by the negative of its kth element.
(5) Strike out column k from the matrix.

This procedure can be applied for any number of constraints. In some cases, the constraints are not all independent, which is encountered in the context of the epicyclic gearbox in Example 9.6.2, in which each one of 10 different gear meshes produced a different constraint. However, there are only eight degrees of freedom in the original system and only six of the constraint equations are independent. In this case, the procedure described here stops after $\mathbf{T}_6$ and $\mathbf{E}_6$ are obtained because all of the columns of $\mathbf{E}_6$ are zero. The fact that the first six columns of $\mathbf{E}_6$ are zero is not surprising; the fact that the remaining columns are also zero signifies the lack of independence in the original constraints. To prevent round-off errors from introducing spurious constraints, the magnitudes of the columns of $\mathbf{E}_0$ should be compared to the magnitudes of $\mathbf{E}_i$. The process is complete when all of the magnitudes are very low compared to their original values.

The following two simple examples illustrate the procedure.

EXAMPLE 9.6.4. Consider the case of a six degrees of freedom system to which a single constraint is applied, with $\mathbf{E}_0$ given by

$$\mathbf{E}_0 = \begin{bmatrix} 2 & -10 & 3 & -4 & 5 & 9 \end{bmatrix}^T$$

Solution. The second entry here has the largest absolute value; hence, $k = 2$. After step (3), in which the second row of the identity matrix is replaced by $\mathbf{E}_0^T$, the matrix is

$$\mathbf{T}_1 = \begin{bmatrix} 1 & 0 & 0 & 0 & 0 & 0 \\ 2 & -10 & 3 & -4 & 5 & 9 \\ 0 & 0 & 1 & 0 & 0 & 0 \\ 0 & 0 & 0 & 1 & 0 & 0 \\ 0 & 0 & 0 & 0 & 1 & 0 \\ 0 & 0 & 0 & 0 & 0 & 1 \end{bmatrix}$$

Dividing the second row by 10 and deleting the second column gives

$$\mathbf{T}_1 = \begin{bmatrix} 1 & 0 & 0 & 0 & 0 \\ 0.2 & 0.3 & -0.4 & 0.5 & 0.9 \\ 0 & 1 & 0 & 0 & 0 \\ 0 & 0 & 1 & 0 & 0 \\ 0 & 0 & 0 & 1 & 0 \\ 0 & 0 & 0 & 0 & 1 \end{bmatrix}$$

EXAMPLE 9.6.5. Consider the case of a six degrees of freedom system to which two constraints are applied, with $\mathbf{E}_0$ given by

$$\mathbf{E}_0 = \begin{bmatrix} 2 & -10 & 3 & -4 & 5 & 9 \\ 4 & 5 & -2 & 8 & 0 & 0 \end{bmatrix}^T$$

Solution. Proceed as in the previous example to find $\mathbf{T}_1$. Then, evaluating $\mathbf{E}_1 = \mathbf{T}_1^T \mathbf{E}_0$ yields

$$\mathbf{E}_1 = \begin{bmatrix} 0 & 0 & 0 & 0 & 0 \\ 5 & -0.5 & 6 & 2.5 & 4.5 \end{bmatrix}^T$$

Evidently, the first constraint requires no further attention. The second column of $\mathbf{E}_1$ is used to determine $\mathbf{T}_2$, which may be obtained, by noting $k = 3$, as

$$\mathbf{T}_2 = \begin{bmatrix} 1 & 0 & 0 & 0 \\ 0 & 1 & 0 & 0 \\ -0.8333 & 0.08333 & -0.4167 & -0.7500 \\ 0 & 0 & 1 & 0 \\ 0 & 0 & 0 & 1 \end{bmatrix}$$

The product $\mathbf{T} = \mathbf{T}_1\mathbf{T}_2$ is then

$$\mathbf{T} = \begin{bmatrix} 1 & 0 & 0 & 0 \\ 0.5333 & 0.2667 & 0.6667 & 1.200 \\ 0 & 1 & 0 & 0 \\ -0.8333 & 0.08333 & -0.4167 & -0.7500 \\ 0 & 0 & 1 & 0 \\ 0 & 0 & 0 & 1 \end{bmatrix}$$

9.7 Axial and Torsional Vibration with External Excitation

Thus far in this chapter, we consider the development of models for free axial and torsional vibration of rotors. Equation (9.2) represents the dynamics of any undamped linear vibrating system not subjected to excitation. We now examine the behavior of these systems in the presence of some excitation. The emphasis here is on important mechanisms of excitation that are specific to torsional or axial vibrations and on the so-called *displacement-driven excitation*, which is particularly important in the context of torsional vibration.

The equation governing forced vibration is

$$\mathbf{M\ddot{q}} + \mathbf{C\dot{q}} + \mathbf{Kq} = \mathbf{Q} \tag{9.25}$$

where $\mathbf{q}(t)$ and $\mathbf{Q}(t)$ represent the displacement and forcing vectors, respectively. If $\mathbf{Q}(t)$ is independent of $\mathbf{q}(t)$, then the excitation can be described as *force-driven*. The lateral response of rotors to unbalance as presented in Chapter 6 is a study of force-driven excitation. Equation (9.25) can be solved in the time domain (see Sections 2.3.6 and 2.6) or in the frequency domain if both the excitation and the response are periodic (see Section 2.3.5).

In the case of displacement-driven excitation, the vector of forces, $\mathbf{Q}(t)$, on the right-hand side of Equation (9.25) is determined in part by a constraint equation involving some known vector of prescribed displacements, $\mathbf{q}_{ref}$, such that

$$\mathbf{E}^\top (\mathbf{q} - \mathbf{q}_{ref}) = \mathbf{0} \tag{9.26}$$

A feature of displacement-driven vibrations is that if changes are made to the system, changes automatically occur in the forces applied. We provide here a broad outline of how displacement-driven problems are solved. Equation (9.25) may be written more generally as

$$\mathbf{M}(\ddot{\mathbf{q}} - \ddot{\mathbf{q}}_{ref}) + \mathbf{C}(\dot{\mathbf{q}} - \dot{\mathbf{q}}_{ref}) + \mathbf{K}(\mathbf{q} - \mathbf{q}_{ref}) = \mathbf{Q} + \mathbf{Q}_{ref} \tag{9.27}$$

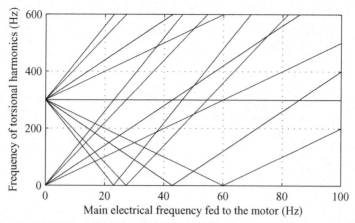

Figure 9.11. Torsional harmonics in the airgap of an induction motor driven by a variable speed drive, following Grieve and McShane (1989).

where the definition of $\mathbf{Q}_{ref}$ is

$$\mathbf{Q}_{ref}(t) = -(\mathbf{M}\ddot{\mathbf{q}}_{ref} + \mathbf{C}\dot{\mathbf{q}}_{ref} + \mathbf{K}\mathbf{q}_{ref}) \tag{9.28}$$

The reformulated problem is described by Equations (9.26) and (9.27), and these are identical in form to Equations (9.10) and (9.11), the solution of which has been discussed. In Section 9.7.3, we show how to derive suitable $\mathbf{q}_{ref}$ in cases where this is not already fully determined.

We now examine some cases of excitation specific to torsional and/or axial vibration, including force- and displacement-driven examples.

9.7.1 Force-Driven Excitation of Torsional Vibration

There are numerous different sources of force-driven torsional excitation and we outline the more important ones in this section. Several of the sources of torsional forcing arise in rotating electrical machines and the underlying causes include the following:

- In most DC electrical machines, the current normally should be constant to achieve a constant torque. Current ripples often occur as a result of the way that the DC is obtained from an AC source; the result of these current ripples is oscillatory components of torque in the airgap.
- In sinusoidal AC electrical machines such as induction motors, a constant torque often requires a balanced set of sinusoidal currents being fed into the three-phase terminals. The presence of higher harmonics can cause torsional oscillations. The power electronics of variable-speed drives have been a notable source of these higher harmonics (Grieve and McShane, 1989; Wolff and Molnar, 1985). Figure 9.11 is a map of torsional excitation frequencies as a function of steady motor-running speed experienced in a large variable speed drive.
- In switched AC electrical machines such as stepper motors and switched-reluctance motors, net rotation is achieved by alternately switching currents on

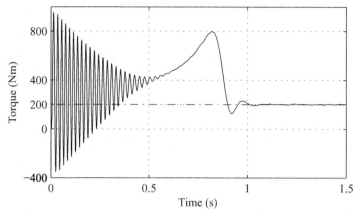

Figure 9.12. The variation in airgap torque during a typical induction motor startup.

and off into particular electrical phases. Stepper motors produce torque harmonics at frequencies many times greater than the rotational speed (i.e., often, more than 100 times). Switched-reluctance machines typically produce between three and six torque pulsations per cycle (depending on the number of stator-pole pairs), and it is not unusual to find that the root mean-square magnitude of the total torque signal is 50 percent higher than the mean torque.

Electrical machines also can produce significant transient torques when they are switched on (Alolah et al., 1999) or when an electrical fault occurs (e.g., a short circuit). An extremely common torsional excitation occurs when a three-phase induction machine is energized. Figure 9.12 shows a typical plot of airgap torque in a machine as a function of time after a 30-kW induction motor is started. The airgap torque settles eventually to a constant value (determined by the mechanical load and the losses); however, in the early stages of the startup, large torque oscillations are seen at supply frequency. These decay as the motor speed rises.

Reciprocating machines such as compressors, pumps, and internal combustion engines are also a notable source of force-driven torsional excitation. All of these machines develop oscillatory components of torque as a combination of the following three main effects:

- fluid loads on the pistons
- nonuniform power loss through friction between pistons and cylinders
- varying effective inertia of the crankshaft

Each cylinder of a reciprocating pump or compressor completes a full cycle of operation in each full rotation of the shaft. Two-stroke internal combustion engines do the same. In all cases, each cylinder of a machine is capable of producing torsional harmonics at multiples of rotational speed. By contrast, in a four-stroke internal-combustion engine, each cylinder completes a full cycle of operation only every two full revolutions of the shaft, and these engines produce torsional forcing at every integer multiple of one-half of the rotational speed.

The torsional forcing produced by a four-stroke internal-combustion engine is examined for illustration purposes. For a given cylinder, at any given crank angle, ϕ, there is a ratio, $s(\phi)$, between the vertical downward velocity of the piston and

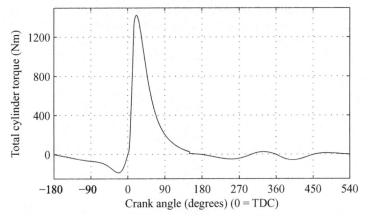

Figure 9.13. The torque for a single cylinder of a typical four-stroke gasoline engine over two complete revolutions of the crankshaft.

the angular velocity of the crankshaft. For a small increment, $\Delta\phi$, in the crankshaft angle, the work done by the gas in the cylinder is $P(\phi)As(\phi)\Delta\phi$, where A is the cross-sectional area of the cylinder and $P(\phi)$ is the gas pressure in the cylinder, above atmospheric. The component of the torque at the crankshaft due to the gas load is therefore $T_G = P(\phi)As(\phi)$. The torque component at the crankshaft associated with varying power loss due to friction can be derived similarly. The friction force between piston and cylinder often may be assumed to be reasonably constant in magnitude, F, and the direction of this force is opposite to the direction of the piston motion within the cylinder. Employing a work argument, the torque component at the crankshaft due to piston friction is found to be $T_F = -Fs(\phi)\,\text{sign}(s(\phi))$.

The third main component of variable torque is due to variation in the effective inertia seen at the crankshaft. Letting m_P represent the mass of one piston, the effective inertia of that piston at the crankshaft is $m_P s(\phi)^2$, which reaches zero twice in every revolution of the crankshaft. The inertia due to the crankshaft is fixed. The effective inertia of the connecting rods also varies with ϕ, but this is typically small compared to the inertia contributed by the piston. The total effective inertia at one cylinder of a reciprocating machine consists of a mean value, I, and a zero mean, variable component $I_{pn}(\phi)$.

If the machine rotates at a constant speed, then the inertia variation provides a component of torque given by $T_{pn} = -\dfrac{1}{2}\dfrac{\mathrm{d}I_{pn}(\phi)}{\mathrm{d}\phi}\Omega^2$. Often, the inertia of the reciprocating parts is much less than the nonreciprocating parts. Neglecting higher-order terms, the acceleration term in the equations of motion is approximated by $I\ddot{\phi}$ and the forcing term T_{pn} appears on the right side. The net torque produced by one cylinder then is given by

$$T_{\text{net}} = P(\phi)As(\phi) - Fs(\phi)\,\text{sign}(s(\phi)) - \frac{1}{2}\frac{\mathrm{d}I_{pn}(\phi)}{\mathrm{d}\phi}\Omega^2 \qquad (9.29)$$

Figure 9.13 shows a plot of net torque against crankshaft angle for a single cylinder of a typical four-stroke gasoline engine. In this case, the mean values of the gas, friction, and varying-inertia torques are 85 Nm, -14 Nm, and 0, respectively. The gas torque has a peak that is 1,383 Nm higher than its mean. The friction torque varies

Table 9.2. *Fourier components of the torque given in Figure 9.13, related to the crankshaft angle*

Harmonic	Magnitude (Nm)(peak)	Phase (rad)
0	70.8	0.0
1/2	174.1	−2.0573
1	180.8	2.2305
3/2	173.5	0.3922
2	128.2	−1.2805
5/2	134.5	−3.1321
3	99.7	1.4381
7/2	96.2	−0.2701
4	79.3	−1.9720
9/2	68.1	2.5815
5	57.0	0.8868

between −23 Nm and 0, and the variable inertia torque oscillates between −41 Nm and +41 Nm.

A Fourier analysis of this torque profile is summarized in Table 9.2. The fundamental period for the Fourier analysis is two revolutions of the crankshaft. For the purposes of torsional analysis, a standard approach develops an approximate torsional model in which a separate (i.e., constant) inertia is assumed to be present at each cylinder location. Then, each harmonic is considered separately. This is illustrated by a three-cylinder engine driving a generator in Example 9.7.1.

In many cases, the oscillatory torques are caused by the load driven by a machine. For example, a steel-mill motor experiences severe torsional excitation when a billet of steel enters between the rollers driven by the motor. Escalator drives are subject to uneven loading resulting from changes in the net weight of the passengers being carried. High-speed milling machines encounter torque pulsations every time another edge of the milling cutter begins to cut a new chip from the workpiece. A similar phenomenon occurs in lathes when the workpiece being turned is not a solid of revolution about the main axis of the lathe.

EXAMPLE 9.7.1. A small three-cylinder, four-stroke gasoline engine driving a generator is represented approximately by a five-inertia model, as shown in Figure 9.14. The first three inertias are associated with cylinders 1 through 3 of the engine. The variable inertia is included in the computation of the applied torque, but a constant value of 3.5×10^{-3} kg m^2 is assumed in the inertia terms in the equation of motion. A flywheel, I_{P4}, has inertia 0.15 kg m^2, and the generator rotor is modeled using a single inertia, I_{P5}, with a value of 0.05 kg m^2. The torsional stiffnesses in the model are $k_1 = k_2 = 12$ kNm/rad, $k_3 = 14$ kNm/rad, and $k_4 = 10$ kNm/rad. The engine is turning at 3,000 rev/min and the torque components produced by each cylinder are described in Table 9.2. The firing sequence of the engine is 1–2–3 with a $4\pi/3$ radians delay between cylinders 1 and 2, and so on. The generator exerts a steady-load torque of 212.4 Nm, keeping the engine at a constant speed. The generator does not produce any

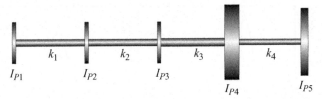

Figure 9.14. A three-cylinder, gasoline-driven generator (Example 9.7.1).

oscillatory components of torque in the airgap. Compute the peak torque in the coupling between the flywheel and the generator in any one cycle.

Solution. The solution begins by assembling a time-invariant model to represent (approximately) the engine–generator set. The stiffness and mass matrices are

$$\mathbf{K} = \begin{bmatrix} 12 & -12 & 0 & 0 & 0 \\ -12 & 24 & -24 & 0 & 0 \\ 0 & -24 & 26 & -14 & 0 \\ 0 & 0 & -14 & 24 & -10 \\ 0 & 0 & 0 & -10 & 10 \end{bmatrix} \times 10^3 \text{ N m/rad}$$

$$\mathbf{M} = \text{diag}\left(\begin{bmatrix} 3.5 & 3.5 & 3.5 & 150 & 50 \end{bmatrix}\right) \times 10^{-3} \text{ kg m}^2$$

The resonances are computed from the generalized eigenvalue problem for $\mathbf{M}$ and $\mathbf{K}$, and the first five of these are 0, 81.21, 141.21, 378.95, and 536.36 Hz.

The prediction of shaft torque at the drive end of the generator is obtained by summing contributions made by each of the different harmonics of half rotational speed. The zeroth harmonic accounts for (70.8×3) Nm of torque. This is the average torque in the generator shaft. The first harmonic has a frequency of 25 Hz. For any one harmonic, the forcing is a complex vector, $\mathbf{Q}$, having five entries of which the final two are zero (i.e., no oscillatory torques are applied directly at the flywheel or the generator). From Table 9.2, the first harmonic of half rotational speed is

$$\mathbf{Q} = \begin{bmatrix} 174.1 & (-87.05 - 150.8_J) & (-87.05 + 150.8_J) & 0 & 0 \end{bmatrix}^\mathsf{T} \text{ Nm}$$

Then, the displacement vector $\mathbf{q}$ is computed for the first harmonic of half rotational speed from

$$\mathbf{q} = (\mathbf{K} - \omega^2 \mathbf{M})^{-1} \mathbf{Q}$$

The difference between the final two entries of $\mathbf{q}$ represents the twist in the shaft between the flywheel and the generator. Multiplying this twist by the stiffness of that shaft produces a prediction of the torque in that shaft. For the first harmonic of half rotational speed, this torque is found to be $(-0.2070 + 0.7833_J)$ Nm. Other harmonics of half rotational speed contribute substantially more. The magnitudes of the contributions of the first 10 harmonics are 0.810, 2.592, 19.67, 10.49, 10.99, 12.03, 2.076, 0.822, 0.393, and 0.190 Nm. Figure 9.15 shows the total torque produced for two revolutions of the engine.

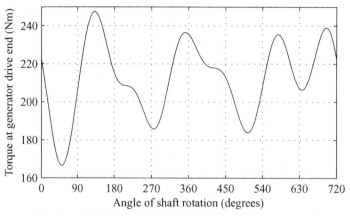

Figure 9.15. Torque for generator drive-end shaft (Example 9.7.1).

The greatest contribution here comes from the third harmonic. Because the cylinders are phased to fire exactly 240° of the crankshaft angle apart, it is not surprising that the third harmonic should be high. Also, the forcing frequency of the third harmonic is 75 Hz, which is sufficiently close to the 81.21 Hz resonance frequency such that substantial dynamic magnification occurs.

EXAMPLE 9.7.2. A major concern for designers of twin-engined aircraft is the possibility that the airplane will completely lose generation capacity. Figure 9.16 shows a seven-inertia idealization of the auxiliary gearbox of an aero-engine. The shafts connecting the electrical generators to the gearbox are deliberately fitted with a weak point (called a *shear-neck*) so that if either generator develops a serious internal mechanical fault, it cannot also cause significant damage to the gearbox. Typically, the shear-neck snaps when the torque in one generator shaft is approximately seven times larger than the rated torque.

Consider the aero-engine gearbox described in Figure 9.16. Inertias I_{g1}, I_{g2}, I_{g3}, and I_{g4} represent individual gears. The inertia values of these gears are 2×10^{-3}, 15×10^{-3}, 15×10^{-3}, and 2×10^{-3} kg m^2, respectively, and the gears have working radii of 35, 100, 100, and 35 mm, respectively. Inertias I_{m1} and I_{m2} represent generators coupled to the gearbox, and these generators each have an inertia of 180×10^{-3} kg m^2. Inertia I_e represents one rotor of an aero-engine, which supplies the power to the gearbox. The inertia of this rotor is 20 kg m^2. Shafts k_{m1} and k_{m2} connect the generators to the gearbox; each has a stiffness of 9,000 Nm/rad. Shaft k_e connects the engine rotor to the gearbox, and this shaft has a stiffness of 400 kNm/rad. Suppose that generator I_{m2} develops a mechanical problem and that it steadily draws more torque along its shaft (k_{m2}) until the shear-neck breaks at 500 Nm. Neglecting all damping, do the following:

(a) Calculate the new natural frequencies of the system after inertia I_{m2} has been disconnected, and calculate the mass-normalized torsional modes of this new system.

(b) Suppose that 500 Nm is being applied to gear inertia I_{g4}. Calculate the required torque on the engine rotor, I_e, so that the system speed does not change.

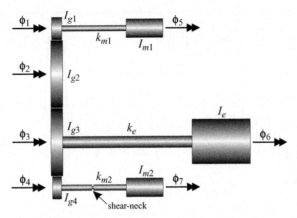

Figure 9.16. Simplified model of an aero-engine gearbox (Example 9.7.2).

(c) When the shear-neck breaks, the torque being applied to gear inertia I_{g4} quickly falls to zero and the magnitude of the torque in shaft k_{m1} increases from zero to a maximum value. Compute that value.

(d) Repeat the previous analysis with a shaft stiffness of $k_e = 50\,\text{kNm/rad}$ rather than $400\,\text{kNm/rad}$.

Solution

(a) With the generator I_{m2} disconnected, there are six degrees of freedom for the unconnected system and three constraints arising from the gear meshes. By referring the mass and stiffness properties to ϕ_3, we have the constraints $\phi_1 = r\phi_3$, $\phi_2 = -\phi_3$, and $\phi_4 = -r\phi_3$, where $r = 100/35$ is the ratio of the gear radii. Thus, the inertia of the gears, referred to as ϕ_3, is $I_{g1}r^2 + I_{g2} + I_{g3} + I_{g4}r^2$. The contribution of the shaft k_{m1} to the stiffness must be written in terms of ϕ_3. Hence, the mass and stiffness matrices, based on the degrees of freedom $\mathbf{q} = \begin{bmatrix} \phi_3 & \phi_5 & \phi_6 \end{bmatrix}^\mathsf{T}$, are

$$\mathbf{M} = \text{diag}\left(\begin{bmatrix} 0.06265 & 0.180 & 20 \end{bmatrix}\right)\,\text{kg}\,\text{m}^2$$

and

$$\mathbf{K} = \begin{bmatrix} 473.469 & -25.714 & -400 \\ -25.714 & 9 & 0 \\ -400 & 0 & 400 \end{bmatrix}\,\text{kNm/rad}$$

The resulting natural frequencies are 0, 33.89, and $438.2\,\text{Hz}$. The corresponding mass-normalized eigenvectors are

$$\mathbf{u}_1 = \begin{Bmatrix} 0.2155 \\ 0.6157 \\ 0.2155 \end{Bmatrix}, \quad \mathbf{u}_2 = \begin{Bmatrix} 0.0743 \\ 2.2739 \\ 0.0587 \end{Bmatrix}, \quad \text{and} \quad \mathbf{u}_3 = \begin{Bmatrix} 3.9886 \\ -0.0757 \\ -0.0105 \end{Bmatrix}\,\text{kg}^{-0.5}\text{m}^{-1}$$

The first mode is a rigid-body mode with zero natural frequency.

(b) For equilibrium, the torque on the gear inertia I_{g3} should be $\tau_{g3} = r\tau_{g4} = 1{,}428.6\,\text{Nm}$, where τ_{g4} is the torque applied to gear inertia I_{g4}. The equilibrium

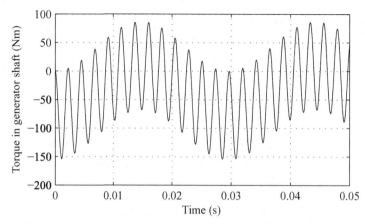

Figure 9.17. The torque response in shaft k_{m1} of the aero-engine gearbox after shear-neck failure (Example 9.7.2).

of the shaft of stiffness k_e means that the torque at the engine inertia, I_e, is $\tau_e = \tau_{g3}$. Hence, the torque exerted on I_e must be 1,428.6 Nm, in the same direction as the torque exerted on I_{g4}.

(c) The time taken for the torque on I_{g4} to fall to zero significantly influences the modes that are excited. Suppose that the torque immediately falls to zero so that the force during the transient response consists only of the constant torque τ_e. To simulate the response, the initial conditions are determined by the steady state based on the deformation due to the constant torques, τ_{g3} and τ_e, from (b). Because the system has a rigid-body mode, the steady-state angle is not unique and we can arbitrarily choose one of the angles. Here, we choose to set ϕ_6 (i.e., the angle at the engine inertia) to be zero. Because the steady-state torque in the shaft k_e is $\tau_e = 1{,}428.6$ Nm from (b), then $\phi_3 = \tau_e/k_e = -0.003571$ rad and, because there is no steady-state torque in shaft k_{m2}, we have $\phi_5 = \phi_4 = r\phi_1 = -0.01020$ rad. The torque in the shaft k_{m1} is given by

$$\tau_{m1} = k_{m1}\left(\phi_5 - \phi_1\right) = k_{m1}\left(\phi_5 - r\phi_3\right)$$

Figure 9.17 shows the transient torque response, τ_{m1}, assuming no damping. It also shows that the maximum torque occurs during the first cycle of the high frequency mode and is approximately 154 Nm.

The major contributor to the torque is the mode with highest frequency. The physical reason for this is that the referred gear inertias are significantly smaller than those of the engine or generators; and hence, the initial motion is mainly confined to the gears, causing a high torque in the shaft of interest, k_{m1}. Therefore, it is possible to get a reasonable estimate of this maximum torque by considering only a single mode. We assume that the response can be approximated by

$$\mathbf{q} = p_2\mathbf{u}_2 + p_3\mathbf{u}_3$$

where the coefficient p_2 is constant. Then, because the third mode is mass-normalized, the equation of motion is

$$\ddot{p}_3 + \omega_3^2 p_3 = P_3 = \mathbf{u}_3^\top \begin{Bmatrix} 0 \\ 0 \\ \tau_e \end{Bmatrix} = -0.0105 \times 1{,}428.6 = -15.0707 \, \text{Nm}$$

where ω_3 is the third natural frequency. The initial conditions are $p_3(0) = P_3/\omega_3^2 = -0.7535 \times 10^{-3}$ and $\dot{p}_3(0) = 0$. Solving this equation gives the modal response

$$p_3 = \frac{P_3}{\omega_3^2} + \left(p_3(0) - \frac{P_3}{\omega_3^2} \right) \cos \omega_3 t$$

The torque in shaft k_{m1} is given by

$$\tau_{m1} = \tau_{c2} + k_{m1} \left(\phi_5 - r\phi_3 \right) = \tau_{c2} + k_{m1} \left(-0.0757 - 3.9886r \right) p_3$$

$$= \tau_{c2} - 1.0324 \times 10^5 p_3$$

where τ_{c2} is the constant torque due to the lower-frequency modes. Because the torque in shaft k_{m1} is zero initially, the maximum torque in shaft k_{m1} is given by the peak-to-peak value of the oscillatory component. Thus, this maximum torque is approximately

$$1.0324 \times 10^5 \times \left| p_3(0) - \frac{P_3}{\omega_3^2} \right| = 155.2 \, \text{Nm}$$

which agrees well with the result from the three degrees of freedom model.

(d) Repeating the previous analysis gives natural frequencies 0, 23.32, and 225.2 Hz. The peak shaft torque in shaft k_{m1} is now 578 Nm, which causes the shear-neck in this shaft to fail also. Thus, the design of the shaft stiffnesses must be considered carefully so that failure of the shear-neck for one generator does not cause a failure in both generator shafts.

9.7.2 Force-Driven Excitation of Axial Vibration

Sustained axial excitation also occurs in reciprocating machines. Because most crankshafts have journal bearings on either side of each piston, the forces exerted through the connecting rod on the crankshaft are reacted mainly at the adjacent journal bearings. These sets of forces cause a bending moment locally in the crankshaft, which causes some curvature. When that curvature is in the same plane as the crankshaft, it tends to either lengthen or shorten the crankshaft axially, as Figure 9.18 illustrates. This source of vibration is best approximated as force-driven excitation because any realistic axial straining of the crankshaft has negligible influence on the forcing applied.

Another source of axial forcing is the vibration of drill-strings in oil and gas exploration. A long shaft (i.e., the drill-string) is assembled piecemeal as a bore hole deepens and a drill-head at the end of the shaft breaks up and moves the material in

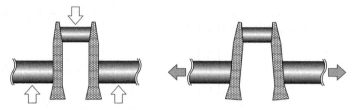

Figure 9.18. The axial excitation in a crankshaft due to transverse loading.

front of it. The drilling action causes significant torsional and axial vibration in the drill-string and can excite resonances of the drill-string. These resonances change as the string becomes increasingly longer.

Transient axial excitation is a frequent occurrence with turbo-machines, such as gas compressors or vacuum pumps, and occurs when there is a sudden change of pressure at the inlet or outlet of the machine. Any net pressure difference across any one disk in a compressor causes a net axial force and, in some cases, these forces can be quite large. Designers of turbo-machines must carefully consider these potentially large axial forces.

Overall, there are far fewer sources of significant axial forcing in rotating machines than there are of torsional forcing. This is primarily because the design intent for virtually all rotating machines is a rotating part that does not move axially. Chapter 10 discusses cases in which torsional vibration is coupled with axial behavior; in these cases, axial vibration results from torsional excitation.

9.7.3 Displacement-Driven Excitation of Torsional Vibration

Section 9.6.2 presents a formal method to model torsional motion of geared systems. In this method, each inertia initially has one independent rotation coordinate and the vector containing all of these rotation coordinates is denoted as $\mathbf{q}_U$. The constraints imposed by the presence of all gear meshes are expressed as Equation (9.12), $\mathbf{E}^\top \mathbf{q}_U = \mathbf{0}$. These constraints are enforced by applying Equation (9.15), $\mathbf{q}_U = \mathbf{T}\mathbf{q}_R$.

With torsional models in particular, it is often necessary to model displacement-driven excitation. The most important example of this is gear-geometry errors, but other cases arise in misaligned coupled rotors. Consider two meshing gears, A and B, with N_A and N_B teeth, respectively. The average ratio of their respective rotational speeds is given by N_B/N_A. If no gear-geometry errors are present, then the ratio $\dot{\phi}_A/\dot{\phi}_B$ is constant and, with suitable definition of angles such that $\phi_A = 0$ when $\phi_B = 0$, it follows that $N_B\phi_A + N_A\phi_B = 0$. Evidently, this is in the form $\mathbf{E}^\top \mathbf{q}_U = 0$.

In the presence of gear-geometry errors, the ratio $\dot{\phi}_A/\dot{\phi}_B$ is not constant and the constraint equation is modified to

$$\mathbf{E}^\top \mathbf{q}_U = \mathbf{e} \tag{9.30}$$

where $\mathbf{e}$ is a vector of mesh errors. Like $\mathbf{q}_U$, $\mathbf{e}$ may be represented as a function of time or frequency depending on the analysis being carried out. Now, a coordinate transformation may be found that inherently ensures that $\mathbf{q}_U$ satisfies Equation (9.30). This transformation is

$$\mathbf{q}_U = \mathbf{T}\mathbf{q}_R + \mathbf{q}_{ref} \tag{9.31}$$

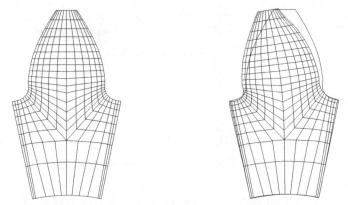

Figure 9.19. A gear tooth deforming under load.

where Equation (9.31) is equivalent to Equation (9.26) if $\mathbf{T}$ satisfies the previously stated condition that $\mathbf{E}^\top\mathbf{T} = \mathbf{0}$ and $\mathbf{Y} = \begin{bmatrix} \mathbf{E} & \mathbf{T} \end{bmatrix}$ satisfies the requirement that $\mathbf{Y}\mathbf{Y}^\top$ is nonsingular. In Equation (9.31), $\mathbf{q}_{ref}$ is selected so that $\mathbf{E}^\top\mathbf{q}_{ref} \equiv \mathbf{e}$. Note that $\mathbf{q}_{ref}$ is not unique, but it is always possible to find a suitable choice. Often, this is determined as the least squares solution

$$\mathbf{q}_{ref} = \mathbf{E}\left(\mathbf{E}^\top\mathbf{E}\right)^{-1}\mathbf{e} \tag{9.32}$$

but other solutions are possible.

Different choices of $\mathbf{q}_{ref}$ result in different solutions for $\mathbf{q}_R$, but the same solution is always obtained for $\mathbf{q}_U$. Substituting Equation (9.31) into Equation (9.10) results in the new equation of motion

$$\mathbf{M}_R\ddot{\mathbf{q}}_R + \mathbf{K}_R\mathbf{q}_R = \mathbf{Q}_R - \mathbf{T}^\top\left(\mathbf{M}_U\ddot{\mathbf{q}}_{ref} + \mathbf{K}_U\mathbf{q}_{ref}\right) \tag{9.33}$$

where $\mathbf{M}_R = \mathbf{T}^\top\mathbf{M}_U\mathbf{T}$, $\mathbf{K}_R = \mathbf{T}^\top\mathbf{K}_U\mathbf{T}$, and $\mathbf{Q}_R = \mathbf{T}^\top\mathbf{Q}_U$ as before. Equation (9.33) can be solved directly to calculate $\mathbf{q}_R$ and, subsequently, $\mathbf{q}_U$ can be found from this using Equation (9.31). Following this, the torques arising in the different gear meshes may be calculated directly.

This analysis assumes that the mesh between two gears is rigid. In reality, no gear mesh is perfectly rigid because of finite flexibility in the gear teeth and elsewhere. Figure 9.19, for example, shows a gear tooth deforming under the influence of bending load. Other pertinent effects are the more widespread deformations of the gear, elastohydrodynamic lubrication effects, and apparent torsional compliance introduced through transverse flexibility of the shaft and the bearing supports. These effects are commonly combined into a single parameter, k_T, representing mesh stiffness, and a corresponding parameter, c_T, for equivalent mesh damping such that a relationship exists between the tangential force in the mesh, f_T, and the corresponding relative tangential movement, x_T, of the gears in the mesh of the form

$$f_T = c_T\dot{x}_T + k_Tx_T \tag{9.34}$$

In this case, the analysis can be conducted again as a force-driven excitation where, instead of applying a coordinate transformation that constrains meshing

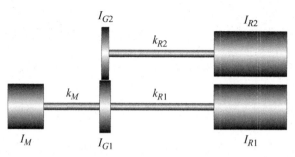

Figure 9.20. A two-rotor model of a vacuum pump.

gears to have related tangential displacements, each gear retains an independent angular coordinate.

We now consider an example with a rigid gear mesh.

EXAMPLE 9.7.3. Figure 9.20 shows two rotors from a simple vacuum pump that are geared together with a 1:1 gear-mesh, where the gears are assumed to be rigid. Inertias I_{G1} and I_{G2} are each 25×10^{-6} kg m^2. Inertias I_{R1} and I_{R2} are each 700×10^{-6} kg m^2. The lower rotor is driven directly by an induction machine, the rotor of which is represented by inertia $I_M = 400 \times 10^{-6}$ kg m^2. The rotors spin with an average speed of 6,000 rev/min. Spring stiffnesses k_{R1} and k_{R2} are each equal to 5,000 Nm/rad and k_M is 10,000 Nm/rad. A small geometry error in the gears causes

$$\phi_{G1} + \phi_{G2} = \left\{600 \times 10^{-6} \cos(200\pi t) + 100 \times 10^{-6} \sin(400\pi t)\right\} \text{ rads}$$

where ϕ_{G1} and ϕ_{G2} are the angular coordinates of the two gears. Calculate the magnitude of the oscillatory torque occurring in the torsional spring k_{R1}, assuming there is no other source of torsional excitation.

Solution. The vector of unconnected angular displacements is

$$\mathbf{q}_U = [\phi_M \ \phi_{G1} \ \phi_{R1} \ \phi_{G2} \ \phi_{R2}]^\top$$

Defining

$$\mathbf{E} = [0 \ 1 \ 0 \ 1 \ 0]^\top$$

The geometry error in the gear mesh is then expressed as

$$\mathbf{e}(t) \equiv \mathbf{E}^\top \mathbf{q}_U = \phi_{G1} + \phi_{G2}$$

$$= \left\{600 \cos(200\pi t) + 100 \sin(400\pi t)\right\} \times 10^{-6} \text{rads}$$

The substitution $\mathbf{q}_U = \mathbf{T}\mathbf{q}_R + \mathbf{q}_{ref}$ now is made using an appropriate transformation matrix

$$\mathbf{T} = \begin{bmatrix} 1 & 0 & 0 & 0 \\ 0 & 1 & 0 & 0 \\ 0 & 0 & 1 & 0 \\ 0 & -1 & 0 & 0 \\ 0 & 0 & 0 & 1 \end{bmatrix}$$

and excitation vector

$$\mathbf{q}_{ref} = \begin{Bmatrix} 0 \\ 300\cos(200\pi t) + 50\sin(400\pi t) \\ 0 \\ 300\cos(200\pi t) + 50\sin(400\pi t) \\ 0 \end{Bmatrix} \times 10^{-6} \text{rads}$$

The equation of motion for the combined system is given by Equation (9.33), where

$$\mathbf{K}_R = \begin{bmatrix} 10,000 & -10,000 & 0 & 0 \\ -10,000 & 20,000 & -5,000 & 5,000 \\ 0 & -5,000 & 5,000 & 0 \\ 0 & 5,000 & 0 & 5,000 \end{bmatrix} \text{Nm}$$

and

$$\mathbf{M}_R = \begin{bmatrix} 400 & 0 & 0 & 0 \\ 0 & 50 & 0 & 0 \\ 0 & 0 & 700 & 0 \\ 0 & 0 & 0 & 700 \end{bmatrix} \times 10^{-6} \text{kg m}^2$$

The natural frequencies can be computed from the reduced mass and stiffness matrices as 0.0, 425.4, 634.1, and 3,247.2 Hz. Because all of the quantities in Equation (9.33) are now known, the complete system response can be computed. The resulting torque in the torsional spring k_{R1} consists of two components and is $(1.5694\cos(200\pi t) + 0.3076\sin(400\pi t))$ Nm.

9.8 Parametric Excitation of Torsional Systems

In Section 9.7.1, it is noted that in a reciprocating machine, the effective inertia of the crankshaft varies due to the movement of the piston(s) relative to it. To a first approximation, this can be accounted for by applying a variable torque to the crankshaft (see Equation (9.29)). However, Equation (9.29) is not adequate to explain the secondary-resonance phenomenon in torsional vibrations that can occur in large reciprocating machines, such as a marine diesel engine, in which the massive pistons cause a significant variation in the effective inertia of the crankshaft. This problem was studied by Pasricha and Carnegie (1976, 1979), Porter (1965), and others. Drew et al. (1999) provide experimental results from a single-cylinder reciprocating engine. The following example is based on a paper by Pasricha and Hassan (1997).

Consider a two-cylinder, in-line reciprocating engine driving a heavy flywheel, as shown in Figure 9.21. Due to the large flywheel, the shaft speed Ω is constant. We neglect gas forces and damping and assume that

- the connecting rods are relatively long so that the reciprocating masses move with simple harmonic motion
- the mass of each connecting rod can be lumped into two masses, one at the crankshaft and one at the piston

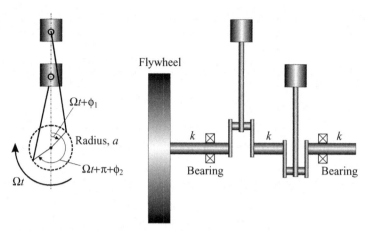

Figure 9.21. The simplified model of a reciprocating machine with two pistons.

The angular position of the crankshaft for the first cylinder consists of a contribution from the steady rotation, Ωt, and the perturbation due to the vibration, ϕ_1. The kinetic energy for the first piston is then

$$T_1 = \frac{1}{2}ma^2(\sin^2 \Phi_1)\dot{\Phi}_1^2 = \frac{1}{4}ma^2(1 - \cos 2\Phi_1)\dot{\Phi}_1^2 \tag{9.35}$$

where $\Phi_1 = \Omega t + \phi_1$, a is the radius of the crankshaft, and m is the effective mass of the reciprocating part for one cylinder including the contribution of the connecting rod. The kinetic energy for the second cylinder is similar, except that the position of the crankshaft is $\Phi_2 = \Omega t + \pi + \phi_2$, where the angle π arises because the two crankshafts are $180°$ out-of-phase. A linear model of the crankshaft stiffness is assumed.

The linearized equations of motion obtained using Lagrange's equation are

$$\left\{I + \frac{1}{2}ma^2 \left(1 - \cos 2\Omega t\right)\right\}\ddot{\phi}_1 + \left(ma^2 \Omega \sin 2\Omega t\right)\dot{\phi}_1$$
$$+ \left(ma^2\Omega^2 \cos 2\Omega t + 2k\right)\phi_1 - k\phi_2 = -\frac{1}{2}ma^2\Omega^2 \sin 2\Omega t \tag{9.36}$$

$$\left\{I + \frac{1}{2}ma^2 \left(1 - \cos 2\Omega t\right)\right\}\ddot{\phi}_2 + \left(ma^2 \Omega \sin 2\Omega t\right)\dot{\phi}_2$$
$$+ \left(ma^2\Omega^2 \cos 2\Omega t + k\right)\phi_2 - k\phi_1 = -\frac{1}{2}ma^2\Omega^2 \sin 2\Omega t \tag{9.37}$$

where I is the moment of inertia of the crankshaft, including the contribution of the connecting rod for each cylinder, and k is the torsional stiffness of each shaft section. These differential equations have nonconstant coefficients and can have unstable solutions for certain combinations of parameters. Regions of instability can exist corresponding to the undamped torsional natural frequencies for the equivalent linear time-invariant system. Other regions of instability occur at half the undamped torsional natural frequencies. Damping has a stabilizing effect.

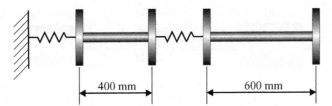

Figure 9.22. A schematic of a two-shaft system (Problem 9.1).

9.9 Summary

We have seen in this chapter that models for the axial and torsional behavior of rotors are simpler in some senses than models for the lateral behavior of rotors but that they have particular features. The simplest torsional and axial models are similar, each having one degree of freedom per node and a simple chain-like structure in the stiffness and damping matrices. Torsional vibration, in particular, often involves a number of distinct machines coupled together to form a single shaft train, and it is invariably distinguished by the fact that the system has one zero-frequency (i.e., rigid-body) mode. Methods for preparing the torsional models of geared systems are discussed and they are useful in understanding displacement-driven excitation, which is especially important in torsional vibrations.

9.10 Problems

9.1 A system consists of two steel rotors coupled together as shown in Figure 9.22. Each disk has a diameter of 300 mm and a thickness of 40 mm. Each shaft has a diameter of 50 mm; one shaft is 400 mm long and the other is 600 mm long. The coupling between the rotors has a torsional stiffness of 0.6 MNm and an axial stiffness of 50 kN/m. At one end of the shaft train, there is a thrust bearing with an axial stiffness of 30 kN/m. Using appropriate software and neglecting the mass of the shafts, verify that the system natural frequencies are 0, 94.58, 146.3, and 366.0 Hz for torsional vibrations and 2.711, 8.204, 1,226, and 1,502 Hz for axial vibrations.

Show that by analyzing each rotor separately, it is possible to estimate the two higher-axial frequencies; then, by treating the two rotors as inextensible bodies, it is possible to estimate the lower pair of axial frequencies. Also show that this approach does not work in the case of torsional vibrations. Assume that $\rho = 7{,}800 \, \text{kg/m}^3$, $E = 200 \, \text{GPa}$, and $G = 80 \, \text{GPa}$.

9.2 The rotor of a turbo-generator set is depicted in Figure 9.23. To a first approximation, the compressor, turbine, and generator each can be represented as a single rigid inertia, and their values are 100, 50, and 80 kg m^2, respectively. Each coupling has an inertia of 10 kg m^2. The shafts have negligible inertia and their torsional stiffnesses are $k_1 = 25 \, \text{MN m/rad}$ and $k_2 = 25 \, \text{MN m/rad}$, respectively. Calculate the first three torsional natural frequencies.

9.3 The system shown in Figure 9.24 consists of two steel rotors coupled together by gears. One rotor has a 400 mm long shaft with a diameter of 50 mm and carries two identical disks at the ends. Each disk has a diameter of 300 mm and a thickness of 40 mm. The other rotor has a 600 mm-long shaft with a diameter

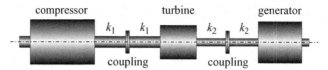

Figure 9.23. A turbo-generator set (Problem 9.2).

of 50 mm and carries two disks at the ends. One disk has a diameter of 300 mm and a thickness of 40 mm; the other has a diameter of 150 mm and a thickness of 40 mm. This disk is in mesh with one of the disks on the other rotor and therefore provides a 2:1 ratio gear coupling between the shafts. Determine the equation of motion for the system and the system natural frequencies. Assume that $\rho = 7{,}800\,\text{kg/m}^3$ and $G = 80\,\text{GPa}$ and neglect the mass of the shafts.

9.4 Consider the aero-engine gearbox system described in Example 9.7.2. Determine the torsional resonance frequencies of the system. Suppose that a gear-mesh error exists between inertias I_{g2} and I_{g3} such that $\phi_2 + \phi_3 = 10^{-5}\sin(3{,}000t)\,\text{rad}$. Without the gear-mesh error, this sum is zero. Calculate the magnitude of the oscillatory torque in the shaft of stiffness k_e and the magnitude of the oscillatory tangential contact force at the common tangent of the working circles of the two central gears of the gearbox.

9.5 Equation (9.31) expresses the vector $\mathbf{q}_U$ in terms of a reduced-dimension vector, $\mathbf{q}_R$, and a vector of offsets, $\mathbf{q}_{ref}$. Many different choices of $\mathbf{q}_{ref}$ are possible. Prove that the same vector $\mathbf{q}_U$ results in all cases.
Hint:
- Using Equation (9.30), show that any two different choices for $\mathbf{q}_{ref}$, denoted $\mathbf{q}_{ref0}$ and $\mathbf{q}_{ref1}$, satisfy $\mathbf{E}^\top\left(\mathbf{q}_{ref1} - \mathbf{q}_{ref0}\right) = \mathbf{0}$.
- Therefore, show that $(\mathbf{q}_{ref1} - \mathbf{q}_{ref0}) = \mathbf{Th}$ for some vector $\mathbf{h}$.
- Use Equation (9.33) to verify that $\mathbf{q}_{R1} + \mathbf{h} = \mathbf{q}_{R0}$, where $\mathbf{q}_{R0}$ and $\mathbf{q}_{R1}$ are the constrained coordinates corresponding to $\mathbf{q}_{ref0}$ and $\mathbf{q}_{ref1}$.
- Then, apply Equation (9.31) to show that the same vector $\mathbf{q}_U$ is obtained.

9.6 Figure 9.25 is a schematic of a five-cylinder diesel generator set. The generator is a large-diameter multipole machine designed to produce 50 Hz of electricity at a rotational speed of 3,000 rev/min. The axial vibrations of this unit are being considered. Masses m_1 and m_6 each represent a section of crankshaft

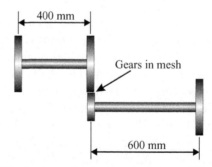

Figure 9.24. A schematic of a two-shaft system (Problem 9.3).

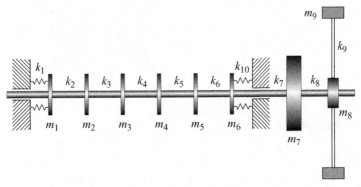

Figure 9.25. A schematic of a five-cylinder diesel generator set (Problem 9.6).

and one crank-web, and these masses are each 90 kg. Masses m_2 through m_5 (inclusive) each represent a section of crankshaft and two crank-webs, which are each 140 kg. Mass m_7 represents a solid coupling between the engine and the generator, with mass 350 kg. Mass m_8 represents the hub of the generator with a mass of 500 kg and m_9 represents the rim of the generator; its mass is 22,000 kg. Stiffnesses k_2, k_3, k_4, k_5, and k_6 each represent an axial stiffness between two consecutive masses, and k_1 and k_{10} represent axial stiffnesses between one mass and ground. The stiffness values are $k_1 = k_{10} = 10\,\mathrm{MN/m}$, $k_2 = k_3 = k_4 = k_5 = k_6 = 2.5\,\mathrm{MN/m}$, $k_7 = k_8 = 40\,\mathrm{MN/m}$, and $k_9 = 3\,\mathrm{MN/m}$. Calculate the first three axial-resonance frequencies of this system and determine the corresponding mode shapes.

9.7 A differential gearbox is used to combine the power output from one large and slow-acting motor (on shaft 1) with the power output from a small and fast-acting motor (on shaft 2) to drive a load fixed to shaft 3. Figure 9.26 is a schematic of the arrangement. The differential gearbox introduces a constraint expressed as $\Omega_1 - \frac{1}{5}\Omega_2 + \Omega_3 = 0$. Treating $\{\phi_1, \phi_2, \phi_3\}$ as the three

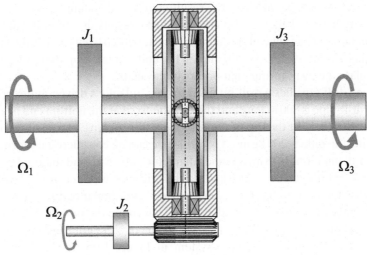

Figure 9.26. A differential gearbox (Problem 9.7). The shaft spin speeds are constrained so that $\Omega_1 - \Omega_2/5 + \Omega_3 = 0$.

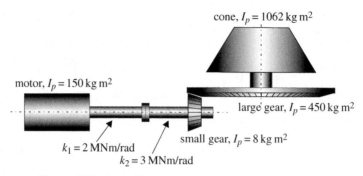

Figure 9.27. A machine for crushing rocks (Problem 9.8).

displacement coordinates in the coordinate vector, $\mathbf{q}_U$, determines a suitable transformation matrix, $\mathbf{T}$, enforcing this constraint. Then, using inertia values $J_1 = 2\,\mathrm{kg\,m}^2$, $J_2 = 5 \times 10^{-4}\,\mathrm{kg\,m}^2$, and $J_3 = 3\,\mathrm{kg\,m}^2$ as the inertias present on each of the three shafts (ignoring any other inertia associated with the gearbox), derive the reduced mass matrix, $\mathbf{M}_R$. Using this mass matrix, determine the angular accelerations of all three shafts when torques of 100 and 6 Nm are applied to shafts 1 and 2, respectively, and with no torque applied to the third shaft.

9.8 A rock-crusher is driven by a large motor directly coupled to the input shaft of the crusher. Figure 9.27 shows an idealization of the crusher, with a pair of bevel gears with the rotational speed of the small bevel gear exactly six times the rotational speed of the large bevel gear. The inertias of various components of the system are given in the figure. The only significant torsional flexibility in the unit is in the two-part shaft. By combining the inertias of the cone and the large gear with the inertia of the pinion (i.e., small gear), find the single equivalent inertia of these objects referred to the pinion. Determine the equivalent torsional stiffness of the two shafts between the motor inertia and the referred inertia of the crusher. Calculate the natural frequency and the mass-normalized shape of the flexible mode of the resulting reduced system.

The action of the cone in crushing rocks results in a broadband torsional excitation over a range of frequencies. For torsional excitation at the cone with frequencies of 10, 20, 30, 40, and 50 Hz and a peak amplitude of 3,000 Nm, calculate the peak torque that occurs in the shaft.

9.9 A long drill-pipe has a drill-bit fixed to one end and is driven at the other end. The drill-pipe is 800 m long and the drill-bit fixed at the bottom end has an inertia of 10 kg m². At the top of the drill-pipe, the effective inertia of the machine driving the pipe is 200 kg m². The drill-pipe can be considered uniform with an outer diameter of 0.78 m, an inner diameter of 0.65 m, and material properties of $G = 75\,\mathrm{GPa}$ and $\rho = 8{,}000\,\mathrm{kg/m}^3$. The drill-pipe is driven at 120 rev/min and, because the drill-bit has four equally spaced and similar cutting edges, a strong torsional excitation is expected at 8 Hz. Determine the torsional resonances of the drill-pipe below 10 Hz using 5, 8, and 50 equal-length elements.

9.10 Consider the diesel-generator system in Figure 9.25. It is subject to periodic axial forcing in which the periodic time is 0.2 s. Each one of the five cylinders fires once per cycle and produces a component of axial force $f(t)$, tending to separate the masses on either side where $f(t) = 6,000\cos(10\pi t) +$ $1,500\cos(30\pi t) - 500\sin(30\pi t)$. The firing sequence is 1-3-5-2-4 and the pistons are phased at exactly $72°$ from one another. Determine the magnitudes of axial response of the generator rim at 5 and 15 Hz and the magnitude of the net axial force between the generator hub and rim for each frequency.

10 More Complex Rotordynamic Models

10.1 Introduction

The purposes of this chapter are primarily to alert readers to the limitations of the analysis provided in previous chapters, to highlight the more complex behavior of certain types of rotor, and to indicate where detailed descriptions of the analysis of these systems can be found.

Chapters 3, 5, and 6 consider the behavior of rotor–bearing systems when the rotor vibrates laterally; that is, the rotor whirls due to the actions of initial radial disturbances or, more important, radial forces. The rotor is modeled as an assembly of shaft elements and rigid disks. The shaft elements include the effects of inertia, bending and shear deflection, rotary inertia, and gyroscopic couples; the disks include the effects of inertia and gyroscopic couples. Bearings and foundations are modeled essentially as assemblies of axial springs and damping elements. Similarly, in Chapter 9, we examine the axial and torsional vibration of rotors and, in each case, we model the system as an assemblage of masses and inertias and axial or torsional springs.

There are three basic assumptions in the aforementioned analysis: (1) the rotor–bearing–foundation system is linear; this assumption is present even in Chapter 7, in which instabilities of various types are examined; (2) although the rotor can deflect laterally, axially, and in torsion, it cannot otherwise deform; the shape of its cross section is fixed and plane cross sections remain plane; and (3) although the rotor can deflect laterally, axially, and in torsion, there is no coupling between these deflections; thus, an axial force can produce an axial deflection of the rotor, but it causes neither a lateral nor a torsional deflection.

Not all rotor–bearing systems can be modeled adequately accepting these above restrictions. A striking example is the helicopter rotor. The vertical rotating drive shaft does not carry a rigid disk but rather a set of very flexible blades, typically mounted on flapping and lagging hinges at the root. In translational flight, the rotor must tilt in the direction of flight to create a propulsive (i.e., thrust) component of lift. The tilting of the rotor disk is achieved by an azimuthal variation in the blade-pitch angle known as *feathering*. Feathering produces an out-of-plane flapping motion that causes the blades to remain on the tilted disk plane as they rotate about the rotor shaft. The flapping motion, in turn, gives rise to an in-plane motion known as *lagging* that conserves angular momentum. Other than the fundamental complexity of the rotordynamic behavior, the aerodynamic forces acting on the blades are highly nonlinear because the rotor must move *edgewise* through the air. In forward flight, for example, the advancing blade tip experiences high local-flow velocities including strong shock waves, whereas the retreating blade experiences a region of reversed flow near the blade root and potentially dynamic stall near the tip.

In the following sections, we briefly examine systems that cannot be analyzed using a shaft-line model. We do not always provide a detailed mathematical analysis of the system.

10.2 Simple Rotating Elastic Systems

The analysis of a general rotating elastic structure must consider all of the accelerations that act throughout the system. Thus, in addition to the accelerations resulting from the elastic deformations of the structure, contributions due to Coriolis and centripetal accelerations are often important. The Coriolis accelerations are manifest in shaft-line models as the gyroscopic terms. Furthermore, the stiffness characteristics of the structure may be modified due to the steady-state internal loads induced by the centrifugal forces. This is true of many rotors; however, the assumption that every section normal to the rotational axis does not deform simplifies the analysis in Chapter 5. Here, we present a more general analysis that focuses initially on a rotating system consisting of a single flexibly supported particle, similar to that studied by Laurenson (1976). This analysis can be extended to an assembly of particles and provides the basis of an FEA, discussed subsequently. The development requires the transformation between stationary and rotating coordinate systems (see Chapter 7).

In Chapter 7, we develop the equation of motion of a rotating structure with respect to a fixed (i.e., inertial) system of axes and a system of axes that is fixed to and rotates with the structure. The choice between using fixed or rotating axes in an analysis depends on the properties of the structure and the foundation. When the rotor is a solid of revolution (or cyclically symmetric), it is often advantageous to base the analysis in the stationary frame, irrespective of the stator properties. Equally, if the rotor is nonsymmetric and the stator is isotropic, then it is usually advantageous to analyze the system in the rotating reference frame (see Chapter 7). If the rotor is nonsymmetric and the stator is anisotropic, then analysis is difficult in either frame of reference. Irrespective of which set of axes is used for analysis, the resonance frequencies and response of the system can be presented relative to either class of axes.

Suppose that a point mass m is connected to a movable framework by springs. The structure rotates about the Oz axis with rotational speed Ω. The springs have stiffnesses $k_{\tilde{x}}$, $k_{\tilde{y}}$, and $k_{\tilde{z}}$ in the $O\tilde{x}$, $O\tilde{y}$, and $O\tilde{z}$ directions, respectively, where the axis set $O\tilde{x}\tilde{y}\tilde{z}$ rotates with the framework about the inertial Oz axis with angular velocity Ω. Thus, Oz is always collinear with $O\tilde{z}$. The position of the particle relative to the origin O is defined as $\tilde{u}$ and $\tilde{v}$ in the $O\tilde{x}$ and $O\tilde{y}$ directions. The position of the point where the particle rests if the structure is not moving is given by $\tilde{u}_0$ and $\tilde{v}_0$. The elastic deformations are denoted as $\tilde{u}_e$ and $\tilde{v}_e$. Thus,

$$\tilde{u} = \tilde{u}_0 + \tilde{u}_e, \qquad \tilde{v} = \tilde{v}_0 + \tilde{v}_e \qquad (10.1)$$

The elastic force on the particle in rotating coordinates is

$$\left\{\begin{matrix} f_{\tilde{x}} \\ f_{\tilde{y}} \end{matrix}\right\}_{\text{elastic}} = -\begin{bmatrix} k_{\tilde{x}} & 0 \\ 0 & k_{\tilde{y}} \end{bmatrix} \left\{\begin{matrix} \tilde{u}_e \\ \tilde{v}_e \end{matrix}\right\} \tag{10.2}$$

Then, from Equation (7.17), because $\dot{\tilde{u}}_0 = \dot{\tilde{v}}_0 = 0$

$$m\left\{\begin{matrix} \ddot{\tilde{u}}_e \\ \ddot{\tilde{v}}_e \end{matrix}\right\} + 2\begin{bmatrix} 0 & -m\Omega \\ m\Omega & 0 \end{bmatrix} \left\{\begin{matrix} \dot{\tilde{u}}_e \\ \dot{\tilde{v}}_e \end{matrix}\right\} - m\Omega^2 \left\{\begin{matrix} \tilde{u}_0 + \tilde{u}_e \\ \tilde{v}_0 + \tilde{v}_e \end{matrix}\right\} = \left\{\begin{matrix} f_{\tilde{x}} \\ f_{\tilde{y}} \end{matrix}\right\}_{\text{elastic}} \tag{10.3}$$

Thus,

$$m\left\{\begin{matrix} \ddot{\tilde{u}}_e \\ \ddot{\tilde{v}}_e \end{matrix}\right\} + 2\begin{bmatrix} 0 & -m\Omega \\ m\Omega & 0 \end{bmatrix} \left\{\begin{matrix} \dot{\tilde{u}}_e \\ \dot{\tilde{v}}_e \end{matrix}\right\} + \begin{bmatrix} k_{\tilde{x}} - m\Omega^2 & 0 \\ 0 & k_{\tilde{y}} - m\Omega^2 \end{bmatrix} \left\{\begin{matrix} \tilde{u}_e \\ \tilde{v}_e \end{matrix}\right\} = m\Omega^2 \left\{\begin{matrix} \tilde{u}_0 \\ \tilde{v}_0 \end{matrix}\right\} \tag{10.4}$$

This equation of motion may be written in matrix terms in the form

$$\mathbf{M}\ddot{\tilde{\mathbf{q}}}_e - 2\Omega\mathbf{J}\mathbf{M}\dot{\tilde{\mathbf{q}}}_e + \left(\mathbf{K} - \Omega^2\mathbf{M}\right)\tilde{\mathbf{q}}_e = \Omega^2\mathbf{M}\tilde{\mathbf{q}}_0 \tag{10.5}$$

where $\tilde{\mathbf{q}}_e = \left\{\begin{matrix} \tilde{u}_e \\ \tilde{v}_e \end{matrix}\right\}$, $\tilde{\mathbf{q}}_0 = \left\{\begin{matrix} \tilde{u}_0 \\ \tilde{v}_0 \end{matrix}\right\}$. The mass and stiffness matrices are given by

$$\mathbf{M} = \begin{bmatrix} m & 0 \\ 0 & m \end{bmatrix}, \qquad \mathbf{K} = \begin{bmatrix} k_{\tilde{x}} & 0 \\ 0 & k_{\tilde{y}} \end{bmatrix} \tag{10.6}$$

and $\mathbf{J}$ is defined in Equation (7.8) as

$$\mathbf{J} = \begin{bmatrix} 0 & 1 \\ -1 & 0 \end{bmatrix} \tag{10.7}$$

A damping matrix defined in the rotating frame, $\mathbf{C}$, also may be included in Equation (10.5). The terms $-2\Omega\mathbf{J}\mathbf{M}\dot{\tilde{\mathbf{q}}}_e$ and $-\Omega^2\mathbf{M}\tilde{\mathbf{q}}_e$ in Equation (10.5) are due to the contributions of the Coriolis acceleration and the centripetal acceleration, respectively. When Oz is the axis of spin, then the Coriolis acceleration couples the vibration in the $O\tilde{x}$ and $O\tilde{y}$ directions. Centripetal accelerations affect only vibration in these directions. The effect of the Coriolis acceleration increases the difference between the natural frequencies, as demonstrated in Example 10.2.1. Inspecting Equation (10.5), we can see that for a given (i.e., non-zero) value of Ω, the effective stiffnesses and, hence, the natural frequencies are reduced when compared with the natural frequencies of the structure at rest. This is often called *spin softening* or *centripetal softening*. The term $\Omega^2\mathbf{M}\tilde{\mathbf{q}}_0$ defines the centrifugal force acting on the particle due to the spin when the particle is in its reference position. If the force–deflection relationship for the structure is linear, then the steady-state deflection of the elastic structure due to the spin may be obtained from Equation (10.5) by setting the time derivatives to zero. Thus, we have

$$\left(\mathbf{K} - \Omega^2\mathbf{M}\right)\tilde{\mathbf{q}}_{ss} = \Omega^2\mathbf{M}\tilde{\mathbf{q}}_0 \tag{10.8}$$

Equation (10.3) considers only the displacement in the $O\tilde{x}$ and $O\tilde{y}$ directions. The equation of motion in the $O\tilde{z}$ direction can be considered in isolation and is

$$m\ddot{\tilde{w}}_e + k_{\tilde{z}}\tilde{w}_e = 0 \tag{10.9}$$

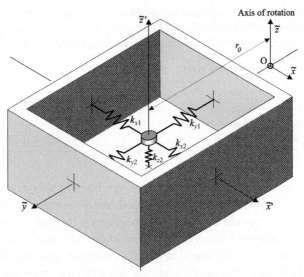

Figure 10.1. A simple rotating elastic structure (Example 10.2.1). Spring $k_{\bar{z}1}$ is not shown.

The solution of the eigenvalue problem corresponding to Equation (10.9) gives the natural frequency $\omega_{\bar{z}} = \sqrt{k_{\bar{z}}/m}$. Thus, for vibration along the $O\bar{z}$ axis, the natural frequency is constant.

EXAMPLE 10.2.1. A point mass of 0.2 kg is mounted in the center of a shallow, rigid box as shown in Figure 10.1. Viewed along the z-axis, the inside of the box has a square cross section 200×200 mm; therefore, the springs $k_{\bar{x}1}$, $k_{\bar{x}2}$, $k_{\bar{y}1}$, and $k_{\bar{y}2}$ have undeformed lengths of 100 mm. The mass is mounted at the exact center of a closed box whose depth in the Oz direction is 100 mm. There are two springs in the Oz direction – one above and one below the mass – with undeformed lengths of 50 mm. Figure 10.1 shows the box with the top plate removed and spring $k_{\bar{z}1}$ is not shown. At rest, the distance, r_0, between the mass and the axis of rotation, $O\bar{z}$, is 200 mm.

The spring stiffnesses are $k_{\bar{y}1} = k_{\bar{y}2} = 100$ kN/m, $k_{\bar{x}1} = k_{\bar{x}2} = 20$ kN/m, and $k_{\bar{z}1} = k_{\bar{z}2} = 25$ kN/m. Determine the equilibrium position of the mass and the three natural frequencies of this system in the rotating frame of reference at speeds of 0, 500, 1,000, and 8,000 rev/min. Ignore any contributions that springs $k_{\bar{x}1}$, $k_{\bar{x}2}$, $k_{\bar{z}1}$, and $k_{\bar{z}2}$ might make to forces in the radial direction. In effect, we assume that the points of attachment of springs $k_{\bar{x}1}$ and $k_{\bar{x}2}$ are free to slide along the walls of the box and that the points of attachment of springs $k_{\bar{z}1}$ and $k_{\bar{z}2}$ are free to slide along the top and base, respectively.

Solution. In this problem, $k_{\bar{x}} = k_{\bar{x}1} + k_{\bar{x}2} = 40$ kN/m, $k_{\bar{y}} = k_{\bar{y}1} + k_{\bar{y}2} = 200$ kN/m, and $k_{\bar{z}} = k_{\bar{z}1} + k_{\bar{z}2} = 50$ kN/m. To determine the steady-state displacement, we use Equation (10.8). The steady-state displacement in the $\bar{x}$ direction is zero (i.e., $\tilde{u}_{ss} = 0$), and in the $\bar{y}$ direction $\tilde{v}_{ss}$ is derived from

$$\left(k_{\bar{y}} - m\Omega^2\right) \tilde{v}_{ss} = m\Omega^2 r_0 \tag{10.10}$$

The equilibrium displacements of the point mass at the four rotational speeds are 0, 0.5498, 2.218, and 470.8 mm. The fourth displacement is very large

Table 10.1. *Natural frequencies in the rotating frame for the in-plane modes (Example 10.2.1)*

Speed (rev/min)	First natural frequency (Hz)	Second natural frequency (Hz)	Eigenvalues
0	71.18	159.15	$\pm 447.2_J$, $\pm 1000.0_J$
500	70.21	160.02	$\pm 441.1_J$, $\pm 1005.4_J$
1,000	67.40	162.51	$\pm 423.5_J$, $\pm 1021.1_J$
8,000	–	259.57	± 237.2, $\pm 1630.9_J$

and would locate the particle outside of the box! In this example, $\sqrt{k_{\bar{y}}/m} = 1,000$ rad/s or 9,549 rev/min. At this speed, the predicted displacement, from Equation (10.10), would be infinite. This places a theoretical upper bound on the rotational speed of the system. In practice, this speed tends to be much higher than the rotational speed at which centrifugal loads would burst the rotor. In a real system, this arises when the internal stress at some point exceeds the yield stress of the material (see Section 6.12). In this example, the springs would fail.

The natural frequency associated with oscillation of the mass in the axial direction is independent of spin speed and can be calculated using $\omega_{\bar{z}} = \sqrt{k_{\bar{z}}/m}$. This frequency is 79.58 Hz.

Equation (10.5) applies here because there is no elastic or viscous coupling between the axial direction and the circumferential and radial directions. This equation of motion is converted to an eigenvalue problem by setting the right side of the equation to zero and using the procedure described in Section 5.8. Table 10.1 shows the resulting natural frequencies and eigenvalues. The damping is zero in this case. At 8,000 rev/min, one eigenvalue pair is real, and the individual eigenvalues of this pair have equal magnitudes but opposite signs. The positive root indicates instability, and, because the system is undamped, the stability boundary occurs when one eigenvalue is zero. From the characteristic equation derived from Equation (10.5), by setting the eigenvalue to zero, the stability boundary occurs when

$$\left(k_{\bar{x}} - m\Omega^2\right)\left(k_{\bar{y}} - m\Omega^2\right) = 0 \tag{10.11}$$

The solution with the lowest rotational speed is $\Omega = \sqrt{k_{\bar{x}}/m}$ and equals 4,271 rev/min. Figure 10.2 shows how the natural frequencies of the system, in the rotating frame of reference, vary with rotational speed. The lowest resonance drops to zero at 4,271 rev/min; above this speed, the equilibrium position of the system is unstable.

10.2.1 Stress and Geometric Stiffening

There are two basic mechanisms for nonlinearity in the force–deflection relationship in a general elastic rotating structure: (1) nontrivial changes in the geometry of the structure, and (2) the presence of significant internal forces and stresses within the structure. These effects are illustrated in the context of a single particle supported by

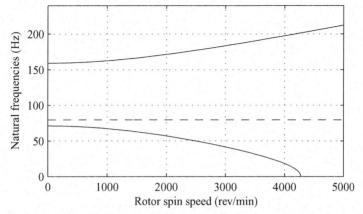

Figure 10.2. Natural frequencies of the system in the rotating frame (Example 10.2.1). The axial mode is shown as a dashed line.

massless springs. Coriolis acceleration and centripetal acceleration are purely kinematic phenomena arising from the fact that the coordinates are fixed to the rotor, whereas stress and geometric stiffness are phenomena of the particular structure and differ among structures. Likins et al. (1973) give three simple examples of rotating mechanisms carrying a single mass. In each example, the geometric stiffness is different but the centripetal acceleration and Coriolis effects are identical. Here, a different example is presented.

Figure 10.3(a) is a plan view of the particle suspended within a rigid box, as indicated in Figure 10.1 with the particle at the rest position. Figure 10.3(b) shows the particle in the steady-state deformed position at spin speed Ω. This conveniently illustrates the two distinct mechanisms. The stiffness associated with the radial deflections of the particle is increased as a result of the change in orientation of the springs $k_{\tilde{x}1}$, $k_{\tilde{x}2}$, $k_{\tilde{z}1}$, and $k_{\tilde{z}2}$. For example, the increase in stiffness due to the change in orientation of $k_{\tilde{x}1}$ is $k_{\tilde{x}1} \sin^2 \alpha$. For small deflections, the angle α is proportional to the steady-state displacement $\tilde{v}_{ss}$; thus, the additional radial stiffness contribution is proportional to the square of the steady-state deflection. As shown in Equation (10.10), for sufficiently small Ω, the steady-state deflection is approximately proportional to Ω^2. Hence, we find here that the *geometric stiffening* effects are proportional to Ω^4. For general systems, the geometric stiffening effects also have a component proportional to Ω^2.

We now consider the restoring force due to the springs when the particle is displaced circumferentially. Figure 10.4(a) shows the particle having undergone a small radial deflection; Figure 10.4(b) shows the particle having a circumferential deflection, $\tilde{u}$, in addition to the radial deflection. The contribution to the restoring force from the radial springs, $k_{\tilde{y}1}$ and $k_{\tilde{y}2}$, in the circumferential direction, $f_{\tilde{x}}$, is

$$f_{\tilde{x}} = f_{\tilde{y}1} \sin \beta_1 - f_{\tilde{y}2} \sin \beta_2 \qquad (10.12)$$

where $f_{\tilde{y}1}$ and $f_{\tilde{y}2}$ are the tensile or compressive forces in the springs $k_{\tilde{y}1}$ and $k_{\tilde{y}2}$, respectively. When $\tilde{u}$ is small, the angles β_1 and β_2 are each approximately proportional to $\tilde{u}$ and the forces are approximately constant with respect to $\tilde{u}$ and given by $f_{\tilde{y}1} = k_{\tilde{x}1}\tilde{v}_{ss}$ and $f_{\tilde{y}2} = -k_{\tilde{x}2}\tilde{v}_{ss}$. Furthermore, if $\tilde{v}_{ss}$ is small compared to the dimensions of the structure, these forces are due to the steady-state displacement and are

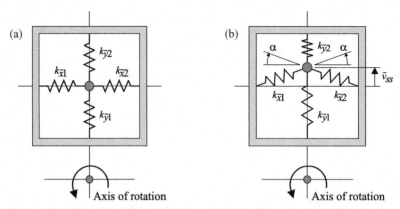

Figure 10.3. Plan view of a particle in a rigid box: (a) at rest, and (b) spinning.

proportional to Ω^2. This is called *stress stiffening* or *centrifugal stiffening*. In general systems, internal stresses play the same role as the forces $f_{\bar{y}1}$ and $f_{\bar{y}2}$ and are also proportional to Ω^2. Therefore, the stress-stiffening effect, both here and in general, is proportional to Ω^2. Thus, the stiffness matrix, $\mathbf{K}$, can be a strong function of spin speed with components proportional to both the square and the fourth power of Ω. The dependence of $\mathbf{K}$ on Ω invariably can be represented sufficiently accurately for any practical purposes by specifying $\mathbf{K}(\Omega)$ in terms of $\mathbf{K}_0$, $\mathbf{K}_2$, and $\mathbf{K}_4$ as

$$\mathbf{K}(\Omega) \approx \mathbf{K}_0 + \Omega^2 \mathbf{K}_2 + \Omega^4 \mathbf{K}_4 \qquad (10.13)$$

EXAMPLE 10.2.2. For the system described in Example 10.2.1, the spring stiffnesses are $k_{\bar{x}1} = k_{\bar{x}2} = 120\,\text{kN/m}$ and $k_{\bar{z}1} = k_{\bar{z}2} = 180\,\text{kN/m}$. At rest, all of the springs have zero internal forces. All of the springs are linear and they are unable to carry any moments. Thus, the forces in the springs necessarily act along the line between their respective endpoints, and these forces are directly proportional to the respective changes in length.

Determine the equilibrium position of the mass and the equivalent net stiffnesses of the point mass in the $O\bar{x}$, $O\bar{y}$, and $O\bar{z}$ directions for rotational speeds between 1,000 and 9,000 rev/min in steps of 1,000 rev/min for (a) $k_{\bar{y}1} = k_{\bar{y}2} = 150\,\text{kN/m}$ and (b) $k_{\bar{y}1} = 200\,\text{kN/m}$ and $k_{\bar{y}2} = 100\,\text{kN/m}$. Then, for the speeds

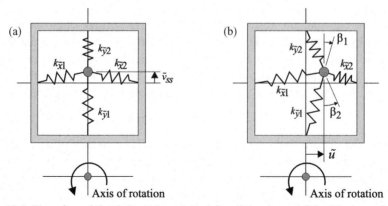

Figure 10.4. Plan view of the deflected particle: (a) radial only, and (b) with circumferential.

Table 10.2. *Radial displacement estimates (Example 10.2.2)*

Speed (rev/min)	$m\Omega^2 r_0$ (N)	Initial estimate of v_{ss} (mm)	Final estimate of v_{ss} (mm)	Converged f_c (N)
1,000	438.65	1.4729	1.4720	0.26778
2,000	1,754.6	6.0248	5.9642	17.653
3,000	3,947.8	14.086	13.398	192.85
4,000	7,018.4	26.494	23.073	906.08
5,000	10,966	44.729	34.177	2,587.0
6,000	15,791	71.440	46.481	5,517.0
7,000	21,494	111.64	60.215	9,900.5
8,000	28,074	175.86	75.829	15,969
9,000	35,531	290.41	93.932	24,038

given, compute the resonance frequencies of the particle both in the plane and out of the plane for (a) and (b).

Solution. Equation (10.10) is extended to account for the influence of the springs $k_{\tilde{x}1}$, $k_{\tilde{x}2}$, $k_{\tilde{z}1}$, and $k_{\tilde{z}2}$. The symmetry of the system means that the only non-zero steady-state deflection is the radial displacement $\tilde{v}_{ss}$. In the radial direction, the equation of motion becomes

$$(f_{\tilde{x}1} + f_{\tilde{x}2})\sin\alpha + (f_{\tilde{z}1} + f_{\tilde{z}2})\sin\gamma + (k_{\tilde{y}} - m\Omega^2)\tilde{v}_{ss} = m\Omega^2 r_0 \qquad (10.14)$$

where γ is the equivalent of the angle α in the Oz direction and $k_{\tilde{y}} = k_{\tilde{y}1} + k_{\tilde{y}2}$. Note that $k_{\tilde{y}} = 300\,\text{kN/m}$ for both (a) and (b). The forces $f_{\tilde{x}1}$, $f_{\tilde{x}2}$, $f_{\tilde{z}1}$, and $f_{\tilde{z}2}$ are the tensile or compressive forces in the corresponding springs. Both the angles, β and γ, and the spring forces depend on the steady-state displacement. For example, considering spring $k_{\tilde{x}1}$, $\tan\alpha = \tilde{v}_{ss}/L_x$, where L_x is the undeformed length of the spring (equal to 100 mm in this example) and $f_{\tilde{x}1} = k_{\tilde{x}1}\left(\sqrt{\tilde{v}_{ss}^2 + L_x^2} - L_x\right)$. Equation (10.14) is a nonlinear equation for the unknown steady-state displacement $\tilde{v}_{ss}$ and may be solved iteratively by splitting the equation into two parts:

$$f_c = (f_{\tilde{x}1} + f_{\tilde{x}2})\sin\alpha + (f_{\tilde{z}1} + f_{\tilde{z}2})\sin\gamma \qquad (10.15)$$

$$(k_{\tilde{y}} - m\Omega^2)\tilde{v}_{ss} = m\Omega^2 r_0 - f_c \qquad (10.16)$$

Initially, letting $f_c = 0$, Equation (10.16) is used to estimate $\tilde{v}_{ss}$, which is equivalent to the approach adopted in Example 10.2.2. This estimated steady-state displacement is then used in Equation (10.15) to obtain a force correction f_c. The correction is applied in Equation (10.16) to update $\tilde{v}_{ss}$, and the process is repeated until convergence.

Table 10.2 lists the initial estimate of particle deflection ignoring geometric stiffening for each speed. The fourth column in Table 10.2 lists the converged estimate of the steady-state displacement. Both geometric and stress stiffening are present in the table. By 4,000 rev/min, these nonlinear stiffening contributions are beginning to become significant. If higher values were given for $k_{\tilde{x}1}$, $k_{\tilde{x}2}$, $k_{\tilde{z}1}$, and $k_{\tilde{z}2}$, the nonlinear effects would be greater.

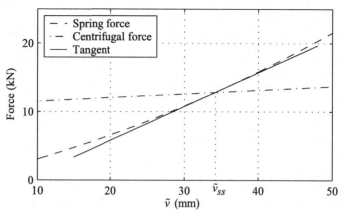

Figure 10.5. The resultant spring radial force and the centrifugal force at rotational speed 5,000 rev/min (Example 10.2.2(a)).

Knowing the steady-state deflections at a given spin speed of the rotor, the internal loads in the springs and their lengths and angles also are known and we can compute the net forces that are required to perturb the point mass by small additional displacements in the circumferential, radial, and axial directions. The term *tangent stiffness* is often applied to this rate of change of restoring force with respect to the particle position. Because of the symmetry in the present example, no coupling arises between the three directions. Figure 10.5 shows the resultant radial forces from the springs together with the centrifugal force on the particle as a function of the radial position of the particle, $\tilde{v}$, for a rotational speed of 5,000 rev/min and zero circumferential displacement ($\tilde{u} = 0$). Where the plots of these forces cross gives the steady-state displacement, $\tilde{v}_{ss}$. Also shown in the figure is the tangent to the spring force at the steady-state displacement, the slope of which gives the radial stiffness.

(a) Table 10.3 shows the three separate stiffness values as function of speeds for the case when $k_{\bar{y}1} = k_{\bar{y}2} = 150\,\text{kN/m}$. It is especially interesting that the tangent stiffness in the circumferential and axial directions becomes negative at speeds of 8,000 rev/min and higher. We can interpret this negative stiffness as an indication that the equilibrium-point positions we computed are unstable. If the particle is caused to move by any amount in the axial or circumferential direction, it continues to move away. This negative stiffness arises as a direct result of *stress stiffening* and it is a useful pointer to the fact that stress stiffening can result in negative stiffness contributions when one or more components are in compression. In the present case, the component in compression is the spring $k_{\bar{y}2}$. From Table 10.2, notice that at 8,000 rev/min, this spring has been compressed from its original length of 100 mm to a final length of 24.17 mm.

The resonance frequencies of the particle can be computed by applying Equations (10.3) and (10.9) with $\tilde{u}(t) = \tilde{u}_0 e^{st}$, $\tilde{v}(t) = \tilde{v}_0 e^{st}$, and $\tilde{w}(t) = \tilde{w}_0 e^{st}$. This produces six characteristic roots that contain the resonance-frequency information. Because the system is undamped, all roots are either purely real or purely imaginary. Moreover, when a pair of real roots exists, the sum of the two roots is zero. Hence, the presence of any real roots in this

Table 10.3. *Tangent stiffnesses for the particle (Example 10.2.2(a))*

Speed (rev/min)	Circumferential stiffness (kN/m)	Radial stiffness (kN/m)	Axial stiffness (kN/m)
1,000	240.04	300.55	359.65
2,000	240.61	308.82	354.34
3,000	242.59	341.88	333.35
4,000	244.44	408.48	291.89
5,000	239.37	494.09	238.58
6,000	214.99	579.59	177.45
7,000	144.66	655.25	87.620
8,000	−73.874	718.53	−135.24
9,000	−1900.3	769.72	−1955.9

case necessarily indicates instability. Figure 10.6 shows that the particle is already unstable at 7,250 rev/min, although none of the tangent stiffnesses is yet negative at that speed. Coriolis forces couple particle movements in the radial and circumferential directions, but movements in the axial direction are decoupled from the others; these resonances can be solved separately and by inspection.

(b) Consider the case in which $k_{\bar{y}1} = 200$ kN/m and $k_{\bar{y}2} = 100$ kN/m. Because $(k_{\bar{y}1} + k_{\bar{y}2})$ remains the same as before (300 kN/m), the equilibrium deflections of Table 10.2 still apply; also, the tangent stiffness in the radial direction is unchanged. However, the tangent stiffnesses in the circumferential and axial directions is now different. Table 10.4 summarizes these tangent stiffnesses. Figure 10.7 provides the natural frequencies as a function of rotational speed and shows that the particle becomes unstable at approximately 7,880 rev/min. Table 10.4 shows that all of the tangent stiffnesses are positive at and below 8,000 rev/min, which demonstrates that stability must be determined from the full eigenvalue problem rather than from the sign of the tangent stiffnesses.

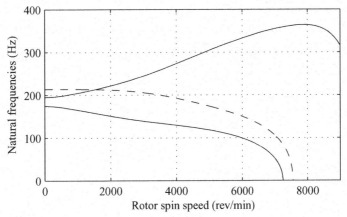

Figure 10.6. Natural frequencies of the point mass in the rotating frame as a function of speed (Example 10.2.2(a)). The axial mode is shown as a dashed line.

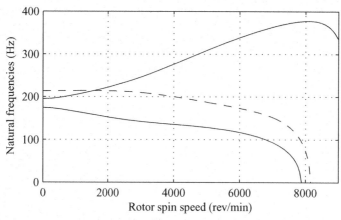

Figure 10.7. Natural frequencies of the point mass in the rotating frame as a function of speed (Example 10.2.2(b)). The axial mode is shown as a dashed line.

10.2.2 Damping in a Spinning Rotor

In Section 10.2, we observe that the rotation of a frame of reference causes the mass of a particle to make contributions to the stiffness and damping terms in an equation of motion. These contributions relate only to degrees of freedom (i.e., particle motions) normal to the axis of rotation. The stiffness contribution is always diagonal and negative, and the damping contribution is always skew-symmetric. In Section 10.2.1, we see that other speed-dependent contributions also apply to the stiffness terms in the equation of motion – namely, stress stiffening and geometric stiffening. It is natural to question whether the presence of some damping in a structure causes any similar effects when the structure rotates. This question is particularly apposite because Chapter 7 explains that internal damping in a rotor can be a root cause of instability. Suppose that a system of particles is connected by parallel spring–dashpot units and spun at a steady speed about an axis. If the particles are oscillating in some given pattern relative to the rotating frame, then the force contributions arising from the dampers are dependent on the spin speed only to the extent that the spinning may cause a small change in the geometry of the system. We refer to such an effect

Table 10.4. *Tangent stiffnesses for the particle (Example 10.2.2(b))*

Speed (rev/min)	Circumferential stiffness (kN/m)	Radial stiffness (kN/m)	Axial stiffness (kN/m)
1,000	241.51	300.55	361.12
2,000	246.60	308.82	360.33
3,000	256.23	341.88	346.99
4,000	268.81	408.48	316.26
5,000	278.06	494.09	277.28
6,000	274.28	579.59	236.75
7,000	239.12	655.25	182.09
8,000	104.55	718.53	43.179
9,000	−1102.2	769.72	−1157.7

as *geometric damping*. Dashpots alone have zero mean force across them so they cannot make any contribution to stress stiffening. In practical circumstances, it is rarely if ever necessary to account for geometric damping; however, for academic accuracy, it is observed that where it occurred, this effect would be proportional to the fourth power of rotational speed, Ω. Thus,

$$\mathbf{C}(\Omega) \approx \mathbf{C}_0 + \Omega^4 \mathbf{C}_4 \tag{10.17}$$

Using the definitions in Equations (10.13) and (10.17) in conjunction with the equations of motion in Equation (10.5) provides a general framework to analyze the small oscillations of a system of discrete particles in a rotating reference frame.

10.3 Finite Element Analysis of Rotors with Deformable Cross Sections

In general, the detailed analysis of a spinning rotor or rotating machine demands the use of FE models. In previous chapters, *shaft-line models* suffice where the elasticity of a rotor is represented by beam elements (or similar) joining nodes on the axis of rotation and where the inertia properties of the rotor are contributed (at least, in part) by discrete disk elements. Although these shaft-line models are one-dimensional FE models, they are not adequate in a number of cases where particular cross sections of the rotor may deform significantly either in or out of the plane of the cross section. In these cases, two- or three-dimensional models are needed with substantially larger numbers of degrees of freedom. For example, Nandi and Neogy (2001) give examples where typically 80 – but, in one case, 625 – three-dimensional 20-node brick elements were required to accurately model particular hollow-tapered rotors. Combescure and Lazarus (2008) compare beam models of large machines with two- and three-dimensional models.

These more extensive FE models for flexible rotors may consist of three-dimensional elements, two-dimensional plate/shell elements for relatively thin plates or shells, and one-dimensional elements for beams. The one-dimensional models are especially useful for blades on the rotors of turbo-machines. In cases in which the rotor is (or can be approximated as) a solid or revolution about the axis of rotation, an efficient two-dimensional analysis can be employed; details of this type of analysis are described herein.

In Section 10.2, insight is given to effects that relate to the behavior of one or more particles in a rotating reference frame. In particular, the Coriolis coupling between radial and circumferential directions and the spin-softening effect are explained. The origins of stress-stiffening and geometric-stiffening are outlined. Although Section 10.2 is focused entirely on a structure consisting of a discrete mass, the extension to continuous distributions of mass is not a major step and is addressed here.

10.3.1 General Finite Element Models

In Chapter 4, a brief derivation of the mass and stiffness matrices for two- and three-dimensional FEA is provided for static structures. Each element represents a finite volume of the structure being analyzed and its boundaries are defined by nodes at corners and (in some cases) at midside locations. Individual elements are related to a parent or master element with a simple shape. A mapping, Φ, associates any given point in the parent element with a unique corresponding point in the actual (i.e., physical) element. For three-dimensional elements, any one point in the parent element is specified by the three spatial coordinates (ξ, η, ζ) and the corresponding point in the actual element is specified by the three spatial (global) coordinates (x, y, z). The mapping, Φ, then has three components:

$$x = \Phi_x(\xi, \eta, \zeta) \tag{10.18}$$

$$y = \Phi_y(\xi, \eta, \zeta) \tag{10.19}$$

$$z = \Phi_z(\xi, \eta, \zeta) \tag{10.20}$$

A second mapping, Θ, defines the point deflections (u, v, w) in the directions (x, y, z), respectively, in terms of the three coordinates of the corresponding point in the parent element (ξ, η, ζ), so that

$$u = \Theta_x(\xi, \eta, \zeta) \tag{10.21}$$

$$v = \Theta_y(\xi, \eta, \zeta) \tag{10.22}$$

$$w = \Theta_z(\xi, \eta, \zeta) \tag{10.23}$$

Both mappings are constructed as linear combinations of smooth functions called shape functions, one for each node. The element is described as *isoparametric* if the same shape functions used to embody Φ are also used to embody Θ. Isoparametric elements are the most common type of elements in FEA, and Section 4.8 discusses the formulation for two-dimensional elements.

The mapping, Φ, is known if the (x, y, z) coordinates are known for each node defining the element. Then, the coordinates of any point within the element can be determined from the coordinates of the corresponding point within the reference element. The mapping is explicit in the forward sense only; that is, that given (ξ, η, ζ), we can compute (x, y, z) directly but not vice versa. The mapping, Θ, is also known if the displacements (u, v, w) are known at each node of the element. Then, the displacements at any point within the element can be determined and the strains at that point also may be determined.

An expression for the total kinetic energy of an individual element within a static structure is given by Equation (4.59). Extending this to three dimensions gives

$$T_e = \frac{1}{2} \iiint \rho(x, y, z) \begin{Bmatrix} \dot{u} \\ \dot{v} \\ \dot{w} \end{Bmatrix}^\top \begin{Bmatrix} \dot{u} \\ \dot{v} \\ \dot{w} \end{Bmatrix} \mathrm{d}x\mathrm{d}y\mathrm{d}z \tag{10.24}$$

The limits of integration here correspond to the boundaries of the element and the integration is evaluated numerically. The details of how this integration is performed are beyond the scope of this book (see Zienkiewicz et al., 2005). The expression for the total strain energy of a two-dimensional element for a static structure is given by Equation (4.68). Extending this to three dimensions gives

$$U_e = \frac{1}{2} \iiint \epsilon(x, y, z)^\top \mathbf{D}(x, y, z)\epsilon(x, y, z)) \, \mathrm{d}x\mathrm{d}y\mathrm{d}z \tag{10.25}$$

where ϵ represents the vector of strains at the point (x, y, z) in the actual element and where $\mathbf{D}$ represents the elasticity relationship between the stress vector, σ, and the strain vector, ϵ. Again, the limits of integration here correspond to the boundaries of the element and the integration is evaluated numerically. The strains within ϵ are determined from the derivatives of point displacements (u, v, w) with respect to the spatial coordinates.

Because the mappings Φ and Θ are formed as linear combinations of known shape functions, the total kinetic energy, T_e, associated with a given element can be written as

$$T_e = \frac{1}{2}\dot{\mathbf{q}}_e^\top \mathbf{M}_e \dot{\mathbf{q}}_e \qquad (10.26)$$

and the total strain energy, U_e, associated with a given element can be written as

$$U_e = \frac{1}{2}\mathbf{q}_e^\top \mathbf{K}_e \mathbf{q}_e \qquad (10.27)$$

where $\mathbf{M}_e$ and $\mathbf{K}_e$ are the element mass and stiffness matrices, respectively (for a nonrotating rotor). The global mass and stiffness matrices for the complete rotor structure, $\mathbf{M}_{R0}$ and $\mathbf{K}_{R0}$, respectively, are obtained by the assembly process described in Chapter 4.

An element damping matrix can be obtained for the element if, in addition to the elastic material properties at every point, the viscous damping properties of the material are also known in the form of a matrix, $\mathbf{E}(x, y, z)$. In that case, the instantaneous power being dissipated as heat within the element is

$$P_e = \iiint \dot{\epsilon}(x, y, z)^\top \mathbf{E}(x, y, z)\dot{\epsilon}(x, y, z)\, dx dy dz \qquad (10.28)$$

The element damping matrix is found by following procedures identical to those used for determining the stiffness matrix except that $\mathbf{E}(x, y, z)$ is inserted in place of $\mathbf{D}(x, y, z)$. The global damping matrix, $\mathbf{C}_{R0}$, is obtained by the same assembly process used to form the global mass and stiffness matrices. This process describes material damping but there are many damping mechanisms possible in structures and machines – for example, in joints, materials, and working fluids – and often these must be analyzed on an individual basis.

We now consider how the effects of rotation are incorporated. The nodal displacement coordinates are now expressed in the rotating frame, and they are relative to the equilibrium positions or the nodes. Each node has one radial-displacement coordinate (i.e., positive outward), one circumferential-displacement coordinate (i.e., positive in the same direction as rotor spin), and one axial coordinate (i.e., such that a right-hand screw fixed to the rotor progresses in a positive sense). The equation of motion for the rotor takes the familiar form

$$\tilde{\mathbf{M}}_R \ddot{\tilde{\mathbf{q}}} + \tilde{\mathbf{C}}_R(\Omega)\dot{\tilde{\mathbf{q}}} + \tilde{\mathbf{K}}_R(\Omega)\tilde{\mathbf{q}} = \mathbf{f} \qquad (10.29)$$

where our convention is adopted with quantities in the rotating frame denoted with a tilde.

It is convenient also to consider that the displacement coordinates are grouped such that all of the radial deflections are contiguous; then, all of the circumferential deflections follow (with the same node ordering); and, finally, all of the axial deflections (with the same node ordering). In this case, the global-mass matrix takes the form

$$\tilde{\mathbf{M}}_R = \begin{bmatrix} \mathbf{M}_n & \mathbf{0} & \mathbf{0} \\ \mathbf{0} & \mathbf{M}_n & \mathbf{0} \\ \mathbf{0} & \mathbf{0} & \mathbf{M}_n \end{bmatrix} \tag{10.30}$$

and the global damping and stiffness matrices depend on Ω according to

$$\tilde{\mathbf{C}}_R(\Omega) = \left(\tilde{\mathbf{C}}_{R0} + \Omega^4 \tilde{\mathbf{C}}_{R4}\right) + \begin{bmatrix} \mathbf{0} & -2\Omega\mathbf{M}_n & \mathbf{0} \\ 2\Omega\mathbf{M}_n & \mathbf{0} & \mathbf{0} \\ \mathbf{0} & \mathbf{0} & \mathbf{0} \end{bmatrix} \tag{10.31}$$

and

$$\tilde{\mathbf{K}}_R(\Omega) = \left(\tilde{\mathbf{K}}_{R0} + \Omega^2 \tilde{\mathbf{K}}_{R2} + \Omega^4 \tilde{\mathbf{K}}_{R4}\right) - \begin{bmatrix} \Omega^2\mathbf{M}_n & \mathbf{0} & \mathbf{0} \\ \mathbf{0} & \Omega^2\mathbf{M}_n & \mathbf{0} \\ \mathbf{0} & \mathbf{0} & \mathbf{0} \end{bmatrix} \tag{10.32}$$

$\mathbf{M}_n$ simultaneously represents the mass matrix for all radial translations at nodes in the rotor, the mass matrix for all circumferential translations at rotor nodes, and the mass matrix for axial translations. The final term in Equation (10.31) represents the Coriolis forces. The final term in Equation (10.32) represents the spin-softening term (see Section 10.2). The Coriolis and spin-softening forces are an extension of Equation (10.4), which addresses a single particle.

Matrices $\tilde{\mathbf{C}}_{R0}$ and $\tilde{\mathbf{K}}_{R0}$ are also the damping and stiffness matrices for the rotor structure when there is no rotation; these are obtained by the methods outlined in this section and in Chapter 4 for the FEA of static structures. Section 10.2 provides insight to the origin of the matrix terms $\tilde{\mathbf{K}}_{R2}$, $\tilde{\mathbf{K}}_{R4}$, and $\tilde{\mathbf{C}}_{R4}$ for rotating systems of discrete particles; we now provide details, about how they are obtained for three-dimensional FE models. For all of these terms, the equilibrium-deformed state of the rotor must be computed for the steady running speed, Ω_0. The forcing in this case is due to the centripetal accelerations. A first approximation to this forcing can be obtained by considering that the deflections of the various rotor nodes are negligible in the rotating frame compared to the original position coordinates. Iterative schemes may be devised that compute the corresponding steady-state nodal deflections and, hence, improve the estimate of the steady-state forcing. The deformed rotor has a different geometry than the undeformed rotor and, if the materials behave linearly, the stiffness and damping matrices for a rotor with this modified shape are $(\tilde{\mathbf{K}}_{R0} + \Omega_0^4 \tilde{\mathbf{K}}_{R4})$ and $(\tilde{\mathbf{C}}_{R0} + \Omega_0^4 \tilde{\mathbf{C}}_{R4})$, respectively. To determine $\tilde{\mathbf{C}}_{R4}$ and $\tilde{\mathbf{K}}_{R4}$, it is most convenient to calculate the equilibrium deflections at two different speeds and the corresponding stiffness and damping matrices.

Determining $\tilde{\mathbf{K}}_{R2}$ requires estimation of the steady-state internal stresses within the rotor. At any point (x, y, z) within the rotor, we can obtain the directions of the principal (internal) stresses in the rotor. Denoting these three principal directions by $(x_L, y_L, \text{ and } z_L)$, respectively, and the corresponding principal stresses by

Figure 10.8. A highly discretized three-dimensional model of a rotor. This figure is reproduced with the permission of Rolls-Royce plc, ©Rolls-Royce plc 2007.

$(\sigma_{xxL}, \sigma_{yyL}, \sigma_{zzL})$, $\tilde{\mathbf{K}}_{R2}$ is given implicitly by

$$\tilde{\mathbf{q}}^{\top} \tilde{\mathbf{K}}_{R2} \tilde{\mathbf{q}} \equiv \iiint \sigma_{xxL} \left(\left(\frac{\partial v_L}{\partial x_L} \right)^2 + \left(\frac{\partial w_L}{\partial x_L} \right)^2 \right)$$

$$+ \sigma_{yyL} \left(\left(\frac{\partial w_L}{\partial y_L} \right)^2 + \left(\frac{\partial u_L}{\partial y_L} \right)^2 \right) + \sigma_{zzL} \left(\left(\frac{\partial u_L}{\partial z_L} \right)^2 + \left(\frac{\partial v_L}{\partial z_L} \right)^2 \right) \mathrm{d}x \mathrm{d}y \mathrm{d}z \quad (10.33)$$

where (u_L, v_L, w_L) are the point deflections in the three principal directions, (x_L, y_L, z_L). Individual contributions to $\tilde{\mathbf{K}}_{R2}$ are derived for each element in turn and assembled to give the global matrix, as is done for the other system matrices.

A detailed three-dimensional model of a rotor may easily consist of tens of thousands of degrees of freedom, or even hundreds of thousands. Figure 10.8 illustrates one example. The number of degrees of freedom actually used in a three-dimensional FE model of a rotor is most often determined by the need to model intricate geometrical detail rather than the requirement to capture effectively the dynamic properties of the rotor. For any given flexible rotor, the requirement to model the dynamic properties accurately does produce constraints on the minimum number of degrees of freedom that may be used, but this minimum limit virtually never determines the initial discretization of the rotor. It is common, therefore, to apply model reduction to the initial discretization such that the number of degrees of freedom retained in the reduced model is significantly smaller than the number in the original model, yet still adequate for the intended analysis purposes. Model reduction is discussed in Section 2.5.

An alternative to performing the model reduction numerically, in effect, is to perform the reduction analytically by constraining the deformation within the element, thereby requiring fewer degrees of freedom per element. Plate and shell elements, beam elements, and axisymmetric elements effectively perform this analytical model reduction. Beam elements were already discussed extensively. These one- and two-dimensional elements also allow larger elements to be used, further reducing the number of degrees of freedom required. Readers are referred to standard textbooks on FEA for explanations of plate and shell elements (Fagan, 1992; Zienkiewicz et al., 2005). Systems with cyclic symmetry (e.g., bladed disks) can be

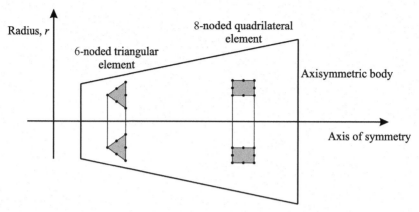

Figure 10.9. Schematic of an axisymmetric body. The axisymmetric elements form a complete ring. Typical eight-node quadrilateral and six-node triangular elements are shown.

analyzed using cyclic symmetry boundary conditions (Chatelet et al., 2005; Kill, 2008). Axisymmetric elements are discussed briefly here because of their particular relevance to the analysis of rotors. It is noteworthy that axisymmmetric rotor models are currently used in early design stages of the most complex rotating machinery and that three-dimensional models tend to be used only in the final stages of design.

10.3.2 Axisymmetric Finite Element Rotor Models

If the rotor geometry is axisymmetric, it is possible to analyze it using axisymmetric solid elements, even if the loads and system displacements are not axisymmetric. Quadrilateral and triangular elements of this class are shown schematically in Figure 10.9.

The model is represented completely by elements in the $\tilde{r}, \tilde{z}$ plane but, mathematically, each element models a complete ring-shaped volume of material. Each node in the model represents an entire circle within the physical object. The displacements and any applied forces are represented using Fourier series. For example, the radial displacement $\tilde{u}(\tilde{\phi})$ at a specific node i is a function of $\tilde{\phi}$ represented by

$$\tilde{u}_i(\tilde{\phi}) = \sum_{N=0}^{\infty} {}_c\tilde{u}_{iN} \cos(N\tilde{\phi}) + \sum_{N=0}^{\infty} {}_s\tilde{u}_{iN} \sin(N\tilde{\phi}) \tag{10.34}$$

Here, ${}_c\tilde{u}_{iN}$ and ${}_s\tilde{u}_{iN}$ are Fourier coefficients. The angle $\tilde{\phi}$ is the angular position and N is the harmonic number (or Fourier order). Similar expressions describe the circumferential and axial displacements at node i, $\tilde{v}_i$, and $\tilde{w}_i$ respectively. The key advantage of axisymmetric models over full three-dimensional models arises from the fact that different harmonics (i.e., Fourier orders) can be analyzed completely independently. Figure 10.10 indicates modes of a simple, slender ring-structure associated with different Fourier orders. The *breathing mode* is only one mode of the ring belonging to Fourier order $N = 0$; two others are rigid-body axial translation and rigid-body rotation about the axis of the ring. The in-plane, rigid-body translation modes of the ring belong to the Fourier order $N = 1$, but there are other modes

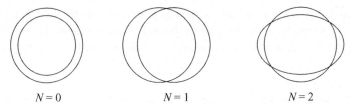

$N = 0$ $N = 1$ $N = 2$

Figure 10.10. Three vibration modes of a simple, slender ring belonging to different Fourier orders.

in this family. Rigid-body tilting of the ring out-of-plane (about any diameter) also belongs to $N = 1$. Similarly, there are in-plane $N = 2$ modes and out-of-plane $N = 2$ modes. As a general rule, the rigid-body modes of any axisymmetric structure all belong to either $N = 0$ (two modes) or to $N = 1$ (four modes).

Consider a full three-dimensional model of a rotor with 100 divisions about the circumference of the model, with 1,000 nodes in each of those planes. This model has 300,000 degrees of freedom in total and can be used to investigate (simultaneously) all Fourier orders from $N = 0$ to $N = 50$. An axisymmetric model of identical resolution consists of 1,000 nodes in a single (i.e., half) cross section and usually uses, only 3,000 degrees of freedom for each different Fourier order. However, in the worst case, in which the natural modes did not have a plane of symmetry within the rotor, 6,000 degrees of freedom is required for each Fourier order. The axisymmetric model can be used to model any Fourier orders of interest. In reality, it is unlikely that all orders up to 50 would be of interest.

In the special case of $N = 0$, the coefficients associated with the $\sin(N\tilde{\phi})$ terms are irrelevant, there are three degrees of freedom per node (i.e., radial, circumferential, and axial translations), and the deformation patterns are axisymmetric. The principal axes for the elastic properties of the rotor material usually coincide with the polar directions $\tilde{r}$, $\tilde{\phi}$, and $\tilde{z}$. Then, with $N = 0$, the stiffness matrix has zero coupling between the circumferential degrees of freedom, $_c\tilde{u}_i$, and either the radial or axial degrees of freedom, $_c\tilde{v}_i$ and $_c\tilde{w}_i$. This is to be expected from symmetry arguments because deflections in the radial and axial directions are symmetric relative to the plane $\tilde{\phi} = 0$, whereas deflections in the circumferential direction are antisymmetric about this plane. In these circumstances, for $\Omega = 0$, some of the $N = 0$ modes of vibration are pure torsional modes in which the only non-zero nodal deflections are in the circumferential direction. The other vibration modes are coupled radial and axial modes. Coriolis effects couple these again through the coefficient matrix for the velocity if the rotor is spinning.

In contrast with all other Fourier orders, modes of vibration of the $N = 0$ modes have no *directionality*. The state of stress of the spinning rotor analyzed for the purposes of setting up the stress-stiffening matrix is a particular case of $N = 0$.

When $N > 0$, both the $\cos(N\tilde{\phi})$ and $\sin(N\tilde{\phi})$ terms are potentially relevant. A pattern of deflection is symmetric about the plane $\tilde{\phi} = 0$ if all of the $\cos(N\tilde{\phi})$ terms of the circumferential deflections are zero and all of the $\sin(N\tilde{\phi})$ terms of the radial and axial deflections are zero. Similarly, a pattern of deflection is antisymmetric about that plane if all of the $\sin(N\tilde{\phi})$ terms of the circumferential deflections and all of the $\cos(N\tilde{\phi})$ terms of the radial and axial deflections are zero.

In the absence of Coriolis forces and any *handedness* in the material properties, it is necessary to model only the symmetric deflection patterns; in these cases, axisymmetric models for harmonic number $N > 0$ require only three degrees of freedom per node. For complete generality, six degrees of freedom are needed for each node (Geradin and Kill, 1984).

Stiffness and damping element matrices for an individual axisymmetric solid element are derived by expressing strains (in polar coordinates) in terms of the nodal displacements and then integrating strain-energy density in the usual way over the volume of the element. Taking $(\tilde{u}, \tilde{v}, \tilde{w})$ to represent point radial, circumferential, and axial deflections, respectively, the six strains are expressed in terms of polar coordinates as

$$
\epsilon_{\tilde{\phi},\tilde{\phi}} = \frac{1}{\tilde{r}}\left(\tilde{u} + \frac{\partial \tilde{v}}{\partial \tilde{\phi}}\right), \qquad \epsilon_{\tilde{r},\tilde{r}} = \frac{\partial \tilde{u}}{\partial \tilde{r}},
$$

$$
\epsilon_{\tilde{z},\tilde{z}} = \frac{\partial \tilde{w}}{\partial \tilde{z}}, \qquad \epsilon_{\tilde{\phi},\tilde{r}} = \frac{1}{\tilde{r}}\left(\frac{\partial \tilde{u}}{\partial \tilde{\phi}} - \tilde{v}\right) + \frac{\partial \tilde{v}}{\partial \tilde{r}}, \qquad (10.35)
$$

$$
\epsilon_{\tilde{r},\tilde{z}} = \frac{\partial \tilde{u}}{\partial \tilde{z}} + \frac{\partial \tilde{w}}{\partial \tilde{x}}, \qquad \epsilon_{\tilde{z},\tilde{\phi}} = \frac{1}{\tilde{r}}\frac{\partial \tilde{w}}{\partial \tilde{\phi}} + \frac{\partial \tilde{v}}{\partial \tilde{z}}.
$$

To obtain the element stiffness matrix, $\tilde{\mathbf{K}}_e$, the strain-energy density is integrated over the entire volume of the element as Equation (10.25) indicates. Two features of the axisymmetric models make this integration easy. First, although there are three integration variables $(\tilde{r}, \tilde{\phi}, \tilde{z})$, the integration with respect to $\tilde{\phi}$ can be done analytically because every quantity of interest has the form $k(\cos^2(N\tilde{\phi}))$, $k(\sin^2(N\tilde{\phi}))$, or $k(\cos(N\tilde{\phi})\sin(N\tilde{\phi}))$ integrated between 0 and 2π. Second, the derivatives of displacement with respect to angle (required in $\epsilon_{\tilde{\phi},\tilde{\phi}}$, $\epsilon_{\tilde{\phi},\tilde{r}}$, and $\epsilon_{\tilde{z},\tilde{\phi}}$) are obtained directly. An element-damping matrix can be formed in a similar way. The only difference between how the stiffness and damping matrices are obtained is that the viscous properties of the element material are used in place of the elastic properties (see Equation (10.28)). The element mass matrix has the same structure as shown in Equation (10.30) and is obtained by integrating kinetic energy density as Equation (10.24) indicates.

In the most general case, every node, i, in an axisymmetric model has the following degrees of freedom:

- $\cos(N\tilde{\phi})$ radial translation $({}_c\tilde{u}_{iN})$
- $\sin(N\tilde{\phi})$ circumferential translation $({}_s\tilde{v}_{iN})$
- $\cos(N\tilde{\phi})$ axial translation $({}_c\tilde{w}_{iN})$
- $\sin(N\tilde{\phi})$ radial translation $({}_s\tilde{u}_{iN})$
- $\cos(N\tilde{\phi})$ circumferential translation $({}_c\tilde{v}_{iN})$
- $\sin(N\tilde{\phi})$ axial translation $({}_s\tilde{w}_{iN})$

For modes that have $\tilde{\phi} = 0$ as a plane of symmetry, only the first three of these degrees of freedom can contain non-zeros at any node. For modes that have $\tilde{\phi} = 0$ as a plane of antisymmetry, only the last three of these degrees of freedom can contain non-zeros at any node.

The process of merging individual elements to form a complete rotor model is the same as described in Chapter 4. Moreover, the axisymmetric model for the complete rotor spinning at a constant speed, Ω, follows the same structure as given for the three-dimensional models of rotors (see Equations (10.30), (10.31), and (10.32)). The same procedure may be applied to obtain the geometric terms, $\tilde{\mathbf{K}}_{R4}$ and $\tilde{\mathbf{C}}_{R4}$; the modified geometry of the rotor spinning at a steady spin speed of Ω_0 is first computed and the global-stiffness and damping matrices for this modified rotor geometry are $\left(\tilde{\mathbf{K}}_{R0} + \Omega_0^4 \tilde{\mathbf{K}}_{R4}\right)$ and $\left(\tilde{\mathbf{C}}_{R0} + \Omega_0^4 \tilde{\mathbf{C}}_{R4}\right)$. The stress-stiffening term, $\tilde{\mathbf{K}}_{R2}$, for these axisymmetric models can be found following procedures parallel to those described previously for three-dimensional rotor models. The directions of the principal stresses due to the steady rotation are first computed at each point of interest in the rotor. Unless the rotor material has a strong handedness, the circumferential direction is always one of these directions. Maintaining consistency with the previous development of $\tilde{\mathbf{K}}_{R2}$, the other two directions are labeled r_L and z_L and the deflections in these directions are u_L and w_L, respectively. Then, $\tilde{\mathbf{K}}_{R2}$ is defined implicitly through

$$\tilde{\mathbf{q}}^\top \tilde{\mathbf{K}}_{R2} \tilde{\mathbf{q}} \equiv \iiint \sigma_{\tilde{\phi}\tilde{\phi}} \frac{1}{\tilde{r}^2} \left(\left(\frac{\partial w_L}{\partial \phi} \right)^2 + \left(\frac{\partial u_L}{\partial \phi} - v \right)^2 \right)$$

$$+ \sigma_{\tilde{r}\tilde{r}L} \left(\left(\frac{\partial v}{\partial r_L} \right)^2 + \left(\frac{\partial w_L}{\partial r_L} \right)^2 \right) + \sigma_{\tilde{z}\tilde{z}L} \left(\left(\frac{\partial v}{\partial z_L} \right)^2 + \left(\frac{\partial u_L}{\partial z_L} \right)^2 \right) dxdydz \quad (10.36)$$

Even when using axisymmetric solid elements, it is sometimes necessary to use a large number of elements. Stephenson et al. (1989) created a model of an experimental three-disk rotor using approximately 750 four-node axisymmetric solid elements and then reduced the model to 23 master degrees of freedom to determine the frequencies of the first five modes of vibration. They modeled the same rotor using 21 12-node axisymmetric solid elements. The latter model had more than 500 degrees of freedom, which was reduced to 25 master degrees of freedom. The results show that models with higher-order axisymmetric elements require fewer elements and degrees of freedom to obtain comparable accuracy. Furthermore, the natural frequencies for models with axisymmetric elements were closer to the measured frequencies than those from shaft-line models.

In Section 10.2, Coriolis forces, stress-stiffening, and spin-softening are discussed. All have an effect on the resonance frequencies of the structure. For harmonic numbers $N > 0$, resonance frequencies of axisymmetric structures occur in identical pairs when the structure has zero rotational speed. As the rotational speed rises, spin-softening tends to reduce resonance frequencies. Coriolis forces split the pairs of resonance frequencies in the rotating frame into forward-traveling modes (i.e., lower resonance frequencies) and backward-traveling modes (i.e., higher resonance frequencies). The resonances most affected by Coriolis forces and spin-softening are those the modes of which involve substantial movements normal to the axis. As the name suggests, stress-stiffening usually tends to increase resonance frequencies. A system consisting of a rotating drum with blades projecting radially inward exhibits the opposite effect.

When an axisymmetric structure is spinning about its axis and resonating in a natural mode, the frequencies of vibration measured by a stationary observer are different than the frequencies of the structure within the rotating frame (see

Section 7.3.1). When observed from the stationary frame, modes that are forward-traveling within the rotating frame have $N\Omega$ added to their resonance frequencies; modes that are backward-traveling within the rotating frame have $N\Omega$ subtracted from their resonance frequencies. The following observations invariably apply:

- Within the rotating frame, modes that are forward-traveling have lower resonance frequencies than modes that are backward-traveling.
- Within the stationary frame, modes that are forward-traveling have higher resonance frequencies than modes that are backward-traveling.
- Modes that are backward-traveling in the rotating frame may be forward-traveling in the stationary frame.

Specific examples of uses of axisymmetric rotor models are explored in subsequent sections.

10.4 Rotor with Flexible Disks

Often, rotor systems can be modeled as flexible shafts carrying rigid disks. In the majority of cases, the assumption that the disks are rigid is valid and it can be shown, for example, that the lowest (i.e., significant) natural frequency of the larger of the two disks carried by the rotor in Example 5.9.1 is greater than 4 kHz if its hub is clamped. Thus, in the range of frequencies shown in Figure 5.28 (i.e., 0 to 150 Hz), we expect the disk to behave as a rigid body.

The assumption that disks are rigid is not valid for all rotor systems. For example, the so-called hard disks used for data storage in computer systems are actually very thin and flexible and have relatively low natural frequencies compared to the shaft to which they are attached. As a consequence, in the analysis of such a system, it is essential to account for the disk flexibility. Papers by Chung et al. (2003), Heo et al. (2003), Jang et al. (2002), Jia and Chun (1997), and Kumar et al. (1997) examine various configurations of a flexible disk on a shaft. Lee and Chun (1998) analyze a rotor carrying multiple flexible disks using an assumed mode method.

10.4.1 Analysis of a Single Flexible Disk

We begin by examining the bending vibrations of an annular disk with a clamped inner edge and a free outer edge. These boundary conditions are chosen because when a disk is fixed to a shaft, it tends to be clamped at its inner edge and is generally free at its outer edge. Figure 10.11 shows a disk vibrating in various bending modes, which have lines and circles of zero displacement called *nodal lines* and *nodal circles* and are an extension of nodes of vibration described in Example 4.5.1. The bending modes are categorized by the number of diametral nodal lines and the number of nodal circles. The notation $(2, 0)$, for example, indicates that the mode has two diametral nodal lines and no circular nodal lines other than the nodal line at the clamped inner edge. This disk is an axisymmetric structure and can be analyzed as such using the methods described in the previous section. In that case, the *harmonic number* or *Fourier order*, N, used in any one analysis, produces results for N nodal diameters.

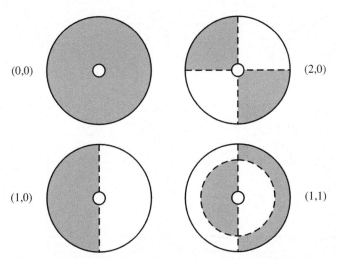

Figure 10.11. Disk vibration showing diametral and circular nodal lines. At a particular instant, the shaded regions of the disk have a positive displacement and the unshaded regions have a negative displacement (i.e., they are out-of-phase with one another). The outer edge of the disk is free.

EXAMPLE 10.4.1. A solid thin disk has an outside diameter of 200 mm, a thickness of 3 mm, and an inside diameter of 20 mm. The material has a density of 2,790 kg/m^3, a Young's modulus of 80 GPa, and a Poisson's ratio of 0.27. If the disk is clamped at its inside edge to a rigid shaft, determine the natural frequencies of the disk for modes having zero, one, two, and three nodal diameters as a function of rotational speed up to 10,000 rev/min. Assume that the shaft cannot undergo any translations or rotations about any axis parallel to the plane of the disk.

Solution. An FE mesh of the disk is created using eight-node axisymmetric quadrilateral elements. The thickness is divided into three layers in the axial direction and the radial distance is divided into eight, as Figure 10.12 indicates. The radial heights of elements close to the center of the disk are deliberately smaller than those of elements close to the edge. Initially, a static calculation is carried out to determine the set of nodal deflections in the disk when the disk is spinning at 1,000 rad/s. The maximum radial deflection computed is 6.225×10^{-6} m. At the outside surface, the disk contracts axially by 5.4×10^{-8} m and the radial and circumferential deflections of every node in the

Figure 10.12. FE mesh of the disk on its own (Example 10.4.1). The dots indicate the nodes and the dot–dash line indicates the shaft centerline.

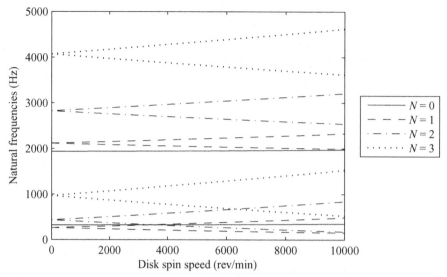

Figure 10.13. The natural frequency map in the stationary frame for the flexible disk (Example 10.4.1). For each pair (with equal natural frequencies for the nonrotating disk), the higher natural frequency is forward-whirling and the lower natural frequency is backward-whirling.

mesh are known. By scaling these displacements by the square of rotational speed, the maximum radial displacement at 10,000 rev/min is approximately 6.826×10^{-6} m. Because these deformations are small compared to the size of the disk, spin-softening and geometric-stiffening are not important in determining the equilibrium deformations of the disk under centrifugal loading. We can assume that the stress state within the disk at any given speed, Ω, is directly proportional to Ω^2.

Using the FE model of the disk, the natural frequencies are computed for $N = 0, 1, 2,$ and 3 and for a range of rotational speeds up to 10,000 rev/min. Figure 10.13 shows the resulting natural frequency map. Figures 10.14 through 10.17 show the first flexural disk modes for each of the cases $N = 0, 1, 2,$ and 3, respectively.

10.4.2 Analysis of Rotor–Disk Assemblies

We now turn to the question of analyzing rotor–disk assemblies. Clearly, these are axisymmetric structures and, as such, they have sets of modes corresponding to all harmonic numbers. For all $N > 1$, a solid shaft tends to have extremely high stiffness. As such, the modes associated with individual harmonic numbers tend to involve significant movement at only one disk at a time and, as such, results for such modes can be generated for individual disks separately.

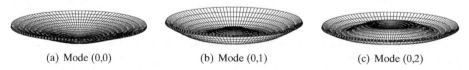

(a) Mode (0,0)   (b) Mode (0,1)   (c) Mode (0,2)

Figure 10.14. Mode shapes of the disk for $N = 0$ (Example 10.4.1).

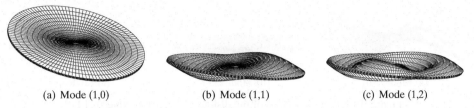

(a) Mode (1,0)   (b) Mode (1,1)   (c) Mode (1,2)

Figure 10.15. Mode shapes of the disk for $N = 1$ (Example 10.4.1).

Lee and Chun (1998) analyzed one system using the *assumed-modes method* (Meirovitch, 1967). The assumed-modes method is closely related to FE modeling of a structure insofar as the total displacement pattern of the structure is considered a linear combination of a finite number of basis-displacement patterns. In the case of FE models, these basis-displacement patterns coincide with the shape functions for the individual elements, whereas for the assumed-modes method, the basis-displacement patterns are specified differently for each problem. Because the user customizes the functions in the assumed-modes method, it is usually possible to obtain good results using significantly fewer degrees of freedom than is needed for an FEA. However, the FE approach has the advantage of versatility. In fact, the procedure of model reduction commonly applied in conjunction with FE models has the net effect of producing a small basis of relatively smooth displacement functions. This removes some of the arbitrary choice that the user of the assumed-modes method otherwise would make.

EXAMPLE 10.4.2. Figure 10.18 shows the rotor used as an example by Lee and Chun (1998). It consists of a shaft 0.8 m long and 20 mm in diameter supported by identical isotropic bearings at each end. The shaft carries two identical disks, 0.3 and 0.5 m from one end of the shaft. The disks are rigidly attached to the shaft and have an outside diameter of 200 mm and a thickness of 3 mm. The shaft and disks are made from a material with a density of 2,790 kg/m^3 and a Young's modulus of 80 GPa. Each one is identical to the disk in Example 10.4.1 except for the boundary condition at the inside diameter. Each isotropic bearing has a stiffness of 3 MN/m and a damping coefficient of 100 Ns/m. Determine the resonance frequencies of this structure as a function of the rotational speed for speeds up to 40,000 rev/min.

Solution. An FE mesh of the rotor is created. The elements used are four-node axisymmetric quadrilateral elements. Figure 10.19 shows the detail of the mesh around the shaft–disk joint. Along the shaft away from the joint, the elements are allowed to become longer in the axial direction. Along the disks in the radial

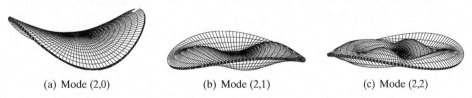

(a) Mode (2,0)   (b) Mode (2,1)   (c) Mode (2,2)

Figure 10.16. Mode shapes of the disk for $N = 2$ (Example 10.4.1).

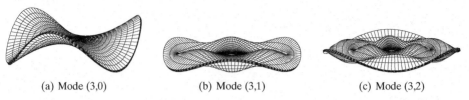

(a) Mode (3,0)   (b) Mode (3,1)   (c) Mode (3,2)

Figure 10.17. Mode shapes of the disk for $N = 3$ (Example 10.4.1).

direction, the element sides are allowed to become larger as the radius increases. A total of 196 elements are present in this model and there are 689 nodes. A model containing 2,067 degrees of freedom (i.e., three per node) is sufficient for the rotor if it is not spinning. Applying model reduction reduces this number to 60 degrees of freedom without seriously impacting the computed resonances. When the rotor is stationary, the first eight resonance frequencies of the system that involve transverse deflections of the shaft center are 44.76, 185.9, 239.6, 266.1, 588.2, 773.5, 1,551, and 1,996 Hz.

In Example 10.4.1, the resonance frequencies of the spinning disk were studied (with its inside diameter clamped). The only disk modes that couple with the shaft lateral dynamics are those with one diametral nodal line (i.e., the modes shown in the lower-left and lower-right diagrams of Figure 10.11 and displayed in oblique form in Figure 10.15). These $N = 1$ modes have resonance frequencies of 265, 2,116, 6,026, and 6,190 Hz when the disk is not spinning. Comparing these disk-resonance frequencies with the top speed of interest for this rotor (i.e., 667 Hz), it is obvious that disk flexibility must be considered. It is useful to analyze the $N = 1$ resonances of one disk for the range of speeds in this case before looking at the characteristics of the rotor as a whole.

Figure 10.20 shows the variation in the frequency of the $(1, 0)$ disk mode when the disk spins about its axis of symmetry for speeds up to 40,000 rev/min. The separation of the pair of frequencies as the speed increases is due to Coriolis coupling, and the increase in frequency of the average of the pair is due to centrifugal stiffening. Although the lower frequency initially decreases with speed due to Coriolis effects, centrifugal stiffness (a term proportional to speed squared) eventually overcomes the Coriolis effects (directly proportional to speed) at a high enough speed and the lower frequency rises. The resonance frequencies in Figure 10.20 are in the stationary frame. The higher-resonance frequency of each pair corresponds to a forward-whirling mode and the lower resonance frequency to a backward-whirling mode.

Having considered the dynamics of an individual flexible disk rigidly clamped at its inner diameter, it is logical now to consider the rotor system, assuming that the disks are rigid. To obtain the rigid disk results, it is assumed that the disk is made from an imaginary material with a density of 2,790 kg/m^3

Figure 10.18. Cross section of the rotor studied by Chun and Lee (1998) (Example 10.4.2).

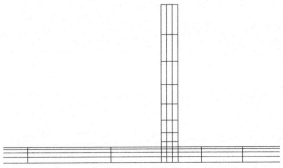

Figure 10.19. Mesh for the example rotor at the shaft–disk junction (Example 10.4.2).

and a Young's modulus of 80,000 GPa (1,000 times higher than the real material from which the flexible disk is made). Figure 10.21 shows the Campbell diagram for the rotor with rigid disks, whereas Figure 10.22 shows the Campbell diagram with flexible disks.

A comparison of Figures 10.21 and 10.22 shows that the flexibility of the disks has a significant effect on the dynamics of the system, and its influence cannot be neglected. To clarify the effect of the disk flexibility, we must look at the mode shapes of the system, shown in Figure 10.23. These mode shapes are for the shaft when stationary (and they do not change greatly with speed). Mode 1 shows virtually no disk vibration, which explains why including disk flexibility in the analysis has virtually no effect on the first mode frequency. There is a small amount of disk vibration in mode 2, and neglecting the disk flexibility changes the frequency of the second mode from 186 to approximately 193 Hz.

Modes 3 and 4 are not present when the disks are assumed to be rigid. In these modes, the disk vibration is significant and close to the frequency of the first mode of vibration of the disk alone (i.e., 269 Hz). Because we have two identical disks, they vibrate out-of-phase in mode 3 at a frequency of 252 Hz and in-phase in mode 4 at a frequency of 279 Hz. In both of these modes, the disk vibration has coupled with the shaft so that the shaft also vibrates. This disk

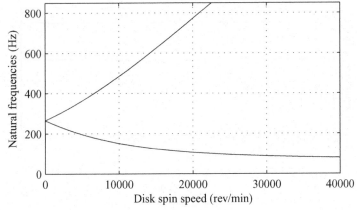

Figure 10.20. The (1,0) disk natural frequencies (forward and backward), in the stationary frame, of a spinning flexible disk on a rigid shaft.

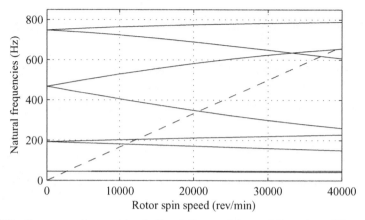

Figure 10.21. Resonance frequencies, in the stationary frame, of the rotor of Example 10.4.2 with rigid disks.

vibration, for example, can cause fatigue problems. Finally, mode 5 is closely related to mode 3 of the shaft with rigid disks, although the natural frequency has increased from 453 to 546 Hz and some disk vibration is evident.

The crossover of resonance frequencies in the region of 550 Hz for speeds between 13,000 and 15,000 rev/min (shown by vertical dashed lines in Figure 10.22) accounts for an interesting change in the nature of the mode shapes. Figure 10.24 shows the four modes in the region in order of frequency at these two particular speeds.

10.5 Detailed Models for Axial Vibration

In some cases, the analysis of the axial vibrations of a rotating machine requires more detail than a shaft-line model can provide. The axial vibration of blade stages is an important example. Here, we illustrate the use of a detailed axisymmetric FE model on a simple example.

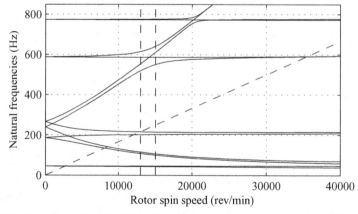

Figure 10.22. Resonance frequencies, in the stationary frame, of the rotor of Example 10.4.2 allowing for disk flexibility. The vertical dashed lines are at 13,000 and 15,000 rev/min.

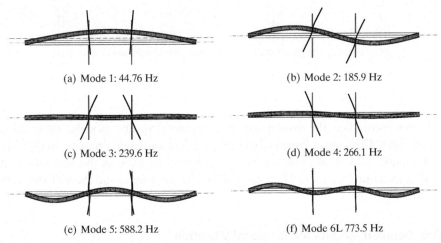

(a) Mode 1: 44.76 Hz

(b) Mode 2: 185.9 Hz

(c) Mode 3: 239.6 Hz

(d) Mode 4: 266.1 Hz

(e) Mode 5: 588.2 Hz

(f) Mode 6L 773.5 Hz

Figure 10.23. First five mode shapes for a stationary rotor with flexible disks (Example 10.4.2).

EXAMPLE 10.5.1. Use an axisymmetric FE model to examine the axial dynamics of the system described in Example 9.2.1.

Solution. Figure 10.25 shows a mesh of the cross section of this rotor. The first four (axial) resonance frequencies computed for this model are 157.03, 682.14, 965.71, and 1,070.56 Hz. From the 12-element shaft-line model of Example 9.3.1, the corresponding natural frequencies are 163.91, 880.57, 1,590.8, and 2,077.2 Hz. As expected, the resonance frequencies computed here are lower than those computed using the shaft-line models. Figure 10.26 shows the computed second mode shape from this model. The disks deform significantly,

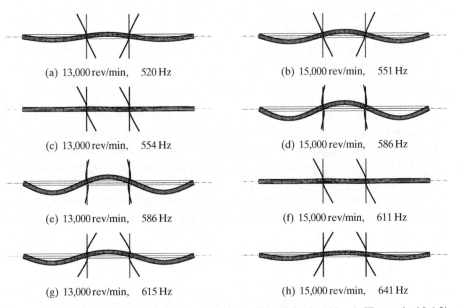

(a) 13,000 rev/min, 520 Hz

(b) 15,000 rev/min, 551 Hz

(c) 13,000 rev/min, 554 Hz

(d) 15,000 rev/min, 586 Hz

(e) 13,000 rev/min, 586 Hz

(f) 15,000 rev/min, 611 Hz

(g) 13,000 rev/min, 615 Hz

(h) 15,000 rev/min, 641 Hz

Figure 10.24. Mode shapes of the rotor with flexible disks for $N = 1$ (Example 10.4.2) at rotor spin speeds of 13,000 rev/min (on the left) and 15,000 rev/min (on the right). The corresponding natural frequency is given in Hz.

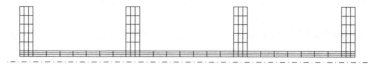

Figure 10.25. Mesh for an axisymmetric finite element model of the rotor in Examples 10.5.1 and 10.6.1.

which contradicts the assumption made in the shaft-line models that the disks are rigid. This causes large errors in the second and higher natural frequencies. Example 9.2.1 shows that the rotor deforms very little in the first mode and that its natural frequency is controlled by the rotor mass and the stiffness of the thrust bearing.

10.6 Detailed Models for Torsional Vibration

Section 10.5 explains that, in at least some cases, it is important to consider more detailed models of axial vibration than can be provided for with shaft-line models. This is also true for torsional vibration.

EXAMPLE 10.6.1. Consider the dynamics of the torsional vibration of the system described in Example 9.5.1 using a detailed axisymmetric FE model. Use the mesh shown in Figure 10.25.

Solution. Using the detailed FE model, the lowest resonance frequencies for the torsional modes are computed as 0, 109.66, 202.06, 263.27, 4,282.7, 4,292.4, and 4,348.5 Hz. From Example 9.5.1, the natural frequencies obtained from a shaft-line model with six elements are 0, 122.81, 226.98, 296.64, 5,869.7, 5,876.7, and 5,883.6 Hz.

Figure 10.27 shows the deformation within one cross section of the rotor for the 263.27 Hz mode. This suggests that the disks are behaving much like rigid objects but the frequency discrepancy indicates that there is significant deformation within the disks. The following example exhibits a more striking degree of nonrigid behavior of the rotor disks.

EXAMPLE 10.6.2. Figure 10.28 shows the cross section of a rotor. The rotor has a uniform inside diameter of 0.070 m and a length of 1.1 m. Its outer profile can be described through a set of pairs of dimensions, each representing outer diameter and length, respectively (in m). These pairs are (0.25, 0.20), (0.10, 0.25), (0.35, 0.10), (0.10, 0.05), (0.18, 0.20), (0.10, 0.20), and (0.30, 0.10). The modulus of rigidity for this material is 78.7 GPa and its density is 7,850 kg/m^3. Calculate the torsional resonance frequencies and modes of this rotor.

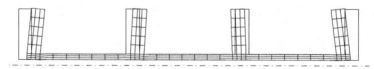

Figure 10.26. Second axial mode shape of the rotor from Example 10.5.1 evaluated using a detailed axisymmetric finite element model.

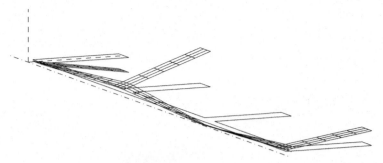

Figure 10.27. Illustration of the 263.27 Hz torsional mode for the rotor of Example 10.6.1.

Solution. Running an axisymmetric model for the mesh described in Figure 10.28 returns the following natural frequencies: 0, 302.21, 432.74, 1,302.3, 5,255.4, and 6,391.1 Hz. If a four-inertia shaft-line model had been set up for this system, where all the torsional flexibility was assumed to reside in the narrower parts of the rotor, then only four resonance frequencies would be computed: 0, 352.15, 501.07, and 1,784.0 Hz. In this case, it is evident that significant amounts of torsional flexibility are contributed by the disks. Figure 10.29 illustrates the torsional mode at 432.74 Hz. In this figure, the distortion of what were originally radial planes within the disks is clear.

10.7 Rotors Consisting of a Flexible Cylinder

Rotors consisting of a flexible cylinder cannot be analyzed with shaft-line models because the cross section of the rotor will distort. We demonstrate typical effects with an example using axisymmetric solid elements.

EXAMPLE 10.7.1. A small experimental electrical machine has a rotor system as shown in Figure 10.30. The rotor has an overall length of 87.5 mm, of which 53.5 mm is a uniform cylinder with an outside diameter 85.5 mm. This uniform cylindrical section consists of a steel shell with a wall thickness of 2.5 mm inside which magnets are fixed (i.e., the shaded gray area in Figure 10.30). The layer of magnets is 4.2 mm thick. A transition section joins the steel cylindrical section to a flat backplate of 3 mm thickness, and the backplate is joined to a shaft that is held firmly in bearings.

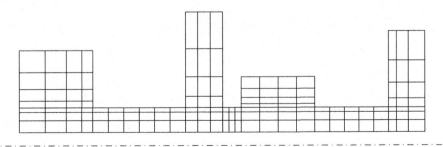

Figure 10.28. Cross section of the rotor and the finite element mesh (Example 10.6.2).

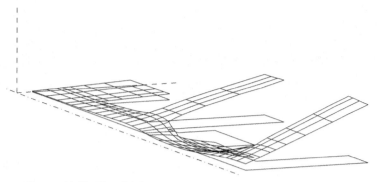

Figure 10.29. The 432.74 Hz mode of the rotor (Example 10.6.2).

The properties for the steel are given by

$$\rho = 7,800 \, \text{kg/m}^3, \quad E = 210 \, \text{GPa}, \quad \nu = 0.285, \quad \text{and} \quad G = E/(2(1+\nu))$$

The set of magnets is not continuous around the inner circumference and an appropriate set of material properties is given by

$$\rho = 4,500 \, \text{kg/m}^3, \quad E_r = E_z = 50 \, \text{GPa}, \quad E_t = 0.05 \, \text{GPa},$$

$$\nu_{rz} = 0.3, \quad \nu_{rt} = \nu_{tz} = 0, \quad G = E_r/(2(1+\nu_{rz}))$$

Subscripts r, t, and z indicate the radial, tangential, and axial directions, respectively.

Find the resonance frequencies and mode shapes for rotational speeds up to 40,000 rev/min.

Solution. The rotor is modeled using 200 8-node isoparametric axisymmetric solid elements, as shown in Figure 10.30. There are 666 nodes, giving the model 1,998 degrees of freedom. In the model, the shaft constraint is represented by springs of stiffness 1,000 GN/m at each one of the nodes at the inside diameter of the shaft. The first stage of the analysis is to compute the state of stress in the rotor when it is spinning at a low speed. Figure 10.31 shows an exaggerated view of the deformations of the rotor under the effects of centrifugal loading. The maximum radial deflection is 5.75×10^{-6} m at 1,000 rad/s. By scaling by Ω^2, the maximum radial deflection at 40,000 rev/min is approximately 0.1 mm. We therefore can assume that spin-softening and geometric-stiffening do not

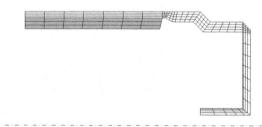

Figure 10.30. Cross section of the rotor of a small experimental electrical machine and the finite element mesh (Example 10.7.1).

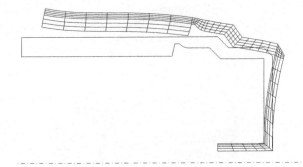

Figure 10.31. Deformation of the rotor under prestress (Example 10.7.1).

affect the steady-state solution and that the stresses are directly proportional to the square of rotational speed. The stress-stiffening matrix therefore can be computed only once and scaled thereafter for different spin speeds.

Standard Guyan model reduction is applied to reduce the number of degrees of freedom to 60 for the calculation of resonances, and the same model-reducing transformation is applied to the stress-stiffening matrix.

Figure 10.32 shows three modes in cross section for each of $N = 0$, $N = 1$, $N = 2$, and $N = 3$ with frequencies reported for zero rotational speed. Figure 10.33 shows the deformed shape for the $N = 3$ mode at 9,251 Hz in a three-dimensional view. This illustrates how powerful the axisymmetric model for

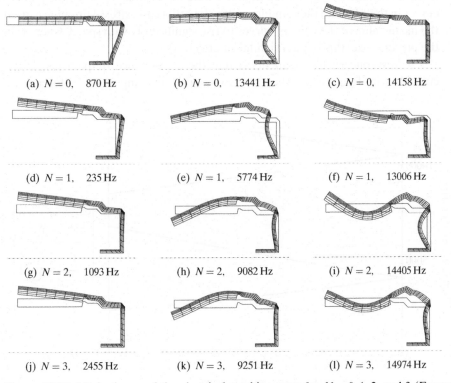

(a) $N = 0$, 870 Hz

(b) $N = 0$, 13441 Hz

(c) $N = 0$, 14158 Hz

(d) $N = 1$, 235 Hz

(e) $N = 1$, 5774 Hz

(f) $N = 1$, 13006 Hz

(g) $N = 2$, 1093 Hz

(h) $N = 2$, 9082 Hz

(i) $N = 2$, 14405 Hz

(j) $N = 3$, 2455 Hz

(k) $N = 3$, 9251 Hz

(l) $N = 3$, 14974 Hz

Figure 10.32. Mode shapes of the electrical machine rotor for $N = 0, 1, 2$, and 3 (Example 10.7.1) for zero rotational speed. The caption gives the values of N and the natural frequency.

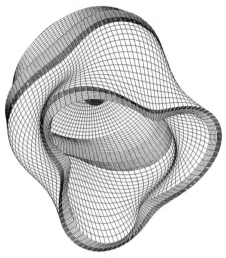

Figure 10.33. A three-dimensional view of the $N = 3$ mode at 9,251 Hz (Example 10.7.1).

such rotors is in terms of being able to represent relatively complex deformation shapes.

For each different value of N, the resonance frequencies (in either the stationary or rotating frame) can be plotted as a function of the rotational speed of the rotor. Figure 10.34 shows the variation of the natural frequencies in the stationary frame for $N = 0, 1, 2, 3,$ and 4; note the contrast between the pairs of curves in this plot and those in Figure 10.22. In that case, stress-stiffening caused the mean value of the pairs of curves to rise significantly with speed, whereas in the present case, this is not a significant effect.

A key concern for engineers is that these resonance frequencies do not coincide with frequencies where the corresponding components of forcing are

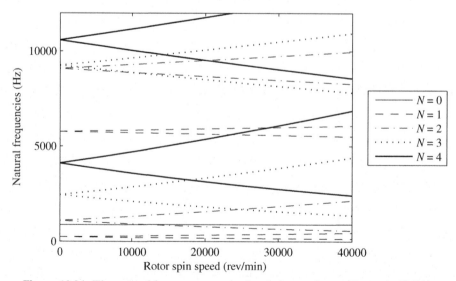

Figure 10.34. The natural frequency map in the stationary frame (Example 10.7.1).

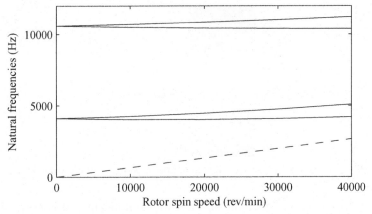

Figure 10.35. Resonance frequencies in the rotating frame of the $N = 4$ modes as a function of rotational speed (Example 10.7.1). The dashed line represents forcing at 4Ω.

substantial. In the present case, for example, the stationary part of the electrical machine (located inside the magnets) exerts a substantial force in proportion to $\sin(4\phi)$, where ϕ is an angle measured in the stationary reference frame. This appears to the rotor as a backward-rotating forcing at frequency 4Ω; it is important that at no point in the speed range should this forcing frequency coincide with the resonance frequency (in the rotating frame) of $N = 4$ modes. Figure 10.35 shows the variation of the rotor resonances with respect to speed for speeds between 0 and 40,000 rev/min.

10.8 Bending Vibrations of Blades Attached to Rotors

A significant proportion of all rotors have blades of some type attached to them. Usually, the blades are mounted on a disk or hub, and we speak of a *bladed-disk assembly*. Steam, gas and water turbines and many types of pumps have blades that range from short and stiff to relatively long and flexible. Wind turbines have particularly long and flexible blades.

The frequency and mode shapes of the blade vibrations are influenced not only by the material and geometric properties of the blade but also by the spin speed, the hub radius, and the setting angle. The setting angle is the angle of inclination of the blade from the plane of rotation, as shown in Figure 10.36. Consider only vibration of the blade due to bending in its most flexible direction. When the vibration is completely out of the plane of rotation ($\alpha = 0$), it is called a *flapwise vibration*. If the vibration is completely in the plane of rotation ($\alpha = 90°$), it is called a *lead-lag* or *spanwise vibration*.

To analyze rotating blades by the FEM, it is necessary to develop the centrifugal-stiffness matrix for a beam element. Hoa (1979) developed this matrix for a uniform Euler-Bernoulli beam element. The paper also accounts for a tip mass. Yokoyama (1988) extended this approach to a Timoshenko beam element. Khulief and Yi (1988) provided tables of frequencies for various blade configurations. Khulief (1989) developed an FE model for a tapered blade with a tip mass based on Euler-Bernoulli theory. Khulief and Bazoune (1992) and Bazoune and Khulief

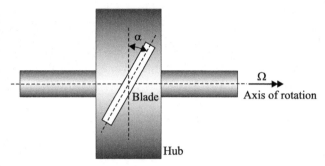

Figure 10.36. Geometry of the blade attachment. α is the setting angle.

(1992) developed and provided the matrices for rotating tapered Timoshenko beam elements and examined the influence of various boundary conditions. Bazoune et al. (1999) provide further modal characteristics of a rotating blade on a hub.

In the analysis that follows, it is assumed that lateral displacements of the blade are small and the hub is rigid. Coriolis effects are insignificant and therefore neglected, but centrifugal-stiffening and spin-softening are considered.

The stiffness of a stationary beam element, including the effect of shear and rotary inertia, is derived from the strain energy due to bending and shear (see Section 5.4.2 and Equation (5.41)). However, for a rotating beam element, we also must include a contribution to the strain energy due to the centrifugal force. In Section 5.4.3, the stiffness matrix due to an axial load is developed. The development of the stiffness matrix due to a centrifugal force, $\tilde{\mathbf{K}}_c^e$, is similar except that the force is not constant but rather proportional to the square of the spin speed, the distance from the axis of rotation, and the distribution of mass. This extra term is sometimes referred to as *stress-stiffening* (or *centrifugal stiffening*). Thus, the total strain energy of a rotating beam element is of the form

$$U^e = \frac{1}{2}\tilde{\mathbf{q}}^{e\top}\left(\tilde{\mathbf{K}}^e + \Omega^2\tilde{\mathbf{K}}_c^e\right)\tilde{\mathbf{q}}^e \tag{10.37}$$

where $\tilde{\mathbf{K}}^e$ is the stiffness matrix for the element and accounts for the bending and shear stiffness. The mass matrix for the rotating element is not affected by the rotation and consists of the two matrices given by Equation (5.45). For the purposes of this analysis, it is necessary to separate the matrices due to translation and rotation, $\tilde{\mathbf{M}}_t^e$ and $\tilde{\mathbf{M}}_r^e$, respectively. Thus, the kinetic energy of the element is given by

$$T^e = \frac{1}{2}\dot{\tilde{\mathbf{q}}}^{e\top}\tilde{\mathbf{M}}^e\dot{\tilde{\mathbf{q}}}^e \tag{10.38}$$

where

$$\tilde{\mathbf{M}}^e = \tilde{\mathbf{M}}_t^e + \tilde{\mathbf{M}}_r^e \tag{10.39}$$

Finally, in developing the matrices for the element, we must include the potential energy due to the centrifugal force per unit volume acting on the beam element. This term affects only the component of vibration in the plane of rotation. Thus, for a blade with a setting angle of α, the resultant element matrix is

$$W^e = -\frac{1}{2}\tilde{\mathbf{q}}^{e\top}\left(\Omega^2\tilde{\mathbf{M}}_t^e\sin^2\alpha\right)\tilde{\mathbf{q}}^e \tag{10.40}$$

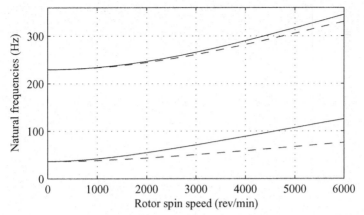

Figure 10.37. Variation of first and second natural frequencies of a cantilever blade with spin speed (Example 10.8.1). Blade-setting angle $\alpha = 0$ (solid line) and $\alpha = \pi/2$ (dashed line).

Assembling these matrices leads to the following equation of motion:

$$\tilde{\mathbf{M}}\ddot{\tilde{\mathbf{q}}} + \left\{ \tilde{\mathbf{K}} + \Omega^2 \left(\tilde{\mathbf{K}}_c - \tilde{\mathbf{M}}_t \sin^2 \alpha \right) \right\} \tilde{\mathbf{q}} = \mathbf{0} \qquad (10.41)$$

The term $\Omega^2 \tilde{\mathbf{M}}_t^e \sin^2 \alpha$ is the spin-softening term and it reduces the stiffness of the system. For flapping vibration, $\alpha = 0$ and, hence, the spin-softening term vanishes. The stress-stiffening is accounted for by the term $\Omega^2 \tilde{\mathbf{K}}_c$. Assuming a harmonic solution, this equation becomes an eigenvalue problem and its solution provides the blade natural frequencies and mode shapes.

Although the application of Equation (10.41) allows the natural frequencies of a rotating blade to be determined, it does not consider any interaction between the flexible blade and the flexible disk and shaft. This was studied by Chun and Lee (1996), who used the assumed-modes method to model the blades, disk, and shaft and to provide examples of modes and frequencies of bladed rotors. Genta and Tonoli (1997) used a harmonic FE approach to model blade arrays.

EXAMPLE 10.8.1. A uniform blade – 20 mm wide, 4 mm thick, and 300 mm long – is mounted on a hub of 150 mm radius. Determine the first and second natural frequencies of the blade as the blade hub spins up to a maximum speed of 6,000 rev/min for blade-setting angles of $\alpha = 0$ and $\alpha = \pi/2$. For the blade, $E = 205$ GPa, $\nu = 0.3$, and $\rho = 7,850$ kg/m^3.

Solution. The blade is modeled using a uniform Timoshenko beam element described by Yokoyama (1988). Twelve elements, each 25 mm long, are used in the analysis.

Figure 10.37 shows how the first and second natural frequencies vary with spin speed. The figure shows that the centrifugal-stiffening increases the natural frequencies as the spin speed increases. When the setting angle $\alpha = \pi/2$, the increase in frequency is reduced (compared to a setting angle of $\alpha = 0$) because of the effect of spin-softening. Figure 10.38 shows how the mode shape also changes as the spin speed is increased. Compared to the mode shape at rest, the spinning blade has a distorted modal shape. There is no difference between

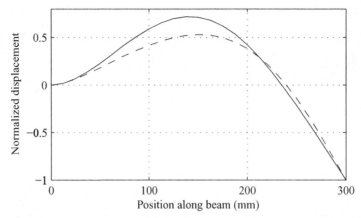

Figure 10.38. Second mode shape of a rotating cantilever beam, with a setting angle $\alpha = \pi/2$ (Example 10.8.1). Beam stationary (solid line) and spinning at 10,000 rev/min (dashed line).

the resonance frequencies for forward- and backward-traveling modes in this case because the high radial stiffness of the blade prevents Coriolis forces from having a significant effect on the bending vibration.

10.9 Coupled Systems

In Chapters 5, 6, and 9, we assume that the lateral, torsional, and axial vibration of a rotor can be considered independent of one another. This is often justified, but there are circumstances in which interactions occur between these different classes of vibration. For example, it is not uncommon for some light coupling to exist between the axial and torsional vibrations so that torsional resonances are detectable in the axial system and vice versa. Interaction between lateral vibrations and either axial or torsional vibrations can occur only when a given rotor is not cyclically symmetrical or is coupled to another component in a way that is not cyclically symmetric. For example:

- If the mass center of a shaft does not coincide with its shear center, coupling occurs between the torsional and lateral vibration (Al-Bedoor, 2001; Meirovitch, 1967; Wu and Yang, 1995). Thus, an open crack in a shaft causes coupling, which has been the subject of numerous research papers. Coupling between lateral and torsional vibration is discussed by Ostachowicz and Krawczuk (1992); between lateral and axial vibration by Papadopoulos and Dimarogonas (1987); and among lateral, torsional, and axial vibrations by Darpe et al. (2004). Turhan and Bulut (2006) considered the coupling between the torsional vibration of the shaft and the bending vibration of the blades in turbo-machines.
- When two shafts are joined together by gears, coupling can occur between the lateral and torsional vibration (Ananda Rao et al., 2003; Lee et al., 2003) and among the lateral, axial, and torsional vibration (Hu et al., 2002; Luo et al., 1996). Rao et al. (1998) considered the lateral response due to torsional excitation of geared rotors.

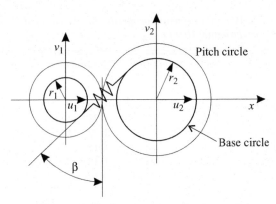

Figure 10.39. Arrangement of gear mesh.

- Rubbing between a rotor and stator couples torsional and lateral vibration (Al-Bedoor, 2000; Edwards et al., 1999).
- A misaligned gear coupling between rotors couples torsional and lateral vibration (Li and Yu, 2001).

We now examine in detail how a pair of parallel spur gears in a gearbox can couple lateral and torsions vibration. The analysis is based on a paper by Rao et al. (1998). In this work, the contact force and displacements of a gear pair are related by representing the teeth in contact by an equivalent stiffness along the pressure line, as shown in Figure 10.39.

Along the pressure line, gear 1 moves $(u_1 \sin \beta + v_1 \cos \beta + r_1\phi_1)$ and gear 2 moves $(u_2 \sin \beta + v_2 \cos \beta - r_2\phi_2)$, where r_1 and r_2 are the base circle radii of the two gears, β is the pressure angle (shown in Figure 10.39), ϕ_1 and ϕ_2 are the torsional displacements of the gears about the z-axis, and – for example – u_2 is the displacement of gear 2 in the x direction. Hence, the relative displacement at the gear mesh is

$$x_g = (u_2 - u_1) \sin \beta + (v_2 - v_1) \cos \beta - (r_1\phi_1 + r_2\phi_2) = \mathbf{N}_g \mathbf{q}_g \qquad (10.42)$$

where

$$\mathbf{N}_g = \begin{bmatrix} -S & -C & 0 & 0 & S & C & 0 & 0 & -r_1 & -r_2 \end{bmatrix}, \qquad (10.43)$$

$S \equiv \sin \beta$, $C \equiv \cos \beta$, and

$$\mathbf{q}_g = \begin{bmatrix} u_1 & v_1 & \theta_1 & \psi_1 & u_2 & v_2 & \theta_2 & \psi_2 & \phi_1 & \phi_2 \end{bmatrix}^\top$$

The definition of $\mathbf{q}_g$ includes the rotations θ_1, ψ_1, θ_2, and ψ_2 to be consistent with the coordinate system in Chapter 5.

The strain energy within the gear mesh is then

$$U_g = \frac{1}{2} k_g x_g^2 = \frac{1}{2} k_g \mathbf{q}_g^\top \mathbf{N}_g^\top \mathbf{N}_g \mathbf{q}_g$$

Table 10.5. *Shaft-section properties for the coupled rotor (Example 10.9.1)*

Nodes	Section length (m)	Section diameter (m)
1–2	0.10	0.30
2–3	4.24	0.30
3–4	1.16	0.22
4–5	0.30	0.22
6–7	0.30	0.15
7–8	5.00	0.15
8–9	0.10	0.15

where k_g is the gear-mesh stiffness. Thus, the stiffness matrix is

$$\mathbf{K}_g = k_g \mathbf{N}_g^{\mathsf{T}} \mathbf{N}_g$$

$$= k_g \begin{bmatrix} S^2 & SC & 0 & 0 & -S^2 & -SC & 0 & 0 & r_1 S & r_2 S \\ SC & C^2 & 0 & 0 & -SC & -C^2 & 0 & 0 & r_1 C & r_2 C \\ 0 & 0 & 0 & 0 & 0 & 0 & 0 & 0 & 0 & 0 \\ 0 & 0 & 0 & 0 & 0 & 0 & 0 & 0 & 0 & 0 \\ -S^2 & -SC & 0 & 0 & S^2 & SC & 0 & 0 & -r_1 S & -r_2 S \\ -SC & -C^2 & 0 & 0 & SC & C^2 & 0 & 0 & -r_1 C & -r_2 C \\ 0 & 0 & 0 & 0 & 0 & 0 & 0 & 0 & 0 & 0 \\ 0 & 0 & 0 & 0 & 0 & 0 & 0 & 0 & 0 & 0 \\ r_1 S & r_1 C & 0 & 0 & -r_1 S & -r_1 C & 0 & 0 & r_1^2 & r_1 r_2 \\ r_2 S & r_2 C & 0 & 0 & -r_2 S & -r_2 C & 0 & 0 & r_1 r_2 & r_2^2 \end{bmatrix} \quad (10.44)$$

It is clear from this matrix that there is coupling between the lateral displacements u_i and v_i and the torsional displacements ϕ_i.

EXAMPLE 10.9.1. The two rotors shown in Figure 10.40 are connected by a pair of spur gears and supported by flexible bearings (Rao et al., 1998). The rotors have the following material properties: $E = 207\,\text{GPa}$, $G = 79.5\,\text{GPa}$, and $\rho = 7{,}800\,\text{kg/m}^3$. Table 10.5 lists the shaft-section properties, Table 10.6 lists the disk properties, and Table 10.7 lists the bearing properties. Bearing damping is modeled but, for clarity, the equivalent dashpots are not shown in Figure 10.40.

Using these data, develop an FE model with nine nodes. Compute the natural frequencies of the system for a generator (node 2) speed of 1,500 rev/min for

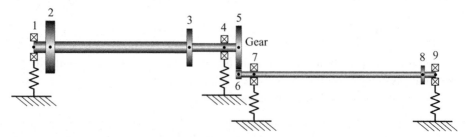

Figure 10.40. Rotors connected by a pair of spur gears (Example 10.9.1). The dots and numbers correspond to the nodes.

Table 10.6. *Disk properties for the coupled rotor (Example 10.9.1) (N is the number of gear teeth)*

Node	Mass (kg)	I_d (kg m^2)	I_p (kg m^2)	r (m)	N
2 (Generator)	525.70	16.10	32.2		
3	116.04	3.115	6.23		
5 (Gear)	726.40	56.95	113.9	0.5086	328
6 (Gear)	5.00	0.002	0.004	0.03567	23
8 (Turbine)	7.45	0.0745	0.149		

contact stiffnesses of 1, 10, and 100 MN/m (1 MN/m is unrealistically small for a gear stiffness but is used here for illustration purposes). Show the Campbell diagram and the response to an unbalance at node 2 for a contact stiffness of 1 MN/m.

Solution. Because each of the nine nodes has five degrees of freedom, the model has 45 degrees of freedom in total. Table 10.8 lists the natural frequencies of the system for the three different contact stiffnesses when the generator runs at 1,500 rev/min.

Figure 10.41 shows the Campbell diagram for this rotor system with a tooth-contact stiffness of $k_g = 10$ MN/m. The frequencies corresponding to modes that are primarily torsional are easily recognized because they are unaffected by the spin speed.

In their paper, Rao et al. (1998) computed the lateral and torsional response due to a short-circuiting torque in the generator. For simplicity in this book, we determine the torsional and lateral response to an out-of-balance in the generator. Figure 10.42 shows the lateral response of the generator and turbine due to unbalance acting on the generator. Figure 10.43 shows the torsional response at the generator and turbine due to this unbalance force. The figures show that due to the coupling in the gears, a lateral force produces a torsional response.

10.10 Rotor–Stator Contact in Rotating Machinery

Unwanted contact between the rotating and stationary parts of a rotating machine – more commonly referred to as *rub* – is a serious problem that is identified regularly as a primary mode of failure in rotating machinery. Rub typically may be caused by mass unbalance, turbine or compressor blade failure, defective bearings and/or

Table 10.7. *Bearing properties for the coupled rotor (Example 10.9.1)*

Node	k_{xx} (MN/m)	k_{yy} (MN/m)	c (Ns/m)
1	183.9	200.4	3,000
4	183.9	200.4	3,000
7	10.10	41.60	3,000
9	10.10	41.60	3,000

Table 10.8. *Natural frequencies (Hz) for the coupled rotor (Example 10.9.1) (The rotor speed at the generator is 1,500 rev/min)*

Mode	$k_g = 1\,\mathrm{MN/m}$	$k_g = 10\,\mathrm{MN/m}$	$k_g = 100\,\mathrm{MN/m}$
1	6.790	11.616	11.641
2	11.682	12.151	12.284
3	12.676	16.816	17.268
4	17.358	17.852	18.458
5	18.528	18.769	23.956
6	36.159	36.769	37.681
7	49.319	49.637	49.889
8	50.791	50.823	50.861
9	56.228	56.383	56.248
10	56.379	57.911	57.752
11	57.946	58.897	59.188
12	59.383	59.608	63.113
13	71.992	72.834	74.203

seals, or rotor misalignment. It therefore is surprising that the quantity of analysis undertaken concerning this relatively common rotordynamic problem is small, in comparison with some of the better-understood faults encountered, such as mass unbalance or cracked shafts. The reason for this discrepancy may be the intrinsically complicated nature of the problem. Several different physical events may occur during a period of contact between rotor and stator: initial impacting stage, frictional behavior between the two contacting parts, and an increase in the stiffness of the rotating system while contact is maintained, to name only three examples. The behavior of the system during this contact period is highly nonlinear and may be chaotic.

An important aspect of the contact event is the generation of heat caused by the friction between the rotor and the stator. Where the rub is primarily caused by vibration due to unbalance, the motion is synchronous and the point of contact

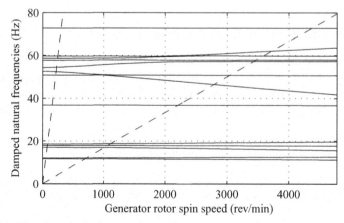

Figure 10.41. The Campbell diagram for the coupled rotor (Example 10.9.1) with tooth-contact stiffness $k_g = 10\,\mathrm{MN/m}$, as a function of generator rotational speed. The dashed lines show the shaft rotational speeds corresponding to the generator (node 2) and turbine (node 8).

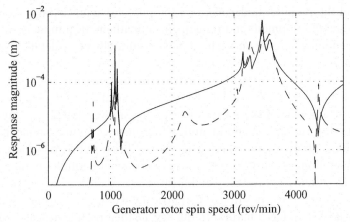

Figure 10.42. Lateral response for the coupled rotor as a function of generator rotational speed (node 2) (Example 10.9.1) at nodes 2 (solid) and 8 (dashed) due to unbalance at node 2. The tooth-contact stiffness is $k_g = 10 \, \text{MN/m}$.

rotates around the stator but is stationary with respect to the rotor. The heat generated causes the rotor to bend, but because this bend is unlikely to be in-phase with the original unbalance, there is a slowly varying synchronous vibration. This was first discussed by Newkirk (1926) and is known as the *Newkirk effect*. The thermal-bending of the rotor causes a slow change in the resultant unbalance on the rotor, which causes a polar plot of the vibration to appear as a spiral; for this reason, such vibrations are sometimes called *spiral*. The time scales involved in these changes can vary considerably; however, on power turbines, they are often about 20 minutes. Kellenburger (1980) undertook a detailed study and his model was extended by Sawicki et al. (2003). Eckert et al. (2006) applied similar arguments to a slightly different situation: the rubbing of slip-rings. They illustrated that in the case of slip-rings (i.e., brush gear), this phenomenon is an inherent part of a machine's operation.

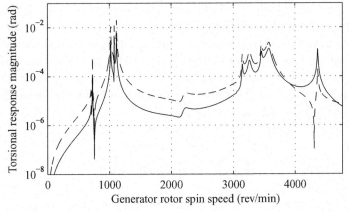

Figure 10.43. Torsional response for the coupled rotor as a function of generator rotational speed (node 2) (Example 10.9.1) at nodes 2 (solid) and 8 (dashed) due to unbalance at node 2. The tooth-contact stiffness is $k_g = 10 \, \text{MN/m}$.

Most of the research undertaken on the purely dynamical analysis of rubbing centers on the development and refinement of mathematical models, allowing the rubbing phenomenon to be understood more accurately. For a detailed overview of the rubbing process and associated literature, readers are referred to the thorough review provided by Muszyńska (1989). Thermal effects, friction, impact, coupling, stiffening, analysis, and vibrational response were included; the paper was the first comprehensive review of the subject. Choy and Padovan (1987) developed a simplified analytical rub model and identified four distinct regimes: noncontact stage, rub initiation, rub interaction, and separation. Ehrich (1988) considered the subharmonic vibration response. Large responses at integer multiples of the natural frequencies of a system are caused by nonlinearities, typically due to asymmetry of the radial stiffness in a rotor positioned eccentrically within its bearings. For instance, if a nonlinear system has a natural frequency, ω, then pseudocritical responses are seen at forcing frequencies of 2ω, 3ω, 4ω, and so on. Ehrich (1992) used the same model to study the chaotic and subcritical super-harmonic response.

Goldman and Muszyńska (1994a) contrasted three different modeling approaches: a classical restitution coefficient-based approach, an inelastic impact with a zero restitution coefficient and sliding, and a discontinuous piecewise approach. They concluded that the resulting dynamics depended significantly on the impact model used. Goldman and Muszyńska (1994b) investigated the effects of order and chaos in a system with contacting components. Experimental data and predicted responses both displayed characteristics of orderly periodic and fully chaotic responses. Further analysis was provided by Choy et al. (1988, 1993), Chu and Zhang (1997, 1998), and von Groll and Ewins (2001, 2002). Edwards et al. (1999) investigated the effect of torsional flexibility on the lateral response.

The analytical model utilized in this work is a development of that provided by Ehrich (1988, 1992), who demonstrated good correlation with previous experimental work by Bentley (1974) and Muszyńska (1984) and with two different sets of real data collected from operational aircraft engines. The basis of the model is a Jeffcott rotor model (see Section 3.9). The rotor is housed within the clearance of the stator, and the rubbing phenomenon of interest arises when contact occurs between the rotor and stator. The local stiffness of the stator is taken to be much higher than that of the rotor, and the system is nonlinear because of this piecewise linear-spring stiffness. The friction that occurs during contact produces a moment that opposes the rotation of the machine, but here it is assumed that the rotor speed remains constant. Edwards et al. (1999) considered the case when this assumption is not valid.

The equations of motion for the system shown in Figure 10.44 are given by the standard equations for a Jeffcott rotor as

$$
\begin{aligned}
m\ddot{u} + c\dot{u} &= m\varepsilon\Omega^2 \cos(\Omega t) + f_x \\
m\ddot{v} + c\dot{v} &= m\varepsilon\Omega^2 \sin(\Omega t) + f_y
\end{aligned}
\tag{10.45}
$$

where f_x and f_y denote the force in the Ox and Oy directions due to the shaft stiffness and the contact and $m\varepsilon$ is the unbalance (i.e., the product of mass and eccentricity). The shaft centerline is displaced vertically downward from the stator

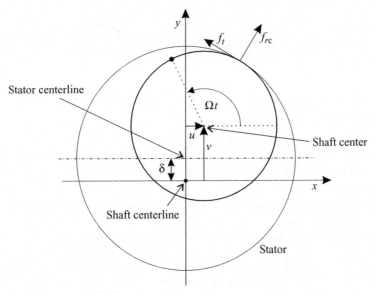

Figure 10.44. The contacting nonlinear rotor–stator system.

centerline by δ, as shown in Figure 10.44. The shaft centerline represents the stationary position of the shaft. The force on the rotor arises from two sources: (1) the stiffness of the rotor, given by k_s; and (2) stiffness due to the rotor–stator contact, given by k_c. This is shown diagrammatically in Figure 10.45. The friction force due to the contact, $-f_t$, is neglected.

The force due to the shaft stiffness is easily derived as $-k_s u$ and $-k_s v$ in the Ox and Oy directions. The force due to the rotor–stator contact is more difficult to derive. The radial displacement of the rotor from the stator centerline is

$$r = \sqrt{(v - \delta)^2 + u^2} \tag{10.46}$$

We assume that when the rotor is at rest, it is not in contact with the stator (i.e., $\delta \leq r_0$). During contact, the stator deforms, which allows the radial displacement, r, to exceed the radial clearance inside the stator, r_0. Then, the force on the rotor due

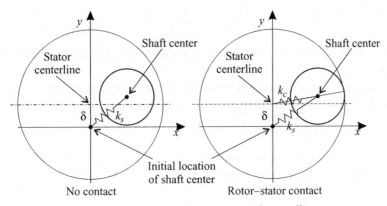

Figure 10.45. The equivalent spring models for the contacting nonlinear rotor–stator system.

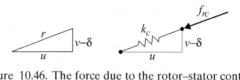

Figure 10.46. The force due to the rotor–stator contact.

to the stator has a magnitude

$$f_{rc} = -k_c (r - r_0) \tag{10.47}$$

The direction of this force is normal to the tangent of the stator and rotor surface at the contact, and the force therefore is directed toward the stator centerline. Hence, the force components in the x and y directions are $f_{rc}u/r$ and $f_{rc}(v - \delta)/r$, as shown in Figure 10.46.

Thus, the forces on the rotor in the Ox and Oy directions are

$$f_x = \begin{cases} -k_s u & \text{for } r \le r_0 \\ -k_s u - k_c (r - r_0) u/r & \text{for } r > r_0 \end{cases} \tag{10.48}$$

$$f_y = \begin{cases} -k_s v & \text{for } r \le r_0 \\ -k_s v - k_c (r - r_0) (v - \delta)/r & \text{for } r > r_0 \end{cases} \tag{10.49}$$

Suppose that we define the force f_r as

$$f_r = \begin{cases} -k_s r & \text{for } r \le r_0 \\ -k_s r_0 - (r - r_0) (k_c + k_s) & \text{for } r > r_0 \end{cases} \tag{10.50}$$

By substituting f_r defined by Equation (10.50) into Equations (10.48) and (10.49), the forces in the x and y directions can be expressed as

$$f_x = f_r \frac{u}{r}$$
$$f_y = f_r \frac{(v - \delta)}{r} - k_s \delta \tag{10.51}$$

The detailed derivation of Equation (10.51) is left as an exercise for readers.

The degree of system nonlinearity is determined by the ratio of local stiffness values when the rotor is free and when the rotor is in contact, given by

$$\beta = \frac{k_s}{k_c + k_s} \tag{10.52}$$

If $k_c = 0$, then the stator stiffness does not affect the dynamics and the system is linear ($\beta = 1$).

The equations may be nondimensionalized using mass eccentricity, ε. Let a caret ($\hat{\ }$) represent a nondimensional quantity so that the displacements become

$$\hat{u} = u/\varepsilon, \quad \hat{v} = v/\varepsilon, \quad \hat{\delta} = \delta/\varepsilon, \quad \hat{r} = r/\varepsilon, \quad \hat{r}_0 = r_0/\varepsilon \tag{10.53}$$

Time is nondimensionalized using the natural frequency of the rotor in contact, $\omega_c^2 = (k_s + k_c)/m$, so that $\tau = \omega_c t$ and

$$\hat{f}_r = \frac{f_r}{(k_s + k_c)\varepsilon}, \quad \hat{\zeta} = \frac{c}{2\sqrt{m(k_s + k_c)}}, \quad \hat{\Omega} = \frac{\Omega}{\omega_c} \tag{10.54}$$

The equations of motion then become

$$\frac{\mathrm{d}^2\hat{u}}{\mathrm{d}\tau^2} + 2\hat{\zeta}\frac{\mathrm{d}\hat{u}}{\mathrm{d}\tau} = \hat{\Omega}^2\cos\left(\hat{\Omega}\tau\right) + \frac{\hat{u}}{\hat{r}}\hat{f}_r$$

$$\frac{\mathrm{d}^2\hat{v}}{\mathrm{d}\tau^2} + 2\hat{\zeta}\frac{\mathrm{d}\hat{v}}{\mathrm{d}\tau} = \hat{\Omega}^2\sin\left(\hat{\Omega}\tau\right) - \beta\hat{\delta} + \frac{\hat{v}-\hat{\delta}}{\hat{r}}\hat{f}_r \tag{10.55}$$

where

$$\hat{f}_r = \begin{cases} -\beta\hat{r} & \text{for } \hat{r} \le \hat{r}_o \\ -\beta\hat{r}_o - (\hat{r}-\hat{r}_o) & \text{for } \hat{r} > \hat{r}_o \end{cases} \tag{10.56}$$

and

$$\hat{r} = \sqrt{\left(\hat{v}-\hat{\delta}\right)^2 + \hat{u}^2} \tag{10.57}$$

The running speeds are expressed using the parameter s, where

$$s = \frac{1}{2}\hat{\Omega}\left(1 + \beta^{-1/2}\right) \tag{10.58}$$

These equations of motion are now used to simulate the nonlinear response of an example system with a rotor-stator rub.

EXAMPLE 10.10.1. Investigate the nonlinear dynamics of a Jeffcott rotor with parameters $\beta = 0.04$, $\hat{\zeta} = 0.01$, $\hat{r}_0 = 10$, and $\hat{\delta} = 8$.

Solution. To demonstrate the effect of including the simple model of the rotor–stator interaction, a numerical simulation is performed and the corresponding steady-state responses are obtained. Runge-Kutta integration is applied to obtain the numerical solutions of Equation (10.55) over a range of running speeds, with 1,000 rotations at each distinct speed. A systematic study was carried out beforehand to verify that this number of rotations was sufficient to eliminate transient effects. The data for the last 200 rotations are stored for postprocessing.

Bifurcation diagrams are an effective means by which results obtained from systems exhibiting nonlinear or chaotic behavior may be presented (see Section 2.7). The diagrams are constructed by plotting the response at wholenumber multiples of the periodic time of the shaft rotation against the desired variable. In this way, periodic behavior results in the repeated plotting of a single point (or points) over time, for each speed increment. Any nonperiodic behavior results in multiple response values being plotted for the speed increment considered. Figure 10.47 shows the bifurcation diagram for vertical displacement as a function of frequency ratio, where the increment in s is 0.01. In addition, Figure 10.48 shows orbit plots obtained at the four speed ratios, $s = 1.2, 1.7, 2.6$, and 2.8.

The trajectory plots included in Figure 10.48 serve to stress the importance of the study. The orbit at $s = 1.2$ is almost a circular, periodic orbit that would not be unusual for a rotor operating near its first critical speed in a linear system. At $s = 1.7$, the corresponding orbit has a significant component at the second harmonic; and at $s = 2.6$, the third harmonic dominates. The response at $s = 2.8$ appears chaotic with no obvious periodic motion.

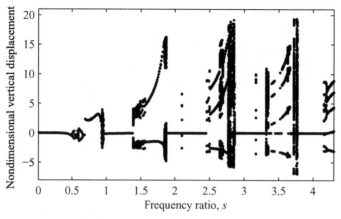

Figure 10.47. Bifurcation plot with varying frequency ratio.

10.11 Alignment

A significant issue when commissioning or recommissioning large multibearing rotating machines, such as turbo-generators, is the problem of alignment. Although it is important to align the bearings of even a simple two-bearing horizontal rotor correctly to allow the bearings to function well, the issue becomes more acute when connecting machines or parts of machines. The question of the *correct* alignment then arises.

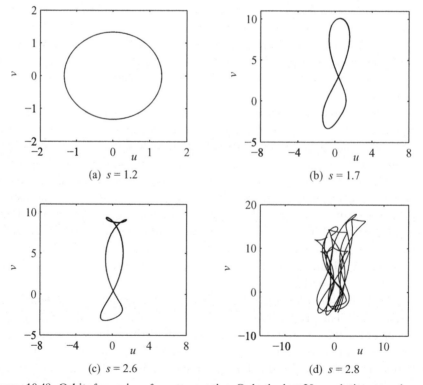

Figure 10.48. Orbits for various frequency ratios. Only the last 20 revolutions are shown.

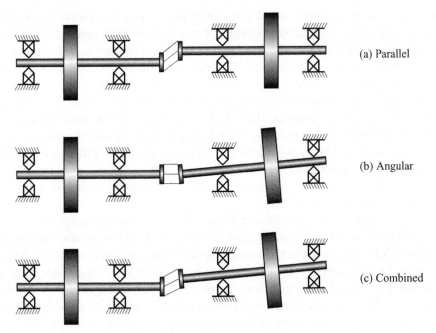

(a) Parallel

(b) Angular

(c) Combined

Figure 10.49. Types of misalignment.

Multibearing horizontal machines may be considered as a series of individual parts the rotors of which are connected. In all cases, the primary requirement is that each bearing takes an appropriate proportion of both the static and dynamic loads, helping to ensure adequate dynamic properties. The distribution of loads between the bearings of such a machine depends strongly on the relative vertical position of the bearings. This alone represents a problem, but the situation is further complicated by relative movements that often occur in service due to changes of temperature and, to a lesser extent, pressure.

Figure 10.49 shows two rotor components that are to be joined via a rigid coupling. In Figure 10.49(a), there is a transverse displacement between the centers of the coupling faces before they are joined. In Figure 10.49(b), there is an angular error between the coupling faces before they are joined. Figure 10.49(c) shows a combination of the two errors. All three illustrations portray *misalignment*. The situation in Figure 10.49(a) is often referred to as *static misalignment* and in Figure 10.49(b) as *dynamic misalignment* (Ertas and Vance, 2002). In reality, either the parallel misalignment of Figure 10.49(a) or the angular misalignment of Figure 10.49(b) justifiably can be termed *static* if the error seen from the stationary frame remains constant while both rotors turn steadily with the same speed. Similarly, either of these misalignments justifiably can be termed *dynamic* if the error seen from the rotating frame remains constant while both rotors turn steadily.

Suppose that individually, the rotor components are geometrically perfect with zero out-of-balance forces. Before the two component rotors are joined, there are transverse displacement and rotation errors at the flanges shown schematically in Figure 10.49. After the component rotors are joined together, some error between the transverse displacement and rotation at the flanges may remain, although these errors are usually reduced. If a coupling has features to force the two flanges to be concentric – such as dowels, fitted-bolts, or spigots – then the transverse displacement and angular errors between coupling faces are zero after joining. In this case, the bearings apply forces on the rotor that are constant relative to a stationary reference frame. These static forces are different than the forces on the rotor at the same bearings prior to joining. Some bearing loads increase and some reduce. Oil-film bearings with a significantly reduced load are a particular concern because this can lead to bearing instabilities. If the transverse displacement and angular errors between coupling faces are the same before and after joining, then the rotor is essentially equivalent to a *bent rotor*. However, there is a significant difference: Because the reason for the misalignment is the position of the bearings, this must be reflected in the machine model. In all real situations, the misalignment present in a machine is a combination of these effects.

In large, high-power machines such as turbo-generators, individual rotors often are rigidly connected by bolting flanges at the ends of each rotor. In this case, we assume that the errors in the transverse displacements and rotations at the flanges are zero. The assembled rotor can be readily modeled as a single rotor with appropriate provision for the mass and inertia of the flange coupling. Nevertheless, there are some problems to examine concerning the alignment of the overall machine. The transverse positions of the bearings are selected to eliminate the cyclic stresses in the shaft and also to provide a good distribution of bearing loads. In turbo-generators, the rotors are both heavy and flexible, implying that each rotor will sag due to gravity. Usually, the uncoupled shafts fail to meet and, if the rotor is coupled in this configuration, high shear forces and bending moments are induced in the shafts. Furthermore, because the resulting stresses are stationary in space, they are cyclic on the rotor, which produces fatigue problems. The forces and moments change the distribution of bearing loads and, hence, alter the dynamic characteristics of the bearings. There are several ways to address this problem. In the case of a solidly coupled system (common in large machines), the vertical levels of the bearings are adjusted so that the location and slopes of neighboring shafts in the uncoupled state are equal. Thus, the coupling faces are parallel and may be bolted together.

An alternative approach is to estimate the load on each bearing and the position of the rotor for a given location of the bearing supports. The bearing supports then may be moved to minimize the cyclic stresses or the response level. If rolling-element bearings are used and are sufficiently stiff to be modeled as constraints, then the deformation in the rotor may be calculated from the rotor model and the load on the bearings calculated. Many large machines are supported on fluid-film bearings and the calculation of misalignment is complicated by the bearing nonlinearities. The approach is demonstrated for short bearings using the models of the rotor and bearings described in Chapter 5. The standard model for short bearings assumes that the load on the bearing is vertical and the magnitude is known. Short-bearing theory then is used to calculate the equilibrium position of the journal in the bearing bush and the stiffness matrix for small perturbations from this position. When the bearing load is unknown, the full nonlinear relationship must be used.

Suppose that the bearings are mounted on a rigid foundation so that all of the degrees of freedom are located on the shaft-line of the rotor. Furthermore, assume that all of the bearings have a finite stiffness so that no degrees of freedom are constrained. Then, the steady-state equation for the rotor is

$$\mathbf{Kq} = \mathbf{Q}_g + \mathbf{Q}_b \qquad (10.59)$$

where $\mathbf{Q}_b$ is the force exerted on the rotor by the bearings and $\mathbf{K}$ represents the effective stiffness matrix for the spinning rotor in isolation. $\mathbf{Q}_g$ is the force due to the weight of the rotor and is

$$\mathbf{Q}_g = \mathbf{Mg} \qquad (10.60)$$

where $\mathbf{M}$ is the mass matrix of the machine, $\mathbf{g}$ is the gravity vector, $\mathbf{g} = -\begin{bmatrix} 0 & g & 0 & 0 & 0 & g & 0 & 0 & \dots \end{bmatrix}^\mathsf{T}$, and g is the acceleration due to gravity. There are two unknown vectors in Equation (10.59): the steady-state nodal displacement vector, $\mathbf{q}$, and the bearing forces, $\mathbf{Q}_b$. These quantities are also linked by the bearing model. Assuming that each bearing is located at a node of the shaft line, then the degrees of freedom related to the ith bearing are selected using the transformation $\mathbf{T}_{bi}$, so that

$$\mathbf{q}_{bi} = \mathbf{T}_{bi}\mathbf{q} \qquad (10.61)$$

The force on the rotor from the ith bearing is then

$$\mathbf{Q}_{bi} = \mathbf{f}_i\,(\mathbf{q}_{bi} - \mathbf{q}_{si}) \qquad (10.62)$$

where $\mathbf{f}_i$ is a nonlinear function and $\mathbf{q}_{si}$ is the offset displacement of the ith bearing. If this nonlinear function were known explicitly, then the result would be a nonlinear algebraic equation for the steady-state displacement, $\mathbf{q}$. However, the model of fluid-film bearings in Section 5.5.1 is the inverse of this function because the load is provided as input and the position of the journal and the stiffness matrix are provided as output. Thus, one possible solution procedure is as follows:

(i) Initially, the bearing loads are calculated assuming that the bearings are very stiff compared to the rotor; therefore, the relative displacements at the bearings are zero. Hence, the rotor displacements at the bearings are equal to the offset displacements of the bearing bushes. Specifying the journal displacements provides constraints, and the bearing loads are obtained as the forces required to ensure these constraints are satisfied using the methods in Section 2.5.

(ii) The model of the rotor and any linear bearings are reduced to those degrees of freedom associated with the fluid-film bearings using dynamic reduction.

(iii) For a given bearing load, the linearized stiffness matrix is calculated using the approach described in Section 5.5.1. The bearing load is not necessarily vertical, so the axes for the stiffness matrix must be rotated.

(iv) The linearized-stiffness matrices of the bearings and the rotor-stiffness matrix are combined to give the full stiffness matrix. The steady-state displacement of the rotor is calculated from this stiffness matrix and the gravity force given by Equation (10.60).

(v) The bearing loads are estimated from the relative displacement at the bearings, using the linearized bearing-stiffness matrix and the bearing load assumed initially. This new estimate of bearing load is used in step (iii) and iterations are performed until the estimated bearing loads converge.

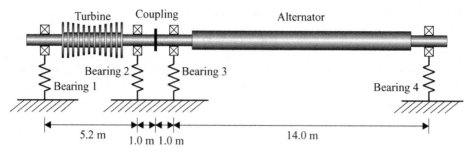

Figure 10.50. The turbine and alternator with four fluid bearings (Example 10.11.1).

(vi) After convergence, the reduction transformation is used to expand the displacements to the full shaft line.

Although this procedure converges if the initial estimate is close to the solution, strategies often are required to obtain convergence from initial estimates that are farther away from the solution. One such strategy is the reduction of the size of the step in early iterations. The use of the linearized-stiffness matrix is equivalent to a Newton-Raphson approach (Lindfield and Penny, 2000) to the optimization.

Adams (2001), Ehrich (1999), and Muszyńska (2005) discussed aspects of misalignment. Gibbons (1976) gave the forces and moments due to parallel misalignment. Al-Hussain and Redmond (2002) considered rigid couplings. Al-Hussain (2003), Sekhar and Prabhu (1995), and Xu and Marangoni (1994a, 1994b) investigated the effects of flexible couplings. Sinha et al. (2004) provided a method to estimate misalignment from rundown data.

> **EXAMPLE 10.11.1.** A single-stage turbine is connected to a large alternator, as shown in Figure 10.50. The bearings are identical with a length of 0.3 m, a diameter of 0.6 m, an oil viscosity of 0.04 Pa s, and a clearance of 0.3 mm. Each rotor in this example is heavy and flexible, implying that both will sag due to gravity. Determine the vertical positions of the bearings to align the rigid coupling.
>
> *Solution.* We begin by calculating the static deflection from the mass and stiffness matrices for the uncoupled system and the bearings are initially at the same (vertical) level. Bearings 1 and 2 are each assumed to carry half of the weight of the turbine, 192 kN, and bearings 3 and 4 are each assumed to carry half of the weight of the alternator, 274 kN. Figure 10.51 shows the steady-state vertical deflection, and the two shafts fail to meet. In the case considered, the turbine rotor bends far less than the alternator. If the rotors are coupled in this configuration, then there will be a very high curvature of the shaft near the coupling, giving high bending moments (and, consequently, high stresses). The resulting steady-state vertical deflection of the shaft is shown in Figure 10.52, which shows that the maximum deflection has reduced slightly. The loads on bearings 1 through 4 are 172, 129, 375, and 256 kN, respectively. This shows that bearing 2 has a significantly reduced load and bearing 3 has a significantly increased load.
>
> The easiest approach to align the rotors for a solid coupling is based on the displacement and slope of the neighboring rotors in the uncoupled state. The

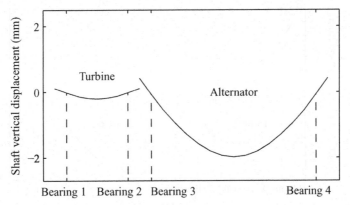

Figure 10.51. The steady-state vertical deflection of the uncoupled turbine and alternator (Example 10.11.1).

aim is to position the rotors so that their coupling faces are at the same height and their cross sections are parallel. This is achieved by varying the heights of the four bearing supports. One option, which is not the best solution in practice, is to fix the displacement and rotation of the coupling to be zero. However, this solution is not unique, and the complete assembly may be translated and rotated. Figure 10.53 shows one solution, in which the sum of squares of the changes in bearing-support positions are minimized. Although the total deflections are considerable, the bending moments in the rotors – when coupled – are reduced significantly.

The discussion thus far considers only the vertical displacement because it is the direction of the force due to gravity. However, there also is a horizontal displacement because of the horizontal forces on the journal arising from the oil film. Figure 10.54 shows the steady-state horizontal displacement when the turbine and alternator are coupled, but the bearing supports are not offset. This horizontal displacement corresponds to the vertical displacement shown in Figure 10.52 and highlights that the machine should be aligned in the horizontal direction as well as vertically.

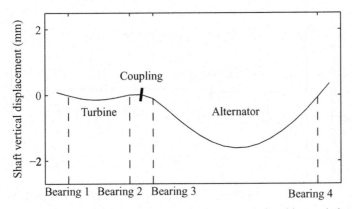

Figure 10.52. The steady-state vertical deflection of the coupled turbine and alternator (Example 10.11.1).

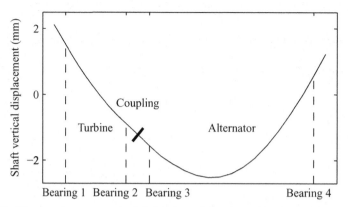

Figure 10.53. The steady-state vertical deflection of the aligned and coupled turbine and alternator (Example 10.11.1).

10.12 Nonlinear Bearings, Oil Whirl, and Oil Whip

In reality, all physical phenomena are nonlinear to some extent, and linear models are simply approximations to this reality. However, linear models frequently provide acceptable (and sometimes excellent) predictions for the behavior of real systems. Situations arise in which the nonlinear effects are so significant that linear models are either inadequate or, at least, do not tell the full story. Nonlinearities can arise for a number of reasons including large strains, large deflections, joint behavior, faults such as cracked rotors and rotor–stator rubs, and devices such as squeeze-film dampers.

The damping and stiffness matrices for a rotor–bearing–foundation system can become nonlinear functions of displacement and velocity, giving

$$\mathbf{M}\ddot{\mathbf{q}} + \left\{ \Omega\mathbf{G} + \mathbf{C}\left(\mathbf{q}, \dot{\mathbf{q}}\right) \right\} \dot{\mathbf{q}} + \mathbf{K}\left(\mathbf{q}\right)\mathbf{q} = \mathbf{Q}_{ub} - \mathbf{Q}_g \qquad (10.63)$$

where $\mathbf{Q}_{ub}$ and $\mathbf{Q}_g$ are vectors of out-of-balance and gravitational forces, respectively.

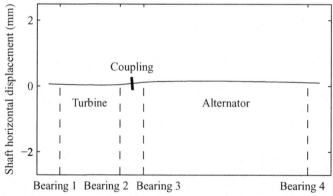

Figure 10.54. The steady-state horizontal displacement of the coupled turbine and alternator (Example 10.11.1).

Solving equations of this type can be a relatively slow and expensive process. An approach used frequently is to seek a solution in the time domain using a time-stepping procedure (see Section 2.6.1). An alternative is the harmonic balance approach that assumes a solution in the form of a series of harmonic terms, the coefficients of which must be determined (see Section 2.7).

Hydrodynamic journal bearings are subject to a form of self-excited vibration that results in an instability called *oil whirl*. The frequency of oil whirl is observed to be slightly less than half the rotor spin speed; hence, this instability is often called *half-speed whirl*. Oil whirl and oil (or resonant) whip are generally discussed together because half-speed whirl often leads to the more serious condition of *oil whip*. When the rotor speed is approximately twice a system natural frequency, the frequency of the half-speed whirl coincides with and becomes locked to the natural frequency. If oil whip then occurs, its frequency is equal to this natural frequency. After this, the vibration frequency remains constant irrespective of the rotor speed. The amplitude of oil whip can be significant and potentially destructive.

Oil whip has been observed since the early 1920s (Newkirk and Taylor, 1925). Verstege (1998) presents a case study of a machine that started to exhibit oil whirl and oil whip when the bearings were changed. Generally, oil whirl and oil whip are more likely to occur in high-speed, lightly loaded journal bearings running with a very low value of eccentricity. Oil whirl can be overcome by reducing the length (and, if possible, the diameter) of the bearing, by increasing bearing clearance, by using lower-viscosity oil, or by introducing oil grooves in the bearing. Alternatively, a different type of bearing can be used, such as a tilting-pad bearing or a lemon-bore bearing (i.e., one with a slightly elliptical bore).

Before giving a specific example of a nonlinear bearing model, we consider a linear model for oil whirl.

10.12.1 Oil Whirl

Although oil-film bearings are generally nonlinear, linear models can be developed for small vibrations (see Section 5.5.1). We consider the phenomenon of half-speed whirl using a series of simple linear models. The analysis of a nonlinear model is provided in Section 10.12.2. For many rotors supported by oil-film bearings, half-speed whirl can be predicted by a simple analysis.

Consider a symmetric, light, horizontal, rigid rotor, supported at the ends by identical oil-film bearings and with a mass of $2m$. Because the rotor mass is small, the bearings are lightly loaded and are operating with small eccentricity. Thus, by letting $\varepsilon \to 0$ in Equation (5.87), the stiffness and damping matrices for each bearing become

$$\mathbf{K}_e = \frac{f}{c}\begin{bmatrix} 0 & 1/\varepsilon \\ -1/\varepsilon & 0 \end{bmatrix} \quad \text{and} \quad \mathbf{C}_e = \frac{f}{c\Omega}\begin{bmatrix} 2/\varepsilon & 0 \\ 0 & 2/\varepsilon \end{bmatrix} \tag{10.64}$$

where $f = mg$. Because the mass of the rotor is small, the inertia forces are negligible compared to the stiffness and damping forces at low frequencies. Thus, for each bearing

$$\mathbf{C}_e \begin{Bmatrix} \dot{u} \\ \dot{v} \end{Bmatrix} + \mathbf{K}_e \begin{Bmatrix} u \\ v \end{Bmatrix} \approx \begin{Bmatrix} 0 \\ 0 \end{Bmatrix} \tag{10.65}$$

Letting $u = u_0 e^{st}$ and $v = v_0 e^{st}$ and substituting in the previous equation gives

$$\left(\frac{f}{c\Omega} \begin{bmatrix} 2/\varepsilon & 0 \\ 0 & 2/\varepsilon \end{bmatrix} s + \frac{f}{c} \begin{bmatrix} 0 & 1/\varepsilon \\ -1/\varepsilon & 0 \end{bmatrix} \right) \begin{Bmatrix} u_0 \\ v_0 \end{Bmatrix} = \begin{Bmatrix} 0 \\ 0 \end{Bmatrix} \tag{10.66}$$

Thus,

$$\det \begin{bmatrix} 2s/\varepsilon\Omega & 1/\varepsilon \\ -1/\varepsilon & 2s/\varepsilon\Omega \end{bmatrix} = 0 \tag{10.67}$$

and, hence, $\left(\dfrac{2s}{\varepsilon\Omega} \right)^2 + \left(\dfrac{1}{\varepsilon} \right)^2 = 0$, giving $s = \pm {}_J\Omega/2$ or $\omega_n = \Omega/2$. Thus, the natural frequency, or frequency of the whirl, is half the shaft speed. The associated eigenvectors are

$$\begin{Bmatrix} u_0 \\ v_0 \end{Bmatrix} = \begin{Bmatrix} 1 \\ \mp {}_J \end{Bmatrix}$$

and, therefore, the mode shape is backward-whirling and circular (see Equation (3.34) and the associated discussion).

This is a simple model but does demonstrate that the half-speed whirl is primarily a phenomenon resulting from the fluid bearing. We assume that the bearing is short and can be linearized and that the shaft is infinitely rigid. In this simple model, the whirl speed is exactly half the shaft speed and the system is neither damped nor unstable. Neither of these facts accord exactly with experimental observation.

The following example repeats this analysis for a specific system. This analysis does not assume that the inertia forces are negligible, nor does it assume that $\varepsilon \rightarrow 0$.

EXAMPLE 10.12.1. A symmetric, rigid rotor of mass m is supported at the ends by two identical oil-film bearings. Considering only symmetric motion, where the displacement at the bearings are identical, calculate the natural frequencies for rotational speeds up to 25,000 rev/min. The systems have the following properties: (a) bearing parameters $L = 0.2$ m, $D = 0.4$ m, $c = 0.3$ mm, $\eta = 0.030$ Pa s; rotor mass m equal to 200, 400, 1,000, 2,000, 4,000 or 10,000 kg; and (b) bearing parameters $L = 15$ mm, $D = 60$ mm, $c = 0.2$ mm, $\eta = 2.25 \times 10^{-2}$ Pa s; rotor mass m equal to 0.4, 0.6, 1.0, 1.6, 2.4, and 4 kg.

Solution. We solve

$$\begin{bmatrix} m/2 & 0 \\ 0 & m/2 \end{bmatrix} \begin{Bmatrix} \ddot{u} \\ \ddot{v} \end{Bmatrix} + \mathbf{C}_e \begin{Bmatrix} \dot{u} \\ \dot{v} \end{Bmatrix} + \mathbf{K}_e \begin{Bmatrix} u \\ v \end{Bmatrix} = \begin{Bmatrix} 0 \\ 0 \end{Bmatrix}$$

for the range of masses, where $\mathbf{C}_e$ and $\mathbf{K}_e$ are given by Equation (5.87). The downward force on each bearing is $mg/2$.

(a) Figure 10.55 shows the lower whirl frequency, which is close to half-speed over a wide range of rotor masses. The second whirl frequency, not shown, is a forward-whirling mode at a higher frequency and is highly damped. With a rotor mass of 200 kg, the frequency is close to half-speed up to about

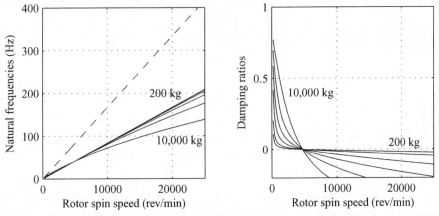

Figure 10.55. The lower natural frequency and associated damping factor for a rigid rotor supported by fluid bearings (Example 10.12.1(a)). The dashed line represents $\omega = \Omega$.

25,000 rev/min and even when the rotor mass is increased to 10,000 kg, the frequency is close to half-speed up to about 8,000 rev/min. For some systems at certain rotational speeds, the natural frequency is in the experimentally observed range of 0.42 to 0.48 times the rotor speed. Also, irrespective of the mass of the rotor, this system becomes unstable at about 5,000 rev/min.

(b) Figure 10.56 shows that for a range of rotor masses, the lower natural frequency is close to half rotor speed; this system also becomes unstable at approximately 6,000 rev/min for a wide range of rotor masses.

The previous example illustrates the behavior of a rigid rotor in oil-film bearings. In the following example, we consider the effect of oil-film bearings on flexible rotors.

EXAMPLE 10.12.2. The flexible rotor described in Example 5.9.6 is supported at each end on an oil-film bearing with the following properties: $L = 40\,\text{mm}$, $D = 80\,\text{mm}$, $c = 60\,\mu\text{m}$, and $\eta = 0.03\,\text{Pa}\,\text{s}$. These bearings were chosen explicitly to

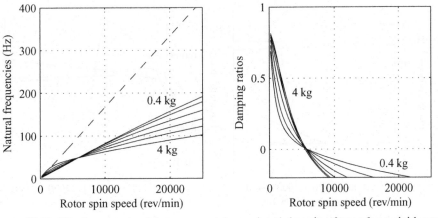

Figure 10.56. The lower natural frequency and associated damping factor for a rigid rotor supported by fluid bearings (Example 10.12.1(b)). The dashed line represents $\omega = \Omega$.

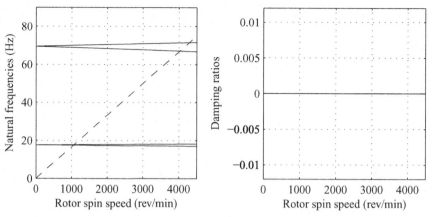

Figure 10.57. Campbell diagram and damping map for the rotor in plain bearings approximated by pinned-pinned supports (Example 10.12.2). The dashed line represents $\omega = \Omega$.

illustrate oil whirl and are unnecessarily large for this rotor. The loads on the bearings arise solely from the total mass of the rotor, which is 107 kg. Calculate the natural frequencies and damping ratios for the system.

Solution. Before considering the system described in the example, we solve the equations of motion for the same rotor simply supported in stiff, self-aligning bearings. Figure 10.57 shows that two natural frequencies, at approximately 18 and 70 Hz, are present in the Campbell diagram. In this simply supported system, the damping factor is zero because there is neither damping nor any source of instability.

Solving the equations of motion for this system supported in the oil-film bearings described in the example, over a range of rotor speeds, gives the Campbell diagram and damping map shown in Figure 10.58. The figure shows that there is a natural frequency equal to half the rotor speed. Using a linear model, vibration at this frequency could not be excited by out-of-balance forces. The actual bearing is nonlinear and out-of-balance force can excite a variety of

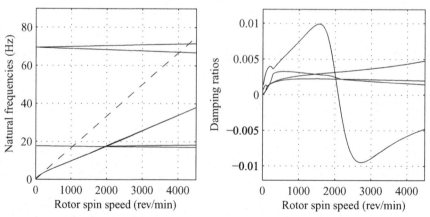

Figure 10.58. Campbell diagram and damping map for the rotor supported by oil-film bearings (Example 10.12.2). The dashed line represents $\omega = \Omega$.

frequency components. However, the linear model can estimate the onset of instability and its initial frequency. The Campbell diagram also shows two frequencies in the region of 18 and 70 Hz, identical to the simply supported model. At approximately 1,970 rev/min, the half-speed natural frequencies are close to the first pair of simply supported natural frequencies of the rotor; above 2,090 rev/min, the system is unstable.

10.12.2 Nonlinear Bearing Models and Oil Whip

Section 10.12.1 demonstrates that a rotating machine with a linear model of oil-film bearings can have a whirl frequency that occurs at approximately half of the rotor speed. This frequency is present over a wide range of rotor speeds and bearing loads. At a particular speed, the machine can become unstable, although we do not show how such an unstable whirl frequency can be excited or how oil whip can be triggered. To do this, we must examine more accurate nonlinear models. However, the linear model can predict the onset of instability and also give an approximate estimate of the frequency of the resulting response.

A nonlinear-bearing model for a short, oil-film journal bearing is presented by Adiletta et al. (1996). They assumed laminar and isothermal fluid flow and derived expressions for the fluid-film forces in terms of the bearing displacements u and v as

$$\begin{Bmatrix} f_x \\ f_y \end{Bmatrix} = \eta \Omega \frac{D}{2} L \left(\frac{D}{2c} \right)^2 \left(\frac{L}{D} \right)^2 \begin{Bmatrix} \hat{f}_x \\ \hat{f}_y \end{Bmatrix} \tag{10.68}$$

where

$$\begin{Bmatrix} \hat{f}_x \\ \hat{f}_y \end{Bmatrix} = -\frac{c\sqrt{(u\Omega - 2\dot{v})^2 + (v\Omega + 2\dot{u})^2}}{\Omega(c^2 - u^2 - v^2)} \begin{Bmatrix} 3V(u/c) - G\sin\alpha - 2S\cos\alpha \\ 3V(v/c) + G\cos\alpha - 2S\sin\alpha \end{Bmatrix} \tag{10.69}$$

In this equation, u and v are the displacements of the rotor in the x and y directions, respectively; η is the oil viscosity; Ω is the speed of rotation of the shaft: and L, D, and c are the bearing length, diameter, and radial clearance, respectively. In Equation (10.69),

$$\alpha = \tan^{-1}\left(\frac{v\Omega + 2\dot{u}}{u\Omega - 2\dot{v}} \right) - \frac{\pi}{2}\text{sign}\left(\frac{v\Omega + 2\dot{u}}{u\Omega - 2\dot{v}} \right) - \frac{\pi}{2}\text{sign}\left(v\Omega + 2\dot{u} \right)$$

$$G(u, v, \alpha) = \frac{2c}{\sqrt{c^2 - u^2 - v^2}} \left\{ \frac{\pi}{2} + \tan^{-1}\left(\frac{v\cos\alpha - u\sin\alpha}{\sqrt{c^2 - u^2 - v^2}} \right) \right\}$$

$$V(u, v, \alpha) = \frac{2c^2 + c(v\cos\alpha - u\sin\alpha)G}{c^2 - u^2 - v^2}$$

$$S(u, v, \alpha) = \frac{c(u\cos\alpha + v\sin\alpha)}{c^2 - (u\cos\alpha + v\sin\alpha)^2}$$

Jing et al. (2005) and de Castro et al. (2008) use this model to demonstrate oil-whirl and oil-whip phenomena. Ding and Leung (2005) investigated the phenomena experimentally.

Thus, for a rotor–bearing system supported by a nonlinear bearing, Equation (10.63) is simplified and becomes

$$\mathbf{M\ddot{q}} + \Omega\mathbf{G\dot{q}} + \mathbf{Kq} = \mathbf{Q}_b\left(\mathbf{q}, \mathbf{\dot{q}}, t\right) + \mathbf{Q}_{ub}\left(\Omega t\right) - \mathbf{Q}_g \qquad (10.70)$$

where $\mathbf{Q}_b$ is a vector of bearing forces including the damping terms, $\mathbf{Q}_{ub}$ is a vector of out-of-balance forces, and $\mathbf{Q}_g$ is a vector of gravitational forces. The stiffness matrix $\mathbf{K}$ only describes the stiffness properties of the rotor. Thus, the rotor is modeled as if it were a free-free rotor and the constraints due to the bearings are accounted for by the forces $\mathbf{Q}_b$. The vectors $\mathbf{Q}_b$, $\mathbf{Q}_{ub}$, and $\mathbf{Q}_g$ each contain a large number of zero terms; care must be taken to ensure that the bearing and unbalance forces are allocated to the correct degrees of freedom.

To solve Equation (10.70) following the approach in Section 2.6.1, either we can use the Newmark-β method, or alternatively we write these equations in the form

$$\begin{Bmatrix} \mathbf{\dot{q}} \\ \mathbf{\dot{v}} \end{Bmatrix} = \begin{bmatrix} \mathbf{0} & \mathbf{I} \\ -\mathbf{M}^{-1}\mathbf{K} & -\Omega\mathbf{M}^{-1}\mathbf{G} \end{bmatrix} \begin{Bmatrix} \mathbf{q} \\ \mathbf{v} \end{Bmatrix} + \begin{Bmatrix} \mathbf{0} \\ \mathbf{M}^{-1}\left(\mathbf{Q}_b\left(\mathbf{q}, \mathbf{v}, t\right) + \mathbf{Q}_{ub}\left(\Omega t\right) - \mathbf{Q}_g\right) \end{Bmatrix} \quad (10.71)$$

These $2n$ first-order, nonlinear equations can be solved using an appropriate numerical ODE solver such as the fourth-order Runge-Kutta procedure. Often, the equations are stiff and either the equations must be reduced or a stiff-equation solver must be used.

EXAMPLE 10.12.3. The flexible rotor described in Example 5.9.6 and used in Example 10.12.2 is supported at each end on an oil-film bearing with the following properties: $L = 40\,\text{mm}$, $D = 80\,\text{mm}$, $c = 60\,\mu\text{m}$, and $\eta = 0.03\,\text{Pa s}$. Using a nonlinear model for the bearings, determine the response of the system due to an out-of-balance mass located on the smaller diameter disk, given as a 1 mm offset of the disk.

Solution. For illustrative purposes, the rotor is adequately modeled with four nodes – one at each bearing and one at each disk – and three beam elements, resulting in a model with 16 degrees of freedom. This model is then reduced to eight degrees of freedom by Guyan reduction (Section 2.5.1), retaining only the translational degrees of freedom. Thus, the nonlinear model consists of eight second-order, nonlinear differential equations or sixteen first-order, nonlinear differential equations. The derivation of the linear rotor model follows the procedure described in Chapters 5 and 6. Because the model is stiff, the model is solved using a stiff equation solver with a variable step size. This time-stepping procedure produces a time series for each coordinate and because we are interested in the steady-state response, the transient must be allowed to decay before further analysis is performed. Figure 10.59 shows the nondimensional, steady-state time response at bearing 1 in the horizontal direction for rotational speeds of 1,600 and 4,000 rev/min. Each plot shows 10 rotations and the main response at 1,600 rev/min clearly is synchronous; whereas at 4,000 rev/min, there is a significant response at a lower frequency. The displacements are nondimensionalized using the radial clearance; hence, all displacements of the rotor at the bearings have a nondimensional displacement with a magnitude of less than 1. This is shown clearly by the orbits given in Figures 10.60 and 10.61, where the orbit stays inside the bearing bush. The orbit at 1,600 rev/min is

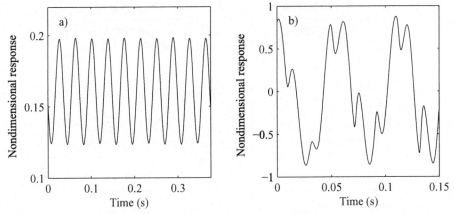

Figure 10.59. Nondimensional steady-state time response in the horizontal direction at the left bearing for the machine with a nonlinear bearing model (Example 10.12.3): (a) 1,600 rev/min, and (b) 4,000 rev/min.

counterclockwise and therefore forward-whirling. The orbit at 4,000 rev/min is predominantly counterclockwise. The response at the disks is much larger than the response at the bearings: at 1,600 rev/min, the disk response is approximately 60 times larger than the bearing response; and at 4,000 rev/min, the disk response is approximately 750 times larger.

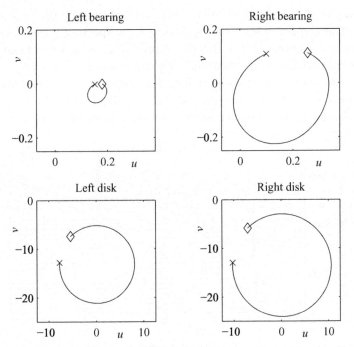

Figure 10.60. Nondimensional steady-state orbits for the machine with a nonlinear bearing model at 1,600 rev/min (Example 10.12.3). The cross denotes the start of the orbit and the diamond denotes the end. The orbit is synchronous and less than one revolution is shown.

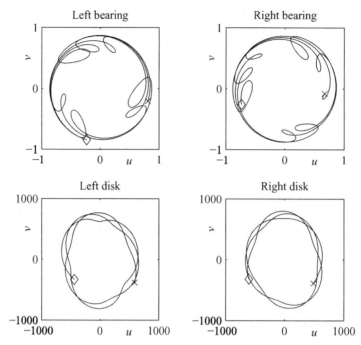

Figure 10.61. Nondimensional steady-state orbits for the machine with a nonlinear bearing model at 4,000 rev/min (Example 10.12.3). The cross denotes the start of the orbit and the diamond denotes the end. The motion for 10 revolutions is shown.

The DFT (see Section 2.6.3) can be applied to these time series to determine their respective frequency spectrum. If the spectra for each of the various speeds are combined, a waterfall plot is produced, as shown in Figure 10.62 for the horizontal direction at bearing 1. The rotor spin speed is increased in increments of 200 rev/min; in a waterfall plot, an appropriately scaled magnitude of the response is added to the spin speed so that the individual responses may be seen clearly. There is a synchronous vibration response at the rotor speed, which

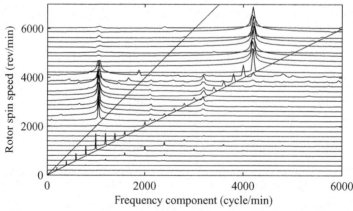

Figure 10.62. Waterfall plot of the horizontal response at bearing 1 for the nonlinear bearing model (Example 10.12.3). The straight lines represent the 0.5X and 1X responses.

becomes large near the critical speeds at approximately 1,000 and 4,000 rev/min. This is predicted by the linear analysis. The figure also shows a large vibration at about 1,000 cycles/min (i.e., 17 Hz) when the rotor spin speed is higher than approximately 2,000 rev/min. This vibration frequency does not change with rotor spin speed and clearly is unrelated to the synchronous excitation. Although no half-speed whirl is apparent in this example, the half-speed whirl appears to cause the oil whip at 1,000 rev/min. For rotor spin speeds above 4,000 rev/min, a second oil-whip response occurs. A small response also appears at the harmonics of the rotor spin speed.

10.13 The Morton Effect

The Morton effect is a synchronous instability that can arise in machines with overhung rotors that are mounted on oil-journal bearings. The phenomenon is basically the excitation of a thermal bend in the shaft; however, whereas the familiar Newkirk effect (see Section 10.10) results from local heating due to a shaft rub, in the Morton effect, heating arises from the viscous shear in the oil film. The effect was first reported by Morton in 1975 (Morton, 2008) and later analyzed more rigorously by Keogh and Morton (1993, 1994). Further work was reported by Balbahadur and Kirk (2002a, 2002b) and Kirk and Guo (2005).

In concept, the effect is straightforward to understand, although precise calculations can be difficult because of the balance of opposing influences. Given some unbalance distribution on a rotor, the orbit at the journals can be readily calculated. Hence, there is some point of minimum clearance that, under some circumstances, rotates around the journal with the shaft. Because the minimum clearance between the rotor and the bearing housing is also the location of greatest shear within the oil film, heat is generated that tends to induce a bend in the shaft. Depending on the relative phases of the unbalance and the deflection, the thermal effects may enhance or reduce the observed vibration response.

The effect has been observed in machines mounted on both journal and tilting-pad bearings. Currently, there is no known general overview of the combination of parameters that exacerbate the effect. Kirk and Guo (2005) concluded that the worst combination is a large circular orbit, with low static load and a small phase-angle difference between the thermal and mechanical sources of unbalance.

10.14 Cracked Rotors

Rotating machines are often subject to high stresses that may cause cracking of the main rotor. Although rotor-cracking is a relatively rare fault, it is potentially dangerous, presenting a severe economic danger in terms of plant integrity as well as a safety issue for the operating staff. Therefore, it is important to understand the characteristics of the vibrational behavior of a cracked rotor, and machines are often monitored to protect against catastrophic failure. Substantial research has been carried out in this area, mainly analyzing the effect of a transverse crack – the most common category – although other crack geometries have been observed.

The behavior of a rotating cracked shaft has been discussed extensively since 1970 with notable early contributions from Davies and Mayes (1984), Dimarogonas (1976), Gasch (1976, 1993), Henry and Okah-Avae (1976), and Mayes and Davies (1976, 1984). Dimarogonas (1996) provided an extensive review of the area. Consider the dynamics of a horizontal rotor with a transverse crack. When the orientation is such that there is a compressive stress across the crack faces due to the self-weight bending (i.e., the shaft deflection due to gravity), the crack tends to be closed and the shaft has its full stiffness (i.e., the crack is effectively absent). However, half a revolution later, the crack tends to open under the action of tensile stresses. The portion of the crack below the rotor's neutral axis is open and this degree of opening determines the dynamic characteristics of the rotor. Thus, as the shaft rotates, its stiffness varies. If operating close to a critical speed, the crack may remain open throughout a full shaft revolution if the unbalance is very high. However, for large machines with modest levels of unbalance and away from critical speeds, the vibration amplitude is significantly less than the shaft deflection due to gravity. With this assumption, the nonlinear equation of motion is converted to a linear equation with time-varying coefficients, thus simplifying the analysis.

The analysis approach first calculates the stiffness of a fully open crack and then incorporates the effect of the crack opening and closing. The reduction in stiffness due to a crack has been modeled in various ways; here, we summarize the analysis of Dimarogonas and Papadopoulos (1983), Papadopoulos and Dimarogonas (1987), and Sekhar (2004), which provided a cracked-shaft element that is easily assembled into the full FE model of the transverse motion of the rotor. The approach uses fracture mechanics to estimate the increased compliance due to the crack in terms of the displacement in the two directions. Others authors included more degrees of freedom; for example, Papadopoulos and Dimarogonas (1987) coupled the longitudinal and bending vibration. Mayes and Davies (1976, 1984) neglected the coupling terms and approximated the increased compliance using the change in second moment of area, validating this assumption with experimental data. Care is needed when applying this technique to study a rotor (e.g., a generator) that has some degree of asymmetry apart from that which is introduced by the crack – a point discussed by Lees and Friswell (2001). Most of the literature is based on a linear analysis; Sinou and Lees (2005) studied the nonlinear behavior of a cracked rotor.

The geometry of a crack is shown in Figure 10.63, in which the crack front is assumed to be straight. The depth of the crack is a and the radius of the shaft section is R. For a fully open symmetric crack, the stiffness in the $\tilde{x}$ and $\tilde{y}$ directions remains uncoupled. The bending compliances in the two directions, $\tilde{x}$ and $\tilde{y}$ (Dimarogonas and Papadopoulos, 1983) are

$$c_{\tilde{x}} = \frac{16(1 - \nu^2)}{\pi E R^8} \iint \tilde{x}^2 \alpha F_1(\alpha/h)^2 \mathrm{d}\tilde{y}\mathrm{d}\tilde{x} \tag{10.72}$$

and

$$c_{\tilde{y}} = \frac{32(1 - \nu^2)}{\pi E R^8} \iint (R^2 - \tilde{x}^2)\alpha F_2(\alpha/h)^2 \mathrm{d}\tilde{y}\mathrm{d}\tilde{x} \tag{10.73}$$

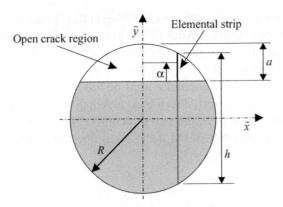

Figure 10.63. The geometry of the cracked shaft.

where the integration is performed over the crack region. We define $\alpha = \tilde{y} - (R - a)$ and $h = 2\sqrt{R^2 - \tilde{x}^2}$, and the Young's modulus is E. The geometric parameters of the crack are

$$F_1(\alpha/h) = \sqrt{\frac{2h}{\pi\alpha}\tan\frac{\pi\alpha}{2h}}\;\frac{0.752 + 2.02(\alpha/h) + 0.37\left[1 - \sin(\pi\alpha/2h)\right]^3}{\cos(\pi\alpha/2h)} \quad (10.74)$$

and

$$F_2(\alpha/h) = \sqrt{\frac{2h}{\pi\alpha}\tan\frac{\pi\alpha}{2h}}\;\frac{0.923 + 0.199\left[1 - \sin(\pi\alpha/2h)\right]^4}{\cos(\pi\alpha/2h)} \quad (10.75)$$

The integration in Equations (10.72) and (10.73) must be performed numerically, although care must be exercised when $a > R$ (Abraham et al., 1994; Dimarogonas, 1994; Papadopoulos, 2004). Figure 10.64 shows the compliance in two directions, nondimensionalized using the factor $\dfrac{(1 - \nu^2)}{ER^3}$, for a range of nondimensional crack depths $\mu = a/R$.

The compliance due to the crack is now added to the compliance of the shaft element. For a fully open symmetric crack, there is no cross-coupling; hence, the

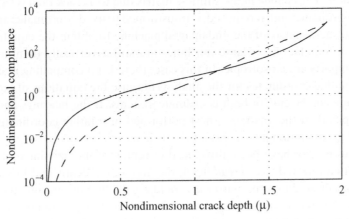

Figure 10.64. The nondimensional direct compliance for a cracked shaft, $c_{\tilde{y}}$ (solid) and $c_{\tilde{x}}$ (dashed), as a function of crack depth $\mu = a/R$.

direct compliance terms in Equations (10.72) and (10.73) may be added directly to
the beam compliance matrix. Sekhar (2004) gave the general approach for a crack
at a node, and Saavedra and Cuitino (2002) considered a crack within an element. If
the coupling terms and the translational compliances are included, then this formal
procedure to include crack compliance must be followed. However, if the element
used to model the crack is short, then a good first approximation is to simply reduce
the stiffness of this element in the two directions (Mayes and Davies, 1984). For
a short element, the majority of the flexibility is due to bending, and the bending
compliance is

$$c_0 = \frac{\ell_e}{E I_0} \tag{10.76}$$

where ℓ_e is the length of the element containing the crack and I_0 is the second moment of area for the uncracked shaft. The total bending compliance of the cracked
shaft element in the $\bar{y}$ direction is obtained as the sum

$$c = c_0 + c_{\bar{y}} \tag{10.77}$$

and the stiffness is easily incorporated into the FE model of the rotor by reducing
the second moment of area in the beam element to

$$I_{\bar{y}} = \frac{\ell_e}{E I_0 c_{\bar{y}} + \ell_e} I_0 \tag{10.78}$$

The procedure in the $\bar{x}$ direction is exactly the same. The length of the section
with reduced properties, ℓ_e, is at the discretion of the modeler within a reasonable
range; however, this choice determines the value of the reduced stiffness. Although
this is a simple model, it incorporates the key reduction in stiffness due to the crack
in the two directions. A real crack has an uncertain geometry and interface mechanics that make it difficult, if not impossible, to model with any degree of accuracy.
Therefore, care must be exercised in using quantitative information from such an
analysis of the dynamics of a machine with a cracked rotor.

The analysis of a rotor–bearing system may be performed in fixed or rotating
coordinates (see Chapter 7). A cracked rotor is asymmetric and the stiffness of the
cracked rotor is more easily determined in coordinates that are fixed to and rotate
with the rotor. The change in the stiffness matrix due to a crack then is calculated in
rotating coordinates and, if required, is transformed to fixed coordinates and merged
with the stiffness matrix of the undamaged machine to obtain the equation of motion. Often, foundations of large machines are stiffer vertically than horizontally
and the supports are anisotropic. In this case, there is no compelling reason to use
fixed or rotating coordinates for the analysis because the equations of motion have
time-varying coefficients in both coordinate systems. If the bearings and supports
are isotropic, then the analysis is best performed in rotating coordinates because
the equations of motion have constant coefficients.

Various models have been introduced to represent the opening and closing of
a crack (often referred to as *breathing* in the literature). Penny and Friswell (2002)
showed that these different approaches yield results that differ only in detail, particularly in the higher harmonic content of the dynamics. Because no two cracks are
identical, the important role of modeling is an understanding of the salient generic
features of a cracked rotor.

Let the stiffness matrix in rotating coordinates for the uncracked rotor be $\tilde{\mathbf{K}}_0$ and the reduction in stiffness due to a crack be $\tilde{\mathbf{K}}_c(\beta)$, where β is the angle between the crack axis and the rotor response at the crack location, which determines the extent to which the crack is open. Thus, the stiffness of the cracked rotor is

$$\tilde{\mathbf{K}}_{cr} = \tilde{\mathbf{K}}_0 - \tilde{\mathbf{K}}_c(\beta) \tag{10.79}$$

To convert this stiffness matrix from rotating to fixed coordinates, we use the transformation matrix, $\mathbf{T}$, which is the generalization of $\mathbf{T}_2$ and $\mathbf{T}_4$ given in Section (7.6). Thus,

$$\mathbf{K}_{cr} = \mathbf{T}(\Omega t)\tilde{\mathbf{K}}_0\mathbf{T}(\Omega t)^\top - \mathbf{T}(\Omega t)\tilde{\mathbf{K}}_c(\beta)\mathbf{T}(\Omega t)^\top = \mathbf{K}_0 - \mathbf{K}_c(\beta, t) \tag{10.80}$$

Let the deflection of the system be $\mathbf{q} = \mathbf{q}_{st} + \mathbf{q}_{dy}$, where $\mathbf{q}_{st}$ is the static deflection of the uncracked rotor due to gravity and $\mathbf{q}_{dy}$ is the dynamic deflection due to the rotating out-of-balance and the effects of the crack. Thus, $\dot{\mathbf{q}} = \dot{\mathbf{q}}_{dy}$ and $\ddot{\mathbf{q}} = \ddot{\mathbf{q}}_{dy}$, and the equation of motion for the rotor in fixed coordinates is

$$\mathbf{M}\ddot{\mathbf{q}}_{dy} + (\mathbf{C} + \Omega\mathbf{G})\dot{\mathbf{q}}_{dy} + (\mathbf{K}_0 - \mathbf{K}_c(\beta, t))(\mathbf{q}_{st} + \mathbf{q}_{dy}) = \mathbf{Q}_u + \mathbf{Q}_g \tag{10.81}$$

where $\mathbf{Q}_u$ and $\mathbf{Q}_g$ are the out-of-balance forces and the gravitational force, respectively. Damping and gyroscopic effects are included as a symmetric positive semidefinite matrix $\mathbf{C}$ and a skew-symmetric matrix $\mathbf{G}$, although they have little direct bearing on the analysis. If there is axisymmetric damping in the rotor, then there also is a skew-symmetric contribution to the undamaged stiffness matrix, $\mathbf{K}_0$. We refer to Equation (10.81) as the *full equations*.

The deflection of the rotor due to gravity varies over each revolution of the rotor because $\mathbf{K}_c$ varies. However, the stiffness reduction due to the crack is usually small, and we may reasonably assume that $\|\mathbf{K}_0\| \gg \|\mathbf{K}_c(\beta, t)\|$. With this assumption, the deflection due to gravity is effectively constant and equal to the static deflection, $\mathbf{q}_{st}$, given by

$$\mathbf{K}_0\mathbf{q}_{st} = \mathbf{Q}_g \tag{10.82}$$

Equation (10.81) then becomes

$$\mathbf{M}\ddot{\mathbf{q}}_{dy} + (\mathbf{C} + \Omega\mathbf{G})\dot{\mathbf{q}}_{dy} + (\mathbf{K}_0 - \mathbf{K}_c(\beta, t))\mathbf{q}_{dy} = \mathbf{Q}_u \tag{10.83}$$

The second approximation commonly used in the analysis of cracked rotors is weight-dominance. If the horizontal rotor system is weight-dominated, it means that the static deflection of the rotor is much greater than the response due to the unbalance or rotating asymmetry; that is, $\|\mathbf{q}_{st}\| \gg \|\mathbf{q}_{dy}\|$. For example, for a large turbine rotor, the static deflection might be of the order of 1 mm; whereas, at running speed, the amplitude of vibration is typically $50\,\mu$m. Even at a critical speed, the allowable level of vibration is only $250\,\mu$m. In this situation, the crack opening and closing are dependent on the rotation only; thus, $\beta = \Omega t + \beta_0$, where Ω is the rotor spin speed and β_0 is a constant. Thus, Equation (10.83) becomes

$$\mathbf{M}\ddot{\mathbf{q}}_{dy} + (\mathbf{C} + \Omega\mathbf{G})\dot{\mathbf{q}}_{dy} + (\mathbf{K}_0 - \mathbf{K}_c(t))\mathbf{q}_{dy} = \mathbf{Q}_u \tag{10.84}$$

where $\mathbf{K}_c$ is now solely a function of time.

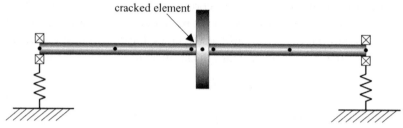

Figure 10.65. The cracked rotor modeled with six elements (Example 10.14.1). The dots indicate the nodes.

For a system with weight dominance, a good approximation to the stiffness-matrix variation of the rotor in the rotating frame is

$$\mathbf{K}_c(t) = f(t)\mathbf{T}(\Omega t)\tilde{\mathbf{K}}_c\mathbf{T}(\Omega t)^\top \qquad (10.85)$$

where $f(t)$ describes the crack opening and $\tilde{\mathbf{K}}_c$ is the change in stiffness due to the fully open crack in the rotating frame. Mayes and Davies (1984) suggested the following function for the crack opening:

$$f(t) = (1 + \cos \Omega t)/2 \qquad (10.86)$$

provided that $t = 0$ defines the downward vertical direction when the crack is fully open. The form of the stiffness variation given by Equations (10.85) and (10.86) means that the rotor responds only at frequencies corresponding to 1X, 2X, and 3X, providing that no other nonlinearity is present (Penny and Friswell, 2002).

Jun et al. (1992) suggested a different approach to calculating the stiffness of a shaft with a breathing crack. They integrated Equations (10.73) and (10.72) only over the part of the crack that is open. Furthermore, there is also a compliance coefficient that couples the two perpendicular directions that must be calculated. The criterion to decide whether a region of the crack surface is open is based on the total stress-intensity factor. Keiner and Gadala (1988) modeled a rotor with a breathing crack using a transient FE analysis with brick elements for the shaft.

Typical effects are now illustrated with an example.

EXAMPLE 10.14.1. Consider a shaft of length 1.5 m and diameter 50 mm that has a disk of diameter 350 mm and thickness 70 mm mounted at the center. The rotor is mounted on flexible supports at each end with a stiffness of 100 kN/m and a damping coefficient of 100 Ns/m. The rotor is modeled with six elements. A crack is located next to the disk, approximately at the midspan. The unbalance, located at the disk, is 0.1 gm. Investigate the response of the rotor for crack depths of $\mu = 0.25, 0.5, 1$. Also consider the effect of the length of the cracked element on the response. The rotor speed is 2,000 rev/min. Assume $E = 211$ GPa and $\rho = 7,810$ kg/m^3.

Solution. Because the bearings and supports are isotropic, the analysis is performed in the rotating coordinates and transformed to the stationary coordinates when required. Figure 10.65 shows the rotor with the six elements. The elements adjacent to the disk have a length of 50 mm, whereas the other elements have a length of 350 mm. This significant difference in the length of adjacent

Table 10.9. *Eigenvalues of the machine at 2,000 rev/min with an undamaged rotor and when the rotor has a fully open crack (Example 10.14.1)*

Mode number	Uncracked (6 elements)	Uncracked (30 elements)	Cracked $\mu = 0.5$
1	$-4.8320 \pm 47.427\jmath$	$-4.8319 \pm 47.427\jmath$	$-4.8035 \pm 47.355\jmath$
2	$-4.8320 \pm 47.427\jmath$	$-4.8319 \pm 47.427\jmath$	$-4.7998 \pm 47.356\jmath$
3	$-49.161 \pm 124.72\jmath$	$-49.156 \pm 124.72\jmath$	$-49.132 \pm 124.70\jmath$
4	$-66.151 \pm 161.56\jmath$	$-66.137 \pm 161.56\jmath$	$-66.151 \pm 161.55\jmath$
5	$-84.831 \pm 478.31\jmath$	$-84.675 \pm 478.15\jmath$	$-84.257 \pm 472.24\jmath$
6	$-84.831 \pm 478.31\jmath$	$-84.675 \pm 478.15\jmath$	$-84.252 \pm 472.28\jmath$

elements is not usually beneficial; however, in this case, the small element at the crack location allows the increased flexibility due to the crack to be positioned accurately. Table 10.9 shows the first six pairs of eigenvalues for this machine modeled with six elements and compares the results to those of a model with 30 elements, each with a length of 50 mm. Clearly, the discretization errors in the six-element model are small. Figure 10.66 shows the Campbell diagram for the uncracked rotor.

We first consider a breathing crack with $\mu = 0.5$. Figure 10.64 shows that the nondimensional compliances are

$$\hat{c}_{\bar{x}}(0.5) = 0.158, \qquad \hat{c}_{\bar{y}}(0.5) = 1.224$$

which enables the calculation of the stiffnesses for the element containing the crack. The stiffness reduction for a fully open crack is 5.3 and 30.4 percent in the $\bar{x}$ and $\bar{y}$ directions, respectively. Table 10.9 also shows the eigenvalues from the six-element model with a fully open crack. The symmetry of the machine and the location of the crack near the center means that the crack predominantly influences the response in the odd-numbered mode pairs (i.e., modes 1, 2, 5, and 6). Furthermore, an open crack with a depth of one-quarter the shaft diameter has only a small effect on the lower natural frequencies.

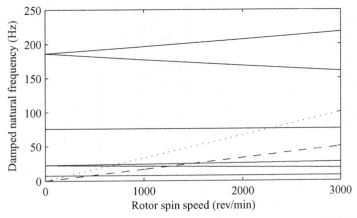

Figure 10.66. The Campbell diagram for the uncracked rotor (Example 10.14.1). The dashed line indicates the rotor spin speed and the dotted line indicates twice the spin speed.

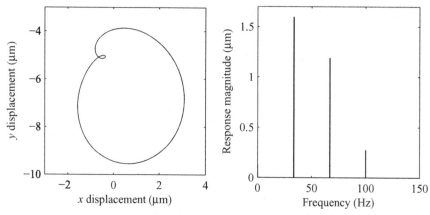

Figure 10.67. The steady-state orbit and frequency response of the cracked rotor (Example 10.14.1), with $\mu = 0.5$.

Now consider the steady-state response. Gravity is modeled as a constant vertical force in the stationary frame, and the force is obtained using the mass matrix. The unbalance force is modeled as a constant force in the rotating frame. The equations of motion in the rotating frame are integrated numerically using a Runge-Kutta fourth-order method, and the response is then transformed to the stationary frame. Figure 10.67 shows the steady-state orbit at the disk in the stationary frame and the frequency content of the horizontal response at the disk. The origin for the orbit is the equilibrium position of the machine with an uncracked shaft. The orbit is always lower than this equilibrium position because the cracked shaft is more flexible than the uncracked shaft. The frequency response highlights that there is a significant response at twice the rotational speed and higher harmonics due to the difference between the stiffness of the cracked shaft in the two directions. This 2X response is often used in practice to diagnose faults in machines.

We now consider a crack that has reached the center of the rotor, $\mu = 1.0$. Figure 10.68 shows the steady-state orbit and the frequency response, which

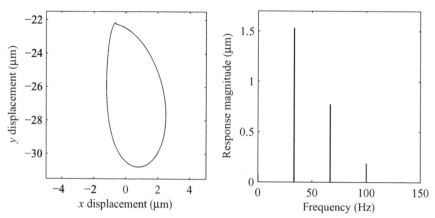

Figure 10.68. The steady-state orbit and frequency response of the cracked rotor (Example 10.14.1), with $\mu = 1.0$.

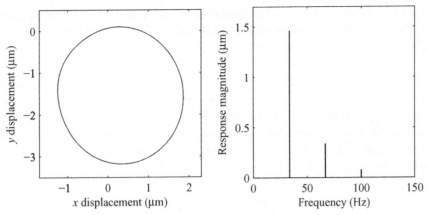

Figure 10.69. The steady-state orbit and frequency response of the cracked rotor (Example 10.14.1), with $\mu = 0.25$.

shows a decrease in the twice-per-revolution vibration. This suprising result is because the stiffness of the cracked shaft is now similar in the two directions (i.e., element stiffness reductions of 61 and 74 percent in the $\tilde{x}$ and $\tilde{y}$ directions, respectively).

Finally, we consider a smaller crack ($\mu = 0.25$). Figure 10.69 shows the steady-state orbit and the frequency response, and it indicates a low relative level of the twice-per-revolution vibration. The orbit is almost circular because the stiffness reductions due to the open crack are small (i.e., 0.42 and 8.0 percent in the $\tilde{x}$ and $\tilde{y}$ directions, respectively). This example highlights that the level of the 2X vibration, relative to the 1X response, is not related in a simple way to crack depth.

Finally, the length of the element containing the crack is increased. Consider the case $\mu = 0.5$, where the response is given by Figure 10.67 for an element of length 50 mm. Now, the shaft is modeled with six elements of equal length (i.e., 0.25 m). The crack is contained in the third element, just to the left of the disk. The stiffness reductions for the longer element are now 1.1 and 8.0 percent in the $\tilde{x}$ and $\tilde{y}$ directions, respectively. The predicted response of the machine changes little; for example, the third pair of eigenvalues is now $-84.508 \pm 474.00\jmath$ and $-84.504 \pm 474.01\jmath$ and the difference with the smaller cracked element shown in Table 10.9 is predominantly due to more refined mesh. The response is visually identical to that shown in Figure 10.67.

10.15 Summary

This chapter is an overview of certain rotor systems that cannot be adequately modeled using a linear shaft-line model. We begin by discussing the dynamics of a single particle, elastically constrained in three dimensions and rotating about an axis in space. This analysis is extended to rotors that cannot be adequately represented by shaft-line models. Such rotors must be modeled using two- or three-dimensional finite elements or by axisymmetric finite elements. For axisymmetric rotors, such elements have the advantage of only requiring degrees of freedom to be specified in one radial plane of the rotor but nonetheless modeling the rotor in three dimensions. We also analyze systems with coupled lateral and torsional vibrations.

We examine several rotor problems, including rotor–stator contact, bearing misalignment, oil whirl and oil whip in oil-film bearings, and cracks in the rotor. A full analysis of all of these phenomena requires the use of a nonlinear model; however, substantial insight may be gained in most cases using a linear model. Nonlinear systems can produce a wide range of responses that may seem inexplicable from a linear analysis. However, most machines are designed to operate in regimes that may be modeled well as linear systems and that venture into highly nonlinear regions only during fault conditions.

Solutions to Problems

Chapter 2

2.1 $a_0 = x_0$, $b_0 = (\dot{x}_0 + \zeta\omega_n x_0)/\omega_d$. Undamped, $x = (\cos 3t)$ mm; damped, $x = (\cos 2.958t + 0.169 \sin 2.958t)$ mm.

2.2 8.23 Hz.

2.3 27.63 kN/m.

2.4 $v_0 = 0.314$ m/s, $a_0 = 98.7$ m/s^2.

2.5 8,796 Ns/m.

2.9 Phase of response relative to force is 180°. (a) Magnitude of response doubles, phase unchanged. (b) Phase changed by $\pi/2$ in the same direction as the phase of the force, magnitude unchanged. (c) Magnitude reduced to 30 percent, phase now 0°. (d) Magnitude reduced to 69 percent, phase now 134°.

2.10 $x(t) = \frac{1}{2} - \sum_{n=1}^{\infty} \frac{1}{n\pi} \sin(n\pi t)$.

2.11 $\omega_1 = 0.8110$ rad/s, $u_2/u_1 = 1.0856$, $\omega_2 = 2.6158$ rad/s, $u_2/u_1 = -0.4606$.

2.13 $\mathbf{u}_{N3} = \begin{bmatrix} 0.22361 & -0.22361 & 0 \end{bmatrix}^\mathsf{T}$.

2.14 (a) $q_1 = q_2 = (6.42 \cos \omega_1 t - 1.42 \cos \omega_2 t)$ mm, $q_3 = (3.81 \cos \omega_1 t + 1.19 \cos \omega_2 t)$ mm. (b) $q_1 = q_2 = -5.84$ mm, $q_3 = 2.46$ mm. Force in spring 1 is 73.8 N, in spring 2 is -83.0 N, in spring 3 is zero, and in spring 4 is 83.0 N. (c) The receptance between coordinates 1 and 3 is -5.84×10^{-5} m/N, the mobility is $-2.20 \times 10^{-3} j$ m/Ns, and the inertance is 8.30×10^{-2} m/Ns^2.

2.15 (a) 6.22 and 23.50 Hz. (b) 4.92 and 13.87 Hz. (c) 5.74 and 13.97 Hz. Procedure (b) gives the best result.

2.16 At 25 Hz, $x_0 = 0.642$ mm, $\phi = 14.75°$. At 55 Hz, $x_0 = 2.49$ mm, $\phi = -14.75°$.

2.17 (a) 2.3, 4.4, 8.0, and 13.2 mm. (b) $\omega > \omega_n$, 50.40 Hz. (c) At 50 Hz, $x_1 = 13.2$ mm, $x_3 = 2.90 \,\mu$m. At 55 Hz, two stable solutions: $x_1 = 80.8$ mm, $x_3 = 0.564$ mm; and $x_1 = -1.24$ mm, $x_3 = -1.9 \times 10^{-6}$ mm (compare to single-term solutions, $x_1 = 81.1$ and -1.24 mm).

Chapter 3

3.1 When the rotor is stationary, 25.16 Hz (twice). At 3,000 rev/min, 23.71 and 26.71 Hz. At 10,000 rev/min, 20.66 and 30.66 Hz. At 3,000 rev/min, the orbit is circular and backward at 23.71 Hz, and circular and forward at 26.71 Hz.

3.2 When the rotor is stationary, 25.16 and 28.69 Hz. At 3,000 rev/min, 24.65 and 29.28 Hz. At 10,000 Hz, 22.09 and 32.69 Hz.

3.3 When the rotor is stationary, the damped natural frequencies are 25.16 Hz (twice) and $\zeta = 0.0395$. At 3,000 rev/min, the damped natural frequencies are 23.71 and 26.71 Hz and $\zeta = 0.0395$. At 10,000 rev/min, the damped natural frequencies are 20.65 and 30.66 Hz and $\zeta = 0.0388$.

3.4 (b) When the rotor is stationary, the natural frequencies are 22.51 and 27.57 Hz; at 3,000 rev/min, they are 22.14 and 28.02 Hz.

3.5 14.24 (twice), 20.19, and 50.19 Hz.

3.6 When the rotor is stationary, the natural frequencies are 15.24 Hz (twice) and 67.15 Hz (twice). The corresponding u/ψ or $-v/\theta$ are 0.3100 and -0.2151, respectively. At 1,000 rev/min, the natural frequencies are 9.959, 22.12, 60.06, and 79.15 Hz. The corresponding u/ψ or $-v/\theta$ are 0.2881, 0.3633, -0.3347, and -0.1269, respectively.

3.7 Case (a), natural frequencies are 28.86, 36.51, and 56.37 Hz. Case (b), natural frequencies are 29.06, 33.83, and 60.73 Hz.

3.8 Stiffness = 50 kN/m, full-model natural frequencies are 13.22 and 34.99 Hz. Rigid-rotor model frequencies are 13.45 and 35.59 Hz. Stiffness = 1 MN/m, full-model natural frequencies are 46.20 and 122.23 Hz. Neither the rigid-rotor model nor the pinned-pinned rotor model is within 30 percent of these frequencies. Stiffness = 100 MN/m, full-model natural frequencies are 71.63 and 189.50 Hz. The pinned-pinned rotor model natural frequencies are 72.14 and 190.86 Hz.

3.9 Isotropic case: natural frequencies are 47.40 Hz (backward), 47.42 Hz (forward), 88.89 Hz (backward), and 154.93 Hz (forward). All orbits are circular (i.e., $\kappa = \pm 1$). Anisotropic case: natural frequencies are 47.41 Hz (translations, $\kappa = -0.0067$ / rotations, $\kappa = -0.2559$), 50.32 Hz ($\kappa = 0.0076$ / -0.2926), 91.62 Hz ($\kappa = -0.4444$ / -0.8888), and 157.88 Hz ($\kappa = 0.4000$ / 0.9347). Negative κ indicates backward whirl; positive κ indicates forward whirl.

3.10 $k = 29.85$ kN/m, $I_p = 0.5988$ kg m^2, $I_d = 0.5384$ kg m^2.

3.12 Two degrees of freedom: (a) 17.4505 and 68.5833 Hz, (b) 6.4971 and 15.8084 Hz. Single degree of freedom: (a) 17.4785 Hz, (b) 6.6074 Hz.

3.13 (a) For combined rotor and fan, center of gravity is 1.2 m from bearing 1, $I_d = 48.333$ kg m^2, $I_p = 26.5$ kg m^2. (b) Yaw, $M_y = 38.851$ kNm, pitch, $M_x = -241.67$ Nm. (c) Vertical loads, bearing 1, 117.7 N; bearing 2, 3.829 kN. Horizontal loads equal and opposite at bearings; magnitude 29.885 kN.

Chapter 4

4.1 Free-free bar, $\omega_1 = 0$, $\omega_2 = 3.4641\sqrt{E/\rho L^2}$. Fixed-free bar, $\omega_1 = 1.6114\sqrt{E/\rho L^2}$.

4.2 $\omega_1 = 3.4641, 3.2863, 3.2228\sqrt{E/\rho L^2}$ using two, three, and four elements.

4.3 $\omega_1 = 1.9027\sqrt{E/\rho L^2}$. Exact solution is $\omega_1 = 1.8366\sqrt{E/\rho L^2}$.

4.4 $\omega_1 = 0.5473\sqrt{E/\rho L^2}$. Exact solution is $\omega_1 = 0.5472\sqrt{E/\rho L^2}$.

4.5 Equal-length elements: $\omega_1 = 1.6114\sqrt{E/\rho L^2}$. Unequal-length elements: $\omega_1 = 1.6157\sqrt{E/\rho L^2}$. Exact solution is $\omega_1 = 1.5708\sqrt{E/\rho L^2}$.

4.6 $\omega_1 = 10.9545\sqrt{EI/\rho AL^4}$.

4.7 $\omega_1 = 22.7359\sqrt{EI/\rho AL^4}$; exact solution $\omega_1 = 22.3733\sqrt{EI/\rho AL^4}$.

4.8 $\omega_1 = 15.5608\sqrt{EI/\rho AL^4}$; exact solution $\omega_1 = 15.4182\sqrt{EI/\rho AL^4}$.

4.9 $\mathbf{M} = \dfrac{\rho AL}{420} \begin{bmatrix} 156 & -22L & 0 \\ -22L & 4L^2 & 0 \\ 0 & 0 & 0 \end{bmatrix} + \begin{bmatrix} 0 & 0 & 0 \\ 0 & 0 & 0 \\ 0 & 0 & m \end{bmatrix}$,

$\mathbf{K} = \dfrac{EI}{L^3} \begin{bmatrix} 12 & -6L & 0 \\ -6L & 4L^2 & 0 \\ 0 & 0 & 0 \end{bmatrix} + \begin{bmatrix} k & 0 & -k \\ 0 & 0 & 0 \\ -k & 0 & k \end{bmatrix}$.

4.10 Using tapered elements, $\omega_1 = 1.8278\sqrt{E/\rho}$. Using uniform elements, $\omega_1 = 1.7933\sqrt{E/\rho}$.

4.11 Using three elements, $\omega_1 = 1.5888\sqrt{E/\rho L^2}$; using one element, $\omega_1 = 1.7321\sqrt{E/\rho L^2}$; using reduction, $\omega_1 = 1.7321\sqrt{E/\rho L^2}$. Exact solution, $\omega_1 = 1.5708\sqrt{E/\rho L^2}$.

4.12 The natural frequencies have the form $\omega_n = \dfrac{C_n}{L^2}\sqrt{\dfrac{EI}{\rho A}}$ for mode n. With three elements, $C_1 = 9.878$, $C_2 = 39.95$, $C_3 = 98.59$. With six elements, $C_1 = 9.870$, $C_2 = 39.51$, $C_3 = 89.18$.

Chapter 5

5.1 Defining percentage difference as $100\,(\omega_{(euler)i}/\omega_{(timo)i} - 1)$, then for the solid shaft, the percentage differences are 0.12, 0.15, 0.55, 0.69, and 0.26. For the hollow shaft, the percentage differences are 0.65, 0.79, 2.99, 3.74, and 1.32.

5.2 The first five natural frequencies using four elements are 25.0929, 30.1380, 59.4160, 65.9957, and 162.5512 Hz. Using eight elements, they are 25.0928, 30.1378, 59.4145, 65.9937, and 162.5205 Hz. Using 16 elements, they are 25.0928, 30.1378, 59.4144, 65.9935, and 162.5182 Hz.

5.3 Natural frequencies using seven tapered and uniform elements are 27.7103, 32.5623, 40.8942, 50.9622, and 93.1906 Hz. Natural frequencies using 11 uniform elements are 27.6201, 32.4496, 40.5212, 50.2079, and 95.7711 Hz. Errors using 11 uniform elements are larger than those using seven tapered and uniform elements, as appropriate.

5.4 At 3,000 rev/min, the Sommerfeld number is 0.3125, radial force is 521.5 N, tangential force is 296.7 N, $\gamma = 29.6°$. Two real eigenvalues, −563.4 and −1,606 s^{-1}; one underdamped mode with damped natural frequency 45.81 Hz; and damping ratio 0.1383. At 6,000 rev/min, the damped natural frequencies are 51.83 and 91.22 Hz, with damping ratios 0.0690 and 0.6605.

5.5 $mk_{sw}^2 = c^2 k$, whirl frequency $= \sqrt{\dfrac{k}{m}}$, maximum power $= \dfrac{\Omega DLc}{\beta}\sqrt{\dfrac{k}{m}}$. $k_{sw} = 11.94$ kN/m, critical damping for stability, $c = 21.99$ Ns/m. When $c = 0$, $\omega_d = 86.3755$ Hz (twice), $\omega_n = 86.3775$ Hz (twice), forward mode, $\zeta = -0.0068$ (unstable), backward mode, $\zeta = 0.0068$. When $c = 20$ Ns/m, $\omega_d = 86.3739$ Hz (twice), forward mode, $\omega_n = 86.3739$ Hz, $\zeta = -0.0006$ (unstable), backward mode, $\omega_n = 86.3811$ Hz, $\zeta = 0.0129$. When $c = 40$ Ns/m, $\omega_d = 86.3690$ Hz (twice), forward mode, $\omega_n = 86.3703$ Hz, $\zeta = 0.0055$, backward mode, $\omega_n = 86.3846$ Hz, $\zeta = 0.0190$.

5.6 Numerical values of frequency and damping are unaffected. Predicted direction of whirl for each mode is reversed.

5.7 Shaft speed, $\Omega = 4{,}522$ rev/min, whirl frequency, $\omega_n = 37.68$ Hz. When added, damper coefficient $c = 80$ Ns/m, shaft speed, $\Omega = 6{,}232$ rev/min, whirl frequency, $\omega_n = 37.09$ Hz. When added, damper coefficient $c = 160$ Ns/m, shaft speed, $\Omega = 7{,}664$ rev/min, whirl frequency, $\omega_n = 35.48$ Hz.

5.8 Simple bearings: 23.25, 24.34, 79.87, 127.7, and 168.4 Hz. Angular contact ball bearings: 22.36, 23.34, 77.81, 123.1, and 163.5 Hz. Shaft tension, 14.9 kN.

5.9 (a) Stationary rotor, natural frequencies are 19.08, 38.74, and 67.24 Hz, (each twice). At 3,000 rev/min, natural frequencies are 18.91, 19.25, 38.49, 38.98, 65.85, and 68.58 Hz. The first, third, and fifth modes have backward circular orbits; the second, fourth, and sixth modes have forward circular orbits.

(b) Stationary rotor, natural frequencies are 18.83, 19.08, 37.00, 38.74, 63.32, and 67.24 Hz. At 3,000 rev/min, natural frequencies are 18.74, 19.16, 36.97, 38.76, 62.94, and 67.57 Hz. At the large disk, the mode orbits are elliptical with κ values of -0.4948, 0.5010, -0.1595, 0.0978, -0.2357, and 0.2983. (The negative sign indicates backward whirl; the positive sign indicates forward whirl.)

(c) Case (i): At 3,000 rev/min, damped natural frequencies are 18.87, 19.67, 35.00, 45.76, 52.70, and 53.10 Hz; damping ratios are 0.0024, 0.0045, 0.0628, 0.0387, 0.3367, and 0.4312; κ values are 0.1463, -0.0047, 0.2970, -0.1739, 0.4842, and 0.4302.

Case (ii): At 3,000 rev/min, damped natural frequencies are 18.94, 19.63, 35.69, 44.27, 45.95, and 49.16 Hz; damping ratios are -0.0016 (unstable), 0.0072, 0.0747, 0.3387, 0.0359, and 0.3617; κ values are 0.2121, -0.0632, 0.4200, 0.5882, -0.1442, and 0.1765.

Case (iii): At 3,000 rev/min, damped natural frequencies are 18.60, 19.53, 32.43, 45.53, 55.76, and 56.52 Hz; damping ratios are -0.0093 (unstable), 0.0148, 0.0560, 0.0650, 0.4457, and 0.2383; κ values are 0.3250, -0.1593, 0.2676, -0.2367, 0.4325, and 0.5638.

The machines for cases (ii) and (iii) are unstable.

5.10 (a) Approximate natural frequency is 17.48 Hz, compared to 16.77 Hz from Problem 5.1. (b) Approximate natural frequency is 30.29 Hz, compared to 25.09 Hz from Problem 5.2.

5.11

	Disk diameter (mm)					
	300	400	500	600	700	800
No shaft mass						
$\Delta\omega_1$ (%)	13.6	5.80	2.94	1.67	1.02	0.667
$\Delta\omega_2$ (%)	31.6	6.58	2.00	0.767	0.341	0.169
With shaft mass						
$\Delta\omega_1$ (%)	0.068	0.016	0.005	0.002	0.001	0.000
$\Delta\omega_2$ (%)	6.835	0.621	0.069	0.011	0.002	0.001

Chapter 6

6.1 Critical speeds 1,467 and 1,557 rev/min. Only the second critical speed is excited by unbalance. Whirl-orbit radius is 3.41 mm.

6.2 Critical speeds 1,714 and 1,844 rev/min. Whirl-orbit radius is 0.646 mm.

6.3 Critical speeds 1,501 and 1,734 rev/min.

6.4. Critical speeds 1,437, 1,674, and 6,397 rev/min (there are only three critical speeds for this system). Maximum whirl-orbit radius is 0.361 mm. Angular acceleration required is 73.8 rad/s^2. Torque required is 11.4 Nm. Maximum angular misalignment is 0.285°.

6.5 Maximum θ and ψ is 0.1944°.

6.6 Whirl amplitude at 1,000 rev/min is 0.624 mm. Balancing mass is 0.232 kg. Response at 1,500 rev/min is 1.13 mm.

6.7 First critical speed 2,296 rev/min. Response at disk 1.62 mm and at bearing 0.065 mm. For no rotation at the support, the first critical becomes 4,340 rev/min.

6.8 1,537 rev/min.

6.9 Campbell diagram (a) is for a machine on isotropic bearings. From this diagram, critical speeds are 2,200 and 5,000 rev/min due to synchronous unbalance; critical speeds are 740 and 1,200 rev/min due to the 3X force; peaks in the response at 10,000 rev/min are 22, 38, 119, 244, and 573 Hz. Diagram (b) is for a machine on anisotropic bearings. From this diagram, critical speeds are 2,200, 2,600, 2,900 and 5,600 rev/min due to synchronous unbalance; critical speeds are 740, 900, 1,000, 1,400, and 5,800 rev/min due to the 3X force; peaks in the response at 10,000 rev/min are 26, 38, 46, 124, 244, and 573 Hz.

6.10 Critical speeds 1,746 and 4,027 rev/min. Mass distribution (a) only weakly excites the first critical speed, and mass distribution (b) only weakly excites the second critical speed.

6.11 Critical speeds (a) 1,016, 1,021, 2,038 and 2,053 rev/min; (b) 1,007, 1,019, 1,967, and 2,046 rev/min; (c) 600, 703, 987, 1,056, 1,825, 2,210, 2,363 and 2,720 rev/min.

6.12 Response at first critical speed due to unbalance and bend is similar.

Chapter 7

7.1 In fixed coordinates, 31.4 Hz (twice), 16.5 Hz, and 110.2 Hz. In rotating coordinates, 18.6, 81.4, 60.2, and 66.5 Hz.

7.2 At 2,400 rev/min, roots are $\pm14.4j$, $\pm365.3j$, $\pm484.6j$, and $\pm662.8j$ rad/s. All roots imaginary, so system is stable. At 2,600 rev/min, roots are $\pm380.7j$, $\pm492.7j$, $\pm678.9j$ rad/s, and $\pm7.57s^{-1}$. One root is real and positive, so system is unstable. At 2,800 rev/min, roots are $\pm17.7j$, $\pm396.5j$, $\pm501.4j$, and $\pm695.5j$ rad/s. All roots imaginary, so system is stable.

7.3 At 1,900 rev/min, roots are $\pm9.465j$, $\pm326.4j$, $\pm355.7j$, and $\pm409.8j$ rad/s (stable). At 2,000 rev/min, roots are ±6.687, $\pm331.4j$, $\pm361.7j$, and $\pm420.2j$ rad/s (unstable). At 2,100 rev/min, roots are $\pm6.331j$, $\pm336.6j$, $\pm367.8j$, and $\pm430.7j$ rad/s (stable).

7.4 $\zeta = 0.005067$ (twice), 0.007476 (twice). At 2,200 rev/min, roots are $-2.084 \pm 127.8j$, $-0.0272 \pm 197.4j$, $-1.973 \pm 197.4j$, and $-1.916 \pm 559.8j$ rad/s (stable). At 2,300 rev/min, roots are $-2.086 \pm 124.3j$, $0.0170 \pm 197.4j$, $-2.017 \pm 197.4j$, and $-1.914 \pm 575.9j$ rad/s (unstable).

7.5 Circular orbits of radii (a) 85.89 μm, (b) 337.0 μm, (c) 245.9 μm.

7.6 1,004 rev/min.

7.8 At 1,500 rev/min, roots are $\pm 306.2j$ and $\pm 3.82j$ rad/s (stable). At 1,540 rev/min, roots are $\pm 310.3j$ and ± 1.61 rad/s (unstable). Gravity critical speed is 755 rev/min. Responses are 0.96 and 2.13 mm, giving an orbit of 2.33 mm radius.

7.9 When stationary, the eigenvalues are $\pm 15.73 \pm 158.89j$ and the natural frequencies are 25.41 Hz (twice). At 3,000 rev/min, the eigenvalues are $15.71 \pm 149.75j$ and $-15.71 \pm 168.60j$ and the natural frequencies are 23.96 and 26.95 Hz. At both rotor spin speeds, the system is unstable.

7.10 4,520 rev/min.

7.11 Unstables speeds: 1,746–1,832 rev/min, 2,944–2,949 rev/min, 4,029–4,226 rev/min. Natural frequencies: 17.47, 19.14, and 75.81 Hz. Pseudonatural frequencies: 32.53, 67.47, 30.86, 69.14, 25.81, and 125.81 Hz.

Chapter 8

8.1 1.245 kg at $-149°$ on disk B; 1.145 kg at $3°$ on disk D.

8.2 (a) Bearing 1, 15.77 N at $-94.27°$; bearing 2, 50.40 N at $159.84°$. (b) Disk 1, 3.282 g m at $-138.80°$; disk 3, 5.044 g m at $151.63°$.

8.3 5.547 g at $124°$.

8.4 0.86 g m at $-126°$; 0.59 g m at $-9°$.

8.5 0.85 g m at $-126°$; 0.60 g m at $51°$.

8.6 Bent rotor.

8.7 1,500 rev/min: 0.335 kg at $17°$; 0.0484 kg at $-143°$. 3,000 rev/min: 0.0932 kg at $162°$; 0.0295 kg at $13°$.

8.8 Using data in both directions: 0.0324 kg at $-138°$; 0.0208 kg at $-8°$. Using data in x direction only: 0.0410 kg at $-147°$; 0.0410 kg at $23°$.

8.9 Using data in both directions: 0.0376 kg at $-141°$; 0.0401 kg at $15°$. Using data in x direction only: 0.0375 kg at $-132°$; 0.0318 kg at $64°$.

8.10 Using data in both directions: 0.305 kg at $177°$; 1.213 kg at $-98°$. Using data in x direction only: 0.765 kg at $180°$; 1.365 kg at $0°$.

8.11 Mode of interest 1, ratio -2.524; mode 2, ratio -2.159. Unbalance correction 0.731 g m at $344°$ on left disk; 0.143 g m at $169°$ on right disk.

8.12 $\mathbf{e}_1 = [0.5391 \ 0.5121]^\top$, $\mathbf{e}_2 = [0.4852 \ -0.5391]^\top$. 0.794 g m at $281°$ on disk 1; 0.075 g m at $273°$ on disk 2. Response at disk 1 at 820 rev/min is 87.6 μm (for exact modes, response is 2.65 μm). Response at 3,075 rev/min is zero.

8.13 Spurious unbalance correction is 29.65 g m at $51.02°$. The corresponding value of $|R(\Omega)|$ is 0.0674 kg^{-1}s^{-1}.

8.15 $\delta_1 = -0.763$ g m, $\mathbf{e}_1 = [0 \ -6.5531 \ 0]^\top$, α_1 is 4.00 g m at $88°$, $\mathbf{b}_{c1} = \alpha_1 \mathbf{e}_1$. $\delta_2 = 5.475$ g m, $\mathbf{e}_2 = [0.91325 \ 0 \ -0.34521]^\top$, α_2 is 10.9 g m at $-168°$, $\mathbf{b}_{c2} = \alpha_2 \mathbf{e}_2$. $\delta_3 = 5.084$ g m, $\mathbf{e}_3 = [0.08061 \ 0.98349 \ 0.11961]^\top$, α_3 is 23.3 g m at $95°$, $\mathbf{b}_{c3} = \alpha_3 \mathbf{e}_3$.

Chapter 9

9.1 The approximate axial natural frequencies are identical to those for the full system. The approximate torsional natural frequencies are 0, 129.2, 158.3, and 247.5 Hz.

9.2 0, 58.71, and 124.85 Hz.

9.3 0, 106.26, and 215.27 Hz.

9.4 Resonance frequencies: 0, 32.59, 35.59, and 471.0 Hz. Oscillatory torque and force: 54.78 Nm and 263.5 N.

9.6 1.548 (mainly generator rim), 11.93 (mainly crankshaft), and 17.92 Hz (mainly crankshaft and hub).

9.7 64.54, 371.13, and 9.691 rad/s^2.

9.8 Referred inertia, 50 kg m^2. Equivalent shaft stiffness, 1.2 MN/rad. Natural frequency, 28.47 Hz; mode shape, $[0.0408, \ -0.1225]^\top$ (order of degrees of freedom: motor, pinion). Peak shaft torques, 427.8, 740.3, 3398.9, 385.0, and 179.9 Nm.

9.9 Five elements: 0, 1.942, 4.072, 6.544 and 9.155 Hz. Eight elements: 0, 1.923, 3.919, 6.062, and 8.418 Hz. Fifty elements: 0, 1.911, 3.823, 5.739, 7.661, and 9.591 Hz.

9.10 5 Hz: response 5.988 μm, force 130.0 N. 15 Hz: response 5.097 μm, force 996.0 N.

APPENDIX 1

Properties of Solids

The parameters used to determine the solid properties are as follows:

ρ density
h length along the axis of symmetry
D outside diameter
d inside diameter
$\bar{z}$ position of center of mass along the cylinder or cone axis of symmetry

The subscripts for the diametral moment of inertia, I_d, are:

C center
E end
B base
A apex

Table A1.1. *Properties of solids*

Section	Mass, M	Polar moment of inertia, I_p	Diametral moment of inertia, I_d
Cylinder	$M = \rho \pi h D^2 / 4$	$I_p = M D^2 / 8$	$(I_d)_C = \dfrac{1}{2} I_p + \dfrac{1}{12} M h^2$
	$\bar{z} = \dfrac{h}{2}$		$(I_d)_E = \dfrac{1}{2} I_p + \dfrac{1}{3} M h^2$
Hollow cylinder	$M = \rho \pi h \left(D^2 - d^2 \right) / 4$	$I_p = M \left(D^2 + d^2 \right) / 8$	$(I_d)_C = \dfrac{1}{2} I_p + \dfrac{1}{12} M h^2$
	$\bar{z} = \dfrac{h}{2}$		$(I_d)_E = \dfrac{1}{2} I_p + \dfrac{1}{3} M h^2$
Right circular cone	$M = \rho \pi h D^2 / 12$	$I_p = 3 M D^2 / 40$	$(I_d)_C = \dfrac{1}{2} I_p + \dfrac{3}{80} M h^2$
	$\bar{z} = \dfrac{h}{4}$ from base		$(I_d)_B = \dfrac{1}{2} I_p + \dfrac{1}{10} M h^2$
			$(I_d)_A = \dfrac{1}{2} I_p + \dfrac{3}{5} M h^2$

APPENDIX 2

Stiffness and Mass Coefficients for Certain Beam Systems

This appendix provides the stiffness coefficients of a shaft at a point along the shaft, which is typically a disk. These coefficients may be used in a simple two or four degrees of freedom model of a machine. The first column in Table A2.1 lists the boundary conditions for the shaft at the two bearing locations, which are at distances of a and b from the disk, for systems 1 through 5. A pinned-boundary condition forces the transverse displacement of the beam to be zero at the boundary but the beam is free to rotate. A clamped-boundary condition forces both the transverse displacement and the rotation of the beam to be zero at the boundary. If the boundary condition is free, then no constraints are applied. Generally, a beam is supported at the two ends and the boundary conditions are used to describe the beam system. For example, a clamped-free beam has one end clamped and the other free.

The stiffness matrices are obtained by a static reduction of the stiffness matrices (see Section 2.5.1), where the disk degrees of freedom are the master degrees of freedom. The stiffness matrix for each part of the shaft is equal to that obtained from the FEM, because a uniform beam with load applied only at the ends has its displacement defined by a cubic.

Table A2.2 lists the mass coefficients for systems 1 through 4 based on the displacement model given by the static deformation used to calculate the stiffness coefficients. These mass coefficients should be added to those of the disk to produce the equations of motion of the system. Table A2.3 lists the mass coefficients for system 5. The same approach can be used for systems 6 through 9, but the resulting expressions are more complicated and not given here. Table A2.4 lists the stiffness coefficients for beam systems 6 through 9.

The stiffness and mass matrices contain terms corresponding to both the translational and rotational degrees of freedom. If the transverse vibration *only* is of interest, then the required stiffness is *not* k_{uu} but rather a further static reduction must be performed to eliminate the rotational degree of freedom, giving a stiffness of $k_{uu} - k_{\psi u}^2 / k_{\psi\psi}$. The same static reduction gives the equivalent translational mass as $m_{uu} - 2m_{\psi u}k_{\psi u}/ k_{\psi\psi} + m_{\psi\psi} k_{\psi u}^2 / k_{\psi\psi}^2$.

Table A2.1. *Stiffness coefficients for beam systems 1 through 5 shown in Figure A2.1*

System	Condition	k_{uu} and k_{vv}	$k_{\psi u}$ and $-k_{\theta v}$	$k_{\psi \psi}$ and $k_{\theta \theta}$
1	Pinned-Pinned	$\dfrac{3EI\left(a^3+b^3\right)}{a^3 b^3}$	$\dfrac{3EI\left(a^2-b^2\right)}{a^2 b^2}$	$\dfrac{3EI\left(a+b\right)}{ab}$
2	Clamped-Clamped	$\dfrac{12EI\left(a^3+b^3\right)}{a^3 b^3}$	$\dfrac{6EI\left(a^2-b^2\right)}{a^2 b^2}$	$\dfrac{4EI\left(a+b\right)}{ab}$
3	Clamped-Pinned	$\dfrac{3EI\left(4b^3+a^3\right)}{a^3 b^3}$	$\dfrac{3EI\left(a^2-2b^2\right)}{a^2 b^2}$	$\dfrac{EI\left(4b+3a\right)}{ab}$
4	Clamped-Free	$\dfrac{12EI}{a^3}$	$-\dfrac{6EI}{a^2}$	$\dfrac{4EI}{a}$
5	Pinned-Pinned-Free	$\dfrac{12EI\left(a+3b\right)}{b^3\left(4a+3b\right)}$	$-\dfrac{6EI\left(2a+3b\right)}{b^2\left(4a+3b\right)}$	$\dfrac{12EI\left(a+b\right)}{b\left(4a+3b\right)}$

Table A2.2. *Mass coefficients for beam systems 1 through 4 shown in Figure A2.1*

System	Condition	m_{uu} and m_{vv}	$m_{\psi u}$ and $-m_{\theta v}$	$m_{\psi \psi}$ and $m_{\theta \theta}$
1	Pinned-Pinned	$\dfrac{17}{35}\rho A\left(a+b\right)$	$\dfrac{3}{35}\rho A\left(b^2-a^2\right)$	$\dfrac{2}{105}\rho A\left(a^3+b^3\right)$
2	Clamped-Clamped	$\dfrac{13}{35}\rho A\left(a+b\right)$	$\dfrac{11}{210}\rho A\left(b^2-a^2\right)$	$\dfrac{1}{105}\rho A\left(a^3+b^3\right)$
3	Clamped-Pinned	$\dfrac{\rho A}{35}\left(13a+17b\right)$	$\dfrac{\rho A}{210}\left(18b^2-11a^2\right)$	$\dfrac{\rho A}{105}\left(a^3+2b^3\right)$
4	Clamped-Free	$\dfrac{13}{35}\rho Aa$	$-\dfrac{11}{210}\rho Aa^2$	$\dfrac{1}{105}\rho Aa^3$

Table A2.3. *Mass coefficients for beam system 5*
(Pinned-Pinned-Free) shown in Figure A2.1

m_{uu} and m_{vv}	$\dfrac{\rho A\left(24a^5+272a^2 b^3+351ab^4+117b^5\right)}{35b^2\left(4a+3b\right)^2}$
$m_{\psi u}$ and $-m_{\theta v}$	$-\dfrac{\rho A\left(16a^5+96a^2 b^3+110ab^4+33b^5\right)}{70b\left(4a+3b\right)^2}$
$m_{\psi \psi}$ and $m_{\theta \theta}$	$\dfrac{\rho A\left(8a^5+32a^2 b^3+33ab^4+9b^5\right)}{105\left(4a+3b\right)^2}$

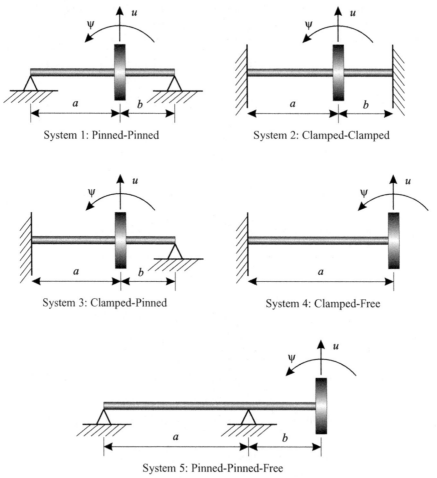

Figure A2.1. Beam systems 1 through 5.

Table A2.4. *Stiffness coefficients for beam systems 6 through 9 shown in Figure A2.2*

System	Coefficients
6	Denominator, D_6 $\left(3EI + a^3 k_1\right)\left(3EI + b^3 k_2\right)$
	k_{uu} and k_{vv} $\dfrac{3EI}{D_6}\left\{3EI\left(k_1 + k_2\right) + \left(a^3 + b^3\right)k_1 k_2\right\}$
	$k_{\psi u}$ and $-k_{\theta v}$ $\dfrac{3EI}{D_6}\left\{3EI\left(-ak_1 + bk_2\right) + ab\left(a^2 - b^2\right)k_1 k_2\right\}$
	$k_{\psi\psi}$ and $k_{\theta\theta}$ $\dfrac{3EI}{D_6}\left\{3EI\left(a^2 k_1 + b^2 k_2\right) + a^2 b^2\left(a + b\right)k_1 k_2\right\}$
7	Denominator, D_7 $a^3\left(3EI + b^3 k\right)$
	k_{uu} and k_{vv} $\dfrac{3EI}{D_7}\left\{12EI + \left(a^3 + 4b^3\right)k\right\}$
	$k_{\psi u}$ and $-k_{\theta v}$ $\dfrac{3EIa}{D_7}\left\{-6EI + b\left(a^2 - 2b^2\right)k\right\}$
	$k_{\psi\psi}$ and $k_{\theta\theta}$ $\dfrac{EIa^2}{D_7}\left\{12EI + b^2\left(3a + 4b\right)k\right\}$
8	Denominator, D_8 $36\left(EI\right)^2 + 12EI\{\left(a + b\right)^3 k_1 + b^3 k_2\} + a^2 b^3\left(4a + 3b\right)k_1 k_2$
	k_{uu} and k_{vv} $\dfrac{12EI}{D_8}\left\{3EI\left(k_1 + k_2\right) + a^2\left(a + 3b\right)k_1 k_2\right\}$
	$k_{\psi u}$ and $-k_{\theta v}$ $-\dfrac{6EI}{D_8}\left\{6EI\left(\left(a + b\right)k_1 + bk_2\right) + a^2 b\left(2a + 3b\right)k_1 k_2\right\}$
	$k_{\psi\psi}$ and $k_{\theta\theta}$ $\dfrac{12EI}{D_8}\left\{3EI\left(\left(a + b\right)^2 k_1 + b^2 k_2\right) + a^2 b^2\left(a + b\right)k_1 k_2\right\}$
9	Denominator, D_9 $12EI + a^3 k$
	k_{uu} and k_{vv} $\dfrac{12EI}{D_9}k$
	$k_{\psi u}$ and $-k_{\theta v}$ $-\dfrac{6EI}{D_9}ak$
	$k_{\psi\psi}$ and $k_{\theta\theta}$ $\dfrac{4EI}{aD_9}\left(3EI + a^3 k\right)$

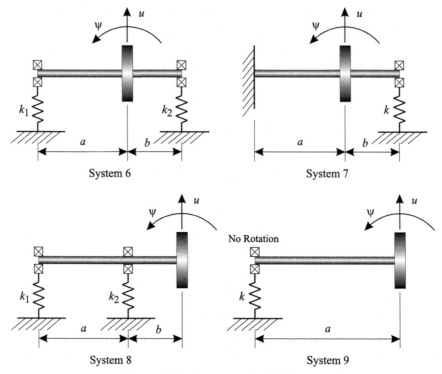

Figure A2.2. Beam systems 6 through 9.

APPENDIX 3

Torsional Constants for Shaft Sections

Table A3.1. *Torsional constants for common sections*

Section	Torsional constant, C
Hollow circle (D, d, outer and inner radii)	$\pi \left(D^4 - d^4 \right) / 32$
Ellipse (a, b, semimajor and semiminor axes)	$\dfrac{\pi a^3 b^3}{a^2 + b^2}$
Equilateral triangle (a, length of one side)	$\dfrac{a^4 \sqrt{3}}{80}$
Square (a, length of one side)	$0.1406 a^4$
Hexagon (a, length of one side)	$1.03 a^4$

Note: Pilkey (2005) provides the torsional constants for more sections.

Bibliography

Abraham, O. N. L., Brandon, J. A., and Cohen, A. M. (1994). Remark on the determination of compliance coefficients at the crack section of a uniform beam with circular cross section, *Journal of Sound and Vibration* 169, 570–574.

Adams, M. L. (2001). *Rotating Machinery Vibration: From Analysis to Troubleshooting* (Marcel Dekker, New York).

Adiletta, G., Guido, A. R., and Rossi, C. (1996). Chaotic motions of a rigid rotor in short journal bearings, *Nonlinear Dynamics* 10, 251–269.

Alauze, C., der Hagopian, J., Gaudiller, L., and Voinis, P. (2001). Active balancing of turbomachinery: Application to large shaft lines, *Journal of Vibration and Control* 7, 249–278.

Al-Bedoor, B. O. (2000). Transient torsional and lateral vibrations of unbalanced rotors with rotor-to-stator rubbing, *Journal of Sound and Vibration* 229, 627–645.

Al-Bedoor, B. O. (2001). Modeling the coupled torsional and lateral vibrations of unbalanced rotors, *Computer Methods in Applied Mechanics and Engineering* 190, 5999–6008.

Alford, J. (1965). Protecting turbomachinery from self-excited rotor whirl, *Journal of Engineering for Power* 87, 333–344.

Al-Hussain, K. M. (2003). Dynamic stability of two rigid rotors connected by a flexible coupling with angular misalignment, *Journal of Sound and Vibration* 266, 217–234.

Al-Hussain, K. M., and Redmond, I. (2002). Dynamic response of two rotors connected by rigid mechanical coupling with parallel misalignment, *Journal of Sound and Vibration* 249, 483–498.

Alolah, R., Badr, M. A., and Abdel-Halim, M. A. (1999). A comparative study on the starting methods of three-phase wound-rotor induction motors: Part I, *IEEE Transactions on Energy Conversion* 14, 918–922.

Ananda Rao, M., Srinivas, J., Rama Raju, V. B. V., and Kumar, K. V. S. S. (2003). Coupled torsional-lateral vibration analysis of geared shaft systems using mode synthesis, *Journal of Sound and Vibration* 261, 359–364.

Arkkio, A., Antila, M., Pokki, K., Simon, A., and Lantto, E. (2000). Electromagnetic force on a whirling cage rotor, *IEE Proceedings: Electric Power Applications* 147, 353–360.

Astley, R. J. (1992). *Finite Elements in Solids and Structures* (Chapman and Hall, London).

Balbahadur, A. C., and Kirk, R. G. (2002a). Part I: Theoretical model for a synchronous thermal instability operating in overhung rotors, *Sixth International Conference on Rotor Dynamics*, IFToMM: Sydney, Australia.

Balbahadur, A. C., and Kirk, R. G. (2002b). Part II: Case studies for a synchronous thermal instability operating in overhung rotors, *Sixth International Conference on Rotor Dynamics*, IFToMM: Sydney, Australia.

Bathe, K.-J., and Wilson, E. L. (1976). *Numerical Methods in Finite Element Analysis* (Prentice-Hall, Englewood Cliffs, N.J.).

Bazoune, A., and Khulief, Y. A. (1992). A finite beam element for vibration analysis of rotating tapered Timoshenko beams, *Journal of Sound and Vibration* 156, 141–164.

Bazoune, A., Khulief, Y. A., and Stephen, N. G. (1999). Further results for modal characteristics of rotating papered Timoshenko beams, *Journal of Sound and Vibration* 219, 157–174.

Bently, D. E. (1974). Forced subrotative speed dynamic action of rotating machinery, *ASME Paper No. 74-Pet-16*.

Bickford, W. B. (1994). *A First Course in Finite Element Analysis*, Second Edition (Richard D. Irwin, Burr Ridge, IL).

Bickford, W. B., and Nelson, H. D. (1985). A conical beam finite element for rotor dynamics analysis, *Journal of Vibrations and Acoustics, Stress and Reliability in Design* 107, 421–430.

Bigret, R. (2004). Balancing. In *Encyclopedia of Vibration*, S. G. Braun, editor-in-chief (Elsevier), pp. 111–124.

Bishop, R. E. D., and Gladwell, G. M. L. (1959). The vibration and balancing of an unbalanced flexible rotor, *Journal of Mechanical Engineering Science* 1, 66–77.

Black, H. F. (1969). Effects of hydraulic forces in annular pressure seals on the vibrations of centrifugal pump rotors, *Journal of Mechanical Engineering Science* 11, 206–213.

Blevins, R. D. (1979). *Formulas for Natural Frequency and Mode Shape* (Van Nostrand Reinhold, New York).

Blough, J. R. (2003). Development and analysis of time-variant discrete Fourier transform order tracking, *Mechanical Systems and Signal Processing* 17, 1185–1199.

Cameron, A. (1976). *Basic Lubrication Theory* (Ellis Horwood, Chichester, England).

Cartmell, M. P. (1990). *Introduction to Linear, Parameteric and Nonlinear Vibrations* (Chapman and Hall, London).

Caughey, T. K., and O'Kelly, M. E. (1965). Classical normal modes in damped linear systems, *Journal of Applied Mechanics* 32, 583–588.

Chatelet, E., D'Ambrosio, F., and Jacquet-Richardet, G. (2005). Toward global modeling approaches for dynamic analyses of rotating assemblies of turbo-machines, *Journal of Sound and Vibration* 282, 163–178.

Childs, D. (1993). *Turbomachinery Rotordynamics: Phenomena, Modeling, and Analysis* (Wiley, New York).

Childs, D. W., Graviss, M., and Rodriguez, L. E. (2007). Influence of groove size on the static and rotordynamic characteristics of short, laminar-flow annular seals, *Journal of Tribology* 129, 398–406.

Choy, F. K., and Padovan, J. (1987). Nonlinear transient analysis of rotor–casing rub events, *Journal of Sound and Vibration* 113, 529–545.

Choy, F. K., Padovan, J., and Li, W. H. (1988). Rub in high-performance turbo-machinery, modeling, solution methodology, and signature analysis, *Mechanical Systems and Signal Processing* 2, 113–133.

Choy, F. K., Padovan, J., and Qian, W. (1993). Effects of foundation excitation on multiple rub interactions in turbo-machinery, *Journal of Sound and Vibration* 164, 349–363.

Chu, F., and Zhang, Z. (1997). Periodic, quasiperiodic, and chaotic vibrations of a rub-impact rotor system supported on oil-film bearings, *International Journal of Engineering Science* 5, 963–973.

Chu, F., and Zhang, Z. (1998). Bifurcation and chaos in a rub-impact Jeffcott rotor system, *Journal of Sound and Vibration* 210, 1–18.

Chun, S.-B., and Lee, C.-W. (1996). Vibration analysis of shaft-bladed disk system by using substructure synthesis and assumed modes method, *Journal of Sound and Vibration* 189, 587–608.

Chung, J., Heo, J. W., and Han, C. S. (2003). Natural frequencies of a flexible spinning disk misaligned with the axis of rotation, *Journal of Sound and Vibration* 260, 763–775.

Chung, J., and Ro, D. S. (1999). Dynamic analysis of an automatic dynamic balancer for rotating mechanisms, *Journal of Sound and Vibration* 228, 1035–1056.

Combescure, D., and Lazarus, A. (2008). Refined finite element modeling for the vibration analysis of large rotating machines: Application to the gas turbine modular helium reactor power conversion unit, *Journal of Sound and Vibration* 318, 1262–1280.

Cook, R. D., Malkus, D. S., Plesha, M. E., and Witt, R. J. (2001). *Concepts and Applications of Finite Element Analysis*, Fourth Edition (John Wiley, N.Y.).

Cookson, R. A., and Kossa, S. S. (1979). The effectiveness of squeeze-film damper bearings supporting rigid rotors without a centralizing spring, *International Journal of Mechanical Sciences* 21, 639–650.

Cookson, R. A., and Kossa, S. S. (1980). The effectiveness of squeeze-film damper bearings supporting flexible rotors without a centralizing spring, *International Journal of Mechanical Sciences* 22, 313–324.

Cooley, J. W., and Tukey, J. W. (1965). An algorithm for the machine calculation of complex Fourier series, *Mathematics of Computation* 19, 297–301.

Cowper, G. R. (1966). The shear coefficient in Timoshenko's beam theory, *Journal of Applied Mechanics* 33, 335–340.

Craig, R. R. (1981). *Structural Dynamics: An Introduction to Computer Methods* (John Wiley, New York).

Craig, R. R., and Bampton, M. C. C. (1968). Coupling of substructures for dynamic analysis, *AIAA Journal* 6, 1313–1319.

Darpe, A. K., Gupta, K., and Chawla, A. (2004). Coupled bending, longitudinal and torsional vibrations of a cracked rotor, *Journal of Sound and Vibration* 269, 33–60.

Davies, W. G. R., and Mayes, I. W. (1984). The vibrational behavior analysis of a multi-shaft, multi-bearing system in the presence of a propagating transverse crack, *Journal of Vibration, Acoustics, Stress, and Reliability in Design* 106, 146–153.

Davis, R., Henshell, R. D., and Warburton, G. B. (1972). A Timoshenko beam element, *Journal of Sound and Vibration* 22, 475–487.

Dawe, D. J. (1984). *Matrix and Finite Element Displacement Analysis of Structures* (Oxford University Press, Oxford, England).

de Castro, H. F., Cavalca, K. L., and Nordmann, R. (2008). Whirl and whip instabilities in rotor-bearing system considering a nonlinear force model, *Journal of Sound and Vibration*, 317, 273–293.

Delamare, J., Rulliere, E., and Yonnet, J. P. (1995). Classification and synthesis of permanent-magnet bearing configurations, *IEEE Transactions on Magnetics* 31, 4190–4192.

Dimarogonas, A. D. (1976). *Vibration Engineering* (West Publishers, St Paul, MN).

Dimarogonas, A. D. (1994). Author's reply to Abraham et al. (1994), *Journal of Sound and Vibration* 169, 575–576.

Dimarogonas, A. D. (1996). Vibration of cracked structures: A State-of-the-art review, *Engineering Fracture Mechanics* 55, 831–857.

Dimarogonas, A. D., and Papadopoulos, C. A. (1983). Vibration of cracked shafts in bending, *Journal of Sound and Vibration* 91, 583–593.

Ding, Q., and Leung, A. Y. T. (2005). Numerical and experimental investigations on flexible multi-bearing rotor dynamics, *Journal of Vibration and Acoustics* 127, 408–415.

Ding, X. J., Yang, Y. L., Chen, W., Huang, S. H., and Zheng, C. G. (2006). Calculation method of efficiency factor in Alfords force, *Journal of Power and Energy* 220, 169–177.

Dorf, R. C., and Bishop, R. H. (2008). *Modern Control Systems*, Eleventh Edition (Pearson Prentice Hall, Upper Saddle River, N.J.).

Drew, S. J., Hesterman, D. C., and Stone, B. J. (1999). The torsional excitation of variable inertia effects in a reciprocating engine. *Mechanical Systems and Signal Processing* 13, 125–144.

Earnshaw, S. (1842). On the nature of the molecular forces which regulate the constitution of the luminiferous ether, *Transactions of the Cambridge Philosophy Society* 7, 97–112.

Eckert, L., Schmied, J., and Ziegler, A. (2006). Case history and analysis of the spiral vibration of a large turbogenerator using three different heat input models, *7th IFToMM Conference on Rotor Dynamics*; Vienna, Austria; 25–28 September.

Edney, S. L., Fox, C. H. J., and Williams, E. J. (1990). Tapered Timoshenko finite elements for rotor dynamics analysis, *Journal of Sound and Vibration* 137, 463–481.

Edwards, S., Lees, A. W., and Friswell, M. I. (1999). The influence of torsion on rotor-stator contact in rotating machinery, *Journal of Sound and Vibration* 225, 767–778.

Ehrich, F. F. (1988). High-order subharmonic response of high-speed rotors in bearing clear-ance, *Journal of Vibration Acoustics Stress and Reliability in Design* 110, 9–16.

Ehrich, F. F. (1992). Observations of subcritical, superharmonic and chaotic response in ro-tordynamics, *Journal of Vibration and Acoustics* 114, 93–100.

Ehrich, F. F. (1999). *Handbook of Rotordynamics* (Krieger Publishing Company, Malabar, FL).

Ertas, B. H., and Vance, J. M. (2002). The effect of static and dynamic misalignment on ball-bearing radial stiffness, *38th AIAA/ASME/SAE/ASEE Joint Propulsion Conference & Exhibit*, 7–10 July 2002, Indianapolis, IN, AIAA 2002–4160.

Fagan, M. J. (1992). *Finite Element Analysis: Theory and Practice* (Longman Scientific and Technical, Harlow, Essex, England).

Fenner, R. T. (1989). *Mechanics of Solids* (Blackwell Scientific Publications, Oxford).

Foiles, W. C., and Allaire, P. E. (2006). Single-plane and multi-plane balancing using only amplitude, *7th IFToMM Conference on Rotordynamics*; Vienna, Austria; Paper Number 182.

Foiles, W. C., Allaire, P. E., and Gunter, E. J. (1998). Review: Rotor balancing, *Shock and Vibration* 5, 325–336.

Foiles, W. C., and Bently, D. E. (1998). Balancing with phase only (single-plane and multi-plane), *Journal of Vibration, Acoustics, Stress, and Reliability in Design* 110, 151–157.

Friswell, M. I., Garvey, S. D., and Penny, J. E. T. (1995). Model reduction using dynamic and iterated IRS techniques, *Journal of Sound and Vibration* 186, 311–323.

Friswell, M. I., Garvey, S. D., and Penny, J. E. T. (1998a). The convergence of the iterated IRS method, *Journal of Sound and Vibration* 211, 123–132.

Friswell, M. I., Garvey, S. D., Penny, J. E. T., and Smart, M. G. (1998b). Computing critical speeds for rotating machines with speed-dependent bearing properties, *Journal of Sound and Vibration* 213, 139–158.

Friswell, M. I., and Penny, J. E. T. (1994). The accuracy of jump frequencies in series solutions of the response of a duffing oscillator, *Journal of Sound and Vibration* 169, 261–269.

Friswell, M. I., Penny, J. E. T., Garvey, S. D., and Lees, A. W. (2001). Damping ratio and natural frequency bifurcations in rotating systems, *Journal of Sound and Vibration* 245, 960–967.

Friswell, M. I., and Mottershead, J. E. (1995). *Finite Element Model Updating in Structural Dynamics* (Kluwer Academic Publishers Dordrecht, Netherlands).

Früchtenicht, J., Jordon, H., and Seinsch, H. O. (1982). Exzentrizitäts felder als Ursache von Laufinstabilitäten bei asynchronmaschinen, *Archiv für Elektrotechnik* 65, 271–292.

Garvey, S. D., Friswell, M. I., Williams, E. J., Lees, A. W., and Care, I. (2002). Robust bal-ancing for rotating machines, *IMechE Journal of Engineering Science* 216, 1117–1130.

Garvey, S. D., Penny, J. E. T. and Friswell, M. I. (1998). The relationship between the real and imaginary parts of complex modes, *Journal of Sound and Vibration* 212, 75–83.

Garvey, S. D., Williams, E. J., Cotter, G., Davies, C., and Grum, N. (2005). Reduction of noise effects for in situ balancing of rotors, *Journal of Vibration and Acoustics* 127, 234–246.

Gasch, R. (1976). Dynamic behavior of a simple rotor with a cross-sectional crack, *IMechE Conference on Vibrations in Rotating Machinery*, Cambridge, UK, 1976, Paper C178/76.

Gasch, R. (1993). A survey of the dynamic behavior of a simple rotating shaft with a trans-verse crack, *Journal of Sound and Vibration* 160, 313–332.

Gasch, R., Markert, R., and Pfützner, H. (1979). Acceleration of unbalanced flexible rotors through the critical speeds, *Journal of Sound and Vibration* 63, 393–409.

Genta, G., and Delprete, C. (1995). Acceleration through critical speeds of an anisotropic, nonlinear, torsionally stiff rotor with many degrees of freedom, *Journal of Sound and Vibration* 180, 369–386.

Genta, G., and Gugliotta, A. (1988). A conical element for finite element rotor dynamics, *Journal of Sound and Vibration* 120, 175–182.

Genta, G., and Tonoli, A. (1997). A harmonic finite element for the analysis of flexural, tor-sional, and axial rotordynamic behavior of bladed arrays, *Journal of Sound and Vibration* 207, 693–720.

Geradin, M., and Kill, N. (1984). A new approach to finite element modeling of flexible rotors, *Engineering Computations* 1, 52–64.

Gibbons, C. B. (1976). Coupling misalignment forces, *Proceedings of the Fifth Turbomachinery Symposium*, College Station, TX, 111–116.

Glasgow, D. A., and Nelson, H. D. (1980). Stability analysis of rotor–bearing systems using component-mode synthesis, *Journal of Mechanical Design* 102, 352–359.

Goldman, P., and Muszyńska, A. (1994a). Chaotic behavior of rotor–stator systems with rubs, *Journal of Engineering for Gas Turbines and Power* 116, 692–701.

Goldman, P., and Muszyńska, A. (1994b). Dynamic effects in mechanical structures with gaps and impacting: Order and chaos, *Journal of Vibration and Acoustics* 116, 541–547.

Golub, G. H., and van Loan, C. F. (1996). *Matrix Computations* (The Johns Hopkins University Press, Baltimore, MD).

Goodman, L. E., and Sutherland, J. G. (1951). Natural frequencies of continuous beams of uniform span length, *Journal of Applied Mechanics* 18, 217–218.

Goodwin, M. J. (1989). *Dynamics of Rotor–Bearing Systems* (Unwin Hyman, London).

Goodwin, M. J., Hooke, C. J., and Penny, J. E. T. (1983). Controlling the dynamic characteristics of a hydrostatic bearing by using a pocket-connected accumulator, *Proceedings of the IMechE*, 197 C, 255–258.

Goodwin, M. J., Penny, J. E. T., and Hooke, C. J. (1984). Variable impedance bearings for turbo-generator rotors, *Proceedings of the Third International Conference on Vibrations in Rotating Machinery*; York, England; September 1984, Paper C288/8, 535–541.

Gordis, J. H. (1992). An analysis of the improved reduced system (IRS) model reduction procedure, *Proceedings of the 10th International Modal Analysis Conference*; San Diego, CA, 471–479.

Green, K., Champneys, A. R., Friswell, M. I., and Munoz, A. M. (2008). Investigation of a multi-ball automatic dynamic balancing mechanism for eccentric rotors, *Royal Society Philosophical Transactions A* 366(1866), 705–728.

Grieve, D. W., and McShane, I. E. (1989). Torque pulsations on inverter-fed induction motors. *Proceedings of the Fourth International Conference on Electrical Machines and Drives*, London, IEE Conference Publication 310, 328–333.

Guyan, R. J. (1965). Reduction of stiffness and mass matrices, *AIAA Journal* 3, 380.

Hamrock, B. J., Schmid, S. R., and Jacobson, B. O. (2004). *Fundamentals of Fluid Film Lubrication* (Marcel Dekker, NJ).

Han, D. J. (2007). Generalized modal balancing for nonisotropic rotor systems, *Mechanical Systems and Signal Processing* 21, 2137–2160.

Han, S. M., Benaroya, H., and Wei, T. (1999). Dynamics of transversely vibrating beams using four engineering theories, *Journal of Sound and Vibration* 225, 935–988.

Harris, T. A. (2001). *Rolling Bearing Analysis*, Fourth Edition (John Wiley and Son, New York).

Henry, T. A., and Okah-Avae, B. E. (1976). Vibrations in cracked shafts, *IMechE Conference on Vibrations in Rotating Machinery*, Cambridge, UK, 15–19.

Henshell, R. D., and Ong, J. H. (1975). Automatic masters for eigenvalue economisation, *Earthquake Engineering and Structural Dynamics*, 3, 375–383.

Heo, J. W., Chung, J., and Choi, K. (2003). Dynamic time responses of a flexible spinning disk misaligned with the axis of rotation, *Journal of Sound and Vibration* 262, 25–44.

Herzog, R., Buhler, P., Gahler, C., and Larsonneur, R. (1996). Unbalance compensation using generalized notch-filters in the multivariable feedback of magnetic bearings, *IEEE Transactions on Control Systems Technology* 4, 580–586.

Hoa, S. V. (1979). Vibration of a rotating beam with tip mass, *Journal of Sound and Vibration* 67, 369–381.

Horn, R. A., and Johnson, C. R. (1985). *Matrix Analysis* (Cambridge University Press England).

Hu, H. Y., Jiang, P. L., and Yu, L. (2002). Coupled axial-lateral-torsional dynamics of a rotor–bearing system geared by spur bevel gears, *Journal of Sound and Vibration* 254, 427–446.

Hutchinson, J. R. (2001). Shear coefficients for Timoshenko beam theory, *Journal of Applied Mechanics* 68, 87–92.

Inman, D. J. (2006). *Vibration with Control* (John Wiley and Sons, Chichester, England).

Inman, D. J. (2008). *Engineering Vibration*, Third Edition (Pearson Prentice Hall, Upper Saddle River, N.J.).

Irons, B. M., and Ahmad, S. (1980). *Techniques of Finite Elements* (Ellis Horwood, Chichester, England).

ISO (1997). ISO 1940-2:1997, Mechanical Vibration: Balance Quality Requirements of Rigid Rotors, Part 2. Balance Errors.

ISO (1998). ISO 11342:1998, Mechanical Vibration: Balancing, Methods and Criteria for the Mechanical Balancing of Flexible Rotors.

ISO (2003). ISO 1940-1:2003, Mechanical Vibration: Balance Quality Requirements for Rigid Rotors in a Constant (Rigid) State, Part 1. Specification and Verification of Balance Tolerances.

ISO (2007). ISO 19499:2007, Mechanical Vibration: Balancing, Guidance on the Use and Application of Balancing Standards.

Jang, G. H., Lee, S. H., and Jung, M. S. (2002). Free vibration analysis of a spinning flexible disk–spindle system supported by ball bearing and flexible shaft using finite element method and substructuring synthesis, *Journal of Sound and Vibration* 251, 59–78.

Jeffcott, H. H. (1919). The lateral vibration of loaded shifts in the neighborhood of a whirling speed: The effects of want of balance, *Philosophical Magazine Series 6*, 37, 304–314.

Jei, Y. G., and Lee, C. W. (1992). Does curve veering occur in the eigenvalue problem of rotors? *Journal of Vibration and Acoustics* 114, 32–36.

Jia, H. S., and Chun, S. B. (1997). Evaluation of the longitudinal coupled vibrations in rotating, flexible disks–spindle systems, *Journal of Sound and Vibration* 208, 175–187.

Jing, J. P., Meng, G., Sun, Y., and Xia, S. B. (2005). On the oil-whipping of a rotor–bearing system by a continuum model, *Applied Mathematical Modeling* 29, 461–475.

Jordan, D. W., and Smith, P. (1977). *Nonlinear Ordinary Differential Equations* (Oxford University Press, Oxford, England).

Jun, O. S., Eun, H. J., Earmme, Y. Y., and Lee, C.-W. (1992). Modeling and vibration analysis of a simple rotor with a breathing crack, *Journal of Sound and Vibration* 155, 273–290.

Kang, Y., Shih, Y.-P., and Lee, A.-C. (1992). Investigation on the steady-state responses of asymmetric rotors, *Journal of Vibration and Acoustics* 114, 194–208.

Keiner, H., and Gadala, M. S. (1988). Comparison of different modeling techniques to simulate the vibration of a cracked rotor, *Journal of Sound and Vibration* 254, 1012–1024.

Kellenburger, W. (1980). Spiral vibrations due to seal rings in turbo-generators: Thermally induced interaction between rotor and stator. *Journal of Mechanical Design* 102, 177–184.

Kellenberger, W., and Rihak, P. (1988). Bimodal (complex) balancing of large turbo-generator rotors having large or small unbalance. *IMechE Conference on Vibrations in Rotating Machinery*, Edinburgh, 479–486, Paper Number C292/88.

Keogh, P. S., and Morton, P. G. (1993). Journal bearing differential heating evaluation with influence on rotordynamic behavior. *Proceedings of the Royal Society: Mathematical and Physical Sciences* 441, 527–548.

Keogh, P. S., and Morton, P. G. (1994). The dynamic nature of rotor thermal bending due to unsteady lubricant shearing within a bearing, *Proceedings of the Royal Society: Mathematical and Physical Sciences* 445, 273–290.

Kessler, C., and Kim, J. (2001). Concept of directional natural mode for vibration analysis of rotors using complex variable descriptions, *Journal of Sound and Vibration* 239, 545–555.

Khulief, Y. A. (1989). Vibration frequencies of a rotating tapered beam with end mass, *Journal of Sound and Vibration* 134, 87–97.

Khulief, Y. A., and Bazoune, A. (1992). Frequencies of rotating tapered Timoshenko beams with different boundary conditions, *Computers and Structures* 42, 781–795.

Khulief, Y. A., and Yi, L. J. (1988). Lead lag vibrational frequencies of a rotating beam with end mass, *Computers and Structures* 29, 1075–1085.

Kill, N. (2008). Application of multistage cyclic symmetry to rotordynamics, *Ninth International Conference on Vibrations in Rotating Machinery*, Exeter, UK, 8–10 September 2008, 267–276, IMechE Paper C663/015/08.

Kirk, R. G., and Guo, Z. (2005). Morton effect analysis: Theory, program and case study. *3rd International Symposium on Stability Control in Rotating Machinery*, Cleveland, OH.

Knospe, C. R., Hope, R. W., Fedigan, S., and Williams, R. (1995). Experiments in the control of unbalance response using magnetic bearings, *Mechanics 5*, 385–400.

Knospe, C. R., Hope, R. W., Tamer, S. M., and Fedigan, S. J. (1996). Robustness of adaptive unbalance control of rotors with magnetic bearings, *Journal of Vibration and Control 2*, 33–52.

Kramer, E. (1993). *Dynamics of Rotors and Foundations* (Springer-Verlag, Berlin, Germany).

Kumar, D. S., Sujatha, C., and Ganesan, N. (1997). Disc flexibility effects in rotor–bearing systems, *Computers and Structures 62*, 715–719.

Lalanne, M., and Ferraris, G. (1999). *Rotordynamics Prediction in Engineering*, Second Edition (John Wiley and Sons, New York).

Laurenson, R. M. (1976). Modal analysis of rotating flexible structures, *AIAA Journal 14*, 1444–1450.

Lee, C.-W. (1993). *Vibration Analysis of Rotors* (Kluwer Academic Publishers, Dordrecht, Netherlands).

Lee, A. S., Ha, J. W., and Choi, D. H. (2003). Coupled lateral and torsional vibration characteristics of a speed increasing geared rotor system, *Journal of Sound and Vibration 263*, 725–742.

Lee, C.-W., and Chun, S.-B. (1998). Vibration analysis of a rotor with multiple flexible disks using assumed-modes method, *Journal of Vibration and Acoustics 120*, 87–94.

Lee, C.-W., Joh, Y.-D., and Kim, Y.-D. (1990). Automatic modal balancing of flexible rotors during operation: Computer-controlled balancing head, *Journal of Mechanical Engineering Science 204*, 19–28.

Lees, A. W., and Friswell, M. I. (2001). The vibration signature of chordal cracks in asymmetric rotors, *19th International Modal Analysis Conference*, Orlando, FL, 124–129.

Lewis, F. M. (1932). Vibration during acceleration through a critical speed, *Transactions of the American Society of Mechanical Engineers 54*, 253–261.

Li, G. X., Lin, Z. L., and Allaire, P. E. (2008). Robust optimal balancing of high-speed machinery using convex optimization, *Journal of Vibration and Acoustics 130*, Article Number 031008.

Li, M., and Yu, L. (2001). Analysis of the coupled lateral torsional vibration of a rotor–bearing system with a misaligned gear coupling, *Journal of Sound and Vibration 243*, 283–300.

Likins, P. W., Barbera, F. J., and Baddeley, V. (1973). Mathematical modeling of spinning elastic bodies for modal analysis, *AIAA Journal 11*, 1251–1258.

Lim, T. C., and Singh, R. (1990). Vibration transmission through rolling-element bearings, Part 1: Bearing stiffness formulation, *Journal of Sound and Vibration 139*, 179–199.

Lim, T. C., and Singh, R. (1994). Vibration transmission through rolling-element bearings, Part V: Effect of distributed contact load in roller-bearing stiffness matrix, *Journal of Sound and Vibration 169*, 547–553.

Lindfield, G. R., and Penny, J. E. T. (2000). *Numerical Methods Using MATLAB* (Prentice Hall Upper Saddle River, New Jersey).

Lum, K. Y., Coppola, V. T., and Bernstein, D. (1996). Adaptive autocentering control for an active magnetic bearing supporting a rotor with unknown mass imbalance, *IEEE Transactions on Control Systems Technology*, 4, 587–597.

Luo, Z., Sun, X., and Fawcett, J. N. (1996). Coupled torsional-lateral–axial vibration analysis of geared shaft systems using substructure synthesis, *Mechanism and Machine Theory 31*, 345–352.

Matsukura, Y., Kiso, M., Inoue, T., and Tomisawa, M. (1979). On the balancing convergence of flexible rotors, with special reference to asymmetric rotors, *Journal of Sound and Vibration 63*, 419–428.

Mayes, I. W., and Davies, W. G. R. (1976). The vibrational behavior of a rotating shaft system containing a transverse crack, *IMechE Conference on Vibrations in Rotating Machinery*, Cambridge, UK, 53–64.

Mayes, I. W., and Davies, W. G. R. (1984). Analysis of the response of a multirotor–bearing system containing a transverse crack, *Journal of Vibration, Acoustics, Stress, and Reliability in Design* 106, 139–145.

Meirovitch, L. (1967). *Analytical Methods in Vibrations* (Macmillan, New York).

Meirovitch, L. (1986). *Elements of Vibration Analysis*, Second Edition (McGraw-Hill, New York).

Merrill, E. F. (1994). Dynamics of AC electrical machines. *IEEE Transactions on Industry Applications* 30, 277–285.

Mohan, S., and Hahn, E. J. (1974). Design of squeeze-film damper supports for rigid rotors, *Journal of Engineering for Industry* 96, 976–982.

Moon, F. C. (2004). *Chaotic Vibrations*, Second Edition (John Wiley and Sons, N.Y.).

Morton, P. G. (2008). Unstable shaft vibrations arising from thermal effects due to oil shearing between stationary and rotating elements, *Ninth International Conference on Vibrations in Rotating Machinery*, Exeter, England, 383–392.

Muszyńska, A. (1984). Partial lateral rotor to stator rubs, *3rd International Conference on Vibrations in Rotating Machinery*; York, UK; Paper C281/84, 327–335.

Muszyńska, A. (1989). Rotor to stationary element rub-related vibration phenomena in rotating machinery: Literature survey, *Shock and Vibration Digest* 21, 3–11.

Muszyńska, A. (2005). *Rotordynamics* (CRC Press, Taylor and Francis, Boca Raton, FL).

NAFEMS (1986). *A Finite Element Primer* (NAFEM, East Kilbride, Glasgow, U.K.).

Nandi, A., and Neogy, S. (2001). Modeling of rotors with three-dimensional solid finite elements, *Journal of Strain Analysis for Engineering Design* 36, 359–371.

Nelson, H. D. (1980). A finite rotating shaft element using Timoshenko beam theory, *Journal of Mechanical Design* 102, 793–803.

Nelson, H. D., and McVaugh, J. M. (1976). The dynamics of rotor–bearing systems using finite elements, *Journal of Engineering for Industry* 98, 593–599.

Newkirk, B. L. (1926). Shaft rubbing: Relative freedom of rotor shafts from sensitiveness to rubbing contact when running above their critical speeds. *Mechanical Engineering* 48, 830–832.

Newkirk, B. L., and Taylor, H. D. (1925). Shaft-whipping due to oil action in journal bearings, *General Electric Review* 28, 559–568.

Newland, D. E. (1984). *An Introduction to Random Vibrations and Spectral Analysis*, Second Edition (Longman Scientific and Technical, Harlow England).

Newland, D. E. (1989). *Mechanical Vibration Analysis and Computation* (Longman Scientific and Technical, Harlow England).

Newmark, N. M. (1959). A method of computation for structural dynamics, *ASCE Journal of Engineering Mechanics* 85, 67–94.

O'Callahan, J. C. (1989). A procedure for an improved reduced system (IRS) model, *Proceedings of the 7th International Modal Analysis Conference*; Las Vegas, NV; 17–21.

O'Callahan, J. C., Avitabile, P., and Riemer, R. (1989). System equivalent reduction expansion process (SEREP), *Proceedings of the 7th International Modal Analysis Conference*; Las Vegas, NV, 29–37.

Ostachowicz, W. M., and Krawczuk, M. (1992). Coupled torsional and bending vibrations of a rotor with an open crack, *Archive of Applied Mechanics* 62, 191–201.

Papadopoulos, C. A. (2004). Some comments on the calculation of the local flexibility of cracked shafts, *Journal of Sound and Vibration* 278, 1205–1211.

Papadopoulos, C. A., and Dimarogonas, A. D. (1987). Coupled longitudinal and bending vibrations of a rotating shaft with an open crack, *Journal of Sound and Vibration* 117, 81–93.

Parkinson, A. G. (1965). The vibration and balancing of shaft rotating in asymmetric bearings, *Journal of Sound and Vibration* 2, 477–501.

Parkinson, A. G. (1966). On balancing of shafts with axial asymmetry, *Proceedings of the Royal Society of London, Series A, Mathematical and Physical Sciences* 294(1436), 66–79.

Parkinson, A. G. (1967). An introduction to the vibration of rotating flexible shafts, *Bulletin of Mechanical Engineering Education* 6, 47–62.

Parkinson, A. G. (1991). Balancing of rotating machinery, *Proceedings of the Institution of Mechanical Engineers, Part C, Journal of Mechanical Engineering Science* 205, 53–66.

Parkinson, A. G., Darlow, M. S., and Smalley, A. J. (1980). A theoretical introduction to the development of a unified approach to flexible rotor balancing, *Journal of Sound and Vibration* 68, 489–506.

Pasricha, M. S., and Carnegie, W. D. (1976). Effects of variable inertia on the damped torsional vibrations of diesel-engine systems, *Journal of Sound and Vibration* 46, 339–345.

Pasricha, M. S., and Carnegie, W. D. (1979). Formulation of the equations of dynamic motion including the effects of variable inertia on the torsional vibrations in reciprocating engines, *Journal of Sound and Vibration* 66, 181–186.

Pasricha, M. S., and Hassan, A. Y. (1997). Effects of damping on secondary resonances in torsional vibrations of a two degree of freedom system – a variable inertia aspect in reciprocating engines, *Sixth International Conference on Recent Advances in Structural Dynamics*, Southampton, UK, 693–707.

Paz, M. (1984). Dynamic condensation, *AIAA Journal* 22, 724–727.

Penny, J. E. T., and Friswell, M. I. (2002). Simplified modeling of rotor cracks, *ISMA 27*; Leuven, Belgium; 607–615.

Perkins, N. C., and Mote, C. D. (1986). Comments on curve veering in eigenvalue problems, *Journal of Sound and Vibration* 106, 451–463.

Petyt, M. (1990). *Introduction to Finite Element Vibration Analysis* (Cambridge University Press).

Pilkey, W. D. (2005). *Formulas for Stress, Strain, and Structural Matrices*, Second Edition (John Wiley & Sons, Inc., Hoboken, NJ).

Porter, B. (1965). Nonlinear torsional vibration of a two-degree-of-freedom system having variable inertia, *Journal of Mechanical Engineering Science* 7, 101–113.

Proctor, M. P., and Gunter, E. J. (2005). Nonlinear whirl response of a high-speed seal test rotor with marginal and extended squeeze-film dampers, *NASA Report, TM-2005-213808*, August 2005, *ISCORMA-3*; Cleveland, OH; 19–23 September 2005.

Qu, Z.-Q. (2004). *Model Order Reduction Techniques: With Applications in Finite Element Analysis* (Springer-Verlag, UK).

Rades, M. (1998). Rotor–bearing model order reduction, *5th IFToMM*, Darmstadt, Germany, 148–159.

Rao, S. S. (1990). *Mechanical Vibrations*, Second Edition (Addison-Wesley, Reading, MA).

Rao, J. S., Shiau, T. N., and Chang, J. R. (1998). Theoretical analysis of lateral response due to torsional excitation of geared rotors, *Mechanism and Machine Theory* 33, 761–783.

Rieger, N. F. (1986). *Balancing of Rigid and Flexible Rotors*, The Shock and Vibration Information Center, Washington, DC.

Saavedra, P. N., and Cuitino, L. A. (2002). Vibration analysis of rotor for crack identification, *Journal of Vibration and Control* 8, 51–67.

Saito, S., and Azuma, T. (1983). Balancing of flexible rotors by the complex modal method, *Journal of Vibrations, Acoustics, Stress, and Reliability in Design* 15, 94–100.

Sawicki, J. T., Montilla-Bravo, A., and Gosiewski, Z. (2003). Thermo-mechanical behavior of rotor with rubbing, *International Journal of Rotating Machinery* 9, 41–47.

Schneider, H. (2000). Exchangeability of rotor modules: A new balancing procedure for rotors in a flexible state, *Seventh International Conference on Vibrations in Rotating Machinery*; Nottingham, UK; Paper Number C576/018/2000, 101–108.

Sekhar, A. S. (2004). Crack identification in a rotor system: A model-based approach, *Journal of Sound and Vibration* 270, 887–902.

Sekhar, A. S., and Prabhu, B. S. (1995). Effects of coupling misalignment on vibrations of rotating machinery, *Journal of Sound and Vibration* 185, 655–671.

Sinha, J. K., Friswell, M. I., and Lees, A. W. (2002). The identification of the unbalance and the foundation model of a flexible rotating machine from a single rundown, *Mechanical Systems and Signal Processing* 16, 255–271.

Sinha, J. K., Lees, A. W., and Friswell, M. I. (2004). Estimating unbalance and misalignment of a flexible rotating machine from a single rundown, *Journal of Sound and Vibration* 272, 967–989.

Sinou, J. J., and Lees, A. W. (2005). The influence of cracks in rotating shafts, *Journal of Sound and Vibration* 285, 1015–1037.

Smith, D. M. (1969). *Journal Bearings in Turbomachinery* (Chapman and Hall, London).

Somervaille, I. J. (1954). Balancing a rotating disk: Simple graphical construction, *Engineering* 177, 241–242.

Stephenson, R. W., Rouch, K. E., and Arora, R. (1989). Modeling rotors with axisymmetric solid harmonic elements, *Journal of Sound and Vibration* 131, 431–443.

Tenhunen, A., Holopainen, T. P., and Arkkio, A. (2003). Impulse method to calculate the frequency response of the electromagnetic forces on whirling-cage rotors. *IEE Proceedings: Electric Power Applications* 150, 752–756.

Thomas, D. L., Wilson, J. M., and Wilson, R. R. (1973). Timoshenko beam finite elements, *Journal of Sound and Vibration* 31, 315–330.

Thomas, H. J., Urlichs, K., and Wohlrab, R. (1976). Rotor instability in thermal turbomachines as a result of gap excitation, *VGB Kraftwerkstechnik* 56, 345–352.

Thompson, J. M. T., and Stewart, H. B. (1986). *Nonlinear Dynamics and Chaos* (John Wiley and Sons Chichester, England).

Thompson, W. T. (1993). *Theory of Vibration with Applications*, Fourth Edition (Prentice Hall, Englewood Cliffs, NJ).

Thomsen, J. J. (1997). *Vibrations and Stability: Order and Chaos* (McGraw-Hill, Maidenhead England).

Tondl, A. (1965). *Some Problems of Rotor Dynamics* (Chapman and Hall, London).

Tondl, A., Ruijgrok, T., Verhulst, F., and Nabergoj, R. (2000). *Autoparametric Resonances in Mechanical Systems* (Cambridge University Press, England).

Turhan, O., and Bulut, G. (2006). Linearly coupled shaft-torsional and blade-bending vibrations in multistage rotor–blade systems, *Journal of Sound and Vibration* 296, 292–318.

Untaroiu, C. D., Allaire, P. E., and Foiles, W. C. (2008). Balancing of flexible rotors using convex optimization techniques: Optimum min-max LMI influence coefficient balancing, *Journal of Vibration and Acoustics* 130, Article Number 021006.

Van de Vegte, J. (1981). Balancing of flexible rotors during operation, *Journal of Mechanical Engineering Science* 23, 257–261.

Van de Vegte, J., and Lake, R. T. (1978). Balancing of rotating systems during operation, *Journal of Sound and Vibration* 57, 225–235.

Verstege, S. (1998). Oil whip in transient operating conditions conditions: Case history, analysis, and remedial action, *Fifth International Conference on Rotor Dynamics*; Darmstadt, Germany; 525–535.

Vold, H., and Leuridan, J. (1995). High-resolution order tracking at extreme slew rates using Kalman tracking filters, *Shock and Vibration* 2, 507–515.

von Groll, G., and Ewins, D. J. (2001). The harmonic balance method with arc-length continuation in rotor–stator contact problems, *Journal of Sound and Vibration* 241, 223–233.

von Groll, G., and Ewins, D. J. (2002). A mechanism of low subharmonic response in rotor–stator contact: Measurements and simulations, *Journal of Vibration and Acoustics* 124, 350–358.

Wauer, J., and Suherman, S. (1998). Vibration suppression of rotating shafts passing through resonances by switching shaft stiffness, *Journal of Vibration and Acoustics* 120, 170–180.

Wilkinson, J. H. (1965). *The Algebraic Eigenvalue Problem* (Clarendon Press, Oxford).

White, M. F. (1979). Rolling-element bearing vibration transfer characteristics: Effect of stiffness, *Journal of Applied Mechanics* 46, 677–684.

Wolff, F. H., and Molnar, A. J. (1985). Variable frequency drives multiply torsional vibration problems, *Power* 129, 83–85.

Wu, J.-S., and Yang, I.-H. (1995). Computer method for torsion-and-flexure-coupled forced vibration of shafting system with damping, *Journal of Sound and Vibration* 180, 417–435.

Xu, M., and Marangoni, R. D. (1994a). Vibration analysis of a motor flexible coupling rotor system subject to misalignment and unbalance. 1. Theoretical-model and analysis, *Journal of Sound and Vibration* 176, 663–679.

Xu, M., and Marangoni, R. D. (1994b). Vibration analysis of a motor flexible coupling rotor system subject to misalignment and unbalance. 2. Experimental validation, *Journal of Sound and Vibration* 176, 681–691.

Yokoyama, T. (1988). Free vibration characteristics of rotating Timoshenko beams, *International Journal of Mechanical Sciences* 30, 743–755.

Zhou, S., and Shi, J. (2001). Active balancing and vibration control of rotating machinery: A survey, *Shock and Vibration Digest* 33, 361–371.

Zienkiewicz, O. C., Taylor, R. L., and Zhu, J. Z. (2005). *The Finite Element Method*, Sixth Edition (Elsevier Butterworth-Heinemann, Oxford).

Zorzi, E. S., and Nelson, H. D. (1980). The dynamics of rotor–bearing systems with axial torque: A finite element approach, *Journal of Mechanical Design* 102, 158–161.

Index